T0039168

METAMATHEMATICS

Foundations & Physicalization

STEPHEN WOLFRAM

METAMATHEMATICS

Foundations & Physicalization

STEPHEN WOLFRAM

Metamathematics: Foundations & Physicalization

Wolfram Media, Inc. | wolfram-media.com

ISBN-978-1-57955-076-9 (hardback)
ISBN-978-1-57955-078-3 (ebook)

Mathematics/Science

Library of Congress cataloging-in-publication data:

Names: Wolfram, Stephen, 1959- author.
Title: Metamathematics : foundations & physicalization / Stephen Wolfram.
Description: First edition. | [Champaign] : Wolfram Media, INC., [2022] | Includes bibliographical references and index.
Identifiers: LCCN 2022030911 (print) | LCCN 2022030912 (ebook) | ISBN 9781579550769 (hardback) | ISBN 9781579550783 (kindle edition)
Subjects: LCSH: Metamathematics. | AMS: General – General and miscellaneous specific topics – Philosophy of mathematics. | Mathematical logic and foundations – Philosophical aspects of logic and foundations. | Mathematical logic and foundations – Philosophical aspects of logic and foundations – Logic in the philosophy of science.
Classification: LCC QA9.8 .W65 2022 (print) | LCC QA9.8 (ebook) | DDC 510.1–dc23/eng/20220708
LC record available at https://lccn.loc.gov/2022030911
LC ebook record available at https://lccn.loc.gov/2022030912

Typeset with Wolfram Notebooks: wolfram.com/notebooks

Printed by Friesens, Manitoba, Canada. ♾ Acid-free paper. First edition. Second printing.

CONTENTS

Preface

Metamathematics is the study of what mathematics is made of, and how it's put together. It's about the foundations of mathematics, and the structure that lies below that most remarkable intellectual enterprise that we call mathematics. If we think of mathematics as an abstraction, then metamathematics is an abstraction of that abstraction—and something that probes deep into foundational questions.

Metamathematics in an organized form is a young field compared to mathematics itself. It first emerged only at the end of the 1800s, at a time when mathematics was expanding in a host of new, abstract directions, and people (like David Hilbert) began to ask global questions about the underlying structure of mathematics, and about what methods might be applied directly to this structure.

Much of what was done ended up under the banner of "mathematical logic" and concerned itself with setting up symbolic expressions that could serve as "axioms", and then using "logical deduction" to progressively establish new "truths of mathematics". For a while it seemed as if thinking at this level might just "solve all of mathematics", but in 1931 Gödel's theorem made it clear that as such, this couldn't work. But the question still remained of what one could say about the underlying structure of mathematics—and about metamathematics.

It was not, realistically, the most popular or accessible of intellectual fields—rapidly tending to become abstruse and highly technical. But from within the field—even a century ago—there began to arise the raw material for a new and very important paradigm: the paradigm of computation.

So what was the relation of this to metamathematics? I myself was in many ways thrust into the center of this some 45 years ago as I began to find ways to use computation to capture the essence of mathematics and to develop practical tools to automate the doing of mathematics. For a long time, though, I basically just used metamathematics as a kind of conceptual backup for things I was inventing.

And indeed I began to view the detailed features of mathematics and metamathematics as some kind of "unnecessary baggage" in the ultimate pure conceptual structure defined by computation—and by the representation of computation that I was developing for what's now the Wolfram Language.

And with computation as my conceptual framework, I began to ask questions of basic science—in particular about what was out there in the computational universe of simple programs. Some of the discoveries I made—like computational irreducibility—had echoes of results from metamathematics, but somehow the metamathematical connection always seemed more like an "interesting footnote" than something of foundational importance.

In the mid-1990s, though, I decided to try to apply what I'd learned from the computational universe directly to metamathematics. And just as I had studied possible programs in the computational universe, I studied possible axiom systems for mathematics. I found many things (described here starting on page 351). But many questions about the nature and global structure of mathematics remained.

A couple of decades passed. And in 2020 came the seemingly unrelated breakthrough of our Physics Project and its discovery that our physical universe can be thought of as being "computational all the way down". But what does this have to do with metamathematics? At the core of the Physics Project is the new paradigm of multicomputation. And it turns out that multicomputation is not only crucial to theories in physics (like general relativity and quantum mechanics), but is also ideally suited to representing things like the pattern of proofs in mathematics, and the structure that needs to be studied in metamathematics.

At first it was a surprise: that ideas born out of a new approach to physics could also apply to metamathematics. But as we began to understand the ruliad, and its unique role as the entangled limit of all possible computations, it began to be clear that in some sense both physics and mathematics arise from samplings of the ruliad by observers like us. It's a remarkable connection and unification. And one of its consequences is that it gives us a new way to think about metamathematics—a "physicalized" view of metamathematics, rooted, like physics, in the ruliad.

It is, I believe, a fresh and very powerful direction—and a serious next step beyond the prevailing view of metamathematics that developed a century ago. There are some significant surprises—as well as some potential conclusions about essentially philosophical questions that have been unresolved since antiquity.

The ideas described in this book are deep and abstract. But because they are fundamentally computational, it's possible to explain and illustrate them very explicitly in what amount to computational diagrams. Looking through this book at all its never-seen-anything-like-that-before pictures, it might seem like it's a window into an alien world. And in some ways it is. Because it's showing us what lies beneath the mathematics that we humans are familiar with.

And, yes, in the end we reach the raw atoms of existence (that we call emes) that make up the ruliad, and from which mathematics and metamathematics—just like physics—are built up.

The core ideas of this book in a sense transcend metamathematics, but metamathematics—with its connection to the somewhat familiar domain of mathematics—provides a convenient environment in which to explore them. And in addition, as we develop these ideas, they tell us about the actualities of metamathematics, and mathematics.

Do you need to know advanced mathematics to read this book? Not as such. Though for example when I talk about the new field of empirical metamathematics, I'll do it in reference to particular areas of mathematics, with some of their key results. But the core of the book—and most of its computational diagrams—can be followed without relying on existing mathematical knowledge.

I view this book as much as anything as a beginning, and in particular a beginning for a new metamathematics. I've tried to give at least an indication of the most obvious directions in which to go. But there is much to be done, and many new results—both theoretical and empirical—to be found. There was great activity around metamathematics roughly a century ago. And it's my hope that the ideas in this book can finally rekindle the excitement of those times. And that with the methods I describe it'll now be possible to make more progress on some of the deepest questions about the character and foundations of mathematics, and to finally be able to convincingly answer questions like "What is mathematics?"

Stephen Wolfram

August 2022

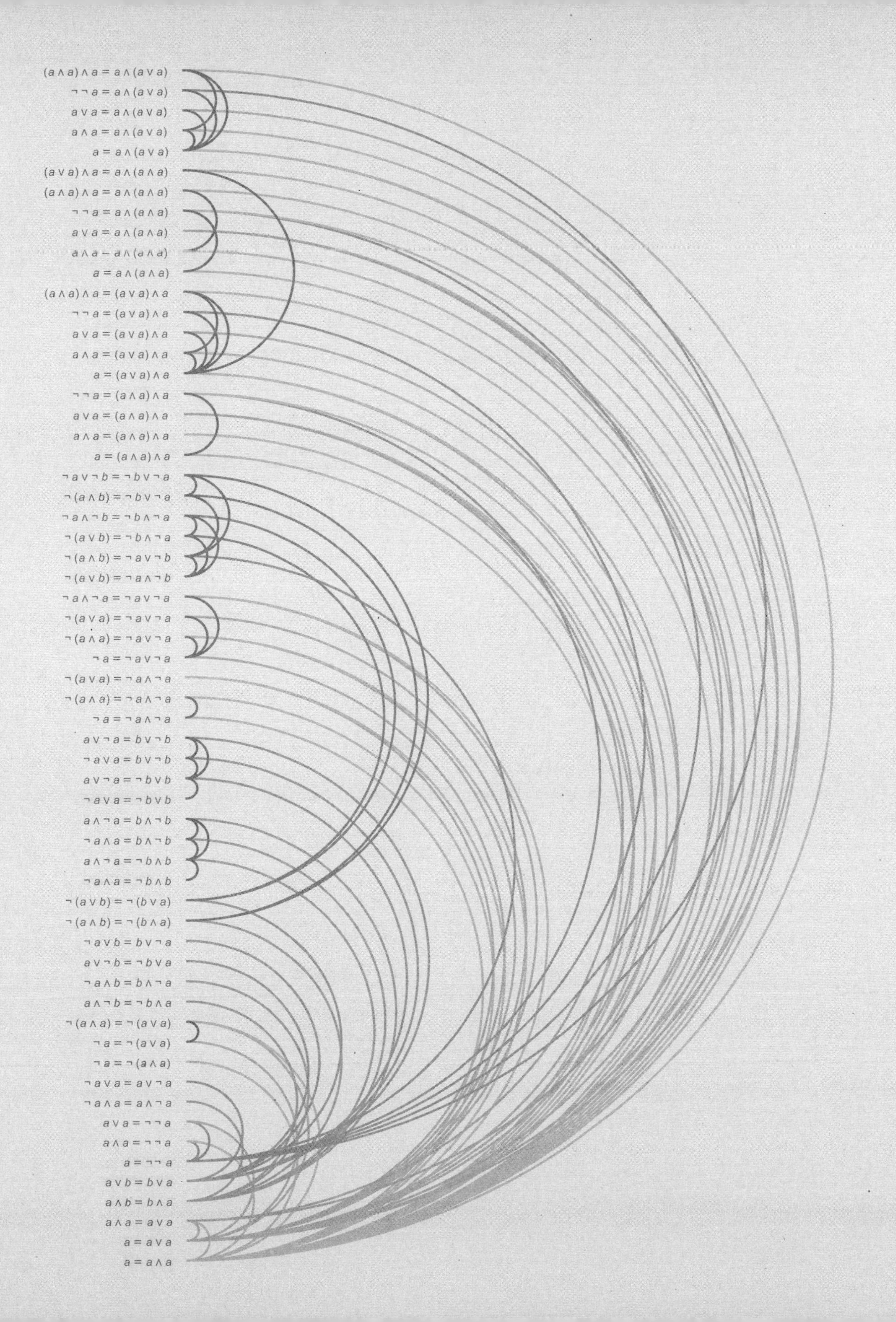

(a ∧ a) ∧ a = a ∧ (a ∨ a)
¬ ¬ a = a ∧ (a ∨ a)
a ∨ a = a ∧ (a ∨ a)
a ∧ a = a ∧ (a ∨ a)
a = a ∧ (a ∨ a)
(a ∨ a) ∧ a = a ∧ (a ∧ a)
(a ∧ a) ∧ a = a ∧ (a ∧ a)
¬ ¬ a = a ∧ (a ∧ a)
a ∨ a = a ∧ (a ∧ a)
a ∧ a = a ∧ (a ∧ a)
a = a ∧ (a ∧ a)
(a ∧ a) ∧ a = (a ∨ a) ∧ a
¬ ¬ a = (a ∨ a) ∧ a
a ∨ a = (a ∨ a) ∧ a
a ∧ a = (a ∨ a) ∧ a
a = (a ∨ a) ∧ a
¬ ¬ a = (a ∧ a) ∧ a
a ∨ a = (a ∧ a) ∧ a
a ∧ a = (a ∧ a) ∧ a
a = (a ∧ a) ∧ a
¬ a ∨ ¬ b = ¬ b ∨ ¬ a
¬ (a ∧ b) = ¬ b ∨ ¬ a
¬ a ∧ ¬ b = ¬ b ∧ ¬ a
¬ (a ∨ b) = ¬ b ∧ ¬ a
¬ (a ∧ b) = ¬ a ∨ ¬ b
¬ (a ∨ b) = ¬ a ∧ ¬ b
¬ a ∧ ¬ a = ¬ a ∨ ¬ a
¬ (a ∨ a) = ¬ a ∨ ¬ a
¬ (a ∧ a) = ¬ a ∨ ¬ a
¬ a = ¬ a ∨ ¬ a
¬ (a ∨ a) = ¬ a ∧ ¬ a
¬ (a ∧ a) = ¬ a ∧ ¬ a
¬ a = ¬ a ∧ ¬ a
a ∨ ¬ a = b ∨ ¬ b
¬ a ∨ a = b ∨ ¬ b
a ∨ ¬ a = ¬ b ∨ b
¬ a ∨ a = ¬ b ∨ b
a ∧ ¬ a = b ∧ ¬ b
¬ a ∧ a = b ∧ ¬ b
a ∧ ¬ a = ¬ b ∧ b
¬ a ∧ a = ¬ b ∧ b
¬ (a ∨ b) = ¬ (b ∨ a)
¬ (a ∧ b) = ¬ (b ∧ a)
¬ a ∨ b = b ∨ ¬ a
a ∨ ¬ b = ¬ b ∨ a
¬ a ∧ b = b ∧ ¬ a
a ∧ ¬ b = ¬ b ∧ a
¬ (a ∧ a) = ¬ (a ∨ a)
¬ a = ¬ (a ∨ a)
¬ a = ¬ (a ∧ a)
¬ a ∨ a = a ∨ ¬ a
¬ a ∧ a = a ∧ ¬ a
a ∨ a = ¬ ¬ a
a ∧ a = ¬ ¬ a
a = ¬ ¬ a
a ∨ b = b ∨ a
a ∧ b = b ∧ a
a ∧ a = a ∨ a
a = a ∨ a
a = a ∧ a

The Physicalization of Metamathematics and Its Implications for the Foundations of Mathematics

(March 7, 2022)

1 | Mathematics and Physics Have the Same Foundations

One of the many surprising (and to me, unexpected) implications of our Physics Project is its suggestion of a very deep correspondence between the foundations of physics and mathematics. We might have imagined that physics would have certain laws, and mathematics would have certain theories, and that while they might be historically related, there wouldn't be any fundamental formal correspondence between them.

But what our Physics Project suggests is that underneath everything we physically experience there is a single very general abstract structure—that we call the ruliad—and that our physical laws arise in an inexorable way from the particular samples we take of this structure. We can think of the ruliad as the entangled limit of all possible computations—or in effect a representation of all possible formal processes. And this then leads us to the idea that perhaps the ruliad might underlie not only physics but also mathematics—and that everything in mathematics, like everything in physics, might just be the result of sampling the ruliad.

Of course, mathematics as it's normally practiced doesn't look the same as physics. But the idea is that they can both be seen as views of the same underlying structure. What makes them different is that physical and mathematical observers sample this structure in somewhat different ways. But since in the end both kinds of observers are associated with human experience they inevitably have certain core characteristics in common. And the result is that there should be "fundamental laws of mathematics" that in some sense mirror the perceived laws of physics that we derive from our physical observation of the ruliad.

So what might those fundamental laws of mathematics be like? And how might they inform our conception of the foundations of mathematics, and our view of what mathematics really is?

The most obvious manifestation of the mathematics that we humans have developed over the course of many centuries is the few million mathematical theorems that have been published in the literature of mathematics. But what can be said in generality about this thing we call mathematics? Is there some notion of what mathematics is like "in bulk"? And what might we be able to say, for example, about the structure of mathematics in the limit of infinite future development?

When we do physics, the traditional approach has been to start from our basic sensory experience of the physical world, and of concepts like space, time and motion—and then to try to formalize our descriptions of these things, and build on these formalizations. And in its early development—for example by Euclid—mathematics took the same basic approach. But beginning a little more than a century ago there emerged the idea that one could build mathematics purely from formal axioms, without necessarily any reference to what is accessible to sensory experience.

And in a way our Physics Project begins from a similar place. Because at the outset it just considers purely abstract structures and abstract rules—typically described in terms of hypergraph rewriting—and then tries to deduce their consequences. Many of these consequences are incredibly complicated, and full of computational irreducibility. But the remarkable discovery is that when sampled by observers with certain general characteristics that make them like us, the behavior that emerges must generically have regularities that we can recognize, and in fact must follow exactly known core laws of physics.

And already this begins to suggest a new perspective to apply to the foundations of mathematics. But there's another piece, and that's the idea of the ruliad. We might have supposed that our universe is based on some particular chosen underlying rule, like an axiom system we might choose in mathematics. But the concept of the ruliad is in effect to represent the entangled result of "running all possible rules". And the key point is then that it turns out that an "observer like us" sampling the ruliad must perceive behavior that corresponds to known laws of physics. In other words, without "making any choice" it's inevitable—given what we're like as observers—that our "experience of the ruliad" will show fundamental laws of physics.

But now we can make a bridge to mathematics. Because in embodying all possible computational processes the ruliad also necessarily embodies the consequences of all possible axiom systems. As humans doing physics we're effectively taking a certain sampling of the ruliad. And we realize that as humans doing mathematics we're also doing essentially the same kind of thing.

But will we see "general laws of mathematics" in the same kind of way that we see "general laws of physics"? It depends on what we're like as "mathematical observers". In physics, there turn out to be general laws—and concepts like space and motion—that we humans can assimilate. And in the abstract it might not be that anything similar would be true in mathematics. But it seems as if the thing mathematicians typically call mathematics is something for which it is—and where (usually in the end leveraging our experience of physics) it's possible to successfully carve out a sampling of the ruliad that's again one we humans can assimilate.

When we think about physics we have the idea that there's an actual physical reality that exists—and that we experience physics within this. But in the formal axiomatic view of mathematics, things are different. There's no obvious "underlying reality" there; instead there's just a certain choice we make of axiom system. But now, with the concept of the ruliad, the story is different. Because now we have the idea that "deep underneath" both physics and mathematics there's the same thing: the ruliad. And that means that insofar as physics is "grounded in reality", so also must mathematics be.

When most working mathematicians do mathematics it seems to be typical for them to reason as if the constructs they're dealing with (whether they be numbers or sets or whatever) are "real things". But usually there's a concept that in principle one could "drill down" and formalize everything in terms of some axiom system. And indeed if one wants to get a global view of mathematics and its structure as it is today, it seems as if the best approach is to work from the formalization that's been done with axiom systems.

In starting from the ruliad and the ideas of our Physics Project we're in effect positing a certain "theory of mathematics". And to validate this theory we need to study the "phenomena of mathematics". And, yes, we could do this in effect by directly "reading the whole literature of mathematics". But it's more efficient to start from what's in a sense the "current prevailing underlying theory of mathematics" and to begin by building on the methods of formalized mathematics and axiom systems.

Over the past century a certain amount of metamathematics has been done by looking at the general properties of these methods. But most often when the methods are systematically used today, it's to set up some particular mathematical derivation, normally with the aid of a computer. But here what we want to do is think about what happens if the methods are used "in bulk". Underneath there may be all sorts of specific detailed formal derivations being done. But somehow what emerges from this is something higher level, something "more human"—and ultimately something that corresponds to our experience of pure mathematics.

How might this work? We can get an idea from an analogy in physics. Imagine we have a gas. Underneath, it consists of zillions of molecules bouncing around in detailed and complicated patterns. But most of our "human" experience of the gas is at a much more coarse-grained level—where we perceive not the detailed motions of individual molecules, but instead continuum fluid mechanics.

And so it is, I think, with mathematics. All those detailed formal derivations—for example of the kind automated theorem proving might do—are like molecular dynamics. But most of our "human experience of mathematics"—where we talk about concepts like integers or morphisms—is like fluid dynamics. The molecular dynamics is what builds up the fluid, but for most questions of "human interest" it's possible to "reason at the fluid dynamics level", without dropping down to molecular dynamics.

It's certainly not obvious that this would be possible. It could be that one might start off describing things at a "fluid dynamics" level—say in the case of an actual fluid talking about the motion of vortices—but that everything would quickly get "shredded", and that there'd soon be nothing like a vortex to be seen, only elaborate patterns of detailed microscopic molecular motions. And similarly in mathematics one might imagine that one would be able to prove theorems in terms of things like real numbers but actually find that everything gets "shredded" to the point where one has to start talking about elaborate issues of mathematical logic and different possible axiomatic foundations.

But in physics we effectively have the Second Law of thermodynamics—which we now understand in terms of computational irreducibility—that tells us that there's a robust sense in which the microscopic details are systematically "washed out" so that things like fluid dynamics "work". Just sometimes—like in studying Brownian motion, or hypersonic flow—the molecular dynamics level still "shines through". But for most "human purposes" we can describe fluids just using ordinary fluid dynamics.

So what's the analog of this in mathematics? Presumably it's that there's some kind of "general law of mathematics" that explains why one can so often do mathematics "purely in the large". Just like in fluid mechanics there can be "corner-case" questions that probe down to the "molecular scale"—and indeed that's where we can expect to see things like undecidability, as a rough analog of situations where we end up tracing the potentially infinite paths of single molecules rather than just looking at "overall fluid effects". But somehow in most cases there's some much stronger phenomenon at work—that effectively aggregates low-level details to allow the kind of "bulk description" that ends up being the essence of what we normally in practice call mathematics.

But is such a phenomenon something formally inevitable, or does it somehow depend on us humans "being in the loop"? In the case of the Second Law it's crucial that we only get to track coarse-grained features of a gas—as we humans with our current technology typically do. Because if instead we watched and decoded what every individual molecule does, we wouldn't end up identifying anything like the usual bulk "Second-Law" behavior. In other words, the emergence of the Second Law is in effect a direct consequence of the fact that it's us humans—with our limitations on measurement and computation—who are observing the gas.

So is something similar happening with mathematics? At the underlying "molecular level" there's a lot going on. But the way we humans think about things, we're effectively taking just particular kinds of samples. And those samples turn out to give us "general laws of mathematics" that give us our usual experience of "human-level mathematics".

To ultimately ground this we have to go down to the fully abstract level of the ruliad, but we'll already see many core effects by looking at mathematics essentially just at a traditional "axiomatic level", albeit "in bulk".

The full story—and the full correspondence between physics and mathematics—requires in a sense "going below" the level at which we have recognizable formal axiomatic mathematical structures; it requires going to a level at which we're just talking about making everything out of completely abstract elements, which in physics we might interpret as "atoms of space" and in mathematics as some kind of "symbolic raw material" below variables and operators and everything else familiar in traditional axiomatic mathematics.

The deep correspondence we're describing between physics and mathematics might make one wonder to what extent the methods we use in physics can be applied to mathematics, and vice versa. In axiomatic mathematics the emphasis tends to be on looking at particular theorems and seeing how they can be knitted together with proofs. And one could certainly imagine an analogous "axiomatic physics" in which one does particular experiments, then sees how they can "deductively" be knitted together. But our impression that there's an "actual reality" to physics makes us seek broader laws. And the correspondence between physics and mathematics implied by the ruliad now suggests that we should be doing this in mathematics as well.

What will we find? Some of it in essence just confirms impressions that working pure mathematicians already have. But it provides a definite framework for understanding these impressions and for seeing what their limits may be. It also lets us address questions like why undecidability is so comparatively rare in practical pure mathematics, and why it is so common to discover remarkable correspondences between apparently quite different areas of mathematics. And beyond that, it suggests a host of new questions and approaches both to mathematics and metamathematics—that help frame the foundations of the remarkable intellectual edifice that we call mathematics.

2 | The Underlying Structure of Mathematics and Physics

If we “drill down” to what we’ve called above the “molecular level” of mathematics, what will we find there? There are many technical details (some of which we’ll discuss later) about the historical conventions of mathematics and its presentation. But in broad outline we can think of there as being a kind of “gas” of “mathematical statements”—like $1+1=2$ or $x+y=y+x$—represented in some specified symbolic language. (And, yes, Wolfram Language provides a well-developed example of what that language can be like.)

But how does the “gas of statements” behave? The essential point is that new statements are derived from existing ones by “interactions” that implement laws of inference (like that q can be derived from the statement p and the statement “p implies q”). And if we trace the paths by which one statement can be derived from others, these correspond to proofs. And the whole graph of all these derivations is then a representation of the possible historical development of mathematics—with slices through this graph corresponding to the sets of statements reached at a given stage.

By talking about things like a “gas of statements” we’re making this sound a bit like physics. But while in physics a gas consists of actual, physical molecules, in mathematics our statements are just abstract things. But this is where the discoveries of our Physics Project start to be important. Because in our project we’re “drilling down” beneath for example the usual notions of space and time to an “ultimate machine code” for the physical universe. And we can think of that ultimate machine code as operating on things that are in effect just abstract constructs—very much like in mathematics.

In particular, we imagine that space and everything in it is made up of a giant network (hypergraph) of “atoms of space”—with each “atom of space” just being an abstract element that has certain relations with other elements. The evolution of the universe in time then corresponds to the application of computational rules that (much like laws of inference) take abstract relations and yield new relations—thereby progressively updating the network that represents space and everything in it.

But while the individual rules may be very simple, the whole detailed pattern of behavior to which they lead is normally very complicated—and typically shows computational irreducibility, so that there’s no way to systematically find its outcome except in effect by explicitly tracing each step. But despite all this underlying complexity it turns out—much like in the case of an ordinary gas—that at a coarse-grained level there are much simpler (“bulk”) laws of behavior that one can identify. And the remarkable thing is that these turn out to be exactly general relativity and quantum mechanics (which, yes, end up being the same theory when looked at in terms of an appropriate generalization of the notion of space).

But down at the lowest level, is there some specific computational rule that's "running the universe"? I don't think so. Instead, I think that in effect all possible rules are always being applied. And the result is the ruliad: the entangled structure associated with performing all possible computations.

But what then gives us our experience of the universe and of physics? Inevitably we are observers embedded within the ruliad, sampling only certain features of it. But what features we sample are determined by the characteristics of us as observers. And what seem to be critical to have "observers like us" are basically two characteristics. First, that we are computationally bounded. And second, that we somehow persistently maintain our coherence—in the sense that we can consistently identify what constitutes "us" even though the detailed atoms of space involved are continually changing.

But we can think of different "observers like us" as taking different specific samples, corresponding to different reference frames in rulial space, or just different positions in rulial space. These different observers may describe the universe as evolving according to different specific underlying rules. But the crucial point is that the general structure of the ruliad implies that so long as the observers are "like us", it's inevitable that their perception of the universe will be that it follows things like general relativity and quantum mechanics.

It's very much like what happens with a gas of molecules: to an "observer like us" there are the same gas laws and the same laws of fluid dynamics essentially independent of the detailed structure of the individual molecules.

So what does all this mean for mathematics? The crucial and at first surprising point is that the ideas we're describing in physics can in effect immediately be carried over to mathematics. And the key is that the ruliad represents not only all physics, but also all mathematics—and it shows that these are not just related, but in some sense fundamentally the same.

In the traditional formulation of axiomatic mathematics, one talks about deriving results from particular axiom systems—say Peano Arithmetic, or ZFC set theory, or the axioms of Euclidean geometry. But the ruliad in effect represents the entangled consequences not just of specific axiom systems but of all possible axiom systems (as well as all possible laws of inference).

But from this structure that in a sense corresponds to all possible mathematics, how do we pick out any particular mathematics that we're interested in? The answer is that just as we are limited observers of the physical universe, so we are also limited observers of the "mathematical universe".

But what are we like as "mathematical observers"? As I'll argue in more detail later, we inherit our core characteristics from those we exhibit as "physical observers". And that means that when we "do mathematics" we're effectively sampling the ruliad in much the same way as when we "do physics".

We can operate in different rulial reference frames, or at different locations in rulial space, and these will correspond to picking out different underlying "rules of mathematics", or essentially using different axiom systems. But now we can make use of the correspondence with physics to say that we can also expect there to be certain "overall laws of mathematics" that are the result of general features of the ruliad as perceived by observers like us.

And indeed we can expect that in some formal sense these overall laws will have exactly the same structure as those in physics—so that in effect in mathematics we'll have something like the notion of space that we have in physics, as well as formal analogs of things like general relativity and quantum mechanics.

What does this mean? It implies that—just as it's possible to have coherent "higher-level descriptions" in physics that don't just operate down at the level of atoms of space, so also this should be possible in mathematics. And this in a sense is why we can expect to consistently do what I described above as "human-level mathematics", without usually having to drop down to the "molecular level" of specific axiomatic structures (or below).

Say we're talking about the Pythagorean theorem. Given some particular detailed axiom system for mathematics we can imagine using it to build up a precise—if potentially very long and pedantic—representation of the theorem. But let's say we change some detail of our axioms, say associated with the way they talk about sets, or real numbers. We'll almost certainly still be able to build up something we consider to be "the Pythagorean theorem"—even though the details of the representation will be different.

In other words, this thing that we as humans would call "the Pythagorean theorem" is not just a single point in the ruliad, but a whole cloud of points. And now the question is: what happens if we try to derive other results from the Pythagorean theorem? It might be that each particular representation of the theorem—corresponding to each point in the cloud—would lead to quite different results. But it could also be that essentially the whole cloud would coherently lead to the same results.

And the claim from the correspondence with physics is that there should be "general laws of mathematics" that apply to "observers like us" and that ensure that there'll be coherence between all the different specific representations associated with the cloud that we identify as "the Pythagorean theorem".

In physics it could have been that we'd always have to separately say what happens to every atom of space. But we know that there's a coherent higher-level description of space—in which for example we can just imagine that objects can move while somehow maintaining their identity. And we can now expect that it's the same kind of thing in mathematics: that just as there's a coherent notion of space in physics where things can for example move without being "shredded", so also this will happen in mathematics. And this is why it's possible to do "higher-level mathematics" without always dropping down to the lowest level of axiomatic derivations.

It's worth pointing out that even in physical space a concept like "pure motion" in which objects can move while maintaining their identity doesn't always work. For example, close to a spacetime singularity, one can expect to eventually be forced to see through to the discrete structure of space—and for any "object" to inevitably be "shredded". But most of the time it's possible for observers like us to maintain the idea that there are coherent large-scale features whose behavior we can study using "bulk" laws of physics.

And we can expect the same kind of thing to happen with mathematics. Later on, we'll discuss more specific correspondences between phenomena in physics and mathematics—and we'll see the effects of things like general relativity and quantum mechanics in mathematics, or, more precisely, in metamathematics.

But for now, the key point is that we can think of mathematics as somehow being made of exactly the same stuff as physics: they're both just features of the ruliad, as sampled by observers like us. And in what follows we'll see the great power that arises from using this to combine the achievements and intuitions of physics and mathematics—and how this lets us think about new "general laws of mathematics", and view the ultimate foundations of mathematics in a different light.

3 | The Metamodeling of Axiomatic Mathematics

Consider all the mathematical statements that have appeared in mathematical books and papers. We can view these in some sense as the "observed phenomena" of (human) mathematics. And if we're going to make a "general theory of mathematics" a first step is to do something like we'd typically do in natural science, and try to "drill down" to find a uniform underlying model—or at least representation—for all of them.

At the outset, it might not be clear what sort of representation could possibly capture all those different mathematical statements. But what's emerged over the past century or so—with particular clarity in Mathematica and the Wolfram Language—is that there is in fact a rather simple and general representation that works remarkably well: a representation in which everything is a symbolic expression.

One can view a symbolic expression such as f[g[x][y, h[z]], w] as a hierarchical or tree structure, in which at every level some particular "head" (like f) is "applied to" one or more arguments. Often in practice one deals with expressions in which the heads have "known meanings"—as in Times[Plus[2, 3], 4] in Wolfram Language. And with this kind of setup symbolic expressions are reminiscent of human natural language, with the heads basically corresponding to "known words" in the language.

And presumably it's this familiarity from human natural language that's caused "human natural mathematics" to develop in a way that can so readily be represented by symbolic expressions.

But in typical mathematics there's an important wrinkle. One often wants to make statements not just about particular things but about whole classes of things. And it's common to then just declare that some of the "symbols" (like, say, x) that appear in an expression are "variables", while others (like, say, Plus) are not. But in our effort to capture the essence of mathematics as uniformly as possible it seems much better to burn the idea of an object representing a whole class of things right into the structure of the symbolic expression.

And indeed this is a core idea in the Wolfram Language, where something like x or f is just a "symbol that stands for itself", while $x\_$ is a pattern (named x) that can stand for anything. (More precisely, _ on its own is what stands for "anything", and $x\_$—which can also be written x:_—just says that whatever _ stands for in a particular instance will be called x.)

Then with this notation an example of a "mathematical statement" might be:

$$x\_ \circ y\_ = (y\_ \circ x\_) \circ y\_$$

In more explicit form we could write this as Equal[f[x_, y_], f[f[y_, x_], y_]]—where Equal (=) has the "known meaning" of representing equality. But what can we do with this statement? At a "mathematical level" the statement asserts that $x\_ \circ y\_$ and $(y\_ \circ x\_) \circ y\_$ should be considered equivalent. But thinking in terms of symbolic expressions there's now a more explicit, lower-level, "structural" interpretation: that any expression whose structure matches $x\_ \circ y\_$ can equivalently be replaced by $(y\_ \circ x\_) \circ y\_$ (or, in Wolfram Language notation, just $(y \circ x) \circ y$) and vice versa. We can indicate this interpretation using the notation

$$x\_ \circ y\_ \longleftrightarrow (y\_ \circ x\_) \circ y\_$$

which can be viewed as a shorthand for the pair of Wolfram Language rules:

$$x\_ \circ y\_ \rightarrow (y \circ x) \circ y, (y\_ \circ x\_) \circ y\_ \rightarrow x \circ y$$

OK, so let's say we have the expression $(a \circ b) \circ a$. Now we can just apply the rules defined by our statement. Here's what happens if we do this just once in all possible ways:

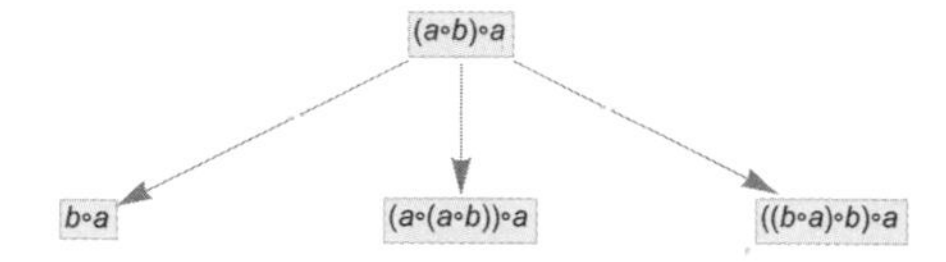

And here we see, for example, that $(a \circ b) \circ a$ can be transformed to $b \circ a$. Continuing this we build up a whole multiway graph. After just one more step we get:

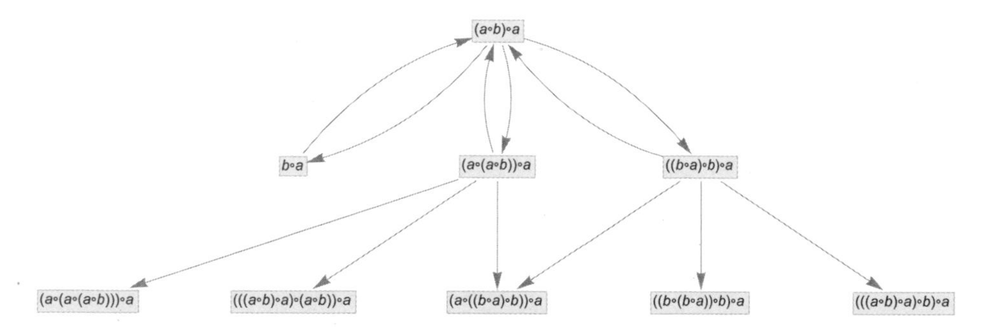

Continuing for a few more steps we then get

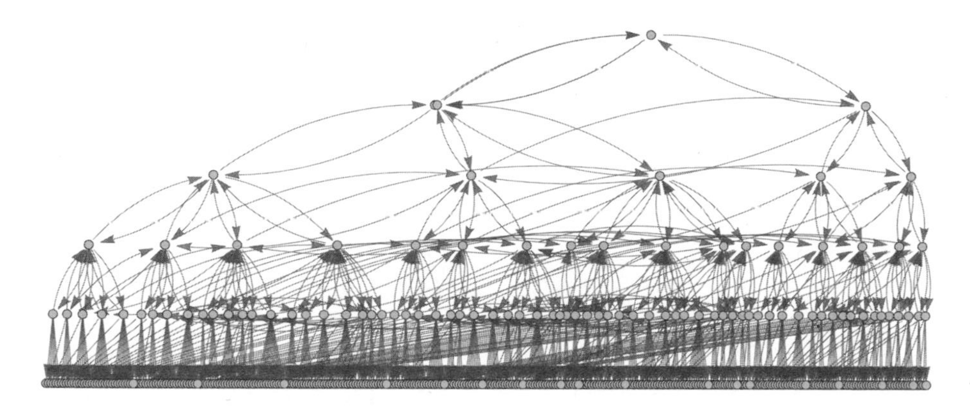

or in a different rendering:

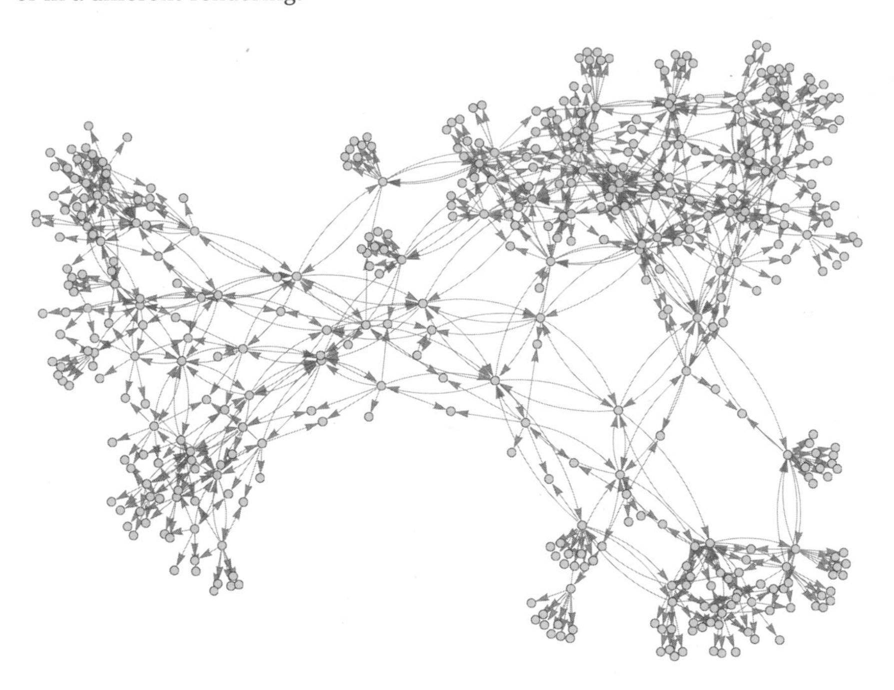

But what does this graph mean? Essentially it gives us a map of equivalences between expressions—with any pair of expressions that are connected being equivalent. So, for example, it turns out that the expressions $(a \circ ((b \circ a) \circ (a \circ b))) \circ a$ and $b \circ a$ are equivalent, and we can "prove this" by exhibiting a path between them in the graph:

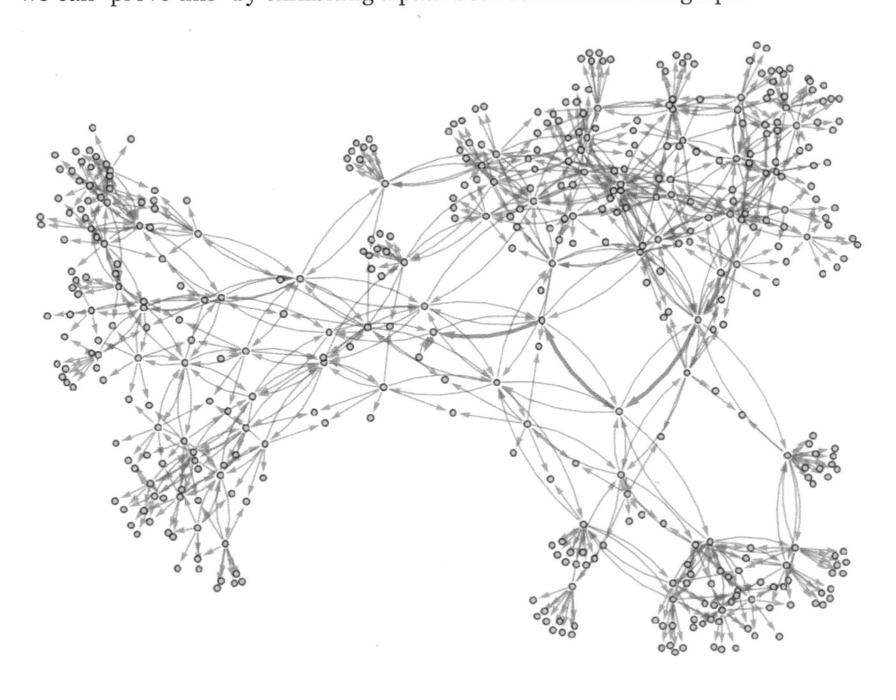

The steps on the path can then be viewed as steps in the proof, where here at each step we've indicated where the transformation in the expression took place:

$(a\circ((b\circ a)\circ(a\circ b)))\circ a$

$(a\circ(((a\circ b)\circ a)\circ(a\circ b)))\circ a$

$(((a\circ b)\circ a)\circ(a\circ b))\circ a$

$(a\circ(a\circ b))\circ a$

$(a\circ b)\circ a$

$b\circ a$

In mathematical terms, we can then say that starting from the "axiom" $x\circ y=(y\circ x)\circ y$ we were able to prove a certain equivalence theorem between two expressions. We gave a particular proof. But there are others, for example the "less efficient" 35-step one

$(a\circ((b\circ a)\circ(a\circ b)))\circ a$

$((b\circ a)\circ(a\circ b))\circ a$

$(((a\circ b)\circ(b\circ a))\circ(a\circ b))\circ a$

$(((a\circ b)\circ((a\circ b)\circ a))\circ(a\circ b))\circ a$

$((((b\circ a)\circ b)\circ((a\circ b)\circ a))\circ(a\circ b))\circ a$

$((((b\circ a)\circ b)\circ((a\circ b)\circ a))\circ((b\circ a)\circ b))\circ a$

$((((b\circ a)\circ b)\circ(((b\circ a)\circ b)\circ a))\circ((b\circ a)\circ b))\circ a$

$((((b\circ a)\circ b)\circ a)\circ((b\circ a)\circ b))\circ a$

$(((((a\circ b)\circ a)\circ b)\circ a)\circ((b\circ a)\circ b))\circ a$

$(((((a\circ b)\circ a)\circ b)\circ a)\circ(((a\circ b)\circ a)\circ b))\circ a$

$((((b\circ a)\circ b)\circ a)\circ(((a\circ b)\circ a)\circ b))\circ a$

$(((a\circ b)\circ a)\circ(((a\circ b)\circ a)\circ b))\circ a$

$(((a\circ b)\circ a)\circ((b\circ a)\circ b))\circ a$

$(((a\circ(a\circ b))\circ a)\circ((b\circ a)\circ b))\circ a$

$(((a\circ((b\circ a)\circ b))\circ a)\circ((b\circ a)\circ b))\circ a$

$(((a\circ((b\circ a)\circ b))\circ a)\circ(a\circ b))\circ a$

$((((b\circ a)\circ b)\circ a)\circ(a\circ b))\circ a$

$((((b\circ(b\circ a))\circ b)\circ a)\circ(a\circ b))\circ a$

$((((b\circ(b\circ a))\circ b)\circ a)\circ((b\circ a)\circ b))\circ a$

$((((b\circ(b\circ a))\circ b)\circ a)\circ((b\circ(b\circ a))\circ b))\circ a$

$(a\circ((b\circ(b\circ a))\circ b))\circ a$

$((b\circ(b\circ a))\circ b)\circ a$

$((((b\circ a)\circ b)\circ(b\circ a))\circ b)\circ a$

$(((((a\circ b)\circ a)\circ b)\circ(b\circ a))\circ b)\circ a$

$(((((a\circ b)\circ a)\circ b)\circ((a\circ b)\circ a))\circ b)\circ a$

$((b\circ((a\circ b)\circ a))\circ b)\circ a$

$(((a\circ b)\circ a)\circ b)\circ a$

$((((b\circ a)\circ b)\circ a)\circ b)\circ a$

$(((a\circ((b\circ a)\circ b))\circ a)\circ b)\circ a$

$(((a\circ(a\circ b))\circ a)\circ b)\circ a$

$(a\circ(((a\circ(a\circ b))\circ a)\circ b))\circ a$

$(a\circ(((a\circ b)\circ a)\circ b))\circ a$

$(a\circ((b\circ a)\circ b))\circ a$

$((b\circ a)\circ b)\circ a$

$(a\circ b)\circ a$

$b\circ a$

corresponding to the path:

For our later purposes it's worth talking in a little bit more detail here about how the steps in these proofs actually proceed. Consider the expression:

$(a \circ ((b \circ a) \circ (a \circ b))) \circ a$

We can think of this as a tree:

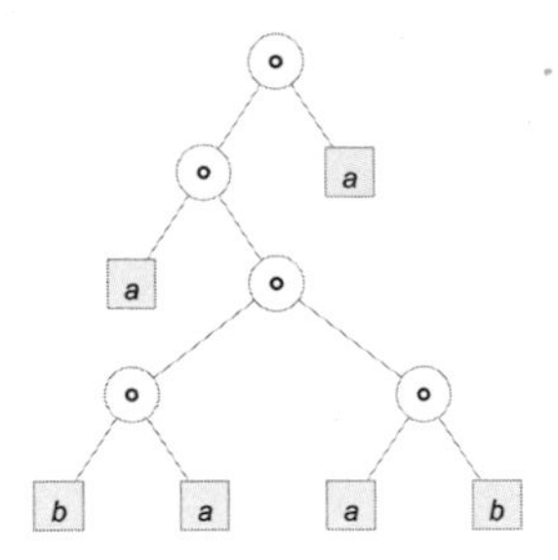

Our axiom can then be represented as:

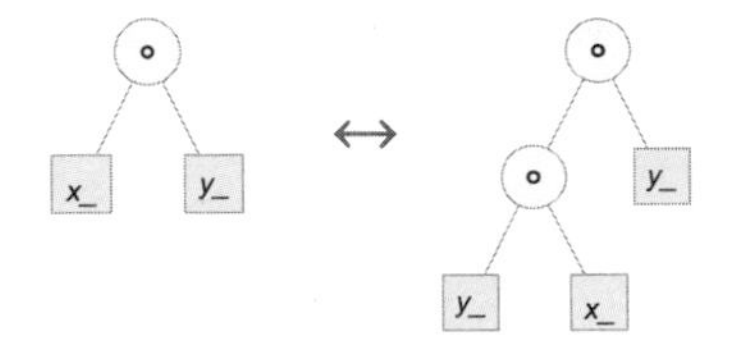

In terms of trees, our first proof becomes

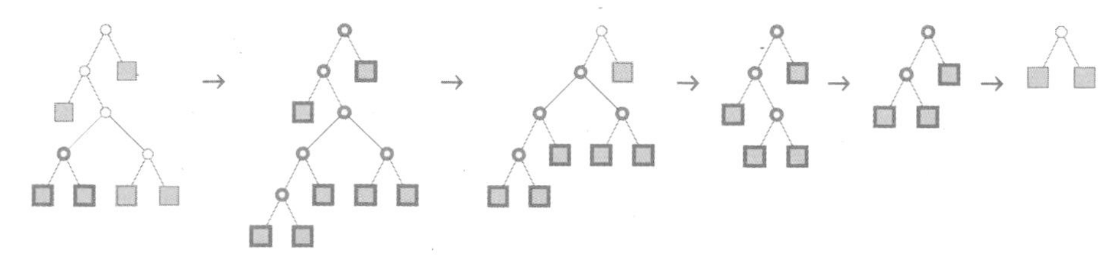

where we're indicating at each step which piece of tree gets "substituted for" using the axiom.

What we've done so far is to generate a multiway graph for a certain number of steps, and then to see if we can find a "proof path" in it for some particular statement. But what if we are given a statement, and asked whether it can be proved within the specified axiom system? In effect this asks whether if we make a sufficiently large multiway graph we can find a path of any length that corresponds to the statement.

If our system was computationally reducible we could expect always to be able to find a finite answer to this question. But in general—with the Principle of Computational Equivalence and the ubiquitous presence of computational irreducibility—it'll be common that there is no fundamentally better way to determine whether a path exists than effectively to try explicitly generating it. If we knew, for example, that the intermediate expressions generated always remained of bounded length, then this would still be a bounded problem. But in general the expressions can grow to any size—with the result that there is no general upper bound on the length of path necessary to prove even a statement about equivalence between small expressions.

For example, for the axiom we are using here, we can look at statements of the form $(a \circ b) \circ a = expr$. Then this shows how many expressions *expr* of what sizes have shortest proofs of $(a \circ b) \circ a = expr$ with progressively greater lengths:

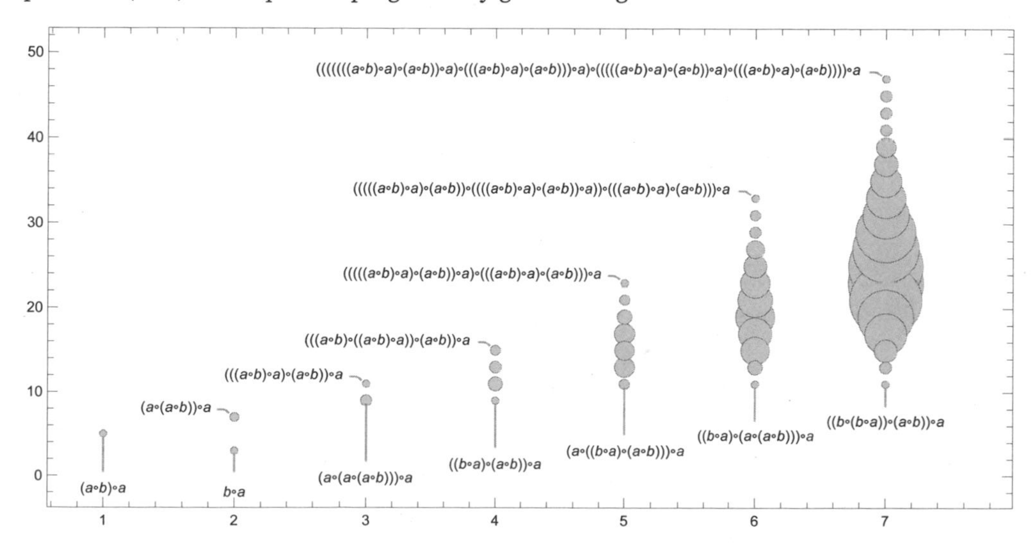

And for example if we look at the statement

$$(a \circ b) \circ a = (a \circ ((b \circ (b \circ a)) \circ (a \circ b))) \circ a$$

its shortest proof is

(a ∘ b) ∘ a
(a ∘ (a ∘ b)) ∘ a
(((a ∘ b) ∘ a) ∘ (a ∘ b)) ∘ a
(((a ∘ b) ∘ ((a ∘ b) ∘ a)) ∘ (a ∘ b)) ∘ a
(a ∘ (((a ∘ b) ∘ ((a ∘ b) ∘ a)) ∘ (a ∘ b))) ∘ a
(a ∘ ((((b ∘ a) ∘ b) ∘ ((a ∘ b) ∘ a)) ∘ (a ∘ b))) ∘ a
(a ∘ ((((b ∘ a) ∘ b) ∘ (b ∘ a)) ∘ (a ∘ b))) ∘ a
(a ∘ ((b ∘ (b ∘ a)) ∘ (a ∘ b))) ∘ a

where, as is often the case, there are intermediate expressions that are longer than the final result.

4 | Some Simple Examples with Mathematical Interpretations

The multiway graphs in the previous section are in a sense fundamentally metamathematical. Their "raw material" is mathematical statements. But what they represent are the results of operations—like substitution—that are defined at a kind of meta level, that "talks about mathematics" but isn't itself immediately "representable as mathematics". But to help understand this relationship it's useful to look at simple cases where it's possible to make at least some kind of correspondence with familiar mathematical concepts.

Consider for example the axiom

$$x\_ \circ y\_ = y\_ \circ x\_$$

that we can think of as representing commutativity of the binary operator $\circ$. Now consider using substitution to "apply this axiom", say starting from the expression $(a \circ b) \circ (c \circ d)$. The result is the (finite) multiway graph:

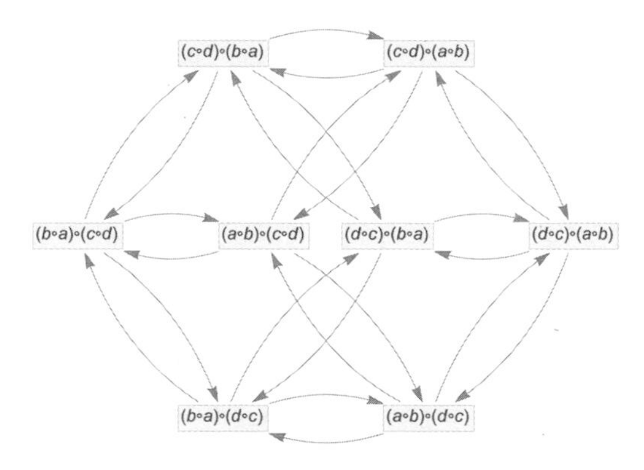

Conflating the pairs of edges going in opposite directions, the resulting graphs starting from any expression involving s $\circ$'s (and $s + 1$ distinct variables) are:

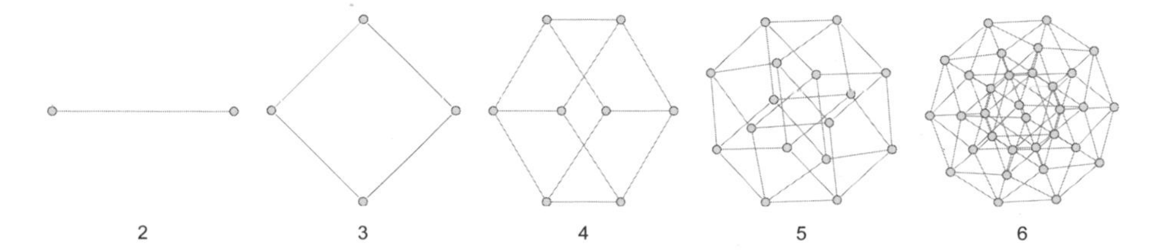

And these are just the Boolean hypercubes, each with 2^s nodes.

If instead of commutativity we consider the associativity axiom

$$x\_ \circ (y\_ \circ z\_) = (x\_ \circ y\_) \circ z\_$$

then we get a simple "ring" multiway graph:

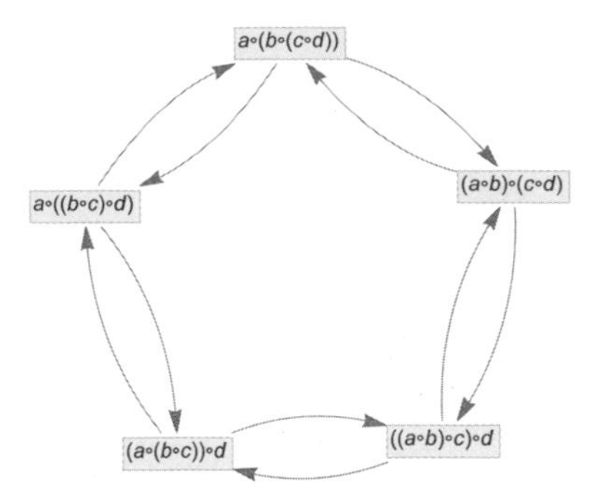

With both associativity and commutativity we get:

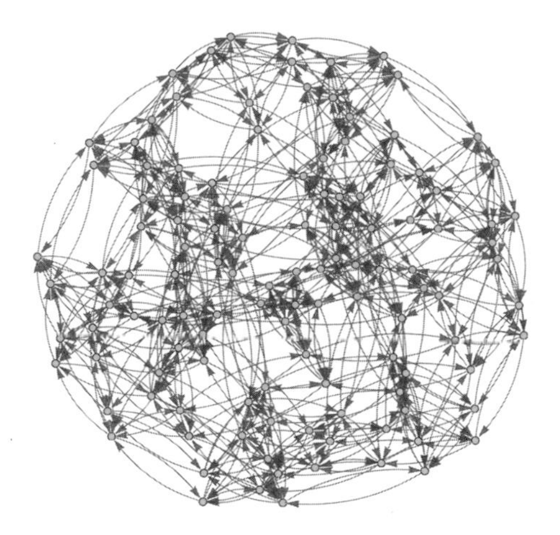

What is the mathematical significance of this object? We can think of our axioms as being the general axioms for a commutative semigroup. And if we build a multiway graph—say starting with $(a \circ b) \circ b$—we'll find out what expressions are equivalent to $(a \circ b) \circ b$ in any commutative semigroup—or, in other words, we'll get a collection of theorems that are "true for any commutative semigroup":

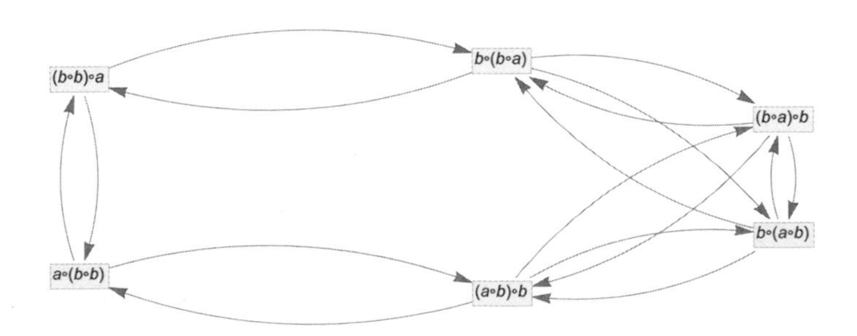

But what if we want to deal with a "specific semigroup" rather than a generic one? We can think of our symbols a and b as generators of the semigroup, and then we can add relations, as in:

$$x\_\circ y\_ = y\_\circ x\_, x\_\circ(y\_\circ z\_) = (x\_\circ y\_)\circ z\_, a\circ a = b\circ b$$

And the result of this will be that we get more equivalences between expressions:

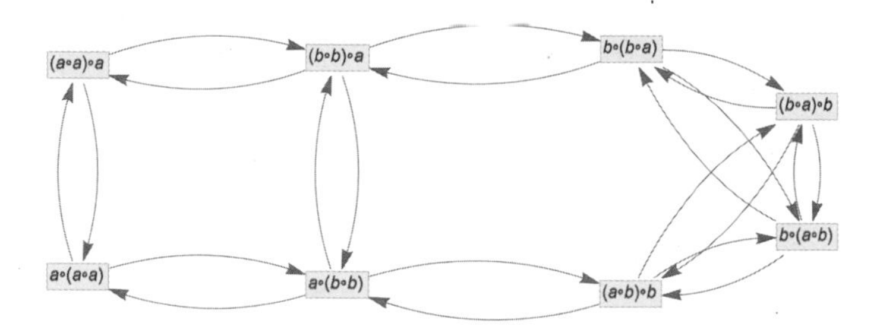

The multiway graph here is still finite, however, giving a finite number of equivalences. But let's say instead that we add the relations:

$$a = b\circ b, b = a\circ b$$

Then if we start from a we get a multiway graph that begins like

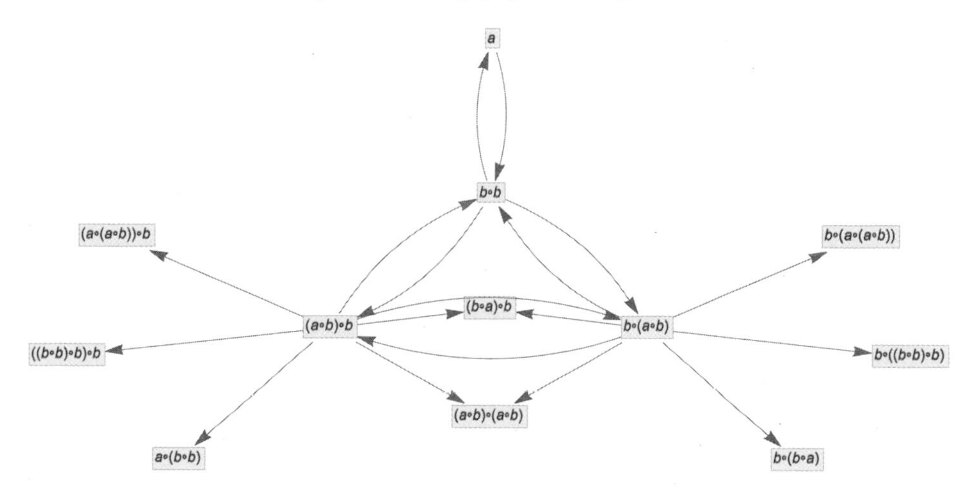

but just keeps growing forever (here shown after 6 steps):

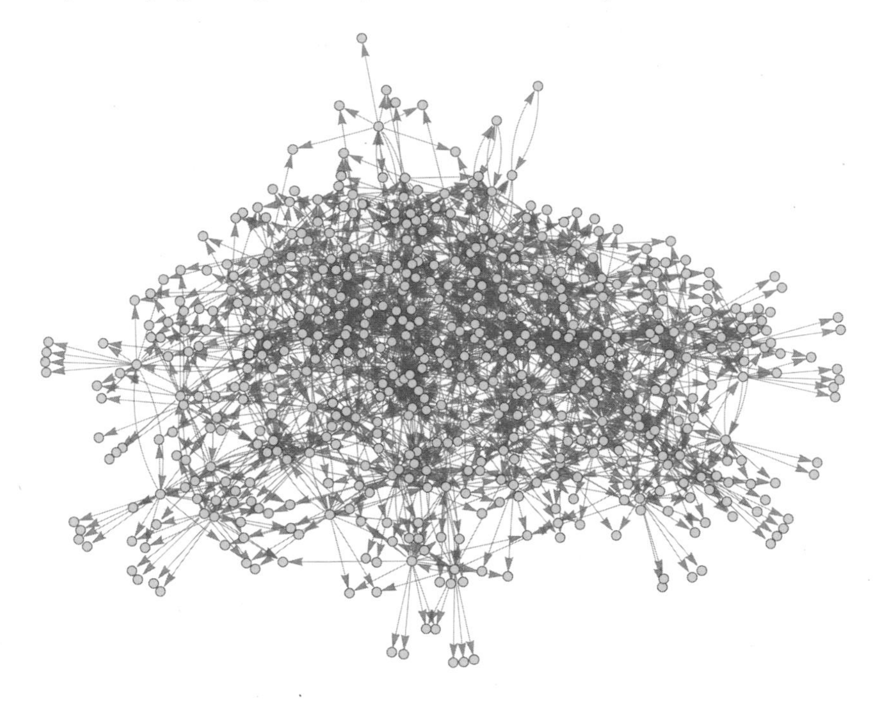

And what this then means is that there are an infinite number of equivalences between expressions. We can think of our basic symbols *a* and *b* as being generators of our semigroup. Then our expressions correspond to "words" in the semigroup formed from these generators. The fact that the multiway graph is infinite then tells us that there are an infinite number of equivalences between words.

But when we think about the semigroup mathematically we're typically not so interested in specific words as in the overall "distinct elements" in the semigroup, or in other words, in those "clusters of words" that don't have equivalences between them. And to find these we can imagine starting with all possible expressions, then building up multiway graphs from them. Many of the graphs grown from different expressions will join up. But what we want to know in the end is how many disconnected graph components are ultimately formed. And each of these will correspond to an element of the semigroup.

As a simple example, let's start from all words of length 2:

The multiway graphs formed from each of these after 1 step are:

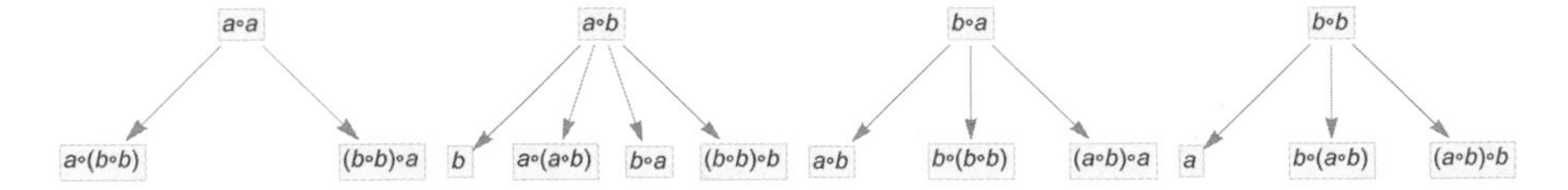

But these graphs in effect "overlap", leaving three disconnected components:

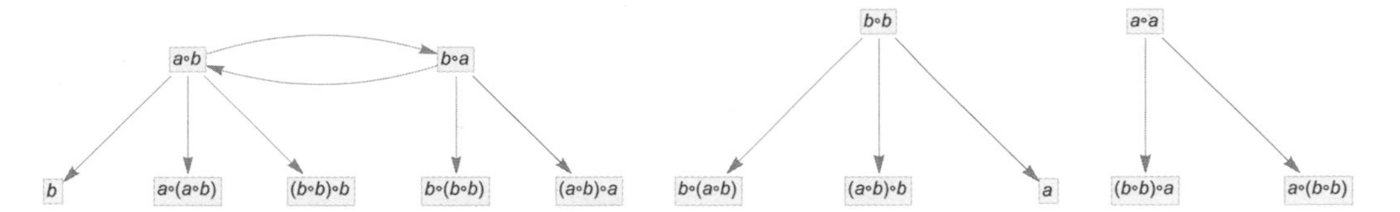

After 2 steps the corresponding result has two components:

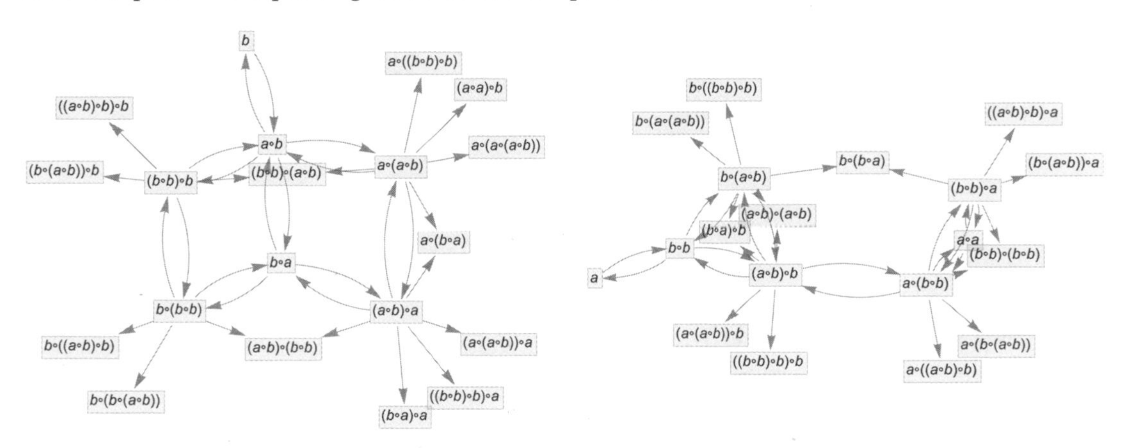

And if we start with longer (or shorter) words, and run for more steps, we'll keep finding the same result: that there are just two disconnected "droplets" that "condense out" of the "gas" of all possible initial words:

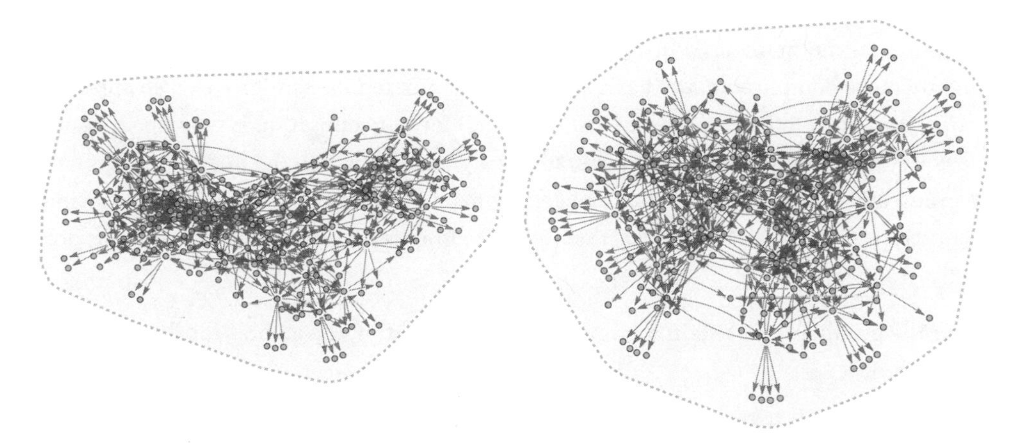

And what this means is that our semigroup ultimately has just two distinct elements—each of which can be represented by any of the different ("equivalent") words in each "droplet". (In this particular case the droplets just contain respectively all words with an odd or even number of *b*'s.)

In the mathematical analysis of semigroups (as well as groups), it's common ask what happens if one forms products of elements. In our setting what this means is in effect that one wants to "combine droplets using ∘". The simplest words in our two droplets are respectively a and b. And we can use these as "representatives of the droplets". Then we can see how multiplication by a and by b transforms words from each droplet:

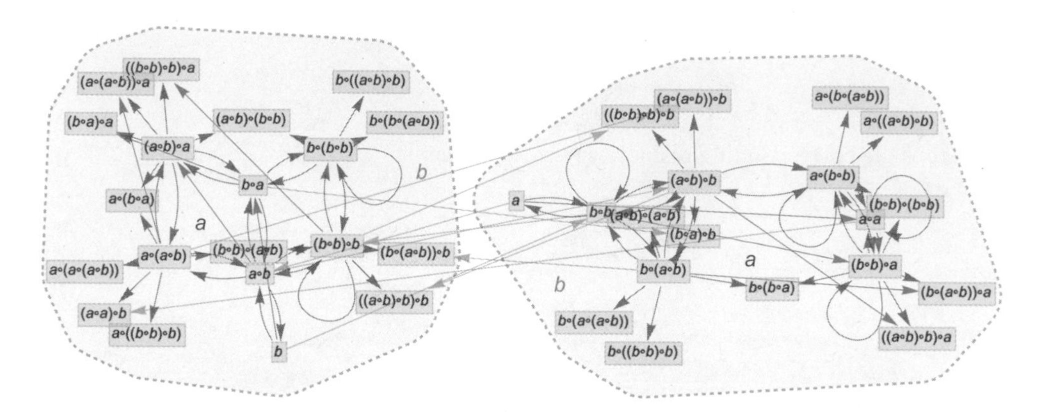

With only finite words the multiplications will sometimes not "have an immediate target" (so they are not indicated here). But in the limit of an infinite number of multiway steps, every multiplication will "have a target" and we'll be able to summarize the effect of multiplication in our semigroup by the graph:

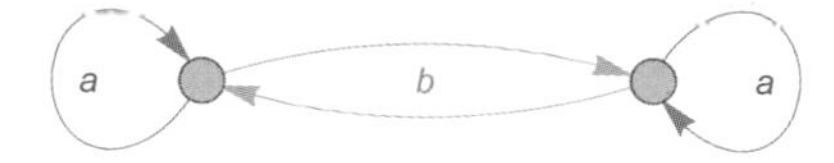

More familiar as mathematical objects than semigroups are groups. And while their axioms are slightly more complicated, the basic setup we've discussed for semigroups also applies to groups. And indeed the graph we've just generated for our semigroup is very much like a standard Cayley graph that we might generate for a group—in which the nodes are elements of the group and the edges define how one gets from one element to another by multiplying by a generator. (One technical detail is that in Cayley graphs identity-element self-loops are normally dropped.)

Consider the group D_2 (the "Klein four-group"). In our notation the axioms for this group can be written:

$$x\_\circ(y\_\circ z\_) = (x\_\circ y\_)\circ z\_,\ x\_\circ y\_ = y\_\circ x\_,\ x\_ = a\circ x\_,\ a = b\circ b,\ a = c\circ c$$

Given these axioms we do the same construction as for the semigroup above. And what we find is that now four "droplets" emerge, corresponding to the four elements of D_2

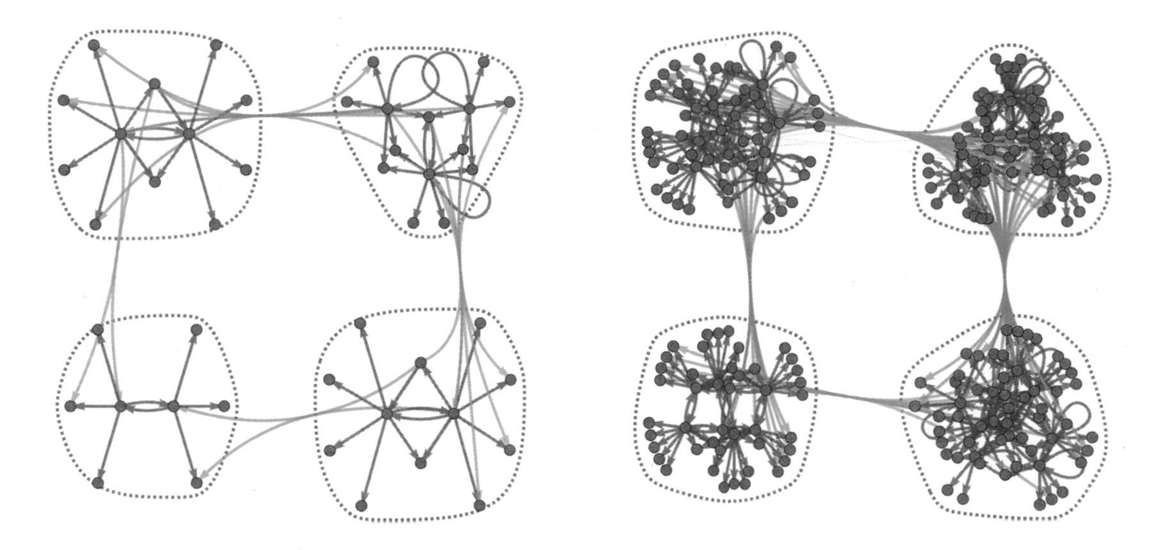

and the pattern of connections between them in the limit yields exactly the Cayley graph for D_2:

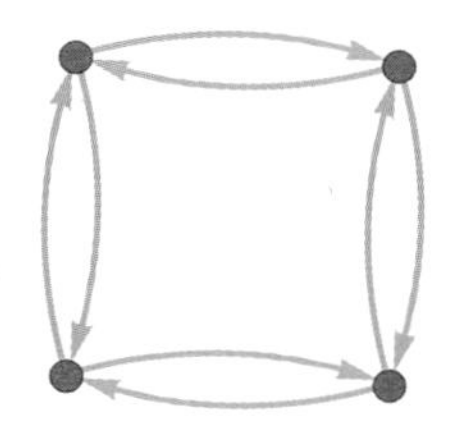

We can view what's happening here as a first example of something we'll return to at length later: the idea of "parsing out" recognizable mathematical concepts (here things like elements of groups) from lower-level "purely metamathematical" structures.

5 | Metamathematical Space

In multiway graphs like those we've shown in previous sections we routinely generate very large numbers of "mathematical" expressions. But how are these expressions related to each other? And in some appropriate limit can we think of them all being embedded in some kind of "metamathematical space"?

It turns out that this is the direct analog of what in our Physics Project we call branchial space, and what in that case defines a map of the entanglements between branches of quantum history. In the mathematical case, let's say we have a multiway graph generated using the axiom:

$x\_ \circ y\_ \leftrightarrow (y\_ \circ x\_) \circ y\_$

After a few steps starting from $a \circ b$ we have:

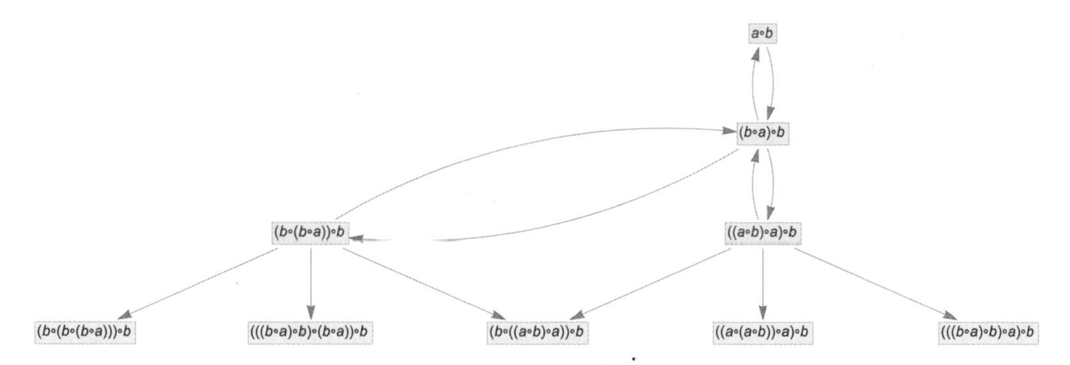

Now—just as in our Physics Project—let's form a branchial graph by looking at the final expressions here and connecting them if they are "entangled" in the sense that they share an ancestor on the previous step:

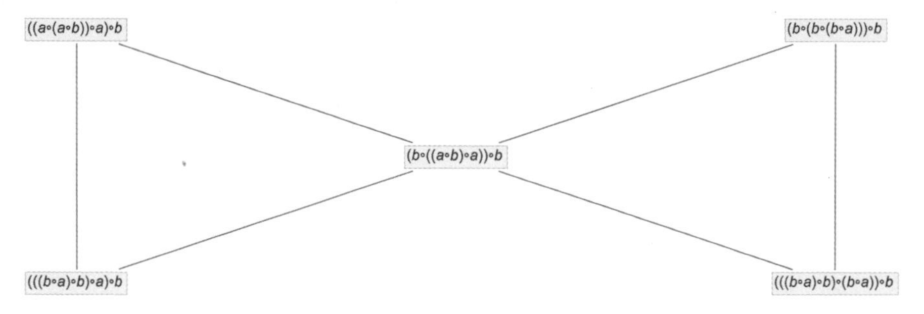

There's some trickiness here associated with loops in the multiway graph (which are the analog of closed timelike curves in physics) and what it means to define different "steps in evolution". But just iterating once more the construction of the multiway graph, we get a branchial graph:

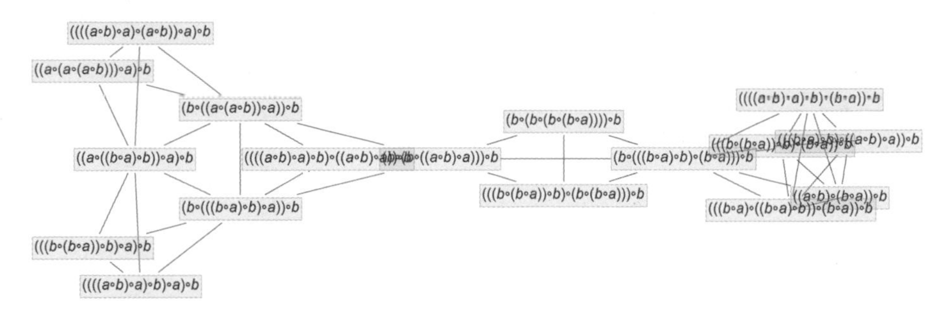

After a couple more iterations the structure of the branchial graph is (with each node sized according to the size of expression it represents):

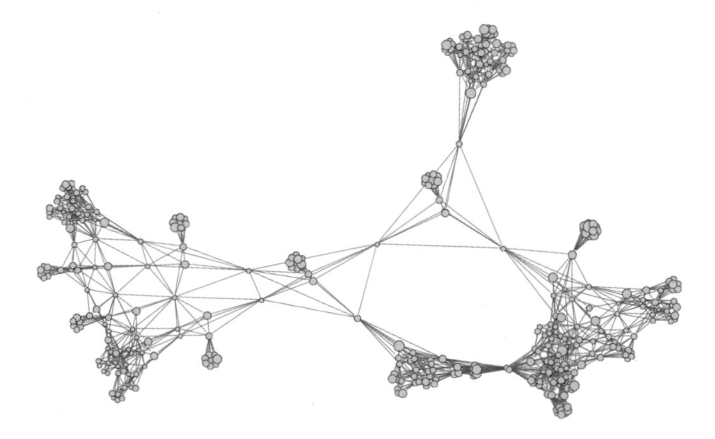

Continuing another iteration, the structure becomes:

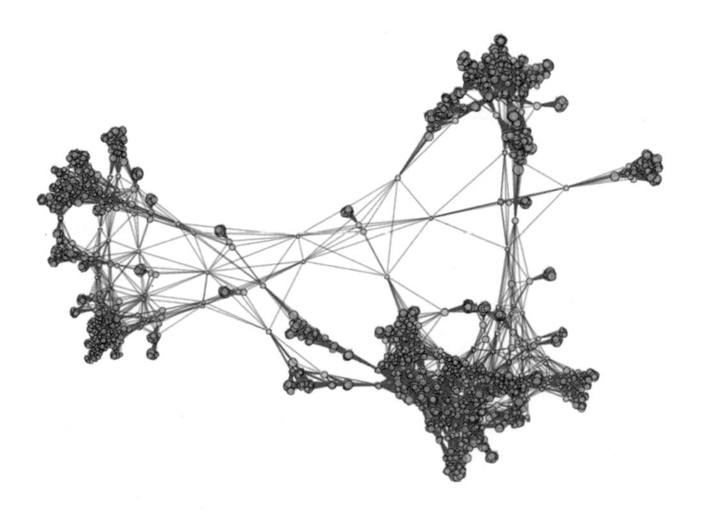

And in essence this structure can indeed be thought of as defining a kind of "metamathematical space" in which the different expressions are embedded. But what is the "geography" of this space? This shows how expressions (drawn as trees) are laid out on a particular branchial graph

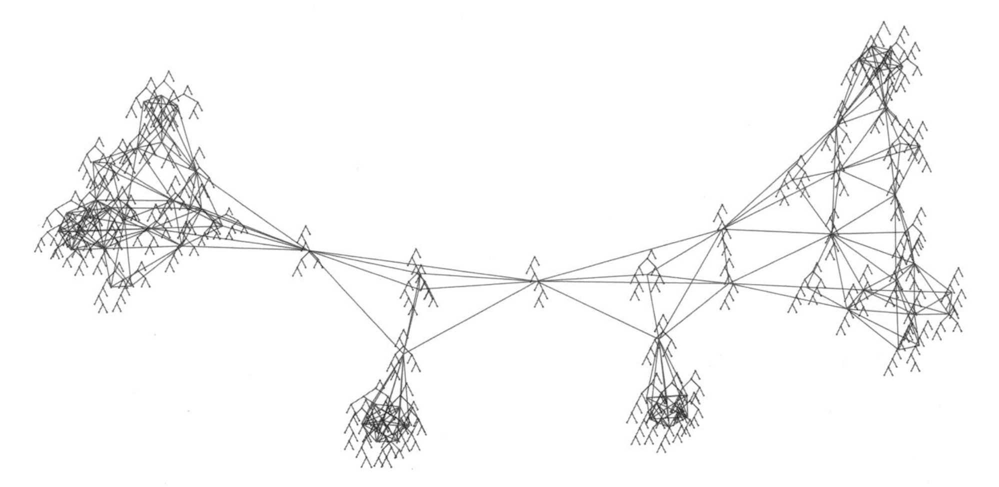

and we see that there is at least a general clustering of similar trees on the graph—indicating that "similar expressions" tend to be "nearby" in the metamathematical space defined by this axiom system.

An important feature of branchial graphs is that effects are—essentially by construction—always local in the branchial graph. For example, if one changes an expression at a particular step in the evolution of a multiway system, it can only affect a region of the branchial graph that essentially expands by one edge per step.

One can think of the affected region—in analogy with a light cone in spacetime—as being the "entailment cone" of a particular expression. The edge of the entailment cone in effect expands at a certain "maximum metamathematical speed" in metamathematical (i.e. branchial) space—which one can think of as being measured in units of "expression change per multiway step".

By analogy with physics one can start talking in general about motion in metamathematical space. A particular proof path in the multiway graph will progressively "move around" in the branchial graph that defines metamathematical space. (Yes, there are many subtle issues here, not least the fact that one has to imagine a certain kind of limit being taken so that the structure of the branchial graph is "stable enough" to "just be moving around" in something like a "fixed background space".)

By the way, the shortest proof path in the multiway graph is the analog of a geodesic in spacetime. And later we'll talk about how the "density of activity" in the branchial graph is the analog of energy in physics, and how it can be seen as "deflecting" the path of geodesics, just as gravity does in spacetime.

It's worth mentioning just one further subtlety. Branchial graphs are in effect associated with "transverse slices" of the multiway graph—but there are many consistent ways to make these slices. In physics terms one can think of the foliations that define different choices of sequences of slices as being like "reference frames" in which one is specifying a sequence of "simultaneity surfaces" (here "branchtime hypersurfaces"). The particular branchial graphs we've shown here are ones associated with what in physics might be called the cosmological rest frame in which every node is the result of the same number of updates since the beginning.

6 | The Issue of Generated Variables

A rule like

$$x\_ \circ y\_ \leftrightarrow (y\_ \circ x\_) \circ y\_$$

defines transformations for any expressions $x\_$ and $y\_$. So, for example, if we use the rule from left to right on the expression $a \circ (b \circ a)$ the "pattern variable" $x\_$ will be taken to be a while $y\_$ will be taken to be $b \circ a$, and the result of applying the rule will be $((b \circ a) \circ a) \circ (b \circ a)$.

But consider instead the case where our rule is:

$$x\_ \circ y\_ \leftrightarrow (y\_ \circ x\_) \circ z\_$$

Applying this rule (from left to right) to $a \circ (b \circ a)$ we'll now get $((b \circ a) \circ a) \circ z\_$. And applying the rule to $a \circ b$ we'll get $(b \circ a) \circ z\_$. But what should we make of those $z\_$'s? And in particular, are they "the same", or not?

A pattern variable like $z\_$ can stand for any expression. But do two different $z\_$'s have to stand for the same expression? In a rule like $z\_ \circ z\_ \leftrightarrow ...$ we're assuming that, yes, the two $z\_$'s always stand for the same expression. But if the $z\_$'s appear in different rules it's a different story. Because in that case we're dealing with two separate and unconnected $z\_$'s— that can stand for completely different expressions.

To begin seeing how this works, let's start with a very simple example. Consider the (for now, one-way) rule

$$a \to x\_$$

where a is the literal symbol a, and $x\_$ is a pattern variable. Applying this to $a \circ a$ we might think we could just write the result as:

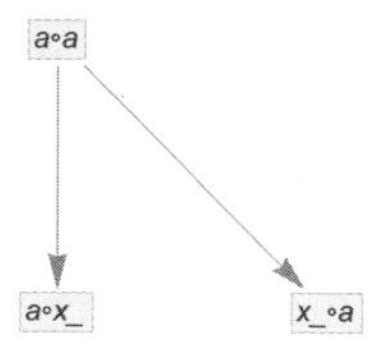

Then if we apply the rule again both branches will give the same expression $x\_ \circ x\_$, so there'll be a merge in the multiway graph:

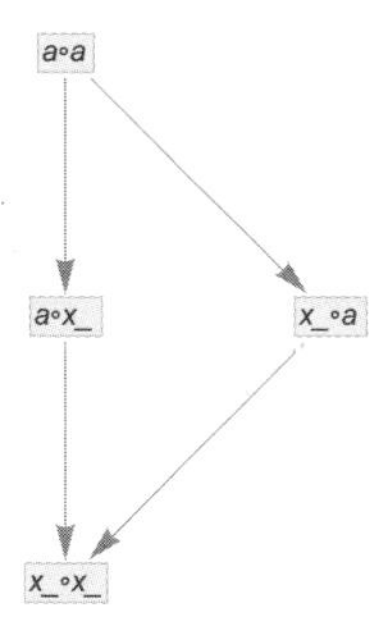

But is this really correct? Well, no. Because really those should be two different $x\_$'s, that could stand for two different expressions. So how can we indicate this? One approach is just to give every "generated" $x\_$ a new name:

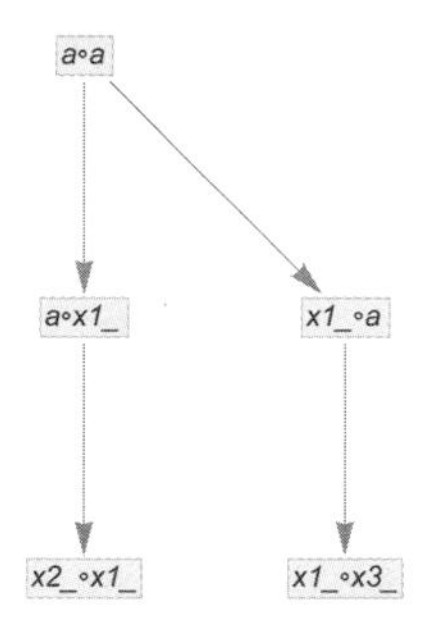

But this result isn't really correct either. Because if we look at the second step we see the two expressions $x1\_ \circ x3\_$ and $x2\_ \circ x4\_$. But what's really the difference between these? The names xi are arbitrary; the only constraint is that within any given expression they have to be different. But between expressions there's no such constraint. And in fact $x1\_ \circ x3\_$ and $x2\_ \circ x4\_$ both represent exactly the same class of expressions: any expression of the form $u\_ \circ v\_$.

So in fact it's not correct that there are two separate branches of the multiway system producing two separate expressions. Because those two branches produce equivalent expressions, which means they can be merged. And turning both equivalent expressions into the same canonical form we get:

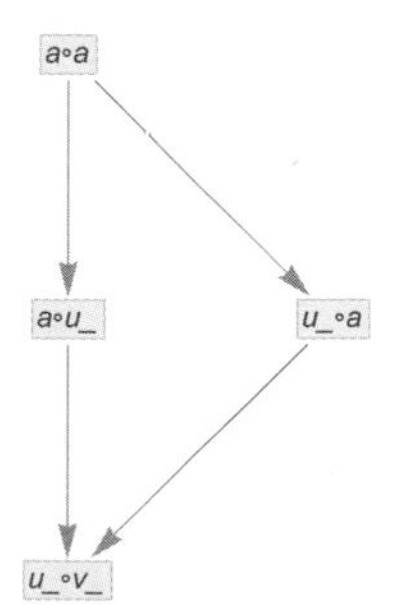

It's important to notice that this isn't the same result as what we got when we assumed that every $x\_$ was the same. Because then our final result was the expression $x\_ \circ x\_$ which can match $a \circ a$ but not $a \circ b$—whereas now the final result is $u\_ \circ v\_$ which can match both $a \circ a$ and $a \circ b$.

This may seem like a subtle issue. But it's critically important in practice. Not least because generated variables are in effect what make up all "truly new stuff" that can be produced. With a rule like $x\_ \circ y\_ \longleftrightarrow (y\_ \circ x\_) \circ y\_$ one's essentially just taking whatever one started with, and successively rearranging the pieces of it. But with a rule like $x\_ \circ y\_ \longleftrightarrow (y\_ \circ x\_) \circ z\_$ there's something "truly new" generated every time $z\_$ appears.

By the way, the basic issue of "generated variables" isn't something specific to the particular symbolic expression setup we've been using here. For example, there's a direct analog of it in the hypergraph rewriting systems that appear in our Physics Project. But in that case there's a particularly clear interpretation: the analog of "generated variables" are new "atoms of space" produced by the application of rules. And far from being some kind of footnote, these "generated atoms of space" are what make up everything we have in our universe today.

The issue of generated variables—and especially their naming—is the bane of all sorts of formalism for mathematical logic and programming languages. As we'll see later, it's perfectly possible to "go to a lower level" and set things up with no names at all, for example using combinators. But without names, things tend to seem quite alien to us humans—and certainly if we want to understand the correspondence with standard presentations of mathematics it's pretty necessary to have names. So at least for now we'll keep names, and handle the issue of generated variables by uniquifying their names, and canonicalizing every time we have a complete expression.

Let's look at another example to see the importance of how we handle generated variables. Consider the rule:

$$x\_ \circ y\_ \longleftrightarrow y\_ \circ z\_$$

If we start with $a \circ a$ and do no uniquification, we'll get:

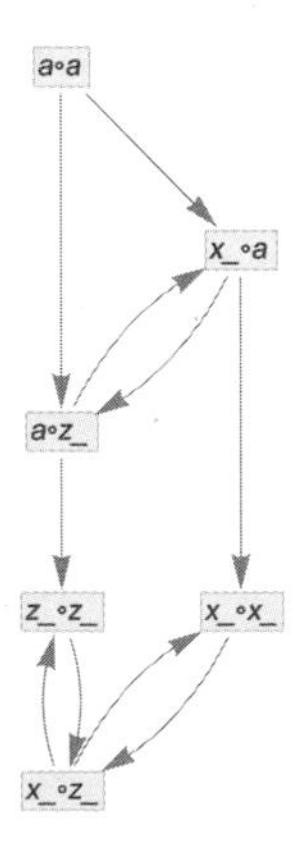

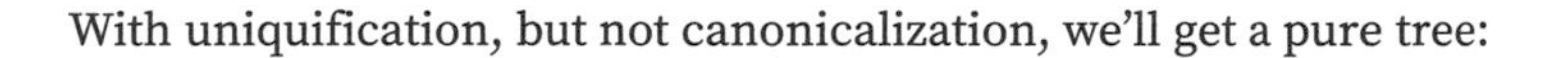

With uniquification, but not canonicalization, we'll get a pure tree:

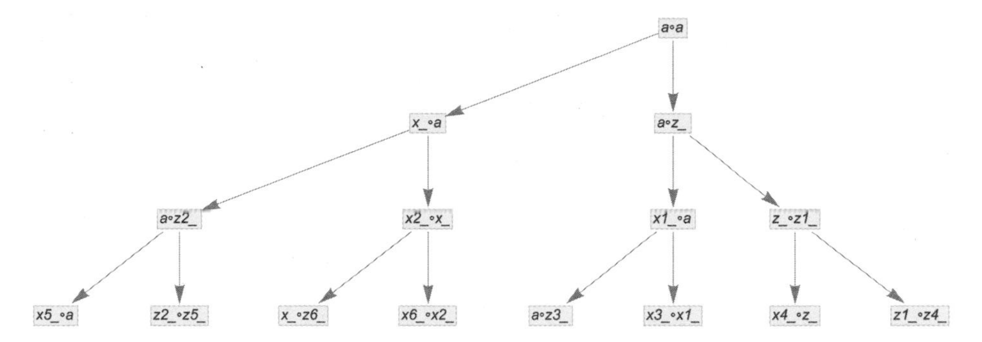

But with canonicalization this is reduced to:

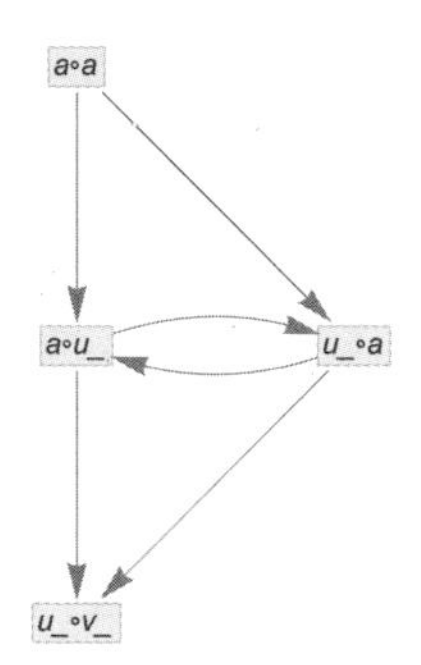

A confusing feature of this particular example is that this same result would have been obtained just by canonicalizing the original "assume-all-x_'s-are-the-same" case.

But things don't always work this way. Consider the rather trivial rule

$x\_ \leftrightarrow y\_$

starting from $a \circ x\_$. If we don't do uniquification, and don't do canonicalization, we get:

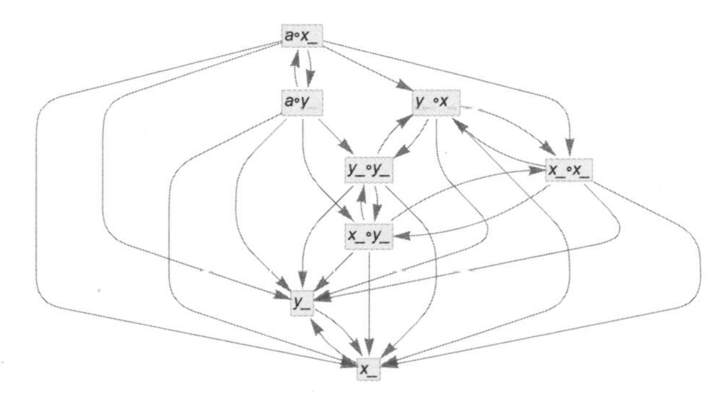

If we do uniquification (but not canonicalization), we get a pure tree:

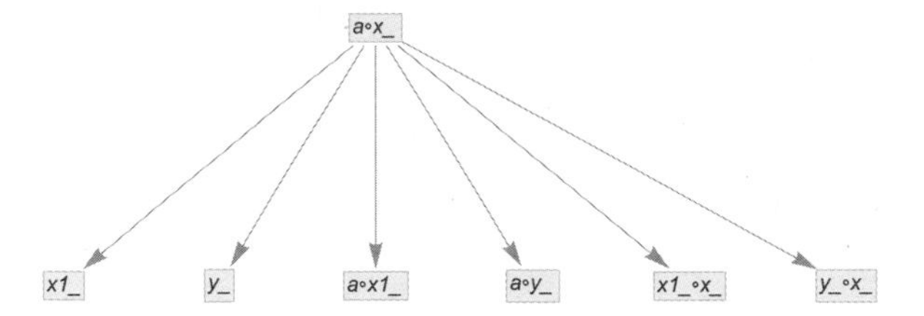

But if we now canonicalize this, we get:

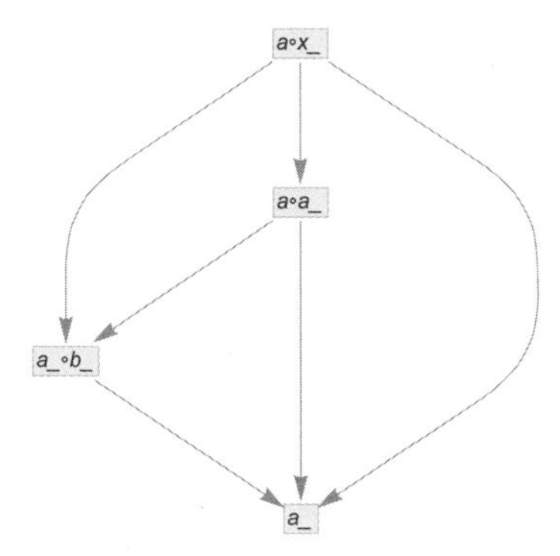

And this is now not the same as what we would get by canonicalizing, without uniquifying:

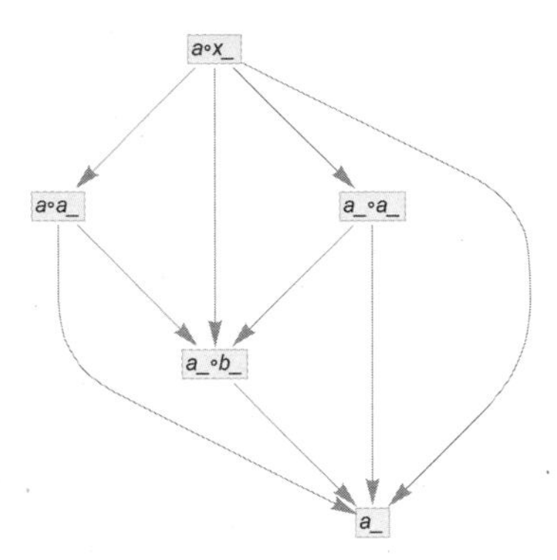

7 | Rules Applied to Rules

In what we've done so far, we've always talked about applying rules (like $x\_ \circ y\_ \longleftrightarrow (y\_ \circ x\_) \circ y\_$) to expressions (like $((b \circ a) \circ a) \circ b$ or $x\_ \circ (y\_ \circ x\_)$). But if everything is a symbolic expression there shouldn't really need to be a distinction between "rules" and "ordinary expressions". They're all just expressions. And so we should as well be able to apply rules to rules as to ordinary expressions.

And indeed the concept of "applying rules to rules" is something that has a familiar analog in standard mathematics. The "two-way rules" we've been using effectively define equivalences—which are very common kinds of statements in mathematics, though in mathematics they're usually written with $=$ rather than with $\longleftrightarrow$. And indeed, many axioms and many theorems are specified as equivalences—and in equational logic one takes everything to be defined using equivalences. And when one's dealing with theorems (or axioms) specified as equivalences, the basic way one derives new theorems is by applying one theorem to another—or in effect by applying rules to rules.

As a specific example, let's say we have the "axiom":

$$x\_ \circ y\_ \longleftrightarrow (y\_ \circ x\_) \circ y\_$$

We can now apply this to the rule

$$a \circ a \longleftrightarrow b \circ b$$

to get (where since $u \longleftrightarrow v$ is equivalent to $v \longleftrightarrow u$ we're sorting each two-way rule that arises)

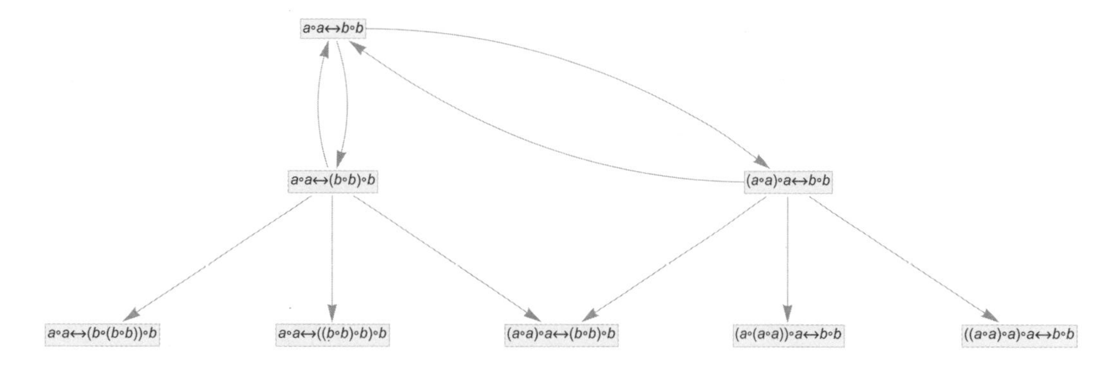

or after a few more steps:

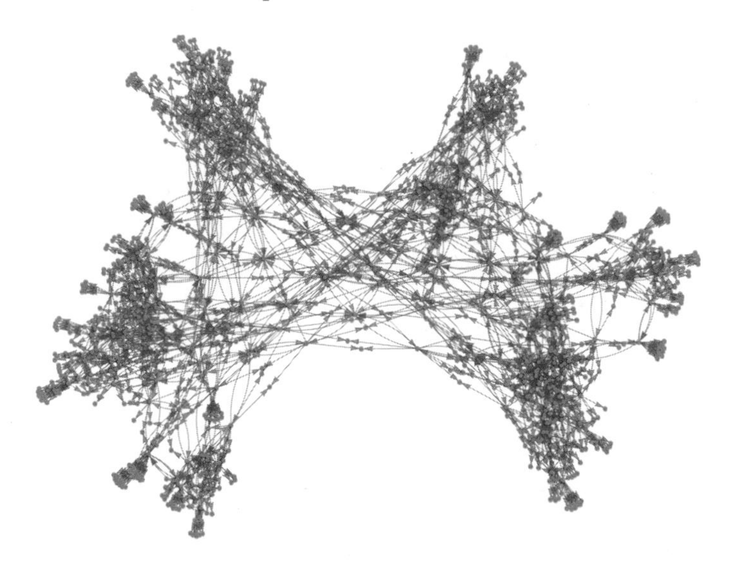

In this example all that's happening is that the substitutions specified by the axiom are getting separately applied to the left- and right-hand sides of each rule that is generated. But if we really take seriously the idea that everything is a symbolic expression, things can get a bit more complicated.

Consider for example the rule:

$x\_ \longleftrightarrow x\_ \circ y\_$

If we apply this to

$a \circ a \longleftrightarrow b \circ b$

then if $x\_$ "matches any expression" it can match the whole expression $a \circ a \longleftrightarrow b \circ b$ giving the result:

$(a \circ a \longleftrightarrow b \circ b) \circ y$

Standard mathematics doesn't have an obvious meaning for something like this—although as soon as one "goes metamathematical" it's fine. But in an effort to maintain contact with standard mathematics we'll for now have the "meta rule" that $x\_$ can't match an expression whose top-level operator is $\longleftrightarrow$. (As we'll discuss later, including such matches would allow us to do exotic things like encode set theory within arithmetic, which is again something usually considered to be "syntactically prevented" in mathematical logic.)

Another—still more obscure—meta rule we have is that *x_* can't "match inside a variable". In Wolfram Language, for example, *a_* has the full form Pattern[a, Blank[]], and one could imagine that *x_* could match "internal pieces" of this. But for now, we're going to treat all variables as atomic—even though later on, when we "descend below the level of variables", the story will be different.

When we apply a rule like $x\_ \longleftrightarrow (x\_ \circ y\_)$ to $a \circ a \longleftrightarrow b \circ b$ we're taking a rule with pattern variables, and doing substitutions with it on a "literal expression" without pattern variables. But it's also perfectly possible to apply pattern rules to pattern rules—and indeed that's what we'll mostly do below. But in this case there's another subtle issue that can arise. Because if our rule generates variables, we can end up with two different kinds of variables with "arbitrary names": generated variables, and pattern variables from the rule we're operating on. And when we canonicalize the names of these variables, we can end up with identical expressions that we need to merge.

Here's what happens if we apply the rule $x\_ \longleftrightarrow (x\_ \circ y\_)$ to the literal rule $a \circ b \longleftrightarrow a$:

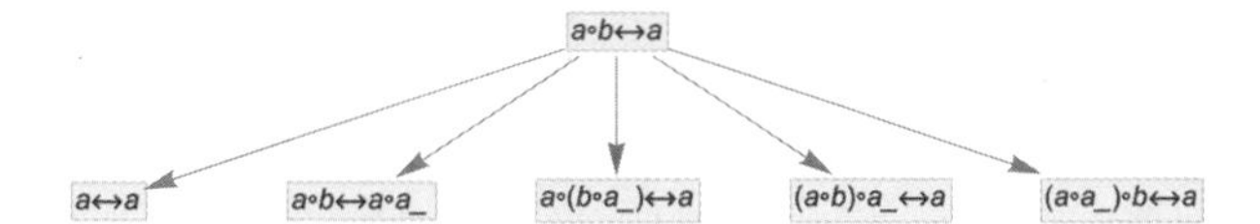

If we apply it to the pattern rule $a\_ \circ b\_ \longleftrightarrow a\_$ but don't do canonicalization, we'll just get the same basic result:

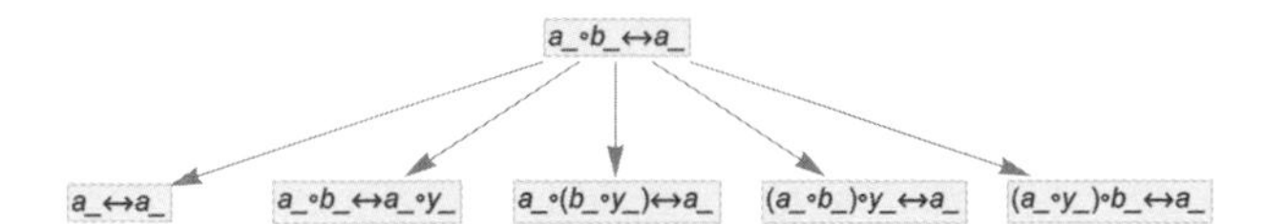

But if we canonicalize we get instead:

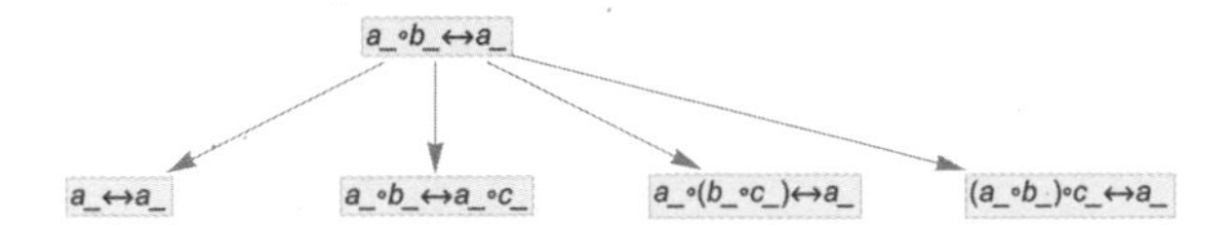

The effect is more dramatic if we go to two steps. When operating on the literal rule we get:

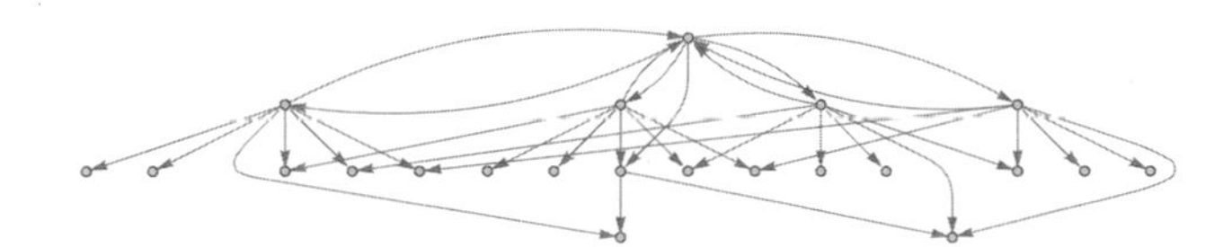

Operating on the pattern rule, but without canonicalization, we get

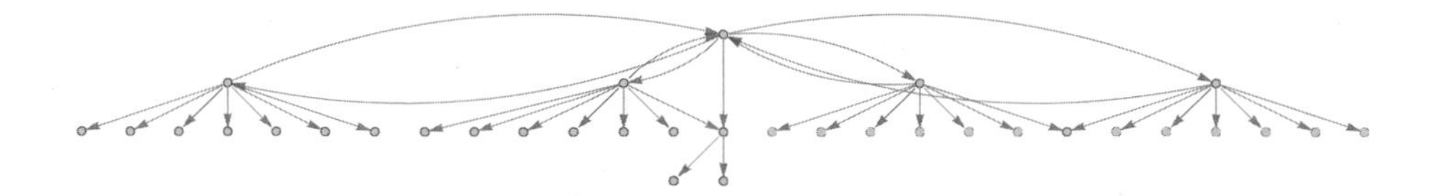

while if we include canonicalization many rules merge and we get:

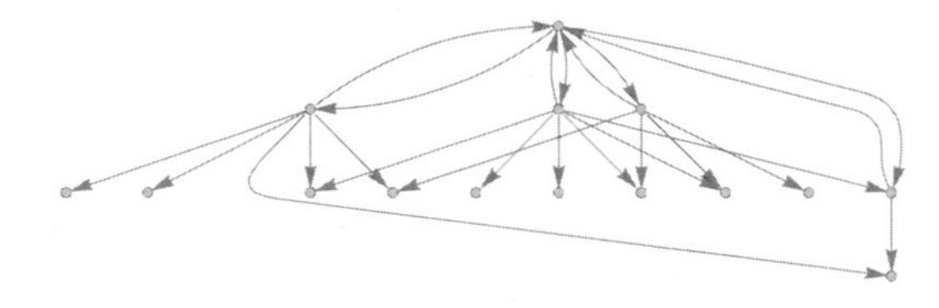

8 | Accumulative Evolution

We can think of "ordinary expressions" like $a \circ b$ as being like "data", and rules as being like "code". But when everything is a symbolic expression, it's perfectly possible—as we saw above—to "treat code like data", and in particular to generate rules as output. But this now raises a new possibility. When we "get a rule as output", why not start "using it like code" and applying it to things?

In mathematics we might apply some theorem to prove a lemma, and then we might subsequently use that lemma to prove another theorem—eventually building up a whole "accumulative structure" of lemmas (or theorems) being used to prove other lemmas. In any given proof we can in principle always just keep using the axioms over and over again—but it'll be much more efficient to progressively build a library of more and more lemmas, and use these. And in general we'll build up a richer structure by "accumulating lemmas" than always just going back to the axioms.

In the multiway graphs we've drawn so far, each edge represents the application of a rule, but that rule is always a fixed axiom. To represent accumulative evolution we need a slightly more elaborate structure—and it'll be convenient to use token-event graphs rather than pure multiway graphs.

Every time we apply a rule we can think of this as an event. And with the setup we're describing, that event can be thought of as taking two tokens as input: one the "code rule" and the other the "data rule". The output from the event is then some collection of rules, which can then serve as input (either "code" or "data") to other events.

Let's start with the very simple example of the rule

$$x \longleftrightarrow y$$

where for now there are no patterns being used. Starting from this rule, we get the token-event graph (where now we're indicating the initial "axiom" statement using a slightly different color):

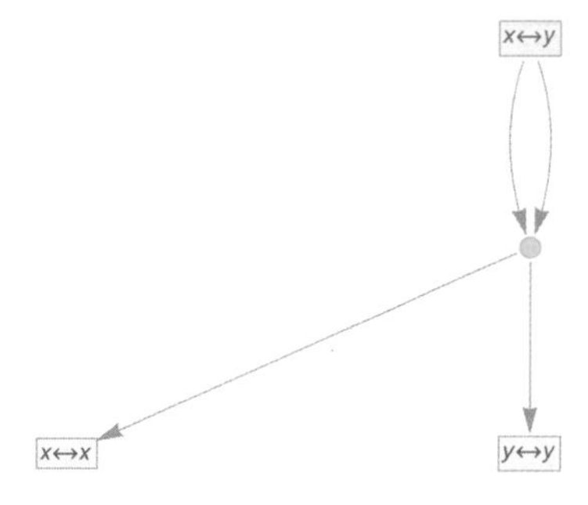

One subtlety here is that the $x \longleftrightarrow y$ is applied to itself—so there are two edges going into the event from the node representing the rule. Another subtlety is that there are two different ways the rule can be applied, with the result that there are two output rules generated.

Here's another example, based on the two rules:

$x \longleftrightarrow y, y \longleftrightarrow z$

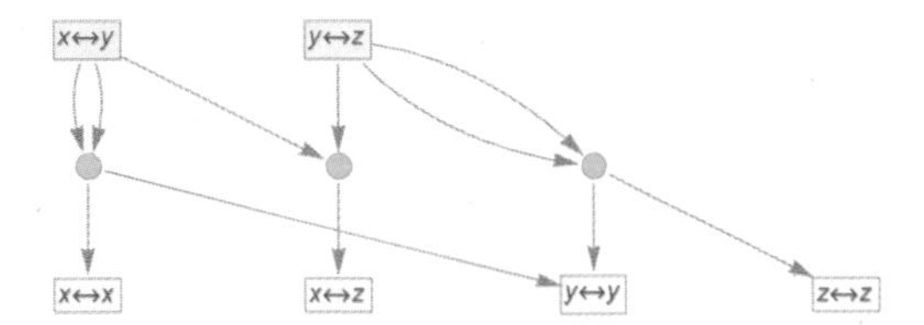

Continuing for another step we get:

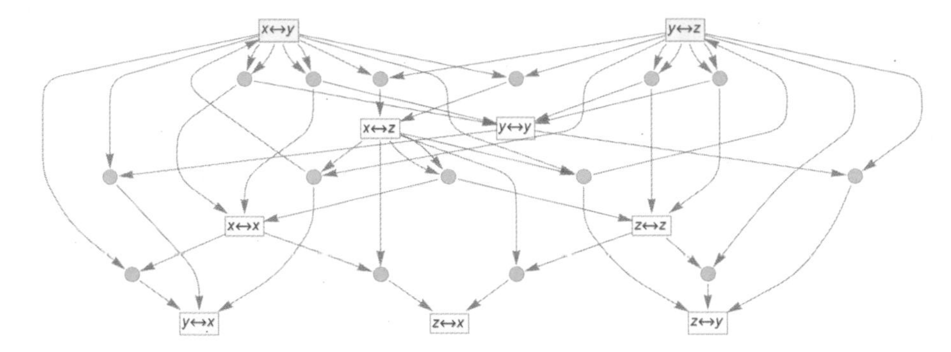

Typically we will want to consider $\longleftrightarrow$ as "defining an equivalence", so that $v \longleftrightarrow u$ means the same as $u \longleftrightarrow v$, and can be conflated with it—yielding in this case:

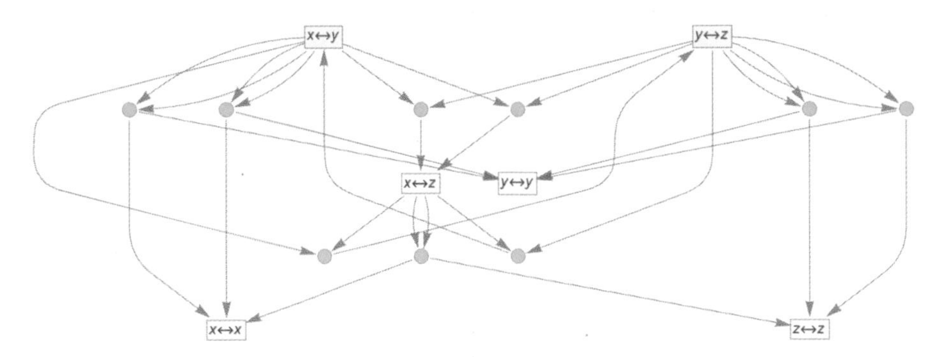

Now let's consider the rule:

$a \circ b \longleftrightarrow b$

After one step we get:

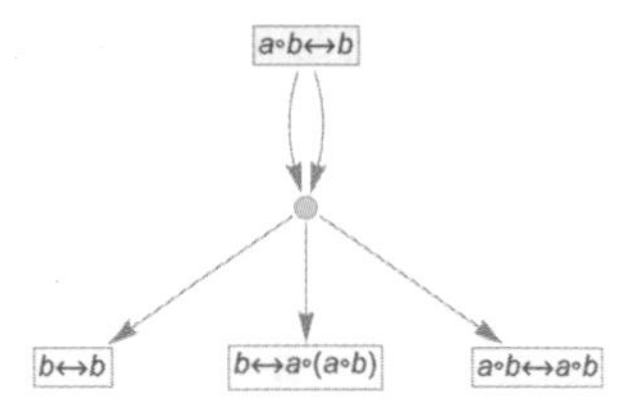

After 2 steps we get:

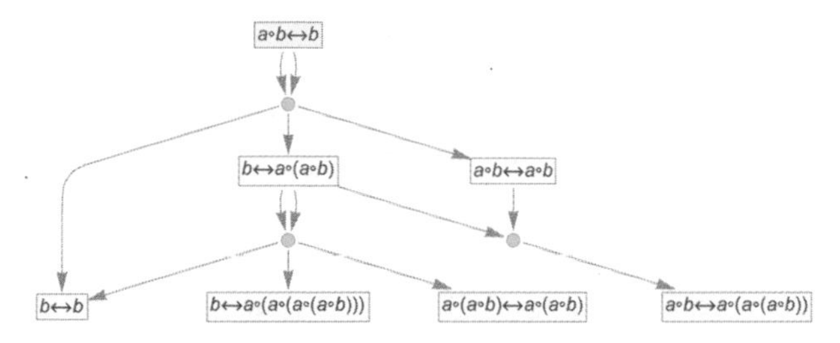

The token-event graphs after 3 and 4 steps in this case are (where now we've deduplicated events):

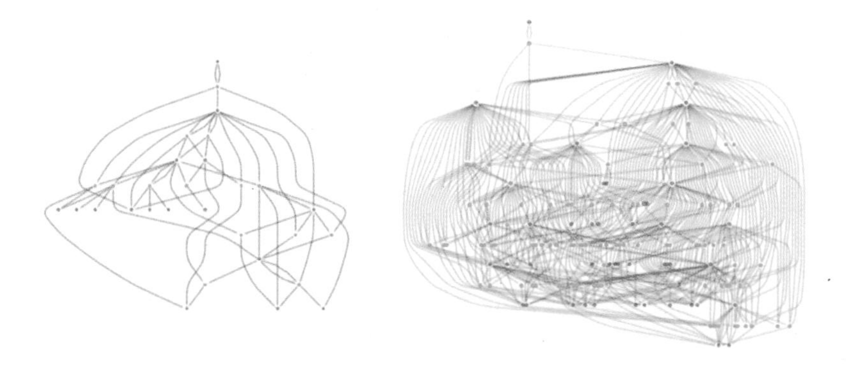

Let's now consider a rule with the same structure, but with pattern variables instead of literal symbols:

$a\_ \circ b\_ \longleftrightarrow b\_$

Here's what happens after one step (note that there's canonicalization going on, so *a_*'s in different rules aren't "the same")

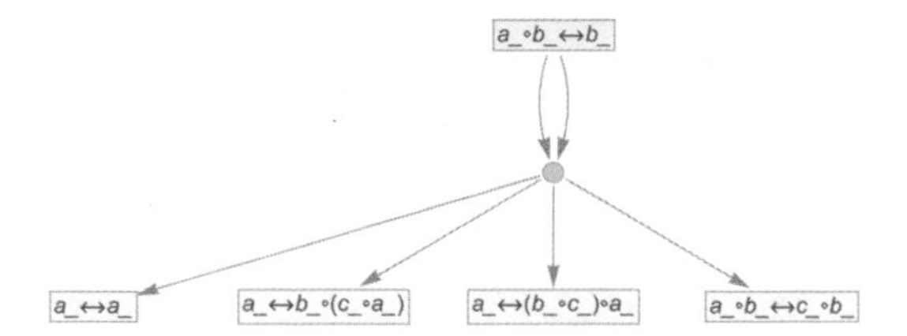

and we see that there are different theorems from the ones we got without patterns. After 2 steps with the pattern rule we get

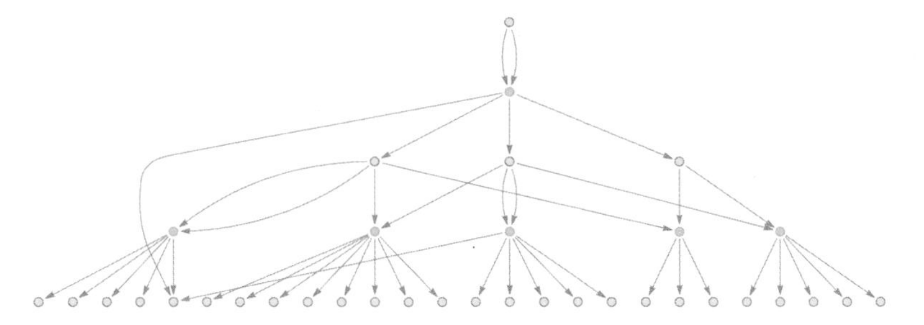

where now the complete set of “theorems that have been derived” is (dropping the _’s for readability)

$a\circ b \leftrightarrow b$	$a \leftrightarrow a$	$a \leftrightarrow b\circ(c\circ a)$	$a \leftrightarrow (b\circ c)\circ a$	$a\circ b \leftrightarrow c\circ b$
$a \leftrightarrow b\circ a$	$a \leftrightarrow b\circ(c\circ(d\circ(e\circ a)))$	$a \leftrightarrow b\circ(c\circ((d\circ e)\circ a))$	$a \leftrightarrow b\circ((c\circ d)\circ(e\circ a))$	$a \leftrightarrow b\circ((c\circ(d\circ e))\circ a)$
$a \leftrightarrow b\circ(((c\circ d)\circ e)\circ a)$	$a \leftrightarrow (b\circ c)\circ(d\circ(e\circ a))$	$a \leftrightarrow (b\circ c)\circ((d\circ e)\circ a)$	$a \leftrightarrow (b\circ(c\circ d))\circ(e\circ a)$	$a \leftrightarrow (b\circ(c\circ(d\circ e)))\circ a$
$a \leftrightarrow (b\circ((c\circ d)\circ e))\circ a$	$a \leftrightarrow ((b\circ c)\circ d)\circ(e\circ a)$	$a \leftrightarrow ((b\circ c)\circ(d\circ e))\circ a$	$a \leftrightarrow ((b\circ(c\circ d))\circ e)\circ a$	$a \leftrightarrow (((b\circ c)\circ d)\circ e)\circ a$
$a\circ b \leftrightarrow c\circ(d\circ(e\circ b))$	$a\circ b \leftrightarrow c\circ((d\circ e)\circ b)$	$a\circ b \leftrightarrow (c\circ d)\circ(e\circ b)$	$a\circ b \leftrightarrow (c\circ(d\circ e))\circ b$	$a\circ b \leftrightarrow ((c\circ d)\circ e)\circ b$
$a\circ(b\circ c) \leftrightarrow d\circ(e\circ c)$	$a\circ(b\circ c) \leftrightarrow (d\circ e)\circ c$	$a\circ(b\circ(c\circ d)) \leftrightarrow e\circ d$	$a\circ((b\circ c)\circ d) \leftrightarrow e\circ d$	$(a\circ b)\circ c \leftrightarrow (d\circ e)\circ c$

or as trees:

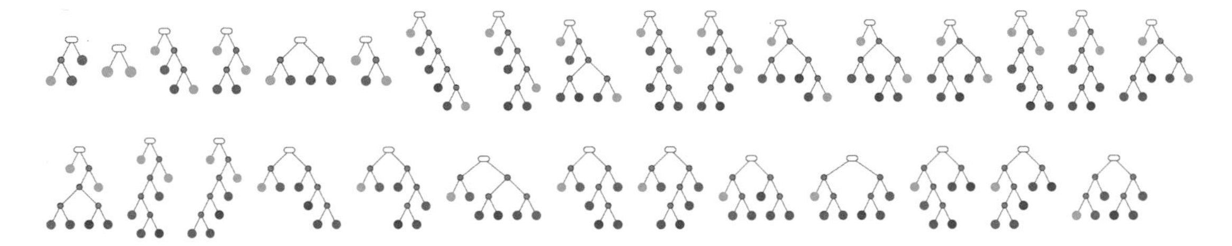

After another step one gets

where now there are 2860 "theorems", roughly exponentially distributed across sizes according to

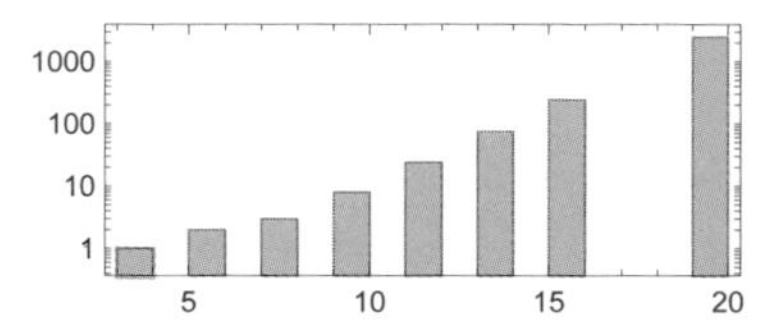

and with a typical "size-19" theorem being:

$$a\_\circ b\_ \leftrightarrow c\_\circ(d\_\circ(e\_\circ(f\_\circ(g\_\circ(h\_\circ(i\_\circ b\_))))))$$

In effect we can think of our original rule (or "axiom") as having initiated some kind of "mathematical Big Bang" from which an increasing number of theorems are generated. Early on we described having a "gas" of mathematical theorems that—a little like molecules—can interact and create new theorems. So now we can view our accumulative evolution process as a concrete example of this.

Let's consider the rule from previous sections:

$$x\_\circ y\_ \leftrightarrow (y\_\circ x\_)\circ y\_$$

After one step of accumulative evolution according to this rule we get:

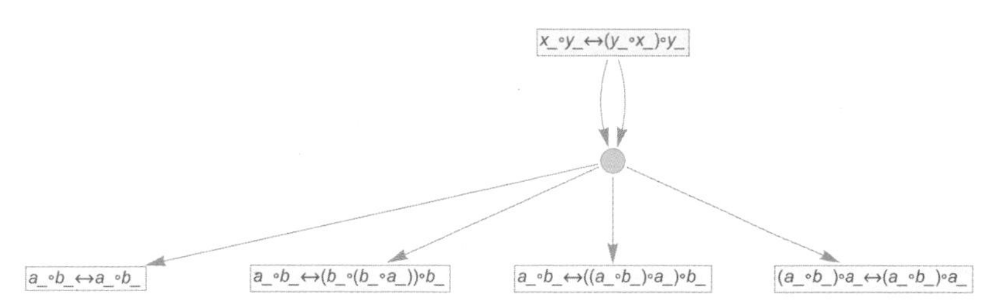

After 2 and 3 steps the results are:

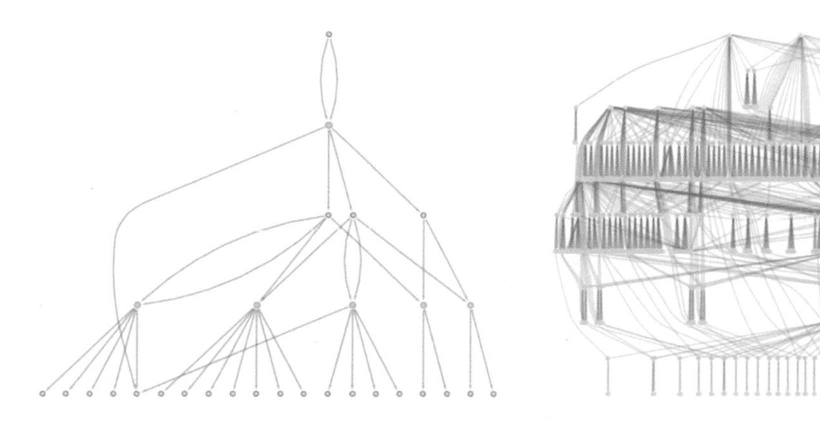

What is the significance of all this complexity? At a basic level, it's just an example of the ubiquitous phenomenon in the computational universe (captured in the Principle of Computational Equivalence) that even systems with very simple rules can generate behavior as complex as anything. But the question is whether—on top of all this complexity—there are simple "coarse-grained" features that we can identify as "higher-level mathematics"; features that we can think of as capturing the "bulk" behavior of the accumulative evolution of axiomatic mathematics.

9 | Accumulative String Systems

As we've just seen, the accumulative evolution of even very simple transformation rules for expressions can quickly lead to considerable complexity. And in an effort to understand the essence of what's going on, it's useful to look at the slightly simpler case not of rules for "tree-structured expressions" but instead at rules for strings of characters.

Consider the seemingly trivial case of the rule:

A ↔ B

After one step this gives

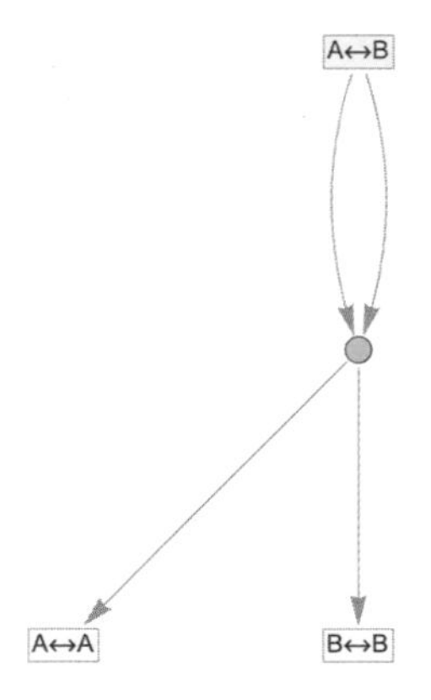

while after 2 steps we get

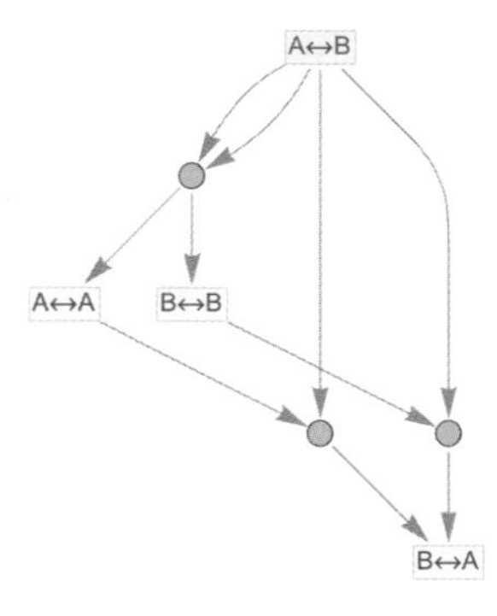

though treating $u \longleftrightarrow v$ as the same as $v \longleftrightarrow u$ this just becomes:

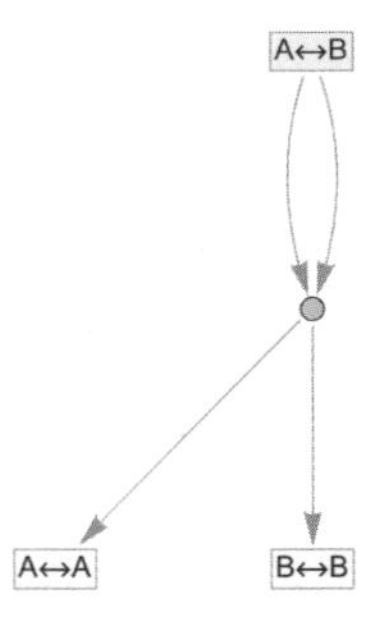

Here's what happens with the rule:

AB ↔ B

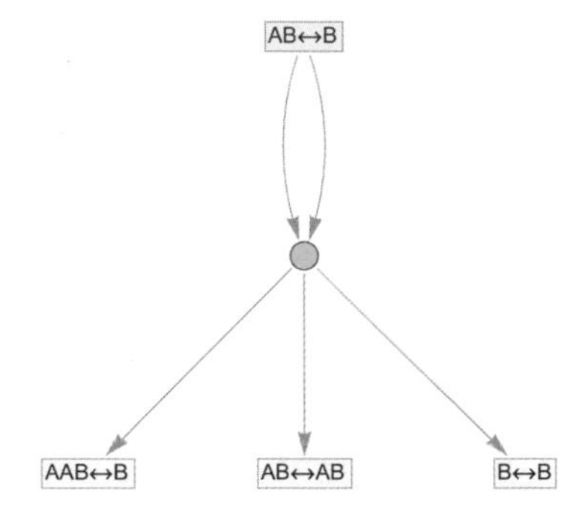

After 2 steps we get

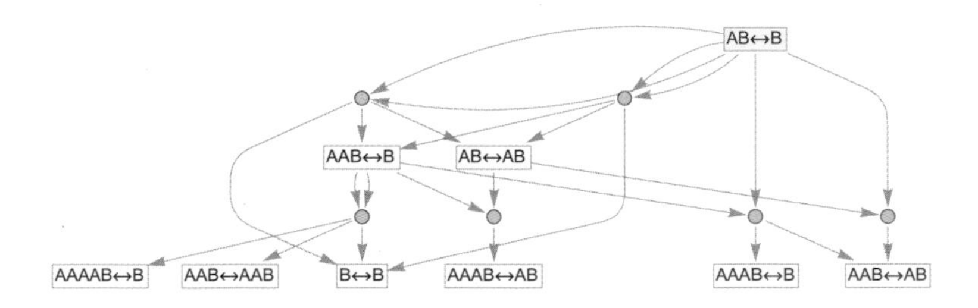

and after 3 steps

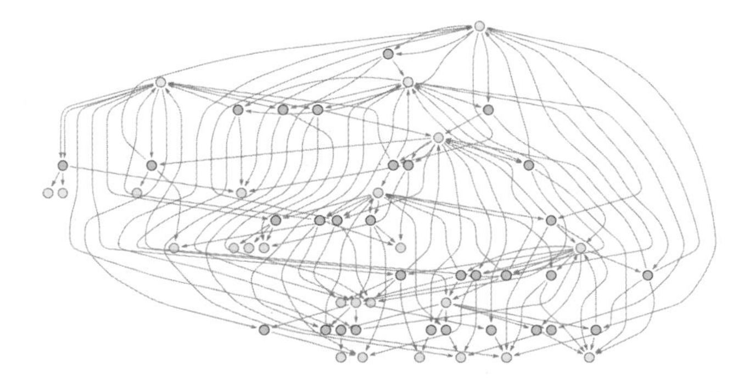

where now there are a total of 25 "theorems", including (unsurprisingly) things like:

AAAAAAB ⟷ AAB

It's worth noting that despite the "lexical similarity" of the string rule AB ⟷ B we're now using to the expression rule $a \circ b \longleftrightarrow b$ from the previous section, these rules actually work in very different ways. The string rule can apply to characters anywhere within a string, but what it inserts is always of fixed size. The expression rule deals with trees, and only applies to "whole subtrees", but what it inserts can be a tree of any size. (One can align these setups by thinking of strings as expressions in which characters are "bound together" by an associative operator, as in A·B·A·A. But if one explicitly gives associativity axioms these will lead to additional pieces in the token-event graph.)

A rule like $a\_ \circ b\_ \longleftrightarrow b\_$ also has the feature of involving patterns. In principle we could include patterns in strings too—both for single characters (as with _) and for sequences of characters (as with __)—but we won't do this here. (We can also consider one-way rules, using → instead of ⟷.)

To get a general sense of the kinds of things that happen in accumulative (string) systems, we can consider enumerating all possible distinct two-way string transformation rules. With only a single character A, there are only two distinct cases

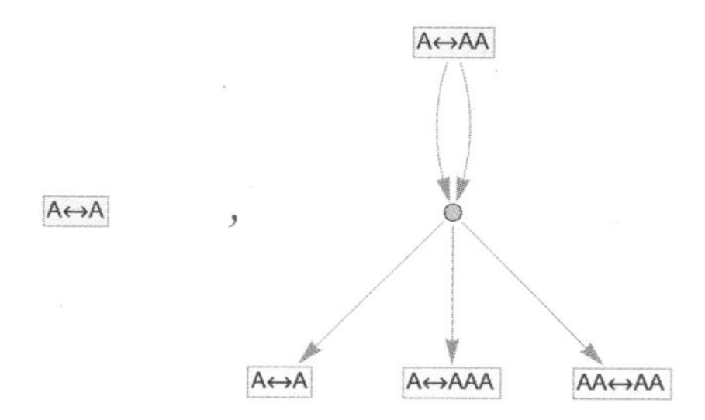

because A ⟷ AA systematically generates all possible $A^n \longleftrightarrow A^m$ rules

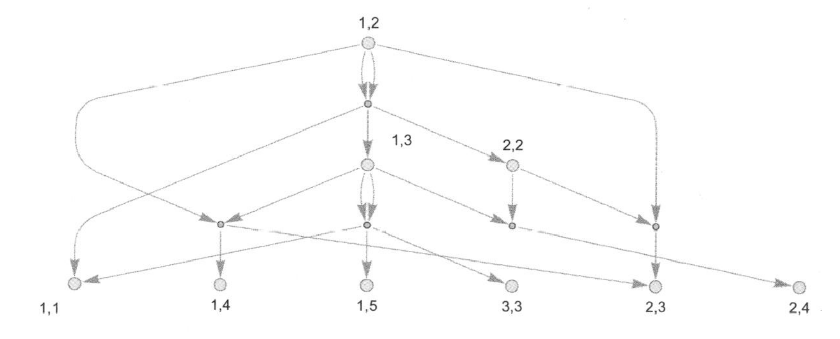

and at t steps gives a total number of rules equal to:

$$(2^{t-1}+1)^2$$

With characters A and B the distinct token-event graphs generated starting from rules with a total of at most 5 characters are:

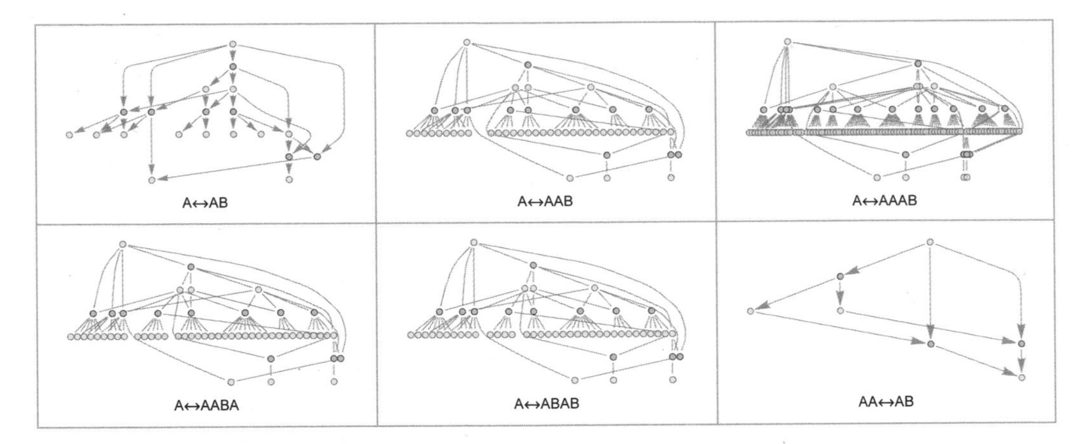

Note that when the strings in the initial rule are the same length, only a rather trivial finite token-event graph is ever generated, as in the case of AA ⟷ AB:

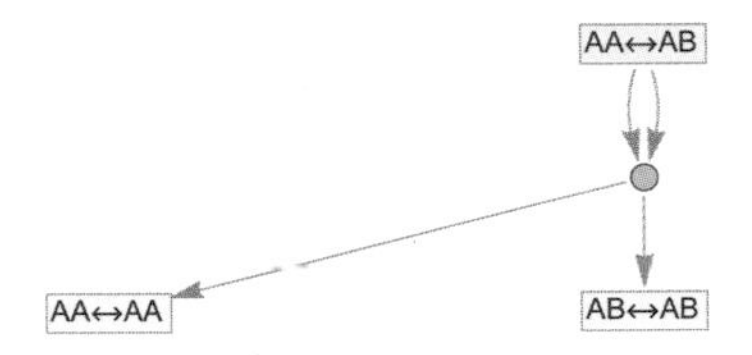

But when the strings are of different lengths, there is always unbounded growth.

10 | The Case of Hypergraphs

We've looked at accumulative versions of expression and string rewriting systems. So what about accumulative versions of hypergraph rewriting systems of the kind that appear in our Physics Project?

Consider the very simple hypergraph rule

$\{\{x, x\}\} \to \{\{x, x\}, \{x, x\}\}$

or pictorially:

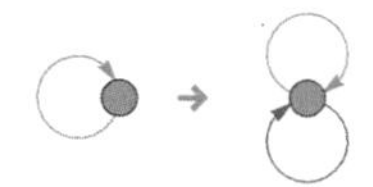

(Note that the nodes that are named 1 here are really like pattern variables, that could be named for example $x\_$.)

We can now do accumulative evolution with this rule, at each step combining results that involve equivalent (i.e. isomorphic) hypergraphs:

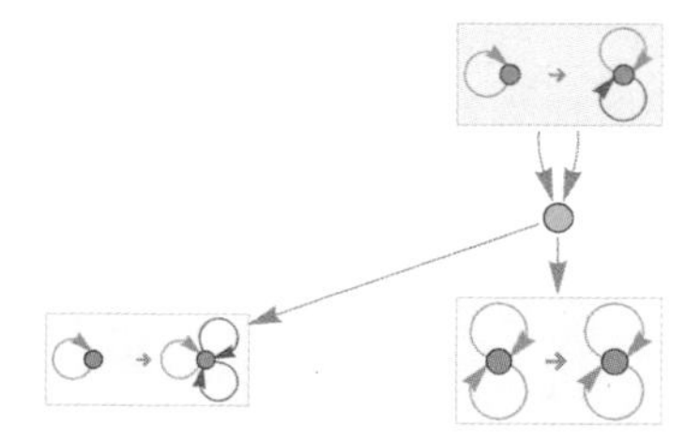

After two steps this gives:

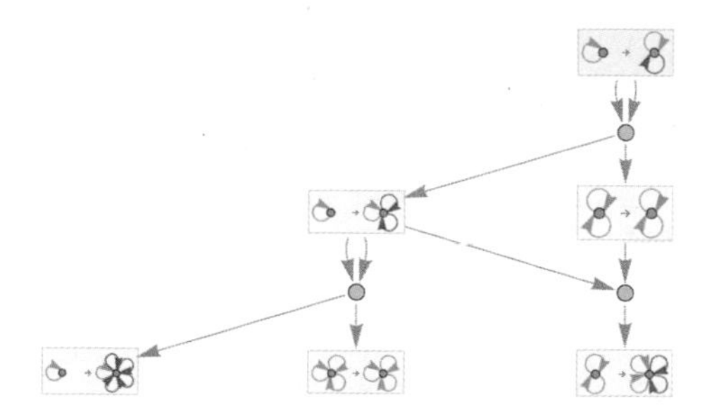

And after 3 steps:

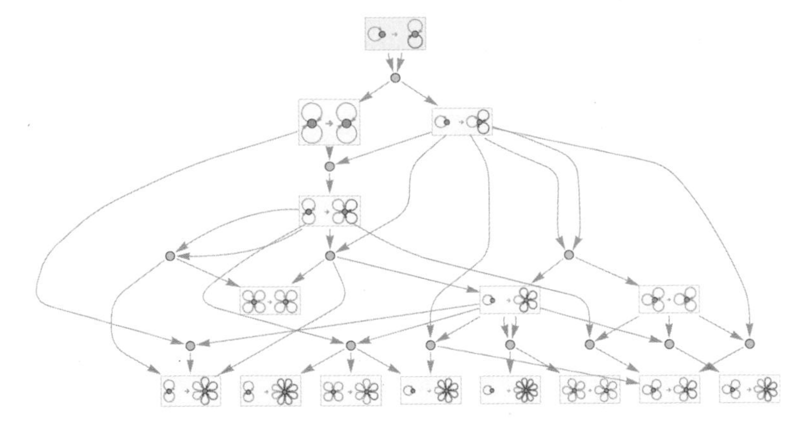

How does all this compare to "ordinary" evolution by hypergraph rewriting? Here's a multiway graph based on applying the same underlying rule repeatedly, starting from an initial condition formed from the rule:

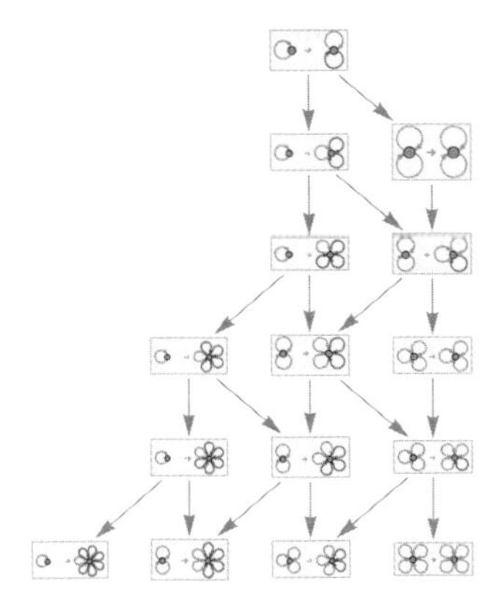

What we see is that the accumulative evolution in effect "shortcuts" the ordinary multiway evolution, essentially by "caching" the result of every piece of every transformation between states (which in this case are rules), and delivering a given state in fewer steps.

In our typical investigation of hypergraph rewriting for our Physics Project we consider one-way transformation rules. Inevitably, though, the ruliad contains rules that go both ways. And here, in an effort to understand the correspondence with our metamodel of mathematics, we can consider two-way hypergraph rewriting rules. An example is the two-way version of the rule above:

$\{\{x, x\}\} \longleftrightarrow \{\{x, x\}, \{x, x\}\}$

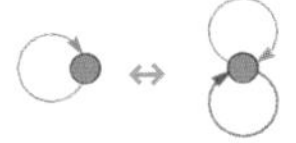

Now the token-event graph becomes

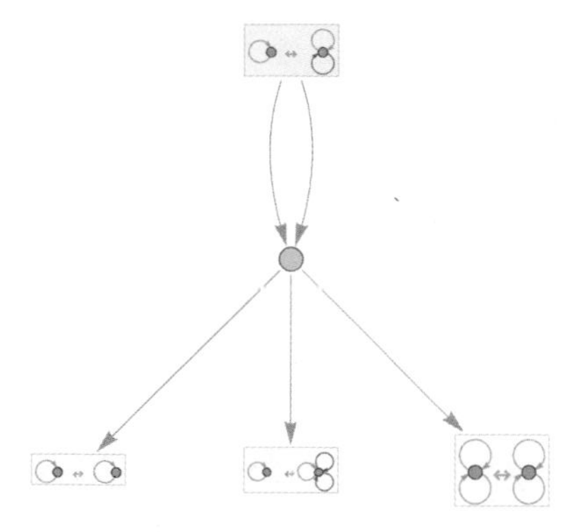

or after 2 steps (where now the transformations from "later states" to "earlier states" have started to fill in):

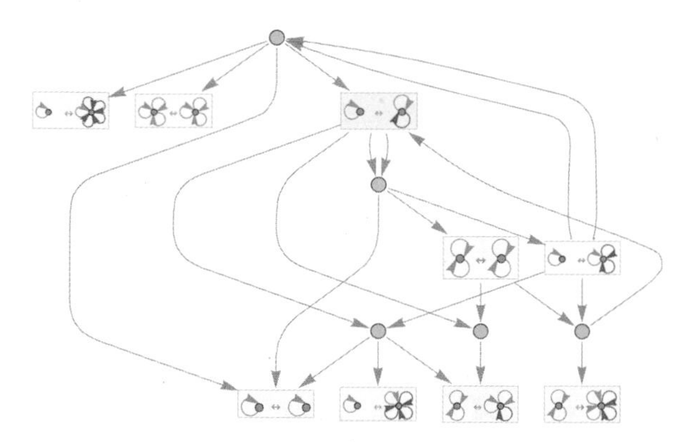

Just like in ordinary hypergraph evolution, the only way to get hypergraphs with additional hyperedges is to start with a rule that involves the addition of new hyperedges—and the same is true for the addition of new elements. Consider the rule:

$\{\{x, y\}, \{x, z\}\} \longleftrightarrow \{\{x, y\}, \{x, w\}, \{y, w\}, \{z, w\}\}$

After 1 step this gives

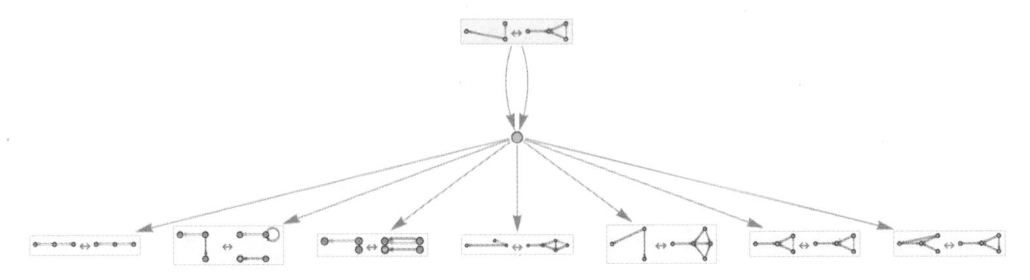

while after 2 steps it gives:

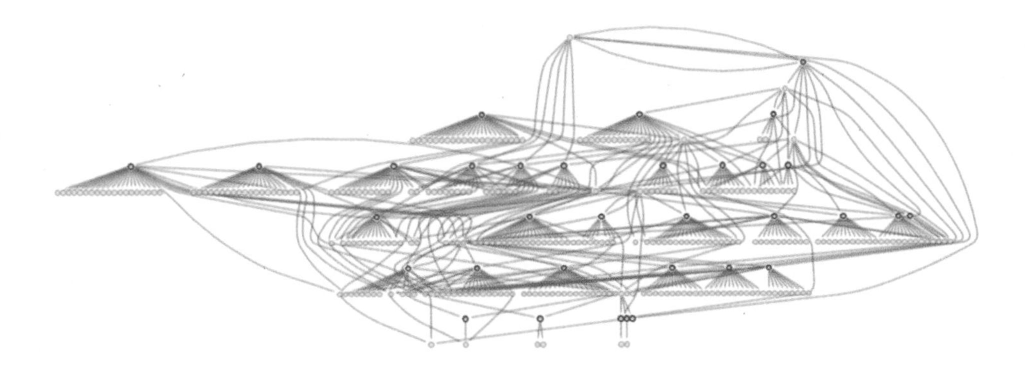

The general appearance of this token-event graph is not much different from what we saw with string rewrite or expression rewrite systems. So what this suggests is that it doesn't matter much whether we're starting from our metamodel of axiomatic mathematics or from any other reasonably rich rewriting system: we'll always get the same kind of "large-scale" token-event graph structure. And this is an example of what we'll use to argue for general laws of metamathematics.

11 | Proofs in Accumulative Systems

In an earlier section, we discussed how paths in a multiway graph can represent proofs of "equivalence" between expressions (or the "entailment" of one expression by another). For example, with the rule (or "axiom")

{A → BBB, BB → A}

this shows a path that "proves" that "BA entails AAB":

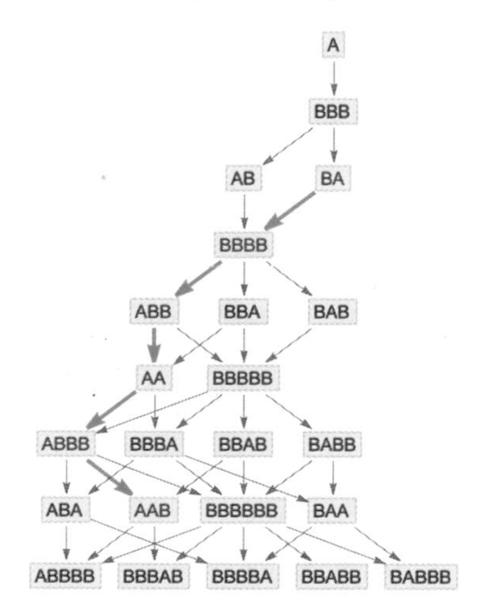

But once we know this, we can imagine adding this result (as what we can think of as a "lemma") to our original rule:

{A → BBB, BB → A, BA → AAB}

And now (the "theorem") "BA entails AAB" takes just one step to prove—and all sorts of other proofs are also shortened:

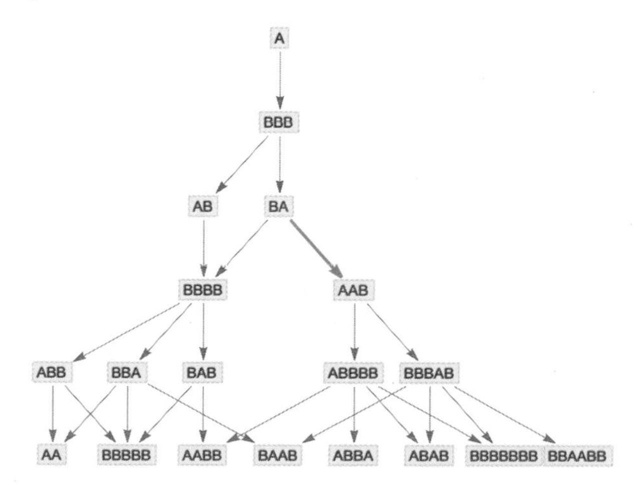

It's perfectly possible to imagine evolving a multiway system with a kind of "caching-based" speed-up mechanism where every new entailment discovered is added to the list of underlying rules. And, by the way, it's also possible to use two-way rules throughout the multiway system:

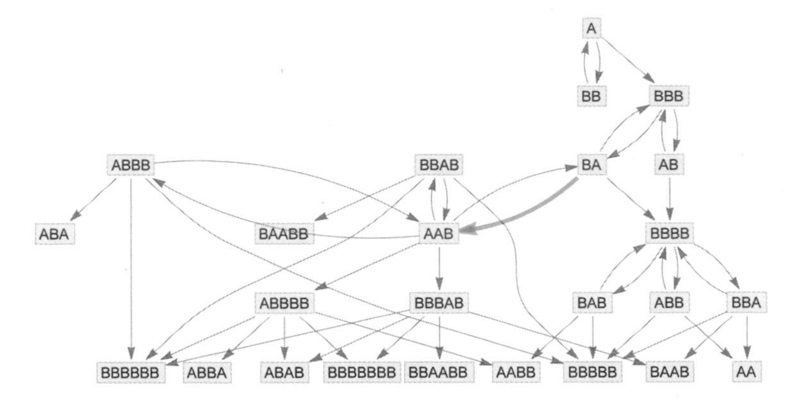

But accumulative systems provide a much more principled way to progressively "add what's discovered". So what do proofs look like in such systems?

Consider the rule:

A ⟷ AB

Running it for 2 steps we get the token-event graph:

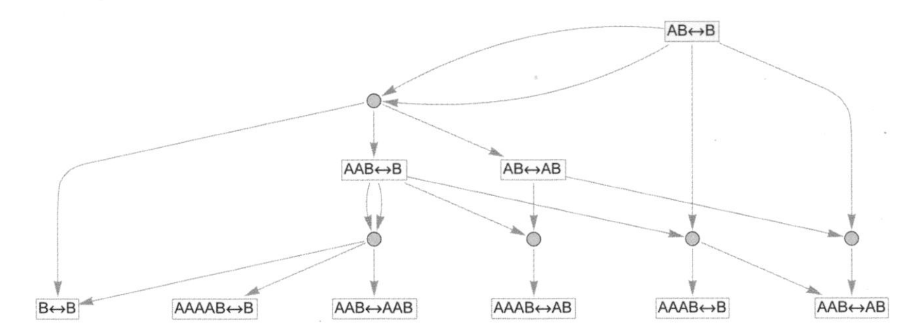

Now let's say we want to prove that the original "axiom" AB ⟷ B implies (or "entails") the "theorem" AAAB ⟷ AB. Here's the subgraph that demonstrates the result:

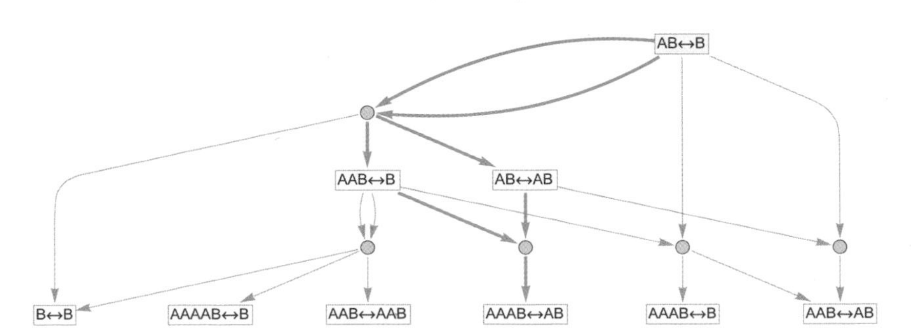

And here it is as a separate "proof graph"

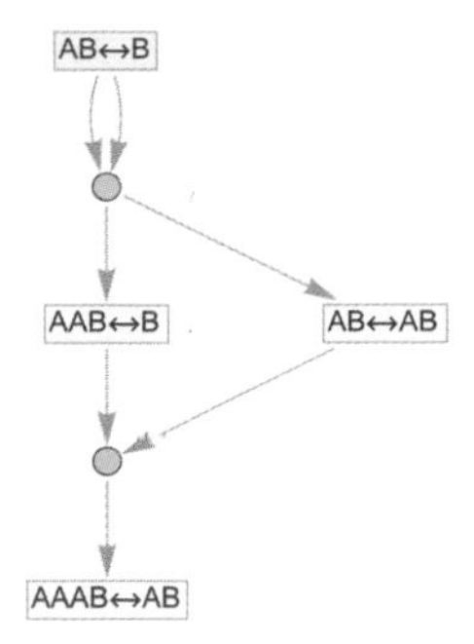

where each event takes two inputs—the "rule to be applied" and the "rule to apply to"—and the output is the derived (i.e. entailed or implied) new rule or rules.

If we run the accumulative system for another step, we get:

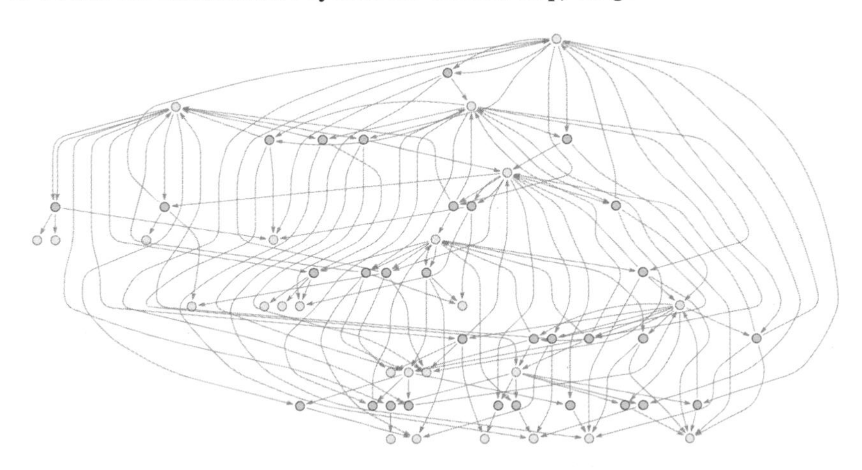

Now there are additional "theorems" that have been generated. An example is:

AAAB ↔ AAB

And now we can find a proof of this theorem:

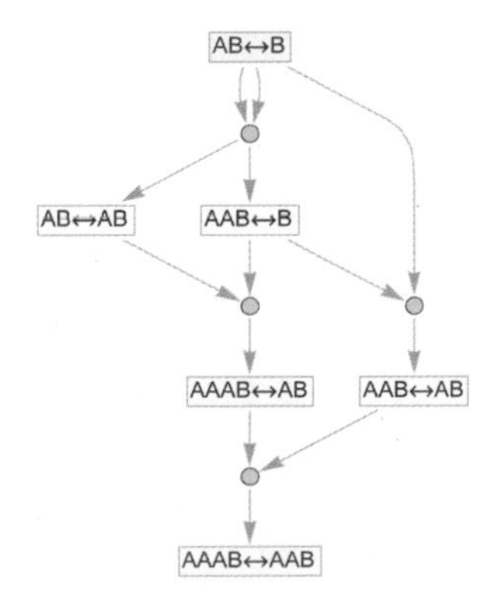

This proof exists as a subgraph of the token-event graph:

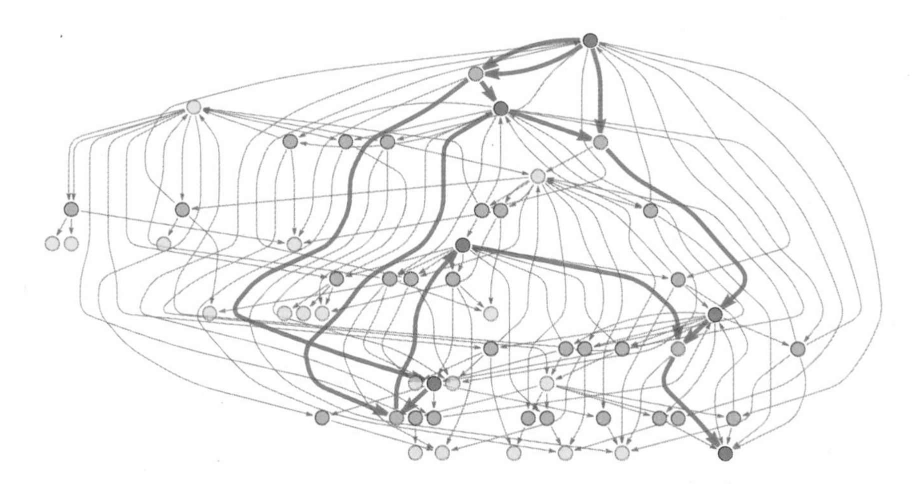

The proof just given has the fewest events—or "proof steps"—that can be used. But altogether there are 50 possible proofs, other examples being:

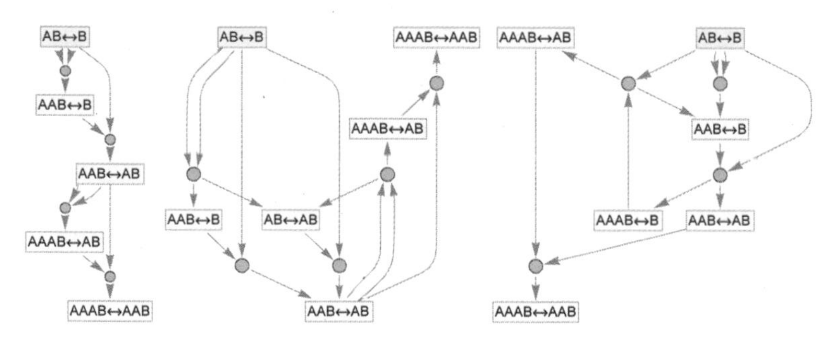

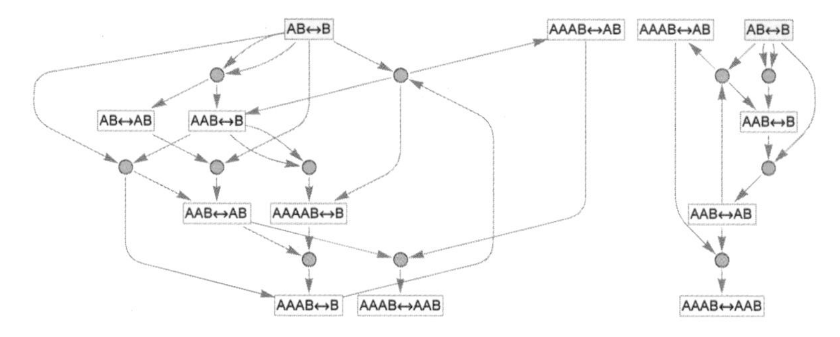

These correspond to the subgraphs:

How much has the accumulative character of these token-event graphs contributed to the structure of these proofs? It's perfectly possible to find proofs that never use "intermediate lemmas" but always "go back to the original axiom" at every step. In this case examples are

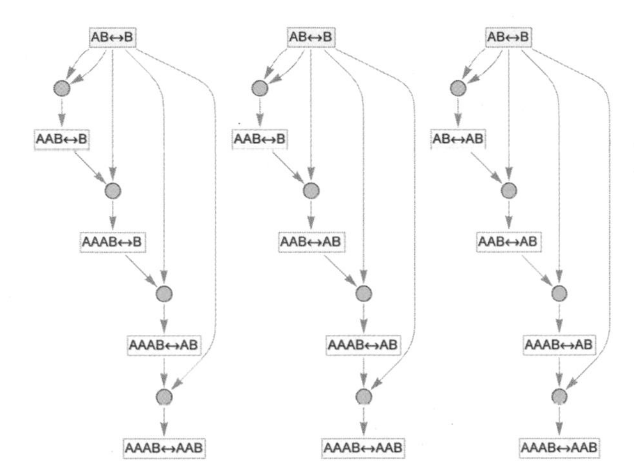

which all in effect require at least one more "sequential event" than our shortest proof using intermediate lemmas.

A slightly more dramatic example occurs for the theorem

AAAAAB ↔ AAAAB

where now without intermediate lemmas the shortest proof is

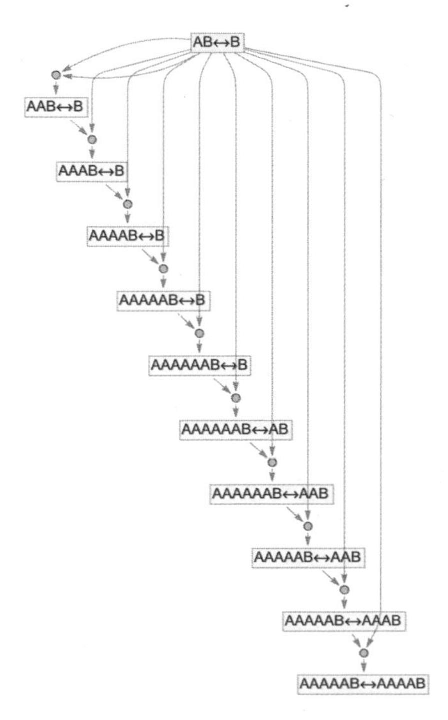

but with intermediate lemmas it becomes:

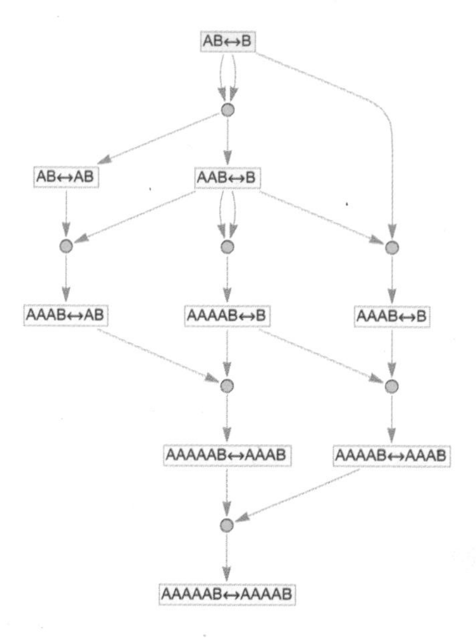

What we've done so far here is to generate a complete token-event graph for a certain number of steps, and then to see if we can find a proof in it for some particular statement. The proof is a subgraph of the "relevant part" of the full token-event graph. Often—in analogy to the simpler case of finding proofs of equivalences between expressions in a multiway graph—we'll call this subgraph a "proof path".

But in addition to just "finding a proof" in a fully constructed token-event graph, we can ask whether, given a statement, we can directly construct a proof for it. As discussed in the context of proofs in ordinary multiway graphs, computational irreducibility implies that in general there's no "shortcut" way to find a proof. In addition, for any statement, there may be no upper bound on the length of proof that will be required (or on the size or number of intermediate "lemmas" that will have to be used). And this, again, is the shadow of undecidability in our systems: that there can be statements whose provability may be arbitrarily difficult to determine.

12 | Beyond Substitution: Cosubstitution and Bisubstitution

In making our "metamodel" of mathematics we've been discussing the rewriting of expressions according to rules. But there's a subtle issue that we've so far avoided, that has to do with the fact that the expressions we're rewriting are often themselves patterns that stand for whole classes of expressions. And this turns out to allow for additional kinds of transformations that we'll call cosubstitution and bisubstitution.

Let's talk first about cosubstitution. Imagine we have the expression f[a]. The rule $a \to b$ would do a substitution for a to give f[b]. But if we have the expression f[c] the rule $a \to b$ will do nothing.

Now imagine that we have the expression f[x_]. This stands for a whole class of expressions, including f[a], f[c], etc. For most of this class of expressions, the rule $a \to b$ will do nothing. But in the specific case of f[a], it applies, and gives the result f[b].

If our rule is f[x_] → s then this will apply as an ordinary substitution to f[a], giving the result s. But if the rule is f[b] → s this will not apply as an ordinary substitution to f[a]. However, it can apply as a cosubstitution to f[x_] by picking out the specific case where x_ stands for b, then using the rule to give s.

In general, the point is that ordinary substitution specializes patterns that appear in rules—while what one can think of as the "dual operation" of cosubstitution specializes patterns that appear in the expressions to which the rules are being applied. If one thinks of the rule that's being applied as like an operator, and the expression to which the rule is being applied as an operand, then in effect substitution is about making the operator fit the operand, and cosubstitution is about making the operand fit the operator.

It's important to realize that as soon as one's operating on expressions involving patterns, cosubstitution is not something "optional": it's something that one has to include if one is really going to interpret patterns—wherever they occur—as standing for classes of expressions.

When one's operating on a literal expression (without patterns) only substitution is ever possible, as in

f[*x*_] → *g*[*x*]: *f*[*a*] → ● → *g*[*a*]

corresponding to this fragment of a token-event graph:

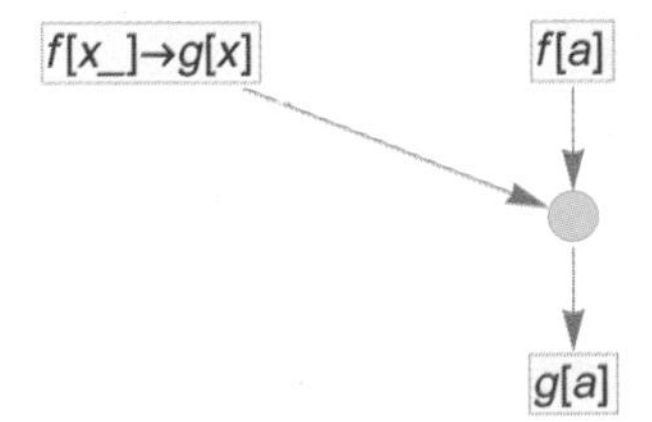

Let's say we have the rule f[a] → s (where f[a] is a literal expression). Operating on f[b] this rule will do nothing. But what if we apply the rule to f[x_]? Ordinary substitution still does nothing. But cosubstitution can do something. In fact, there are two different cosubstitutions that can be done in this case:

f[a] → s: f[x_] → s

f[a] → s: f[x_] → f[s]

What's going on here? In the first case, f[x_] has the "special case" f[a], to which the rule applies ("by cosubstitution")—giving the result *s*. In the second case, however, it's *x_* on its own which has the special case f[a], that gets transformed by the rule to *s*, giving the final cosubstitution result f[s].

There's an additional wrinkle when the same pattern (such as *x_*) appears multiple times:

a → b: f[x_, x_, x_] → f[b, a, a]

a → b: f[x_, x_, x_] → f[a, b, a]

a → b: f[x_, x_, x_] → f[a, a, b]

In all cases, *x_* is matched to *a*. But which of the *x_*'s is actually replaced is different in each case.

Here's a slightly more complicated example:

f[a] → s: f[x_] ↔ f[x_] → s ↔ f[a]

f[a] → s: f[x_] ↔ f[x_] → f[a] ↔ s

f[a] → s: f[x_] ↔ f[x_] → f[s] ↔ f[f[a]]

f[a] → s: f[x_] ↔ f[x_] → f[f[a]] ↔ f[s]

In ordinary substitution, replacements for patterns are in effect always made "locally", with each specific pattern separately being replaced by some expression. But in cosubstitution, a "special case" found for a pattern will get used throughout when the replacement is done.

Let's see how this all works in an accumulative axiomatic system. Consider the very simple rule:

$x\_ \circ y\_ \leftrightarrow y\_ \circ x\_$

One step of substitution gives the token-event graph (where we've canonicalized the names of pattern variables to *a_* and *b_*):

But one step of cosubstitution gives instead:

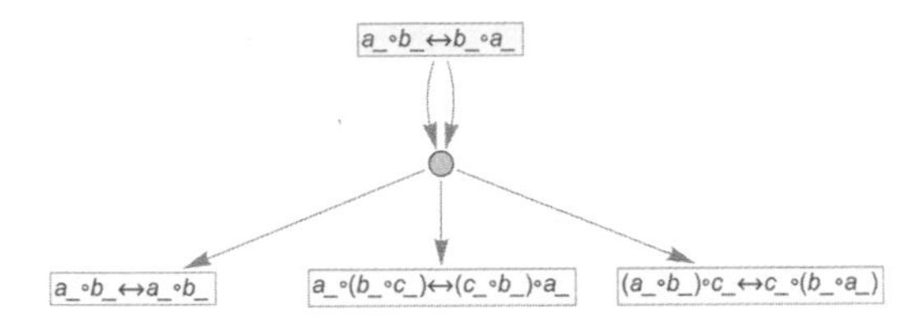

Here are the individual transformations that were made (with the rule at least nominally being applied only in one direction):

x_∘y_ → y∘x: a_∘b_ ⟷ b_∘a_ → y_∘x_ ⟷ y_∘x_

x_∘y_ → y∘x: a_∘b_ ⟷ b_∘a_ → y_∘x_ ⟷ y_∘x_

x_∘y_ → y∘x: a_∘b_ ⟷ b_∘a_ → (y_∘x_)∘b_ ⟷ b_∘(x_∘y_)

x_∘y_ → y∘x: a_∘b_ ⟷ b_∘a_ → a_∘(y_∘x_) ⟷ (x_∘y_)∘a_

x_∘y_ → y∘x: a_∘b_ ⟷ b_∘a_ → a_∘(x_∘y_) ⟷ (y_∘x_)∘a_

x_∘y_ → y∘x: a_∘b_ ⟷ b_∘a_ → (x_∘y_)∘b_ ⟷ b_∘(y_∘x_)

The token-event graph above is then obtained by canonicalizing variables, and combining identical expressions (though for clarity we don't merge rules of the form $a \longleftrightarrow b$ and $b \longleftrightarrow a$).

If we go another step with this particular rule using only substitution, there are additional events (i.e. transformations) but no new theorems produced:

Cosubstitution, however, produces another 27 theorems

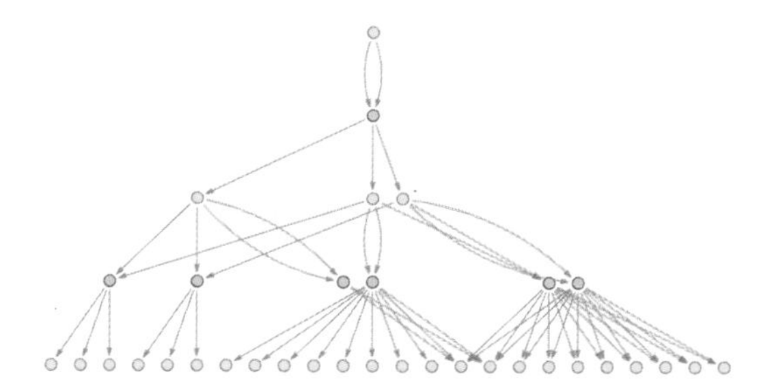

or altogether

$a\circ b\leftrightarrow b\circ a$	$a\circ b\leftrightarrow a\circ b$	$a\circ(b\circ c)\leftrightarrow(c\circ b)\circ a$	$(a\circ b)\circ c\leftrightarrow c\circ(b\circ a)$
$a\circ(b\circ c)\leftrightarrow a\circ(b\circ c)$	$a\circ(b\circ c)\leftrightarrow(b\circ c)\circ a$	$a\circ(b\circ(c\circ d))\leftrightarrow(b\circ(d\circ c))\circ a$	$a\circ(b\circ(c\circ d))\leftrightarrow((c\circ d)\circ b)\circ a$
$a\circ(b\circ(c\circ d))\leftrightarrow((d\circ c)\circ b)\circ a$	$a\circ(b\circ(c\circ(d\circ e)))\leftrightarrow(((e\circ d)\circ c)\circ b)\circ a$	$a\circ(b\circ((c\circ d)\circ e))\leftrightarrow((e\circ(d\circ c))\circ b)\circ a$	$a\circ((b\circ c)\circ d)\leftrightarrow(d\circ(b\circ c))\circ a$
$a\circ((b\circ c)\circ d)\leftrightarrow(d\circ(c\circ b))\circ a$	$a\circ((b\circ c)\circ d)\leftrightarrow((c\circ b)\circ d)\circ a$	$a\circ((b\circ(c\circ d))\circ e)\leftrightarrow(e\circ((d\circ c)\circ b))\circ a$	$a\circ(((b\circ c)\circ d)\circ e)\leftrightarrow(e\circ(d\circ(c\circ b)))\circ a$
$(a\circ b)\circ c\leftrightarrow c\circ(a\circ b)$	$(a\circ b)\circ c\leftrightarrow(a\circ b)\circ c$	$(a\circ b)\circ(c\circ d)\leftrightarrow(c\circ d)\circ(b\circ a)$	$(a\circ b)\circ(c\circ d)\leftrightarrow(d\circ c)\circ(a\circ b)$
$(a\circ b)\circ(c\circ d)\leftrightarrow(d\circ c)\circ(b\circ a)$	$(a\circ b)\circ(c\circ(d\circ e))\leftrightarrow((e\circ d)\circ c)\circ(b\circ a)$	$(a\circ b)\circ((c\circ d)\circ e)\leftrightarrow(e\circ(d\circ c))\circ(b\circ a)$	$(a\circ(b\circ c))\circ d\leftrightarrow d\circ(a\circ(c\circ b))$
$(a\circ(b\circ c))\circ d\leftrightarrow d\circ((b\circ c)\circ a)$	$(a\circ(b\circ c))\circ d\leftrightarrow d\circ((c\circ b)\circ a)$	$(a\circ(b\circ c))\circ(d\circ e)\leftrightarrow(e\circ d)\circ((c\circ b)\circ a)$	$(a\circ(b\circ(c\circ d)))\circ e\leftrightarrow e\circ(((d\circ c)\circ b)\circ a)$
$(a\circ((b\circ c)\circ d))\circ e\leftrightarrow e\circ((d\circ(c\circ b))\circ a)$	$((a\circ b)\circ c)\circ d\leftrightarrow d\circ(c\circ(a\circ b))$	$((a\circ b)\circ c)\circ d\leftrightarrow d\circ(c\circ(b\circ a))$	$((a\circ b)\circ c)\circ d\leftrightarrow d\circ((b\circ a)\circ c)$
$((a\circ b)\circ c)\circ(d\circ e)\leftrightarrow(e\circ d)\circ(c\circ(b\circ a))$	$((a\circ(b\circ c))\circ d)\circ e\leftrightarrow e\circ(d\circ((c\circ b)\circ a))$	$(((a\circ b)\circ c)\circ d)\circ e\leftrightarrow e\circ(d\circ(c\circ(b\circ a)))$	

or as trees:

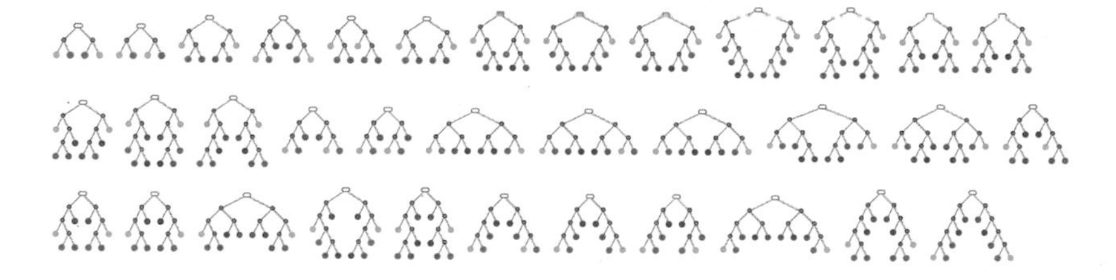

We've now seen examples of both substitution and cosubstitution in action. But in our metamodel for mathematics we're ultimately dealing not with each of these individually, but rather with the "symmetric" concept of bisubstitution, in which both substitution and cosubstitution can be mixed together, and applied even to parts of the same expression.

In the particular case of $a\_\circ b\_\leftrightarrow b\_\circ a\_$, bisubstitution adds nothing beyond cosubstitution. But often it does. Consider the rule:

$x\_\circ a\rightarrow b$

Here's the result of applying this to three different expressions using substitution, cosubstitution and bisubstitution (where we consider only matches for "whole ∘ expressions", not subparts):

$x\_ \circ b \to c$: $\{a \circ b, x\_ \circ y\_, a \circ y\_\} \to \{c, x\_ \circ y\_, a \circ y\_\}$

$x\_ \circ b \to c$: $\{a \circ b, x\_ \circ y\_, a \circ y\_\} \to \{a \circ b, c, a \circ b\}$

$x\_ \circ b \to c$: $\{a \circ b, x\_ \circ y\_, a \circ y\_\} \to \{c, x\_ \circ y\_, a \circ y\_\}$

$x\_ \circ b \to c$: $\{a \circ b, x\_ \circ y\_, a \circ y\_\} \to \{a \circ b, c, a \circ b\}$

$x\_ \circ b \to c$: $\{a \circ b, x\_ \circ y\_, a \circ y\_\} \to \{a \circ b, x\_ \circ b, c\}$

Cosubstitution very often yields substantially more transformations than substitution—bisubstitution then yielding modestly more than cosubstitution. For example, for the axiom system

$$x\_ \circ y\_ \longleftrightarrow (y\_ \circ x\_) \circ y\_$$

the number of theorems derived after 1 and 2 steps is given by:

	step 1	step 2
substitution	5	24
cosubstitution	14	1630
bisubstitution	14	1885

In some cases there are theorems that can be produced by full bisubstitution, but not—even after any number of steps—by substitution or cosubstitution alone. However, it is also common to find that theorems can in principle be produced by substitution alone, but that this just takes more steps (and sometimes vastly more) than when full bisubstitution is used. (It's worth noting, however, that the notion of "how many steps" it takes to "reach" a given theorem depends on the foliation one chooses to use in the token-event graph.)

The various forms of substitution that we've discussed here represent different ways in which one theorem can entail others. But our overall metamodel of mathematics—based as it is purely on the structure of symbolic expressions and patterns—implies that bisubstitution covers all entailments that are possible.

In the history of metamathematics and mathematical logic, a whole variety of "laws of inference" or "methods of entailment" have been considered. But with the modern view of symbolic expressions and patterns (as used, for example, in the Wolfram Language), bisubstitution emerges as the fundamental form of entailment, with other forms of entailment corresponding to the use of particular types of expressions or the addition of further elements to the pure substitutions we've used here.

It should be noted, however, that when it comes to the ruliad different kinds of entailments correspond merely to different foliations—with the form of entailment that we're using representing just a particularly straightforward case.

The concept of bisubstitution has arisen in the theory of term rewriting, as well as in automated theorem proving (where it is often viewed as a particular "strategy", and called "paramodulation"). In term rewriting, bisubstitution is closely related to the concept of unification—which essentially asks what assignment of values to pattern variables is needed in order to make different subterms of an expression be identical.

13 | Some First Metamathematical Phenomenology

Now that we've finished describing the many technical issues involved in constructing our metamodel of mathematics, we can start looking at its consequences. We discussed previously how multiway graphs formed from expressions can be used to define a branchial graph that represents a kind of "metamathematical space". We can now use a similar approach to set up a metamathematical space for our full metamodel of the "progressive accumulation" of mathematical statements.

Let's start by ignoring cosubstitution and bisubstitution and considering only the process of substitution—and beginning with the axiom:

$a\_ \circ b\_ \longleftrightarrow b\_$

Doing accumulative evolution from this axiom we get the token-event graph

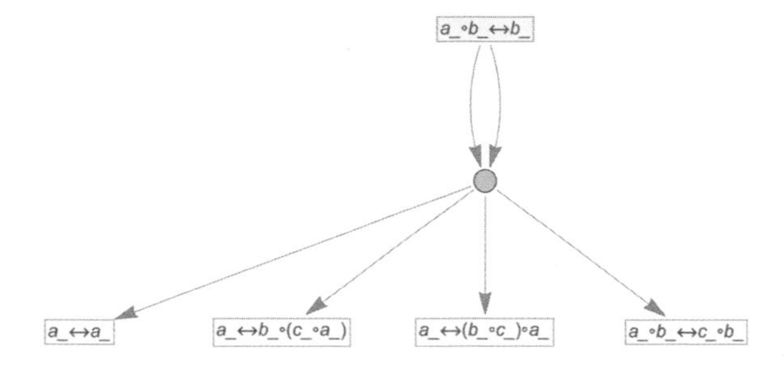

or after 2 steps:

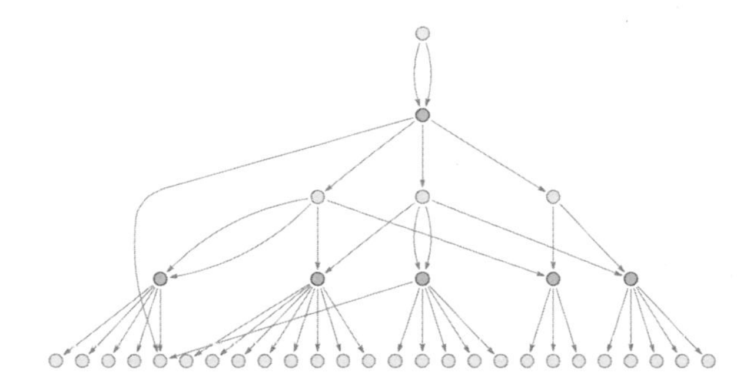

From this we can derive an “effective multiway graph” by directly connecting all input and output tokens involved in each event:

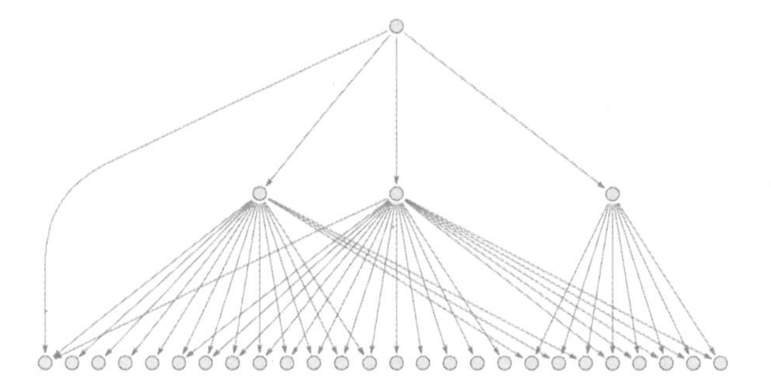

And then we can produce a branchial graph, which in effect yields an approximation to the “metamathematical space” generated by our axiom:

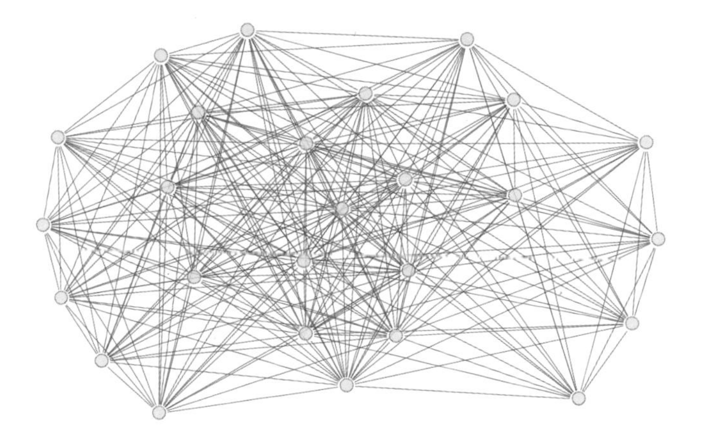

Showing the statements produced in the form of trees we get (with the top node representing ⟷):

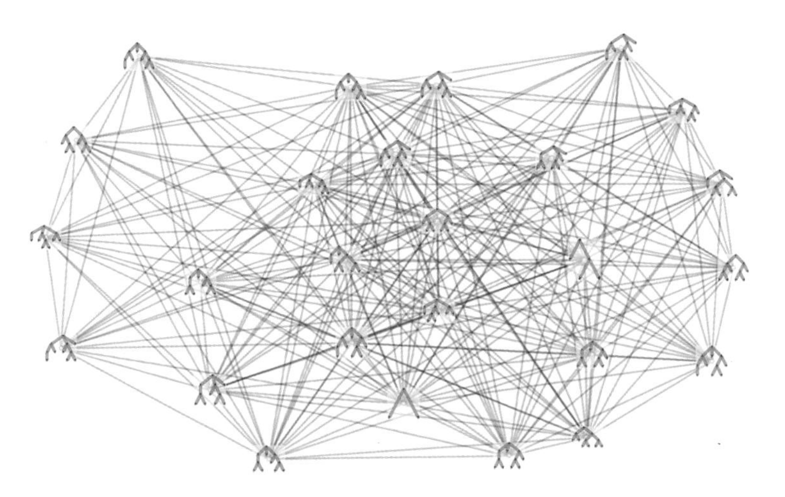

If we do the same thing with full bisubstitution, then even after one step we get a slightly larger token-event graph:

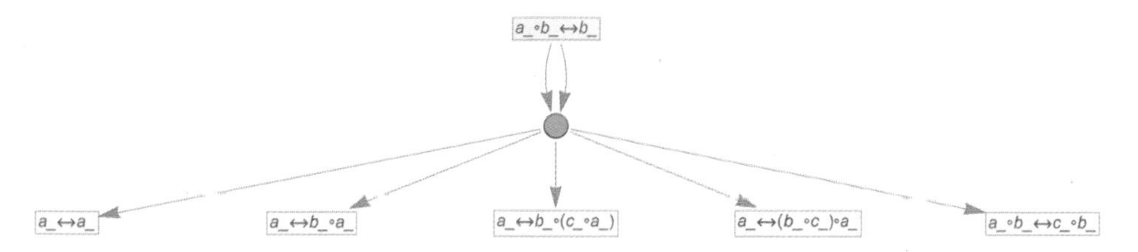

After two steps, we get

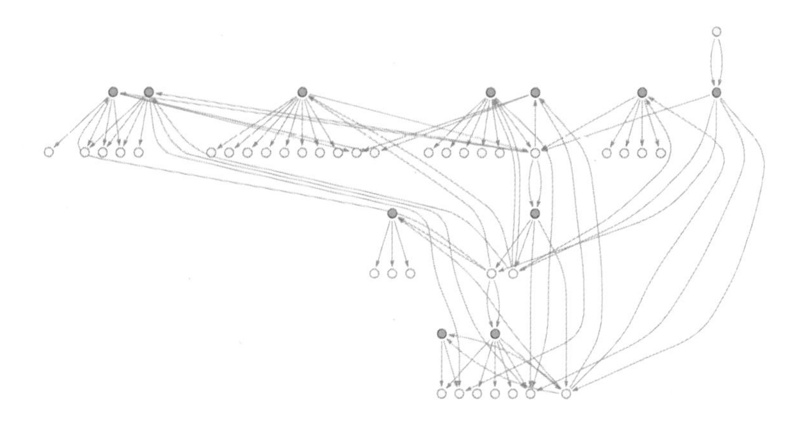

which contains 46 statements, compared to 42 if only substitution is used. The corresponding branchial graph is:

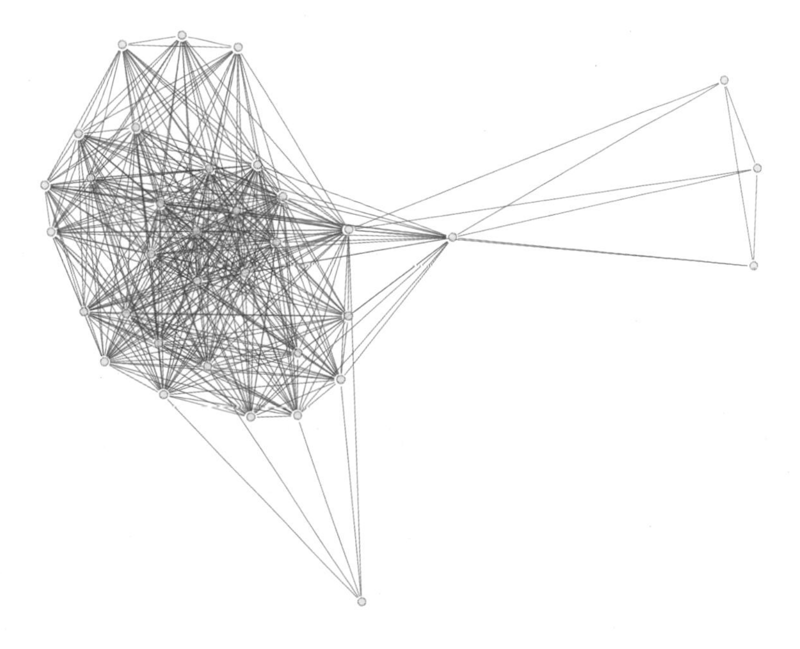

The adjacency matrices for the substitution and bisubstitution cases are then

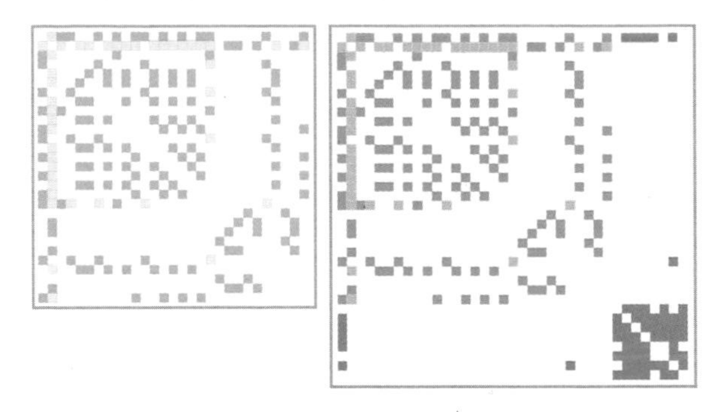

which have 80% and 85% respectively of the number of edges in complete graphs of these sizes.

Branchial graphs are usually quite dense, but they nevertheless do show definite structure. Here are some results after 2 steps:

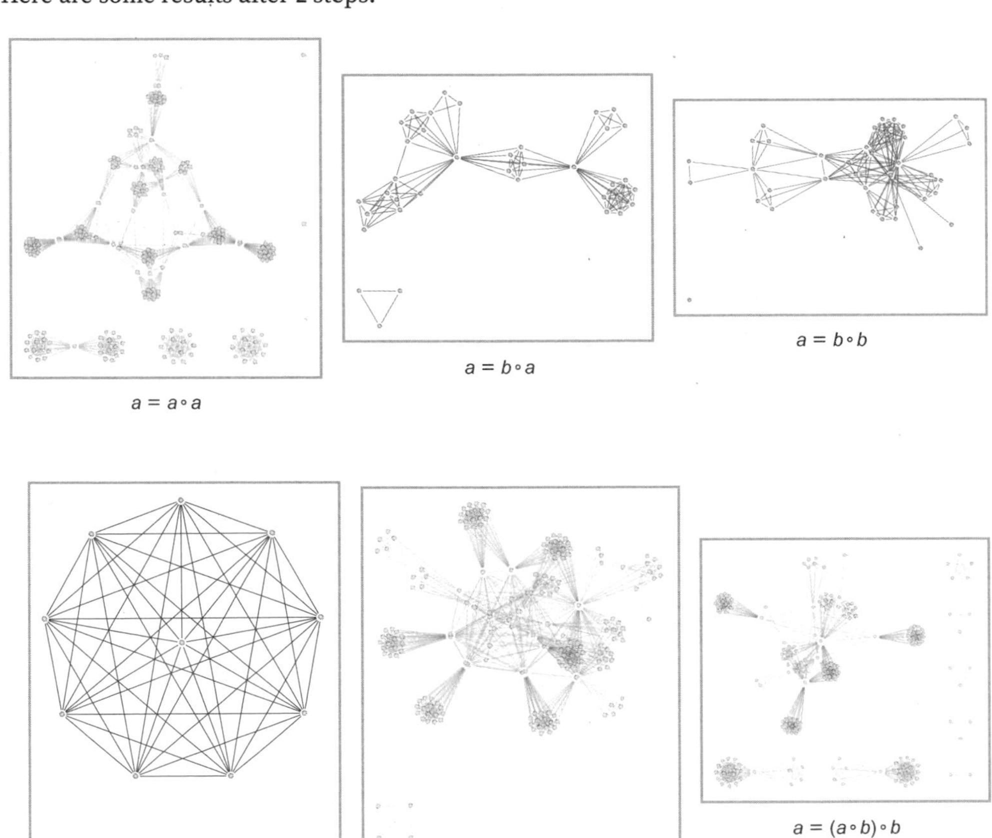

$a = a \circ a$ $a = b \circ a$ $a = b \circ b$

$a \circ a = b \circ b$ $a = a \circ (b \circ b)$ $a = (a \circ b) \circ b$

14 | Relations to Automated Theorem Proving

We've discussed at some length what happens if we start from axioms and then build up an "entailment cone" of all statements that can be derived from them. But in the actual practice of mathematics people often want to just look at particular target statements, and see if they can be derived (i.e. proved) from the axioms.

But what can we say "in bulk" about this process? The best source of potential examples we have right now come from the practice of automated theorem proving—as for example implemented in the Wolfram Language function FindEquationalProof. As a simple example of how this works, consider the axiom

$$x\_\circ y\_ = (y\_\circ x\_)\circ y\_$$

and the theorem:

$$(a\circ((b\circ a)\circ(a\circ b)))\circ a = b\circ a$$

Automated theorem proving (based on FindEquationalProof) finds the following proof of this theorem:

$(a\circ((b\circ a)\circ(a\circ b)))\circ a \longrightarrow ((b\circ a)\circ(a\circ b))\circ a \longrightarrow (((a\circ b)\circ a)\circ(a\circ b))\circ a \longrightarrow (a\circ(a\circ b))\circ a \longrightarrow (a\circ b)\circ a \longrightarrow b\circ a$

Needless to say, this isn't the only possible proof. And in this very simple case, we can construct the full entailment cone—and determine that there aren't any shorter proofs, though there are two more of the same length:

$(a\circ((b\circ a)\circ(a\circ b)))\circ a \longrightarrow (a\circ(((a\circ b)\circ a)\circ(a\circ b)))\circ a \longrightarrow (((a\circ b)\circ a)\circ(a\circ b))\circ a \longrightarrow (a\circ(a\circ b))\circ a \longrightarrow (a\circ b)\circ a \longrightarrow b\circ a$

$(a\circ((b\circ a)\circ(a\circ b)))\circ a \longrightarrow (a\circ(((a\circ b)\circ a)\circ(a\circ b)))\circ a \longrightarrow (a\circ(a\circ(a\circ b)))\circ a \longrightarrow (a\circ(a\circ b))\circ a \longrightarrow (a\circ b)\circ a \longrightarrow b\circ a$

All three of these proofs can be seen as paths in the entailment cone:

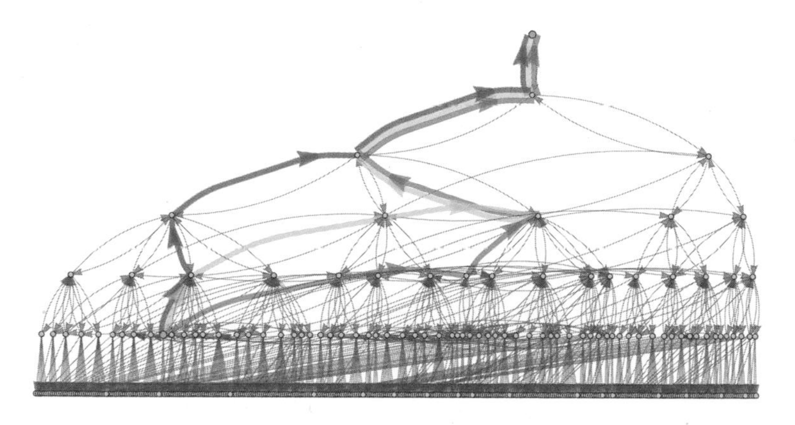

How "complicated" are these proofs? In addition to their lengths, we can for example ask how big the successive intermediate expressions they involve become, where here we are including not only the proofs already shown, but also some longer ones as well:

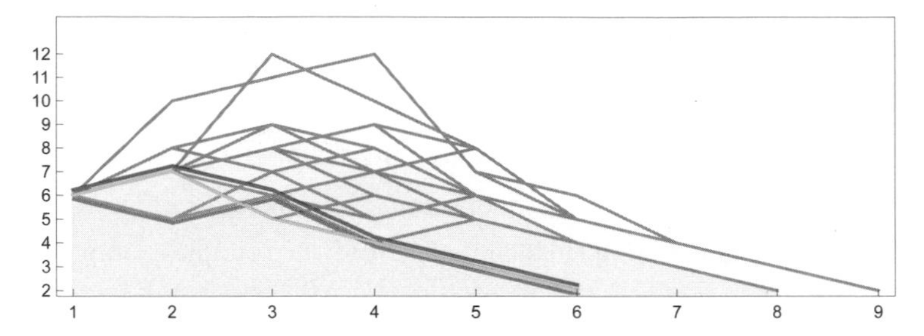

In the setup we're using here, we can find a proof of $lhs = rhs$ by starting with *lhs*, building up an entailment cone, and seeing whether there's any path in it that reaches *rhs*. In general there's no upper bound on how far one will have to go to find such a path—or how big the intermediate expressions may need to get.

One can imagine all kinds of optimizations, for example where one looks at multistep consequences of the original axioms, and treats these as "lemmas" that we can "add as axioms" to provide new rules that jump multiple steps on a path at a time. Needless to say, there are lots of tradeoffs in doing this. (Is it worth the memory to store the lemmas? Might we "jump" past our target? etc.)

But typical actual automated theorem provers tend to work in a way that is much closer to our accumulative rewriting systems—in which the "raw material" on which one operates is statements rather than expressions.

Once again, we can in principle always construct a whole entailment cone, and then look to see whether a particular statement occurs there. But then to give a proof of that statement it's sufficient to find the subgraph of the entailment cone that leads to that statement. For example, starting with the axiom

$$a \circ b = (b \circ a) \circ b$$

we get the entailment cone (shown here as a token-event graph, and dropping _'s):

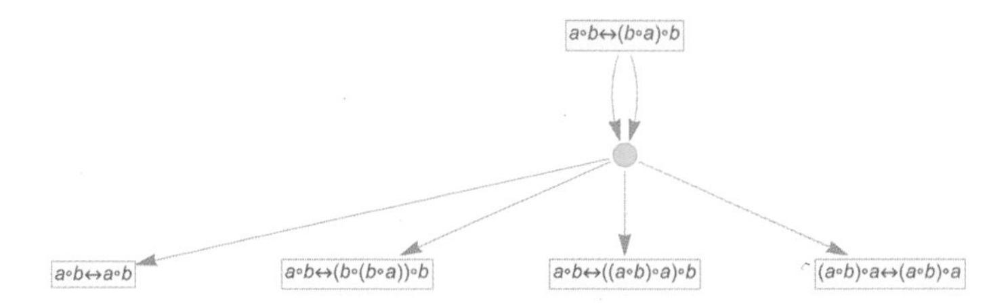

After 2 steps the statement

$$(a \circ b) \circ a = (a \circ (a \circ (a \circ b))) \circ a$$

shows up in this entailment cone

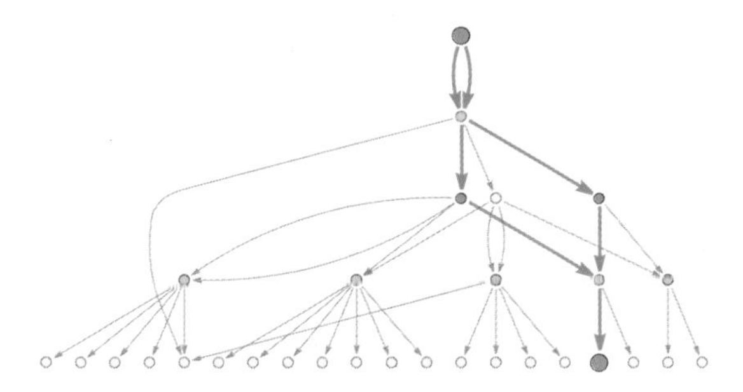

where we're indicating the subgraph that leads from the original axiom to this statement. Extracting this subgraph we get

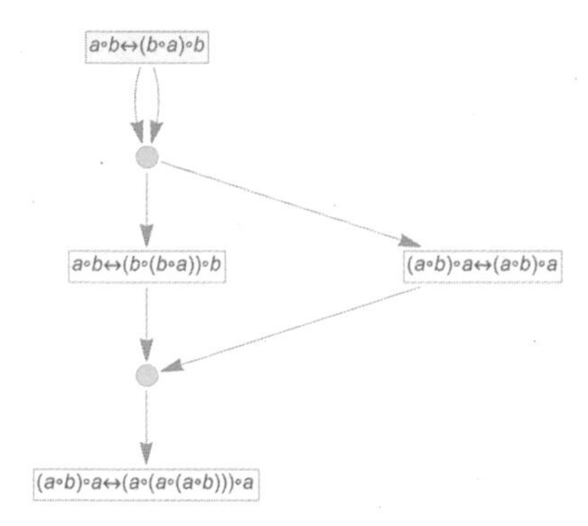

which we can view as a proof of the statement within this axiom system.

But now let's use traditional automated theorem proving (in the form of FindEquationalProof) to get a proof of this same statement. Here's what we get:

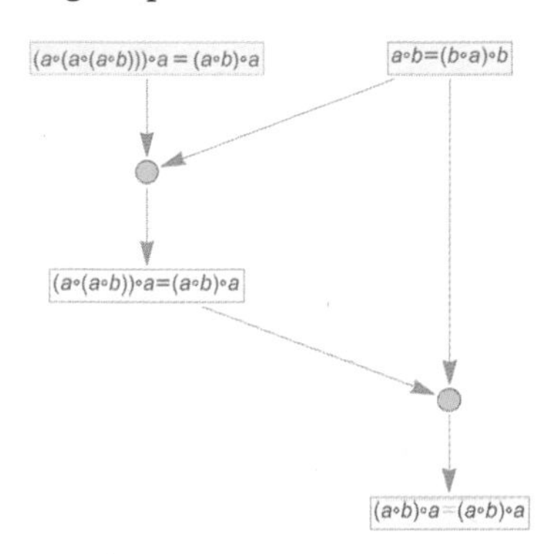

This is again a token-event graph, but its structure is slightly different from the one we "fished out of" the entailment cone. Instead of starting from the axiom and "progressively deriving" our statement we start from both the statement and the axiom and then show that together they lead "merely via substitution" to a statement of the form $x = x$, which we can take as an "obviously derivable tautology".

Sometimes the minimal "direct proof" found from the entailment cone can be considerably simpler than the one found by automated theorem proving. For example, for the statement

$$a \circ b = (((b \circ (b \circ a)) \circ b) \circ (b \circ a)) \circ b$$

the minimal direct proof is

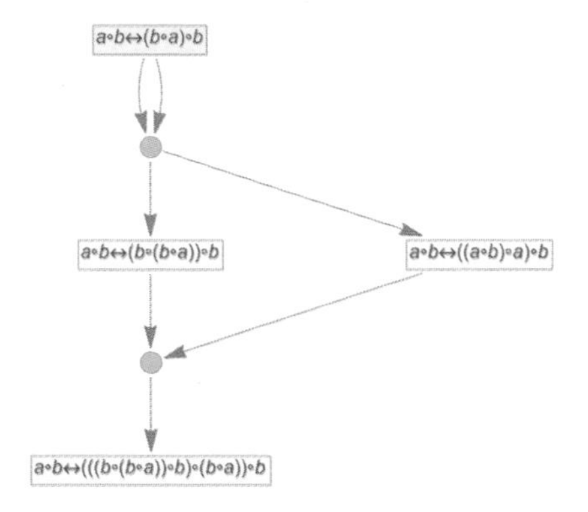

while the one found by FindEquationalProof is:

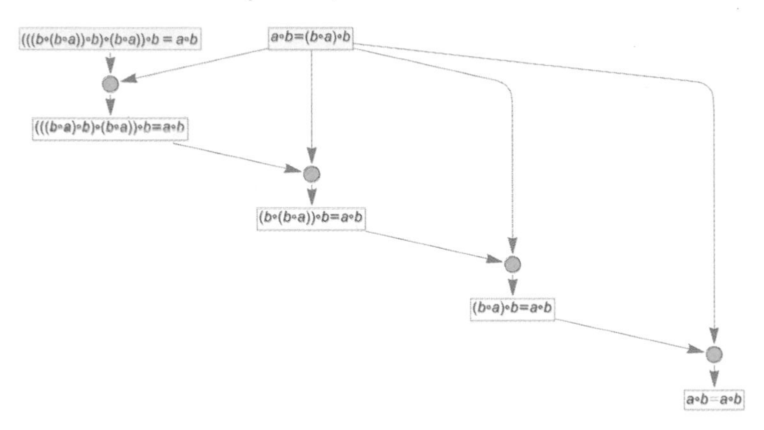

But the great advantage of automated theorem proving is that it can "directedly" search for proofs instead of just "fishing them out of" the entailment cone that contains all possible exhaustively generated proofs. To use automated theorem proving you have to "know where you want to go"—and in particular identify the theorem you want to prove.

Consider the axiom

$$(b \circ a) \circ ((c \circ a) \circ c) = a$$

and the statement:

$$a \circ (a \circ (b \circ b)) = b \circ b$$

This statement doesn't show up in the first few steps of the entailment cone for the axiom, even though millions of other theorems do. But automated theorem proving finds a proof of it—and rearranging the "prove-a-tautology proof" so that we just have to feed in a tautology somewhere in the proof, we get:

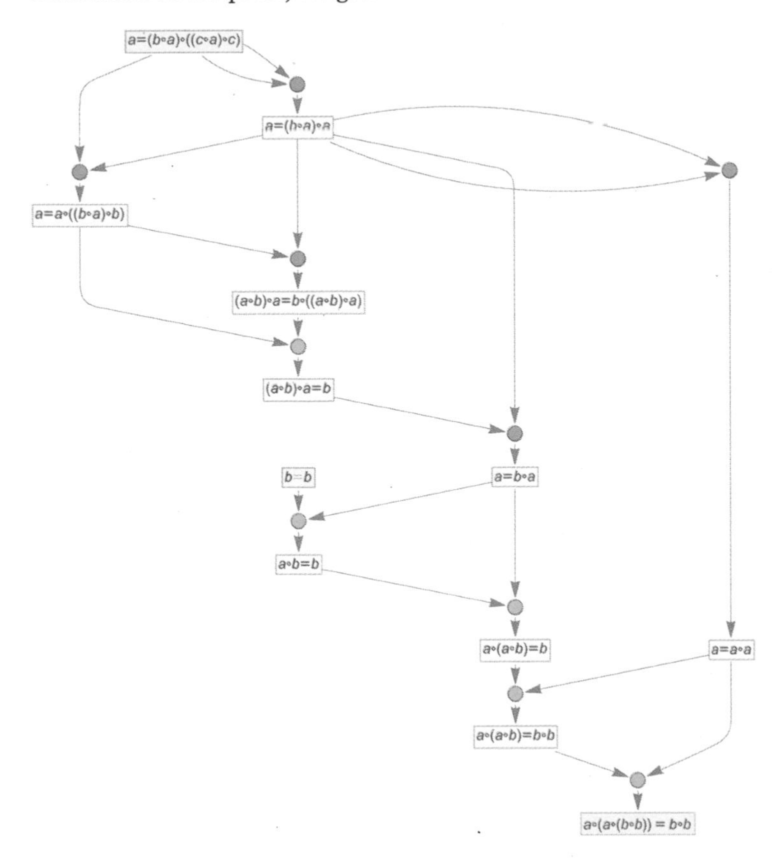

The model-theoretic methods we'll discuss a little later allow one effectively to "guess" theorems that might be derivable from a given axiom system. So, for example, for the axiom system

$$(b \circ a) \circ (c \circ ((c \circ a) \circ c)) = a$$

here's a "guess" at a theorem

$$a \circ ((a \circ b) \circ a) = (b \circ b) \circ (a \circ b)$$

and here's a representation of its proof found by automated theorem proving—where now the length of an intermediate "lemma" is indicated by the size of the corresponding node

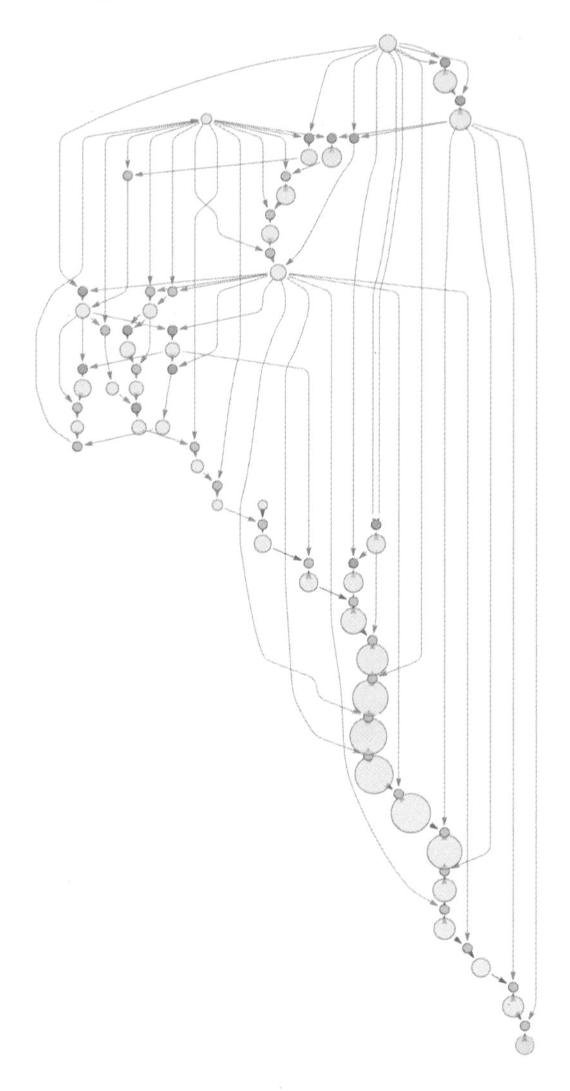

and in this case the longest intermediate lemma is of size 67 and is:

$$(b\circ b)\circ(a\circ b)=((b\circ((b\circ b)\circ b))\circ b)\circ((b\circ((b\circ((b\circ b)\circ b))\circ b))\circ (((b\circ((b\circ b)\circ b))\circ b)\circ(((b\circ b)\circ b)\circ((b\circ((b\circ b)\circ b))\circ(((b\circ b)\circ b)\circ(b\circ((b\circ b)\circ b)))))))$$

In principle it's possible to rearrange token-event graphs generated by automated theorem proving to have the same structure as the ones we get directly from the entailment cone—with axioms at the beginning and the theorem being proved at the end. But typical strategies for automated theorem proving don't naturally produce such graphs. In principle automated theorem proving could work by directly searching for a "path" that leads to the theorem one's trying to prove. But usually it's much easier instead to have as the "target" a simple tautology.

At least conceptually automated theorem proving must still try to "navigate" through the full token-event graph that makes up the entailment cone. And the main issue in doing this is that there are many places where one does not know "which branch to take". But here there's a crucial—if at first surprising—fact: at least so long as one is using full bisubstitution it ultimately doesn't matter which branch one takes; there'll always be a way to "merge back" to any other branch.

This is a consequence of the fact that the accumulative systems we're using automatically have the property of confluence which says that every branch is accompanied by a subsequent merge. There's an almost trivial way in which this is true by virtue of the fact that for every edge the system also includes the reverse of that edge. But there's a more substantial reason as well: that given any two statements on two different branches, there's always a way to combine them using a bisubstitution to get a single statement.

In our Physics Project, the concept of causal invariance—which effectively generalizes confluence—is an important one, that leads among other things to ideas like relativistic invariance. Later on we'll discuss the idea that "regardless of what order you prove theorems in, you'll always get the same math", and its relationship to causal invariance and to the notion of relativity in metamathematics. But for now the importance of confluence is that it has the potential to simplify automated theorem proving—because in effect it says one can never ultimately "make a wrong turn" in getting to a particular theorem, or, alternatively, that if one keeps going long enough every path one might take will eventually be able to reach every theorem.

And indeed this is exactly how things work in the full entailment cone. But the challenge in automated theorem proving is to generate only a tiny part of the entailment cone, yet still "get to" the theorem we want. And in doing this we have to carefully choose which "branches" we should try to merge using bisubstitution events. In automated theorem proving these bisubstitution events are typically called "critical pair lemmas", and there are a variety of strategies for defining an order in which critical pair lemmas should be tried.

It's worth pointing out that there's absolutely no guarantee that such procedures will find the shortest proof of any given theorem (or in fact that they'll find a proof at all with a given amount of computational effort). One can imagine "higher-order proofs" in which one attempts to transform not just statements of the form $lhs = rhs$, but full proofs (say represented as token-event graphs). And one can imagine using such transformations to try to simplify proofs.

A general feature of the proofs we've been showing is that they are accumulative, in the sense they continually introduce lemmas which are then reused. But in principle any proof can be "unrolled" into one that just repeatedly uses the original axioms (and in fact, purely by substitution)—and never introduces other lemmas. The necessary "cut elimination" can effectively be done by always recreating each lemma from the axioms whenever it's needed—a process which can become exponentially complex.

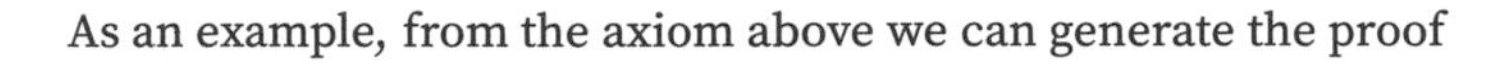

As an example, from the axiom above we can generate the proof

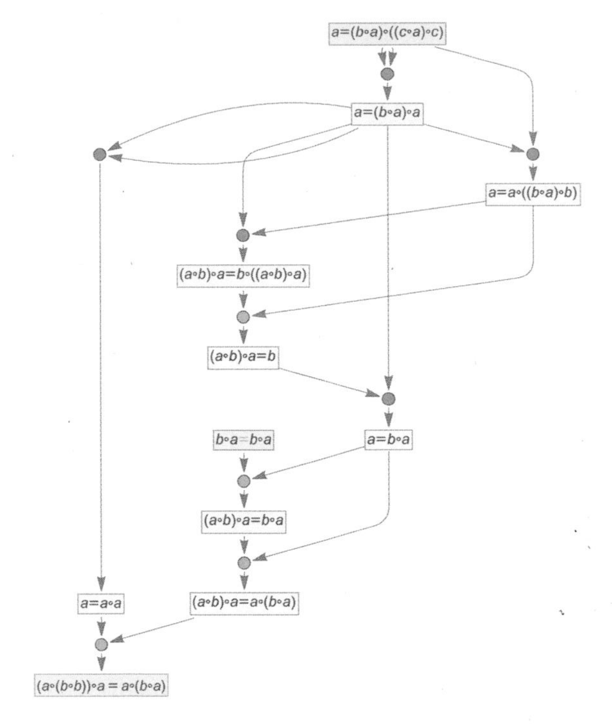

where for example the first lemma at the top is reused in four events. But now by cut elimination we can "unroll" this whole proof into a "straight-line" sequence of substitutions on expressions done just using the original axiom

(a∘(b∘b))∘a

(a∘(((b∘b)∘((((c∘b)∘c)∘b)∘((c∘b)∘c)))∘b))∘a

(a∘(((b∘b)∘b)∘b))∘a

(a∘(((b∘b)∘b)∘((((d∘b)∘d)∘b)∘((d∘b)∘d))))∘a

(a∘b)∘a

((((a∘a)∘a)∘((b∘a)∘b))∘b)∘a

((((a∘a)∘((((e∘a)∘e)∘a)∘((e∘a)∘e)))∘((b∘a)∘b))∘b)∘a

((a∘((b∘a)∘b))∘b)∘a

(((((e∘a)∘a)∘((b∘a)∘b))∘((b∘a)∘b))∘b)∘a

(((((e∘a)∘((((f∘a)∘f)∘a)∘((f∘a)∘f)))∘((b∘a)∘b))∘((b∘a)∘b))∘b)∘a

(((a∘((b∘a)∘b))∘((b∘a)∘b))∘b)∘a

(((a∘((b∘a)∘b))∘((((c∘((b∘a)∘b))∘c)∘((b∘a)∘b))∘((c∘((b∘a)∘b))∘c)))∘b)∘a

(((b∘a)∘b)∘b)∘a

(((b∘a)∘b)∘((((d∘b)∘d)∘b)∘((d∘b)∘d)))∘a

b∘a

(((b∘a)∘a)∘(b∘a))∘((((d∘(b∘a))∘d)∘(b∘a))∘((d∘(b∘a))∘d))

(((b∘a)∘a)∘(b∘a))∘(b∘a)

((a∘(((b∘a)∘a)∘(b∘a)))∘((((c∘(((b∘a)∘a)∘(b∘a)))∘c)∘(((b∘a)∘a)∘(b∘a)))∘((c∘(((b∘a)∘a)∘(b∘a)))∘c)))∘(b∘a)

((a∘(((b∘a)∘a)∘(b∘a)))∘(((b∘a)∘a)∘(b∘a)))∘(b∘a)

((((e∘a)∘((((f∘a)∘f)∘a)∘((f∘a)∘f)))∘(((b∘a)∘a)∘(b∘a)))∘(((b∘a)∘a)∘(b∘a)))∘(b∘a)

((((e∘a)∘a)∘(((b∘a)∘a)∘(b∘a)))∘(((b∘a)∘a)∘(b∘a)))∘(b∘a)

(a∘(((b∘a)∘a)∘(b∘a)))∘(b∘a)

(((a∘a)∘((((e∘a)∘e)∘a)∘((e∘a)∘e)))∘(((b∘a)∘a)∘(b∘a)))∘(b∘a)

(((a∘a)∘a)∘(((b∘a)∘a)∘(b∘a)))∘(b∘a)

a∘(b∘a)

and we see that our final theorem is the statement that the first expression in the sequence is equivalent under the axiom to the last one.

As is fairly evident in this example, a feature of automated theorem proving is that its result tends to be very "non-human". Yes, it can provide incontrovertible evidence that a theorem is valid. But that evidence is typically far away from being any kind of "narrative" suitable for human consumption. In the analogy to molecular dynamics, an automated proof gives detailed "turn-by-turn instructions" that show how a molecule can reach a certain place in a gas. Typical "human-style" mathematics, on the other hand, operates on a higher level, analogous to talking about overall motion in a fluid. And a core part of what's achieved by our physicalization of metamathematics is understanding why it's possible for mathematical observers like us to perceive mathematics as operating at this higher level.

15 | Axiom Systems of Present-Day Mathematics

The axiom systems we've been talking about so far were chosen largely for their axiomatic simplicity. But what happens if we consider axiom systems that are used in practice in present-day mathematics?

The simplest common example are the axioms (actually, a single axiom) of semigroup theory, stated in our notation as:

$$a\_\circ(b\_\circ c\_) \longleftrightarrow (a\_\circ b\_)\circ c\_$$

Using only substitution, all we ever get after any number of steps is the token-event graph (i.e. "entailment cone"):

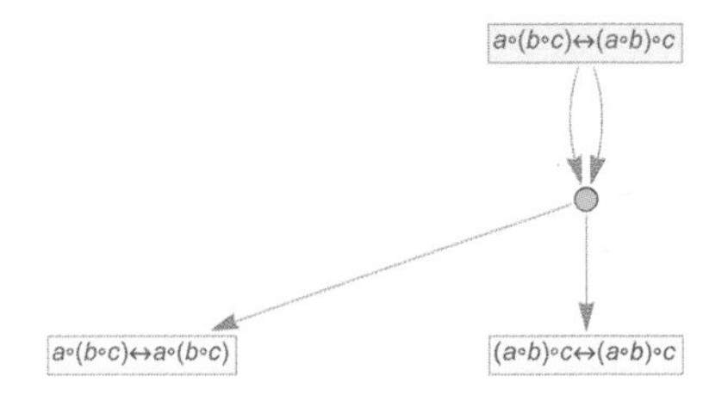

But with bisubstitution, even after one step we already get the entailment cone

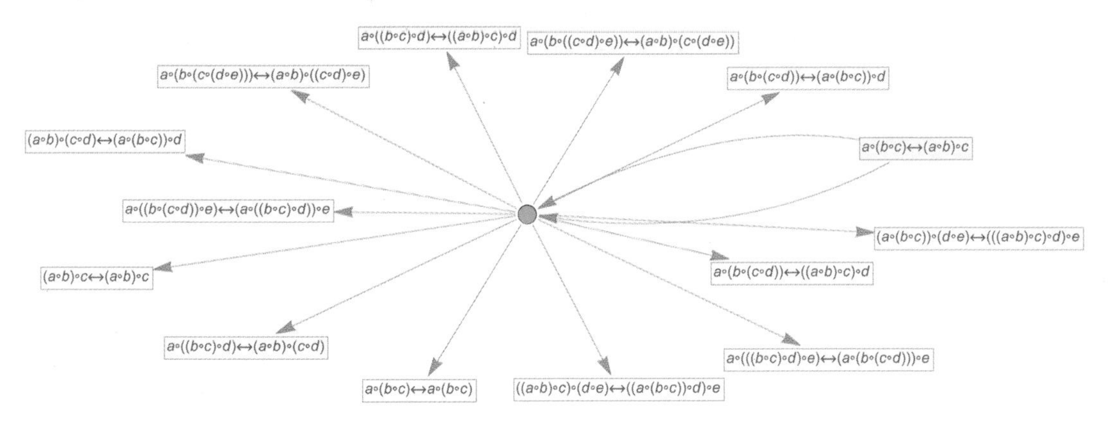

which contains such theorems as:

$$a\circ((b\circ c)\circ d) = (a\circ b)\circ(c\circ d)$$

After 2 steps, the entailment cone becomes

which contains 1617 theorems such as

$a\circ((b\circ(c\circ(d\circ e)))\circ f) = (a\circ((b\circ(c\circ d))\circ e))\circ f$
$(a\circ(b\circ c))\circ(((d\circ e)\circ f)\circ g) = (a\circ(b\circ(c\circ(d\circ(e\circ f)))))\circ g$
$a\circ(b\circ((((c\circ d)\circ e)\circ f)\circ g)) = ((a\circ b)\circ(c\circ(d\circ(e\circ f))))\circ g$
$(a\circ(b\circ c))\circ(d\circ((e\circ(f\circ g))\circ(h\circ i))) = (((a\circ b)\circ c)\circ d)\circ((((e\circ f)\circ g)\circ h)\circ i)$
$((a\circ((b\circ((c\circ d)\circ e))\circ f))\circ g)\circ h = ((a\circ(((b\circ c)\circ d)\circ e))\circ f)\circ(g\circ h)$

with sizes distributed as follows:

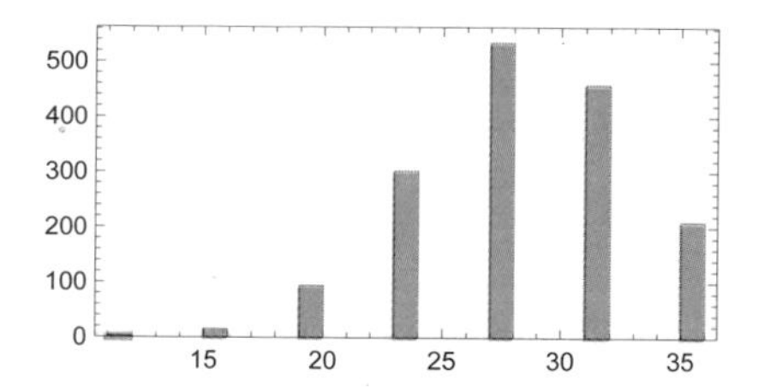

Looking at these theorems we can see that—in fact by construction—they are all just statements of the associativity of $\circ$. Or, put another way, they state that under this axiom all expression trees that have the same sequence of leaves are equivalent.

What about group theory? The standard axioms can be written

$a\circ(b\circ c) = (a\circ b)\circ c$
$a\circ\diamond = a$
$a\circ\overline{a} = \diamond$

where $\circ$ is interpreted as the binary group multiplication operation, overbar as the unary inverse operation, and 1 as the constant identity element (or, equivalently, zero-argument function).

One step of substitution already gives:

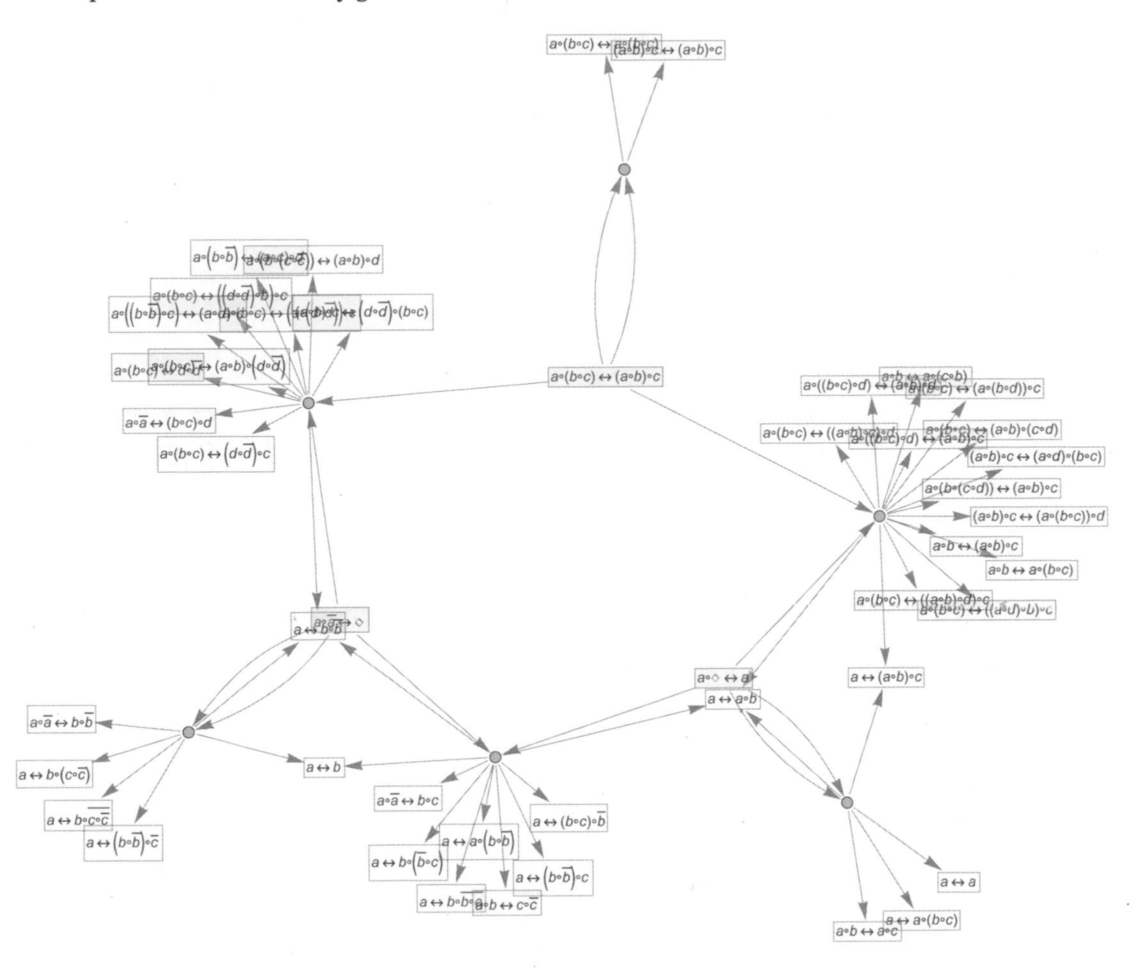

It's notable that in this picture one can already see "different kinds of theorems" ending up in different "metamathematical locations". One also sees some "obvious" tautological "theorems", like $a = a$ and $1 = 1$.

If we use full bisubstitution, we get 56 rather than 27 theorems, and many of the theorems are more complicated:

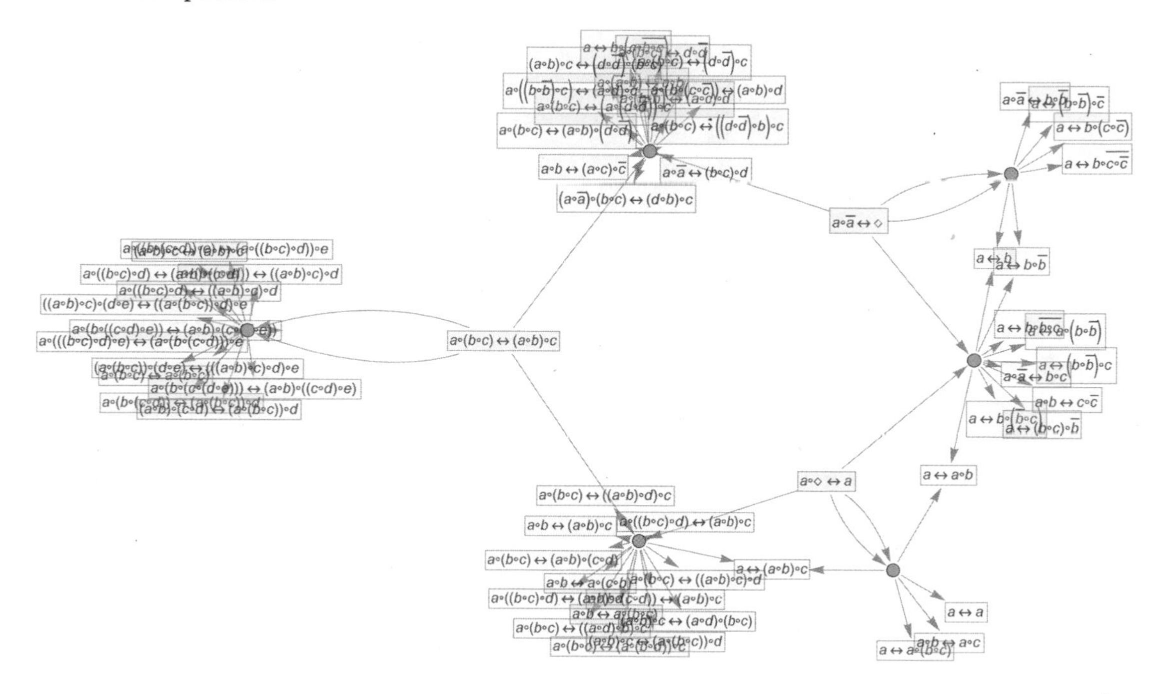

After 2 steps of pure substitution, the entailment cone in this case becomes

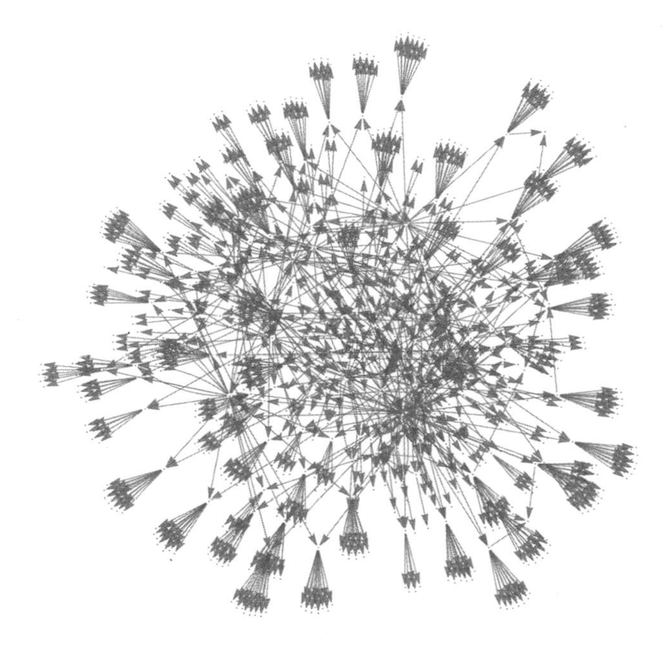

which includes 792 theorems with sizes distributed according to:

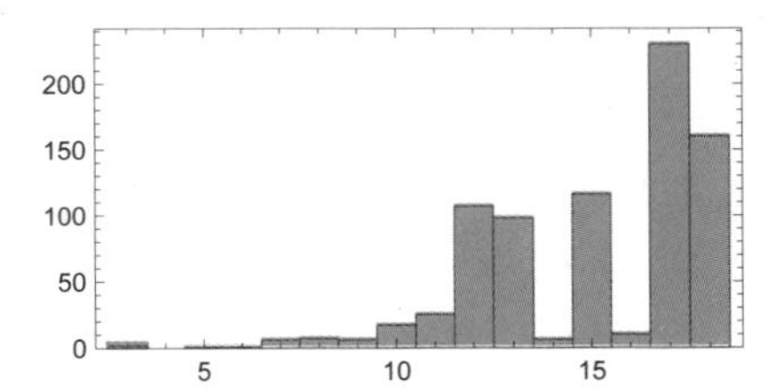

But among all these theorems, do straightforward "textbook theorems" appear, like:

inverse of inverse	$\bar{\bar{a}} = a$
left identity	$\diamond \circ a = a$
left inverse	$\bar{a} \circ a = \diamond$
inverse of composite	$\overline{a \circ b} = \bar{b} \circ \bar{a}$

The answer is no. It's inevitable that in the end all such theorems must appear in the entailment cone. But it turns out that it takes quite a few steps. And indeed with automated theorem proving we can find "paths" that can be taken to prove these theorems—involving significantly more than two steps:

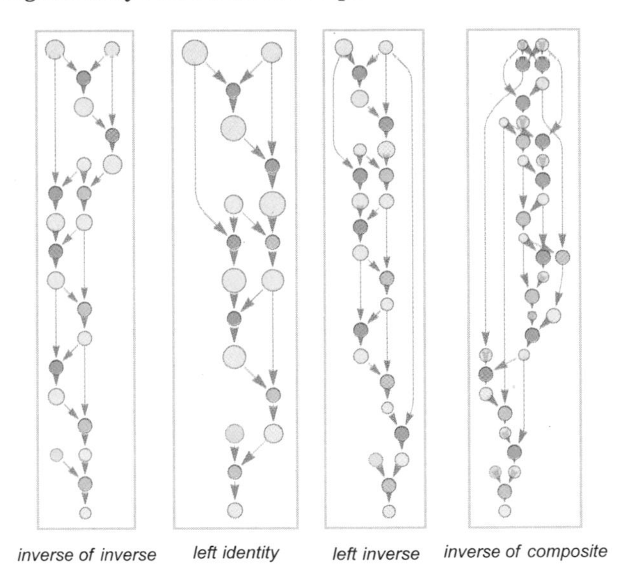

So how about logic, or, more specifically Boolean algebra? A typical textbook axiom system for this (represented in terms of And $\wedge$, Or $\vee$ and Not $\bar{\square}$) is:

$a \vee b = b \vee a$	$a \wedge b = b \wedge a$
$a \vee (b \wedge \bar{b}) = a$	$a \wedge (b \vee \bar{b}) = a$
$a \vee (b \wedge c) = (a \vee b) \wedge (a \vee c)$	$a \wedge (b \vee c) = (a \wedge b) \vee (a \wedge c)$

After one step of substitution from these axioms we get

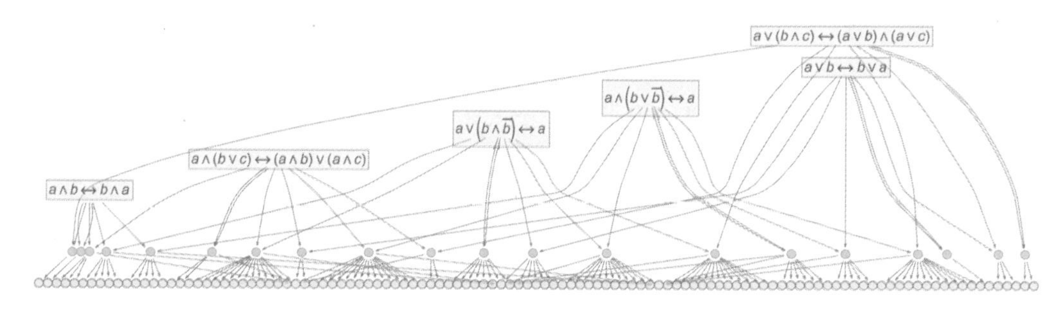

or in our more usual rendering:

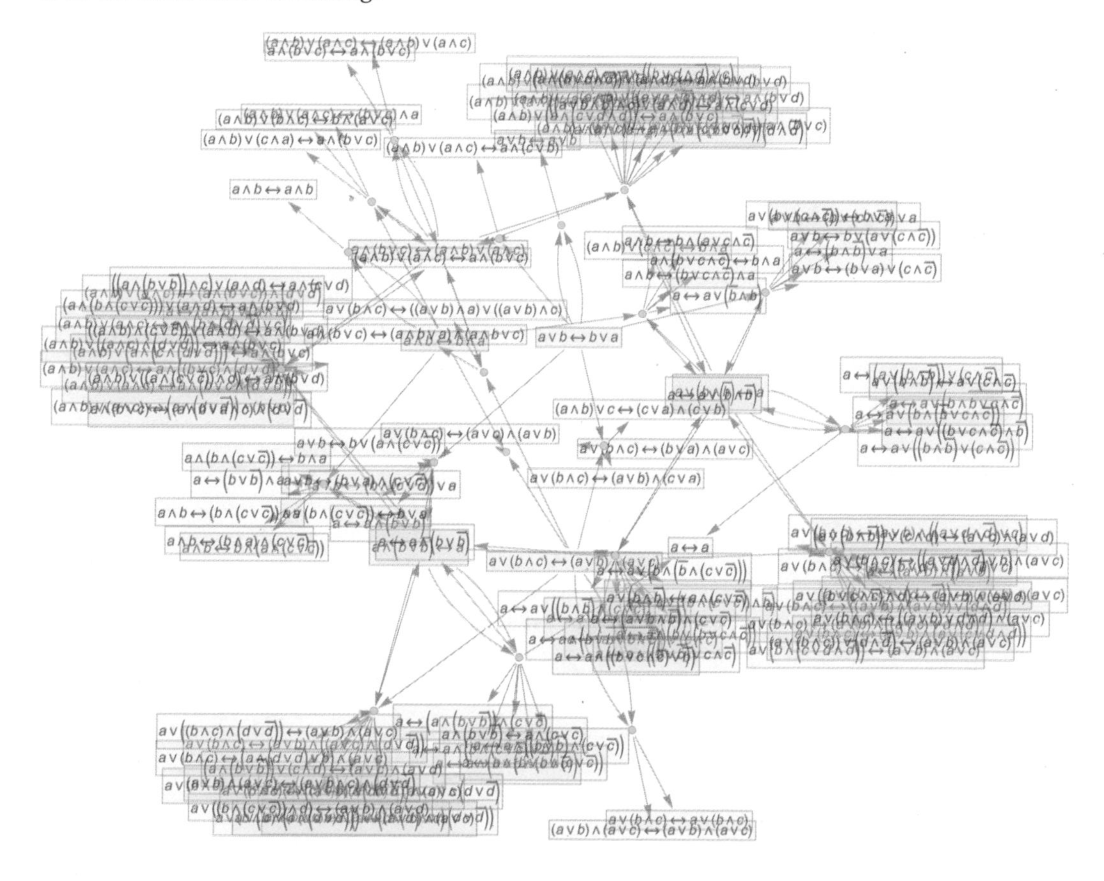

So what happens here with "named textbook theorems" (excluding commutativity and distributivity, which already appear in the particular axioms we're using)?

idempotence of And	$a = a \wedge a$
idempotence of Or	$a = a \vee a$
law of double negation	$a = \overline{\overline{a}}$
law of noncontradiction	$\overline{a} \wedge a = \overline{b} \wedge b$
law of excluded middle	$\overline{a} \vee a = \overline{b} \vee b$

de Morgan law	$\overline{a \vee b} = \overline{a} \wedge \overline{b}$
de Morgan law	$\overline{a \wedge b} = \overline{a} \vee \overline{b}$
absorption law	$a = a \wedge (a \vee b)$
absorption law	$a = a \vee (a \wedge b)$
associativity of And	$(a \wedge b) \wedge c = a \wedge (b \wedge c)$
associativity of Or	$(a \vee b) \vee c = a \vee (b \vee c)$

Once again none of these appear in the first step of the entailment cone. But at step 2 with full bisubstitution the idempotence laws show up

where here we're only operating on theorems with leaf count below 14 (of which there are a total of 27,953).

And if we go to step 3—and use leaf count below 9—we see the law of excluded middle and the law of noncontradiction show up:

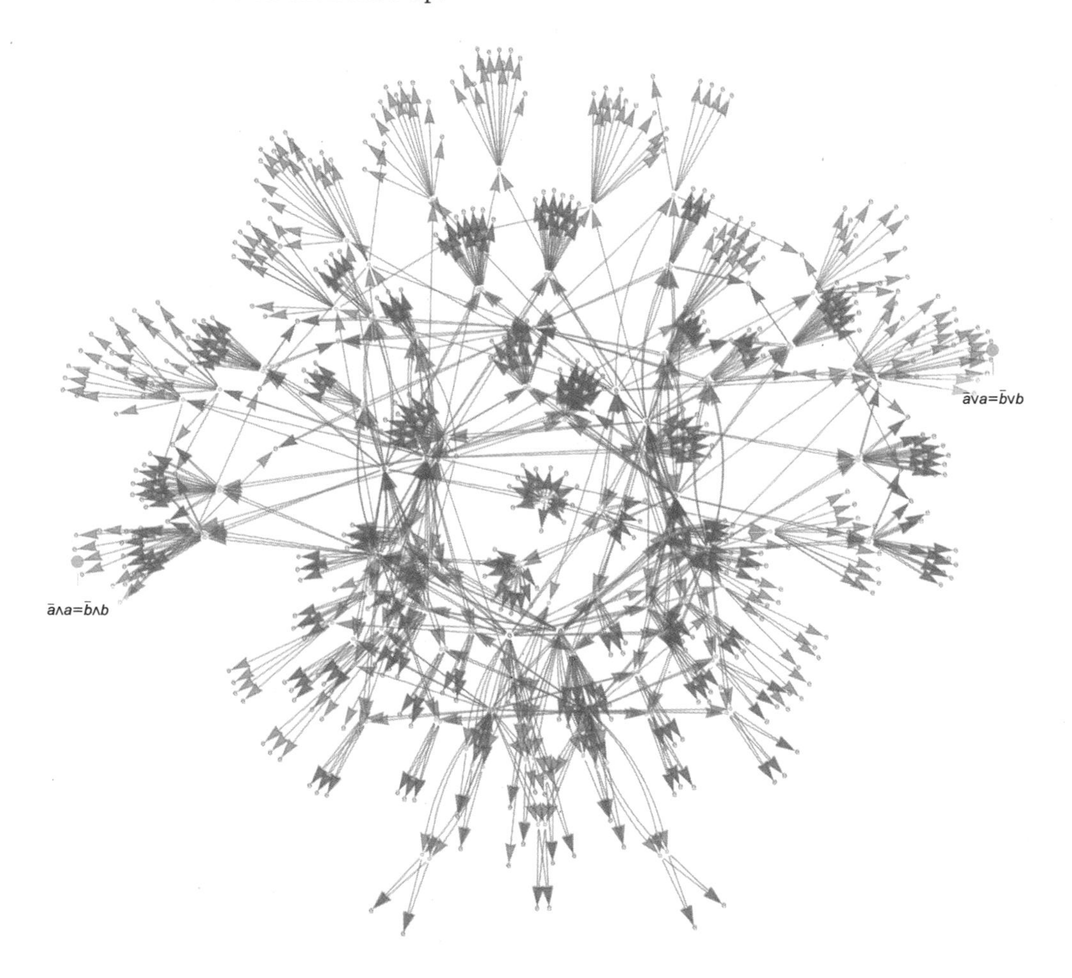

How are these reached? Here's the smallest fragment of token-event graph ("shortest path") within this entailment cone from the axioms to the law of excluded middle:

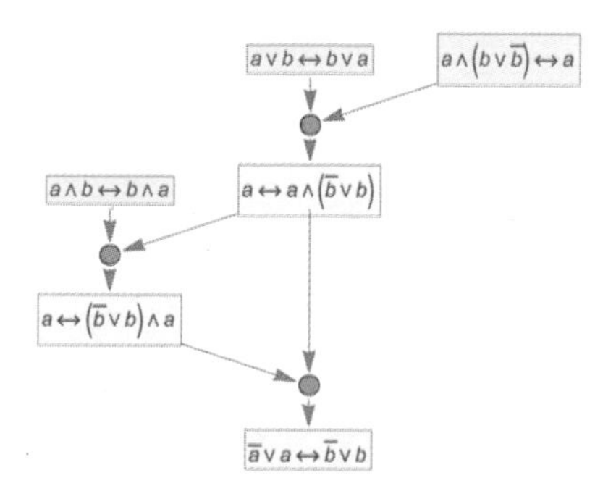

There are actually many possible "paths" (476 in all with our leaf count restriction); the next smallest ones with distinct structures are:

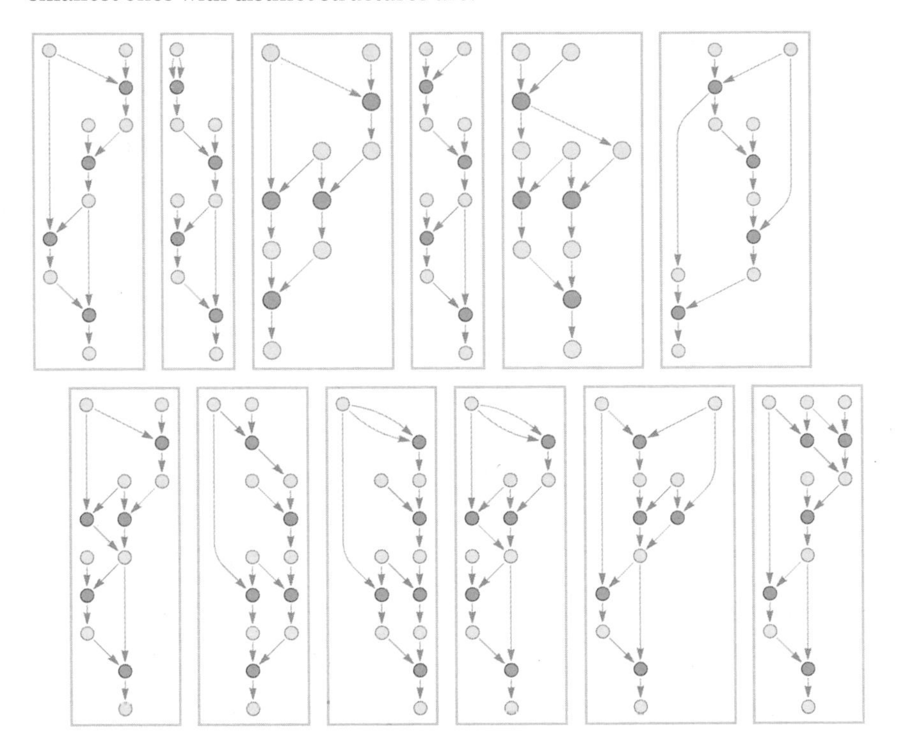

Here's the "path" for this theorem found by automated theorem proving:

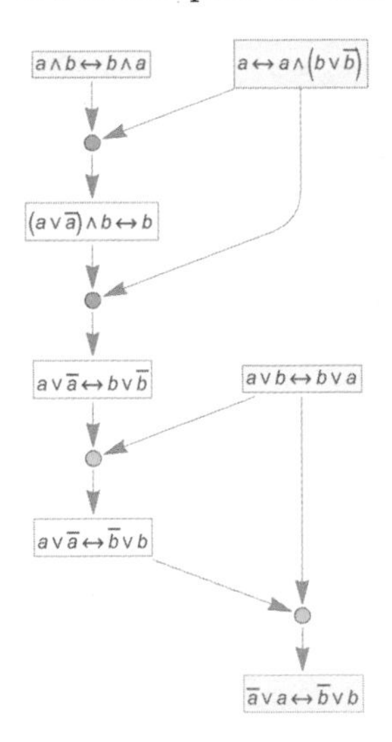

Most of the other “named theorems” involve longer proofs—and so won’t show up until much later in the entailment cone:

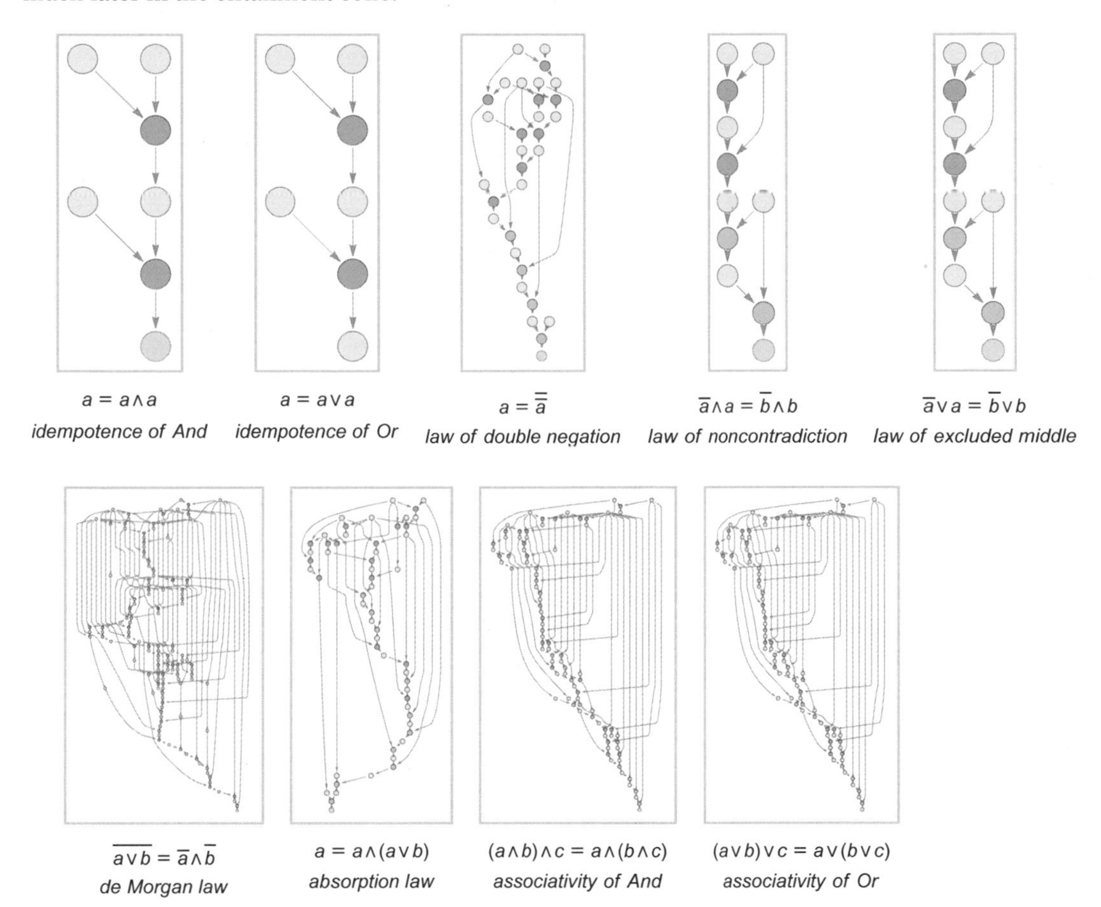

The axiom system we’ve used for Boolean algebra here is by no means the only possible one. For example, it’s stated in terms of And, Or and Not—but one doesn’t need all those operators; any Boolean expression (and thus any theorem in Boolean algebra) can also be stated just in terms of the single operator Nand.

And in terms of that operator the very simplest axiom system for Boolean algebra contains (as I found in 2000) just one axiom (where here $\circ$ is now interpreted as Nand):

$$((b \circ c) \circ a) \circ (b \circ ((b \circ a) \circ b)) = a$$

Here's one step of the substitution entailment cone for this axiom:

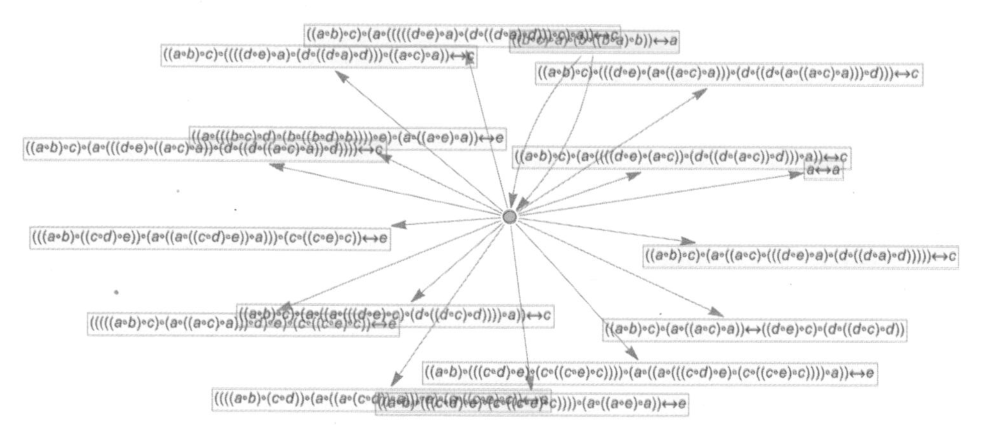

After 2 steps this gives an entailment cone with 5486 theorems

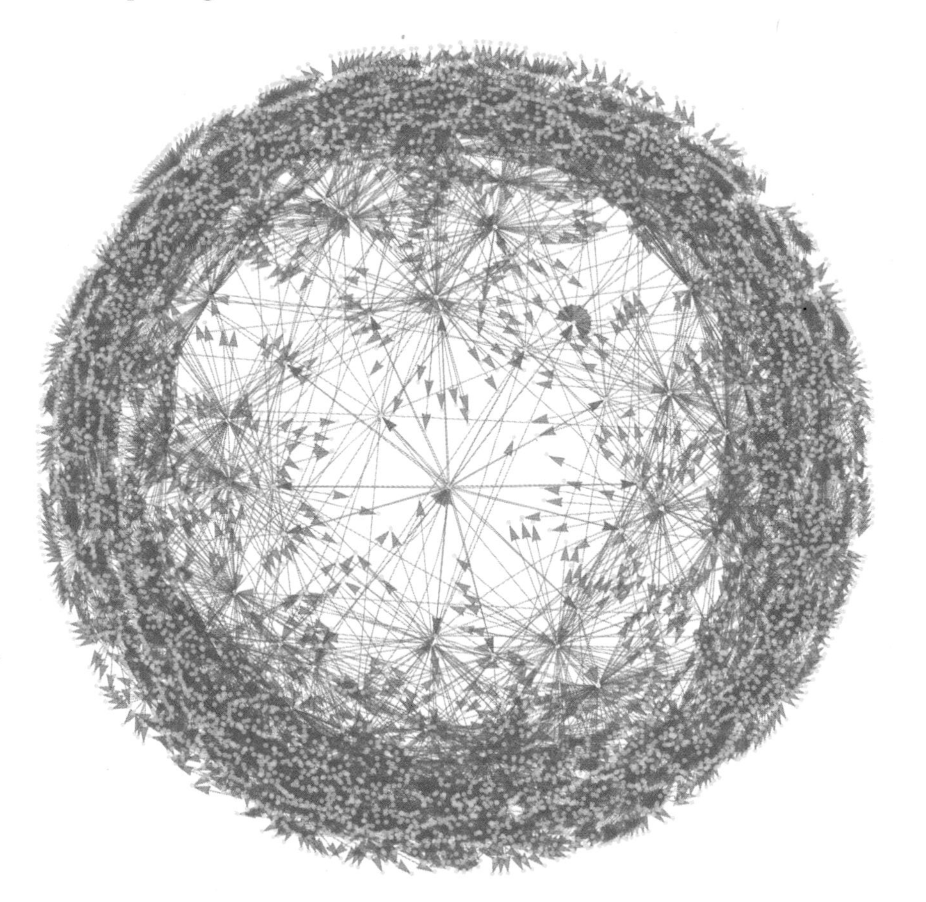

with size distribution:

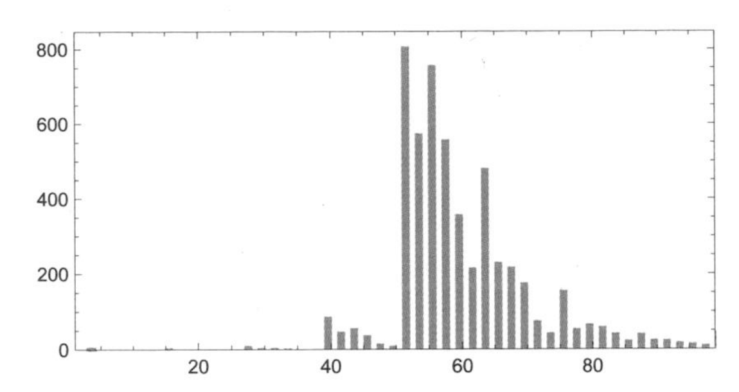

When one's working with Nand, it's less clear what one should consider to be "notable theorems". But an obvious one is the commutativity of Nand:

$$p \circ q = q \circ p$$

Here's a proof of this obtained by automated theorem proving (tipped on its side for readability):

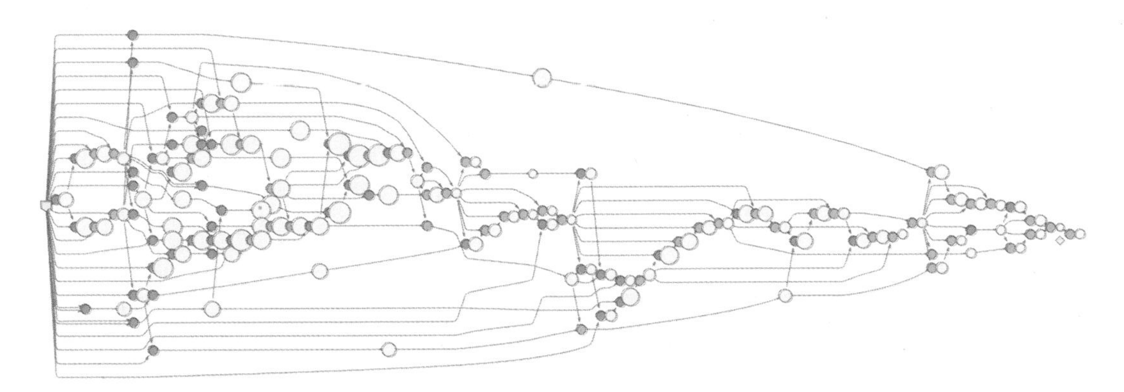

Eventually it's inevitable that this theorem must show up in the entailment cone for our axiom system. But based on this proof we would expect it only after something like 102 steps. And with the entailment cone growing exponentially this means that by the time $p \circ q = q \circ p$ shows up, perhaps 10^{100} other theorems would have done so—though most vastly more complicated.

We've looked at axioms for group theory and for Boolean algebra. But what about other axiom systems from present-day mathematics? In a sense it's remarkable how few of these there are—and indeed I was able to list essentially all of them in just two pages in *A New Kind of Science*:

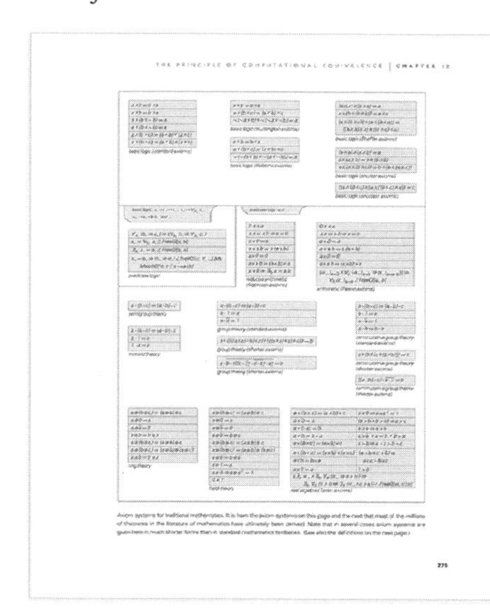

The longest axiom system listed here is a precise version of Euclid's original axioms

$\forall_{\{x,y,z\}}$ implies[congruent[line[x, y], line[z, z]], congruent[x, y]]
$\forall_{\{x,y,z,u,v,w\}}$ implies[and[congruent[line[x, y], line[z, u]], congruent[line[x, y], line[v, w]]], congruent[line[z, u], line[v, w]]]
$\forall_{\{x,y,z\}}$ implies[between[x, y, z], equal[x, y]]
$\forall_{\{x,y,z,u,v\}}$ implies[and[between[x, u, z], between[y, v, z]], $\exists_a$ and[between[u, a, y], between[v, a, x]]]
$\forall_{\{x,y,z,u,v\}}$ implies[and[and[and[congruent[line[x, u], line[x, v]], congruent[line[y, u], line[y, v]]], congruent[line[z, u], line[z, v]]], not[equal[u, v]]], or[or[between[x, y, z], between[y, z, x]], between[z, x, y]]]
$\forall_{\{x,y,z,u,v,w\}}$ implies[and[and[and[between[x, y, w], congruent[line[x, y], line[y, w]]], and[between[x, u, v], congruent[line[x, u], line[u, v]]]], and[between[y, u, z], congruent[line[y, u], line[z, u]]]], congruent[line[y, z], line[v, w]]]
$\forall_{\{x,y,z,a,b,c,u,v\}}$ implies[and[and[and[and[and[not[equal[x, y]], between[x, y, z]], between[a, b, c]], congruent[line[x, y], line[a, b]]], congruent[line[y, z], line[b, c]]], congruent[line[x, u], line[a, v]]], congruent[line[y, u], line[b, v]]], congruent[line[z, u], line[c, v]]]
$\forall_{\{x,y\}}$ implies[equal[x, y], equal[y, x]]
$\forall_{\{x,y,z\}}$ implies[and[equal[x, y], equal[y, z]], equal[x, z]]
$\forall_x$ equal[x, x]
$\forall_{\{a,b\}}$ and[a, b] = and[b, a]
$\forall_{\{a,b\}}$ or[a, b] = or[b, a]
$\forall_{\{a,b\}}$ and[a, or[b, not[b]]] = a
$\forall_{\{a,b\}}$ or[a, and[b, not[b]]] = a
$\forall_{\{a,b,c\}}$ and[a, or[b, c]] = or[and[a, b], and[a, c]]
$\forall_{\{a,b,c\}}$ or[a, and[b, c]] = and[or[a, b], or[a, c]]
$\forall_{\{a,b\}}$ implies[a, b] = or[not[a], b]
$\forall_{\{\alpha,\beta,y,z\}}$ implies[$\exists_x$ implies[and[α[y], β[z]], between[x, y, z]], $\exists_u$ implies[and[α[y], β[z]], between[y, u, z]]]

where we are listing everything (even logic) in explicit (Wolfram Language) functional form. Given these axioms we should now be able to prove all theorems in Euclidean geometry. As an example (that's already complicated enough) let's take Euclid's very first "proposition" (Book 1, Proposition 1) which states that it's possible "with a ruler and compass" (i.e. with lines and circles) to construct an equilateral triangle based on any line segment—as in:

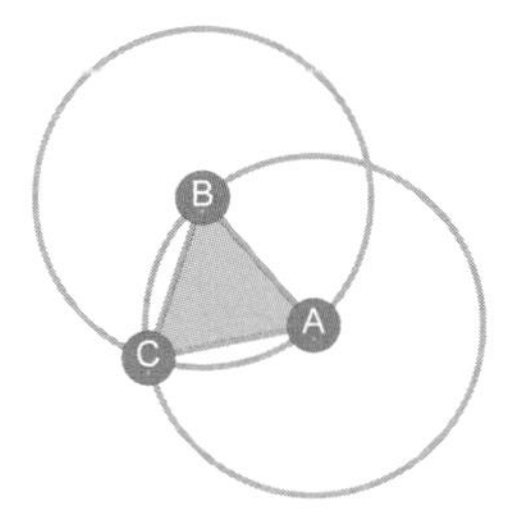

We can write this theorem by saying that given the axioms together with the "setup"

$\forall_{\{a,b,c\}}$ implies[equal[circle[a, b], circle[a, c]], congruent[line[a, b], line[a, c]]]

it's possible to derive:

and[congruent[line[a, b], line[a, c]], congruent[line[b, a], line[b, c]]]

We can now use automated theorem proving to generate a proof

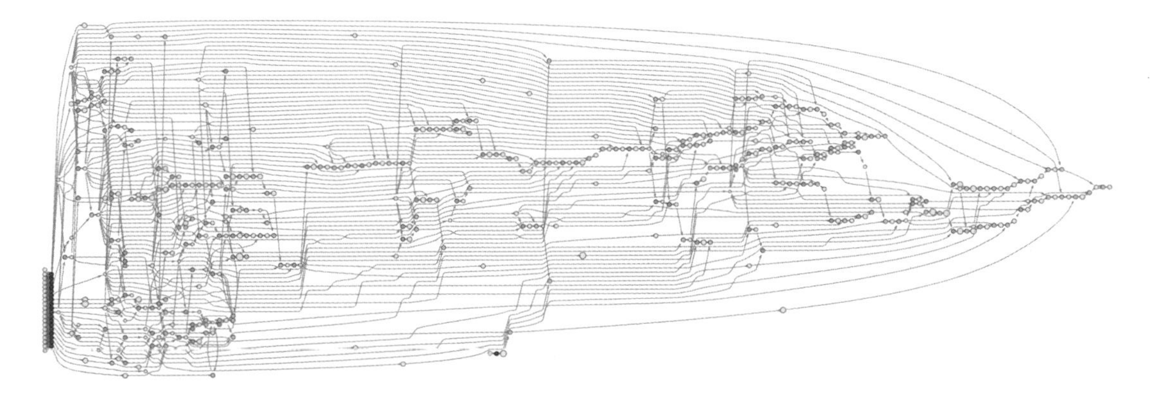

and in this case the proof takes 272 steps. But the fact that it's possible to generate this proof shows that (up to various issues about the "setup conditions") the theorem it proves must eventually "occur naturally" in the entailment cone of the original axioms—though along with an absolutely immense number of other theorems that Euclid didn't "call out" and write down in his books.

Looking at the collection of axiom systems from *A New Kind of Science* (and a few related ones) for many of them we can just directly start generating entailment cones—here shown after one step, using substitution only:

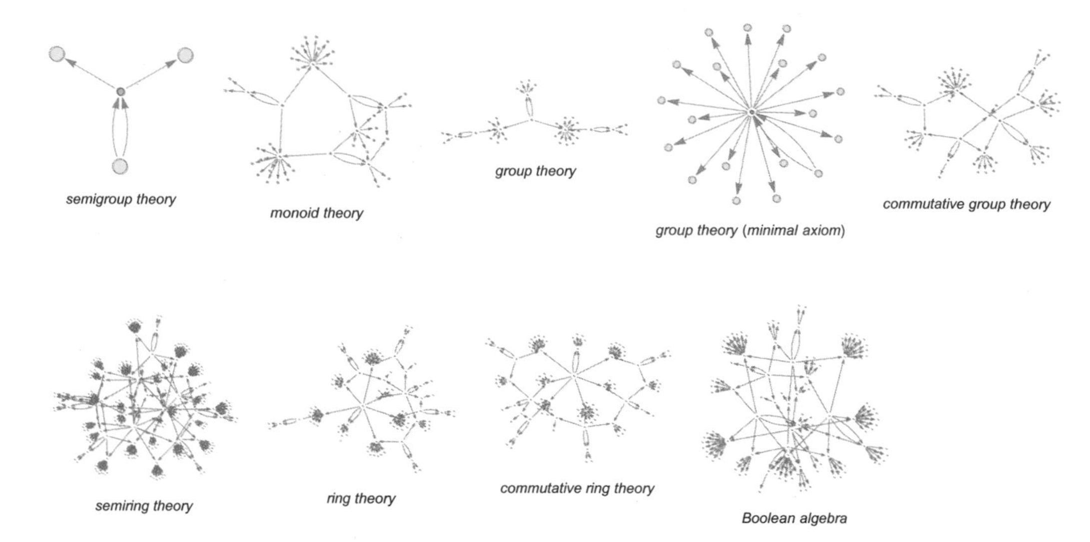

But if we're going to make entailment cones for all axiom systems there are a few other technical wrinkles we have to deal with. The axiom systems shown above are all "straightforwardly equational" in the sense that they in effect state what amount to "algebraic relations" (in the sense of universal algebra) universally valid for all choices of variables. But some axiom systems traditionally used in mathematics also make other kinds of statements. In the traditional formalism and notation of mathematical logic these can look quite complicated and abstruse. But with a metamodel of mathematics like ours it's possible to untangle things to the point where these different kinds of statements can also be handled in a streamlined way.

In standard mathematical notation one might write

$$\forall_a \forall_b \, a \circ b = b \circ a$$

which we can read as "for all a and b, $a \circ b$ equals $b \circ a$"—and which we can interpret in our "metamodel" of mathematics as the (two-way) rule:

$$a\_ \circ b\_ \leftrightarrow b\_ \circ a\_$$

What this says is just that any time we see an expression that matches the pattern $a\_ \circ b\_$ we can replace it by $b\_ \circ a\_$ (or in Wolfram Language notation just $b \circ a$), and vice versa, so that in effect $a\_ \circ b\_$ can be said to entail $b\_ \circ a\_$.

But what if we have axioms that involve not just universal statements ("for all...") but also existential statements ("there exists...")? In a sense we're already dealing with these. Whenever we write $a\_ \circ b\_$—or in explicit functional form, say o[a_, b_]—we're effectively asserting that there exists some operator o that we can do operations with. It's important to note that once we introduce o (or ∘) we imagine that it represents the same thing wherever it appears (in contrast to a pattern variable like $a\_$ that can represent different things in different instances).

Now consider an "explicit existential statement" like

$$\exists_a \, a \circ a = a$$

which we can read as "there exists something a for which $a \circ a$ equals a". To represent the "something" we just introduce a "constant", or equivalently an expression with head, say, α, and zero arguments: $\alpha[]$. Now we can write our existential statement as

$$\alpha[] \circ \alpha[] \longleftrightarrow \alpha[]$$

or:

$$o[\alpha[], \alpha[]] \longleftrightarrow \alpha[]$$

We can operate on this using rules like $a\_ \circ b\_ \longleftrightarrow b\_ \circ a\_$, with $\alpha[]$ always "passing through" unchanged—but with its mere presence asserting that "it exists".

A very similar setup works even if we have both universal and existential quantifiers. For example, we can represent

$$\forall_a \, \exists_b \, a \circ b = a$$

as just

$$a\_ \circ \beta[a\_] \leftrightarrow a\_$$

where now there isn't just a single object, say $\beta[]$, that we assert exists; instead there are "lots of different β's", "parametrized" in this case by a.

We can apply our standard accumulative bisubstitution process to this statement—and after one step we get:

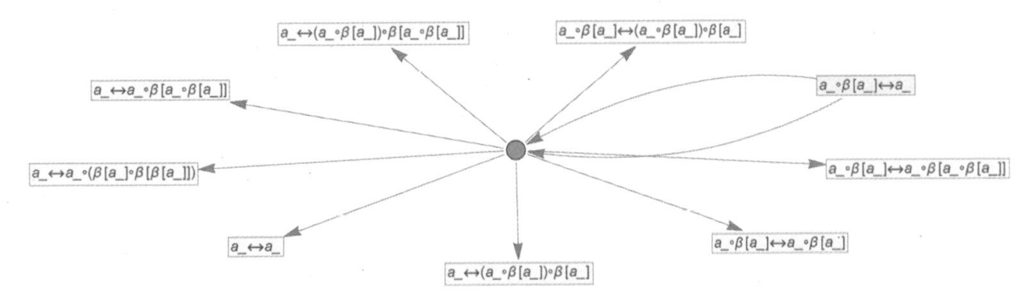

Note that this is a very different result from the one for the "purely universal" statement:

$a\_ \circ b\_ \leftrightarrow a\_$

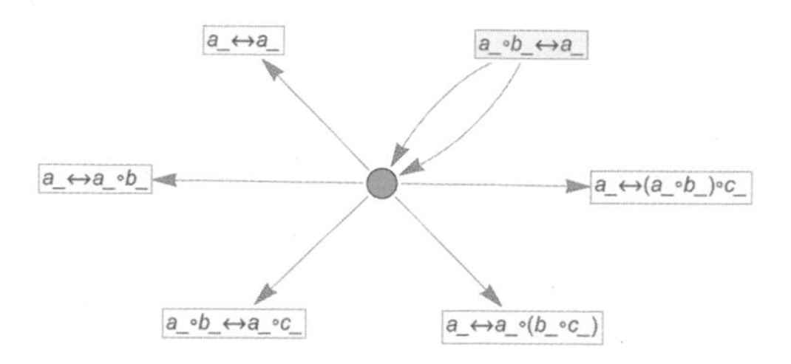

In general, we can "compile" any statement in terms of quantifiers into our metamodel, essentially using the standard technique of Skolemization from mathematical logic. Thus for example

$\forall_a \forall_c \exists_b\, a \circ b = c \circ b$

can be "compiled into"

$a\_ \circ \beta[a\_, c\_] \leftrightarrow c\_ \circ \beta[a\_, c\_]$

while

$\forall_a \exists_b \forall_c\, a \circ b = c$

can be compiled into:

$a\_ \circ \beta[a\_] \leftrightarrow c\_$

If we look at the actual axiom systems used in current mathematics there's one more issue to deal with—which doesn't affect the axioms for logic or group theory, but does show up, for example, in the Peano axioms for arithmetic. And the issue is that in addition to quantifying over "variables", we also need to quantify over "functions". Or formulated differently, we need to set up not just individual axioms, but a whole "axiom schema" that can generate an infinite sequence of "ordinary axioms", one for each possible "function".

In our metamodel of mathematics, we can think of this in terms of "parametrized functions", or in Wolfram Language, just as having functions whose heads are themselves patterns, as in f[n_][a_].

Using this setup we can then "compile" the standard induction axiom of Peano Arithmetic

$\forall_f \forall_y (f[0, y] \wedge \forall_x (f[x, y] \Rightarrow f[s[x], y]) \Rightarrow \forall_x f[x, y])$

into the (Wolfram Language) metamodel form

$f\_[0, y\_] \wedge (f\_[x\_, y\_] \rightarrow f\_[s[x\_], y\_]) \rightarrow f\_[z\_, y\_]$

where the "implications" in the original axiom have been converted into one-way rules, so that what the axiom can now be seen to do is to define a transformation for something that is not an "ordinary mathematical-style expression" but rather an expression that is itself a rule.

But the important point is that our whole setup of doing substitutions in symbolic expressions—like Wolfram Language—makes no fundamental distinction between dealing with "ordinary expressions" and with "rules" (in Wolfram Language, for example, $a \to b$ is just Rule[a, b]). And as a result we can expect to be able to construct token-event graphs, build entailment cones, etc. just as well for axiom systems like Peano Arithmetic, as for ones like Boolean algebra and group theory.

The actual number of nodes that appear even in what might seem like simple cases can be huge, but the whole setup makes it clear that exploring an axiom system like this is just another example—that can be uniformly represented with our metamodel of mathematics—of a form of sampling of the ruliad.

16 | The Model-Theoretic Perspective

We've so far considered something like

$$x \circ y = (y \circ x) \circ y$$

just as an abstract statement about arbitrary symbolic variables x and y, and some abstract operator $\circ$. But can we make a "model" of what x, y, and $\circ$ could "explicitly be"?

Let's imagine for example that x and y can take 2 possible values, say 0 or 1. (We'll use numbers for notational convenience, though in principle the values could be anything we want.) Now we have to ask what $\circ$ can be in order to have our original statement always hold. It turns out in this case that there are several possibilities, that can be specified by giving possible "multiplication tables" for $\circ$:

∘	0	1
0	0	0
1	0	0

∘	0	1
0	0	0
1	0	1

∘	0	1
0	0	1
1	0	1

∘	0	1
0	0	1
1	1	1

∘	0	1
0	1	0
1	1	0

∘	0	1
0	1	1
1	1	1

(For convenience we'll often refer to such multiplication tables by numbers FromDigits[Flatten[m], k], here 0, 1, 5, 7, 10, 15.) Using let's say the second multiplication table we can then "evaluate" both sides of the original statement for all possible choices of x and y, and verify that the statement always holds:

x	y	x∘y	(y∘x)∘y
1	1	1	1
1	0	0	0
0	1	0	0
0	0	0	0

If we allow, say, 3 possible values for x and y, there turn out to be 221 possible forms for $\circ$. The first few are:

∘	0	1	2
0	0	0	0
1	0	0	0
2	0	0	0

∘	0	1	2
0	0	0	0
1	0	0	0
2	0	0	2

∘	0	1	2
0	0	0	0
1	0	0	2
2	0	1	0

∘	0	1	2
0	0	0	0
1	0	0	2
2	0	1	2

∘	0	1	2
0	0	0	0
1	0	0	2
2	0	2	2

∘	0	1	2
0	0	0	0
1	0	1	0
2	0	0	0

∘	0	1	2
0	0	0	0
1	0	1	0
2	0	0	2

∘	0	1	2
0	0	0	0
1	0	1	1
2	0	1	0

...

As another example, let's consider the simplest axiom for Boolean algebra (that I discovered in 2000):

$$((b \circ c) \circ a) \circ (b \circ ((b \circ a) \circ b)) = a$$

Here are the "size-2" models for this

∘	0	1
0	1	0
1	0	0

∘	0	1
0	1	1
1	1	0

and these, as expected, are the truth tables for Nand and Nor respectively. (In this particular case, there are no size-3 models, 12 size-4 models, and in general $\frac{2^n!}{n!}$ models of size 2^n—and no finite models of any other size.)

Looking at this example suggests a way to talk about models for axiom systems. We can think of an axiom system as defining a collection of abstract constraints. But what can we say about objects that might satisfy those constraints? A model is in effect telling us about these objects. Or, put another way, it's telling what "things" the axiom system "describes". And in the case of my axiom for Boolean algebra, those "things" would be Boolean variables, operated on using Nand or Nor.

As another example, consider the axioms for group theory:

$$\{a \circ (b \circ c) = (a \circ b) \circ c,\ a \circ \diamond = a,\ a \circ \overline{a} = \diamond\}$$

Here are the models up to size 3 in this case:

∘	0	1
0	0	1
1	1	0

∘	0	1	2
0	0	1	2
1	1	2	0
2	2	0	1

∘	0	1	2	3
0	0	1	2	3
1	1	0	3	2
2	2	3	0	1
3	3	2	1	0

∘	0	1	2	3
0	0	1	2	3
1	1	0	3	2
2	2	3	1	0
3	3	2	0	1

∘	0	1	2	3
0	0	1	2	3
1	1	3	0	2
2	2	0	3	1
3	3	2	1	0

∘	0	1	2	3
0	0	1	2	3
1	1	2	3	0
2	2	3	0	1
3	3	0	1	2

Is there a mathematical interpretation of these? Well, yes. They essentially correspond to (representations of) particular finite groups. The original axioms define constraints to be satisfied by any group. These models now correspond to particular groups with specific finite numbers of elements (and in fact specific representations of these groups). And just like in the Boolean algebra case this interpretation now allows us to start saying what the models are "about". The first three, for example, correspond to cyclic groups which can be thought of as being "about" addition of integers mod k.

For axiom systems that haven't traditionally been studied in mathematics, there typically won't be any such preexisting identification of what they're "about". But we can still think of models as being a way that a mathematical observer can characterize—or summarize—an axiom system. And in a sense we can see the collection of possible finite models for an axiom system as being a kind of "model signature" for the axiom system.

But let's now consider what models tell us about "theorems" associated with a given axiom system. Take for example the axiom:

$$x = (x \circ y) \circ x$$

Here are the size-2 models for this axiom system:

∘	0	1
0	0	0
1	1	1

∘	0	1
0	0	1
1	0	0

∘	0	1
0	0	1
1	0	1

∘	0	1
0	1	1
1	0	0

∘	0	1
0	1	1
1	0	1

Let's now pick the last of these models. Then we can take any symbolic expression involving ∘, and say what its values would be for every possible choice of the values of the variables that appear in it:

a	b	a∘a	a∘b	b∘a	b∘b	(a∘a)∘a	(a∘a)∘b	(a∘b)∘a	(a∘b)∘b	(b∘a)∘a	(b∘a)∘b	(b∘b)∘a	(b∘b)∘b	a∘(a∘a)	a∘(a∘b)
0	0	1	1	1	1	0	0	0	0	0	0	0	0	1	1
0	1	1	1	0	1	0	1	0	1	1	1	0	1	1	1
1	0	1	0	1	1	1	0	1	1	1	0	1	0	1	0
1	1	1	1	1	1	1	1	1	1	1	1	1	1	1	1
3	5	15	13	11	15	3	5	3	7	7	5	3	5	15	13

The last row here gives an "expression code" that summarizes the values of each expression in this particular model. And if two expressions have different codes in the model then this tells us that these expressions cannot be equivalent according to the underlying axiom system.

But if the codes are the same, then it's at least possible that the expressions are equivalent in the underlying axiom system. So as an example, let's take the equivalences associated with pairs of expressions that have code 3 (according to the model we're using):

$\{a, (a \circ a) \circ a, (a \circ b) \circ a, (b \circ b) \circ a\}$

So now let's compare with an actual entailment cone for our underlying axiom system (where to keep the graph of modest size we have dropped expressions involving more than 3 variables):

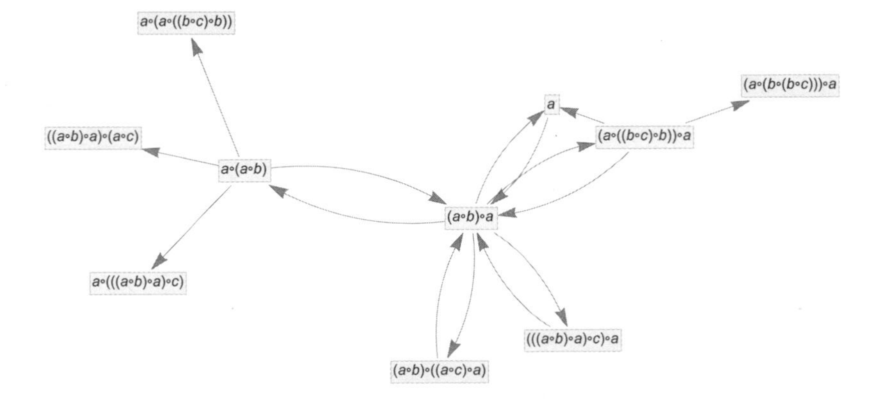

So far this doesn't establish equivalence between any of our code-3 expressions. But if we generate a larger entailment cone (here using a different initial expression) we get

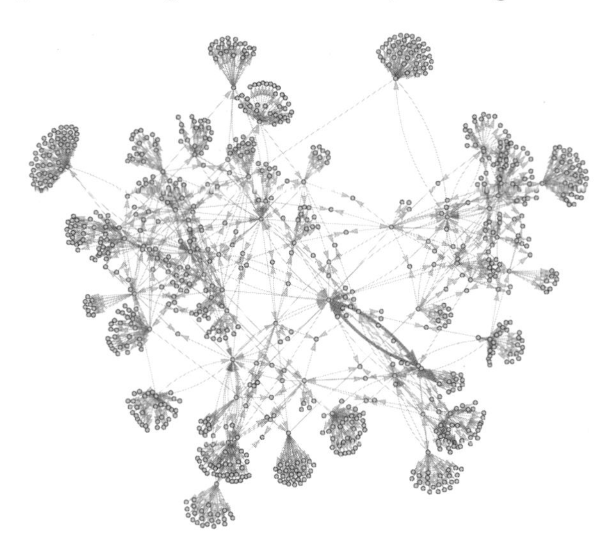

where the path shown corresponds to the statement

$(a \circ a) \circ a = (a \circ b) \circ a$

demonstrating that this is an equivalence that holds in general for the axiom system.

But let's take another statement implied by the model, such as:

$(a \circ b) \circ a = (b \circ b) \circ a$

Yes, it's valid in the model. But it's not something that's generally valid for the underlying axiom system, or could ever be derived from it. And we can see this for example by picking another model for the axiom system, say the second-to-last one in our list above

∘	0	1
0	1	1
1	0	0

and finding out that the values for the two expressions here are different in that model:

a	b	(a ∘ b) ∘ a	(b ∘ b) ∘ a
0	0	0	0
0	1	0	1
1	0	1	0
1	1	1	1
3	5	3	5

The definitive way to establish that a particular statement follows from a particular axiom system is to find an explicit proof for it, either directly by picking it out as a path in the entailment cone or by using automated theorem proving methods. But models in a sense give one a way to "get an approximate result".

As an example of how this works, consider a collection of possible expressions, with pairs of them joined whenever they can be proved equal in the axiom system we're discussing:

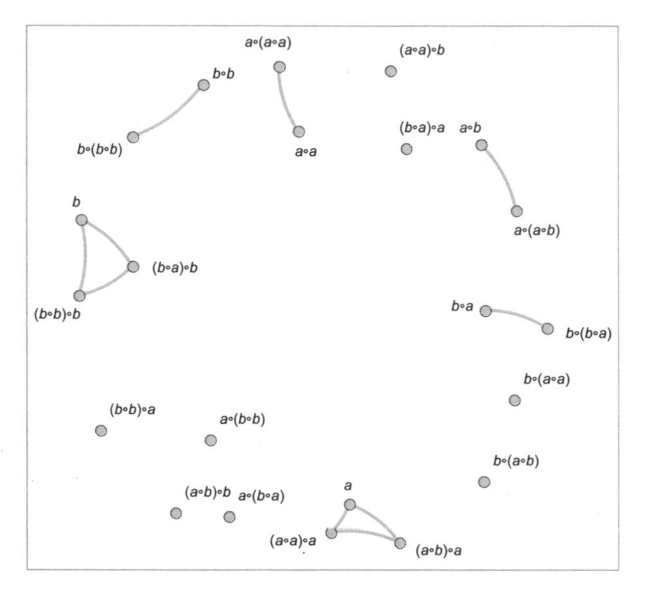

Now let's indicate what codes two models of the axiom system assign to the expressions:

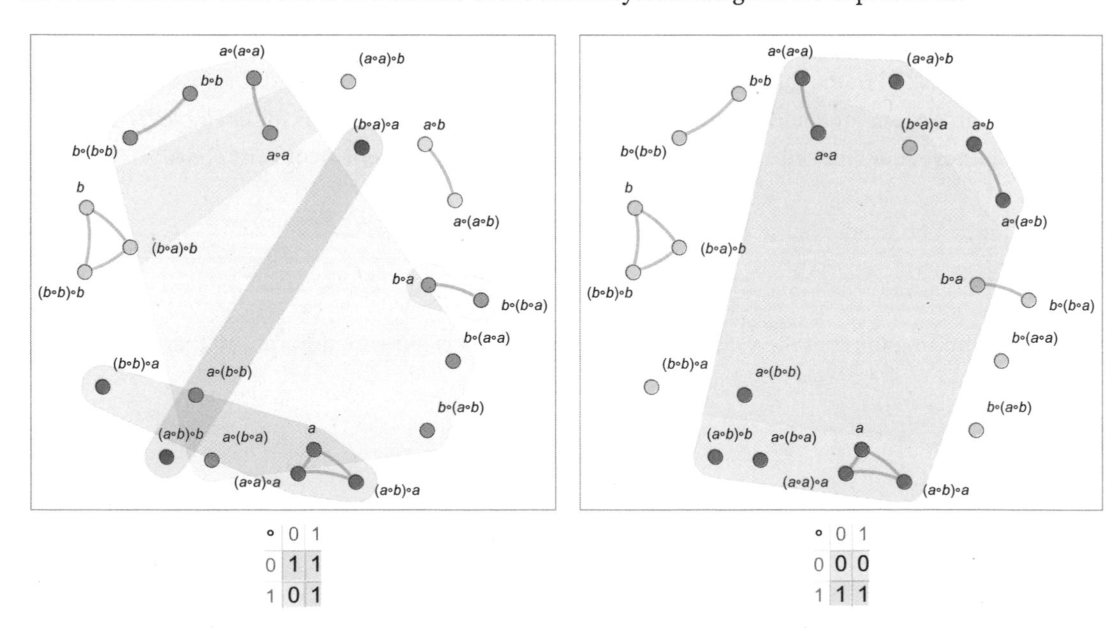

∘	0	1
0	1	1
1	0	1

∘	0	1
0	0	0
1	1	1

The expressions within each connected graph component are equal according to the underlying axiom system, and in both models they are always assigned the same codes. But sometimes the models "overshoot", assigning the same codes to expressions not in the same connected component—and therefore not equal according to the underlying axiom system.

The models we've shown so far are ones that are valid for the underlying axiom system. If we use a model that isn't valid we'll find that even expressions in the same connected component of the graph (and therefore equal according to the underlying axiom system) will be assigned different codes (note the graphs have been rearranged to allow expressions with the same code to be drawn in the same "patch"):

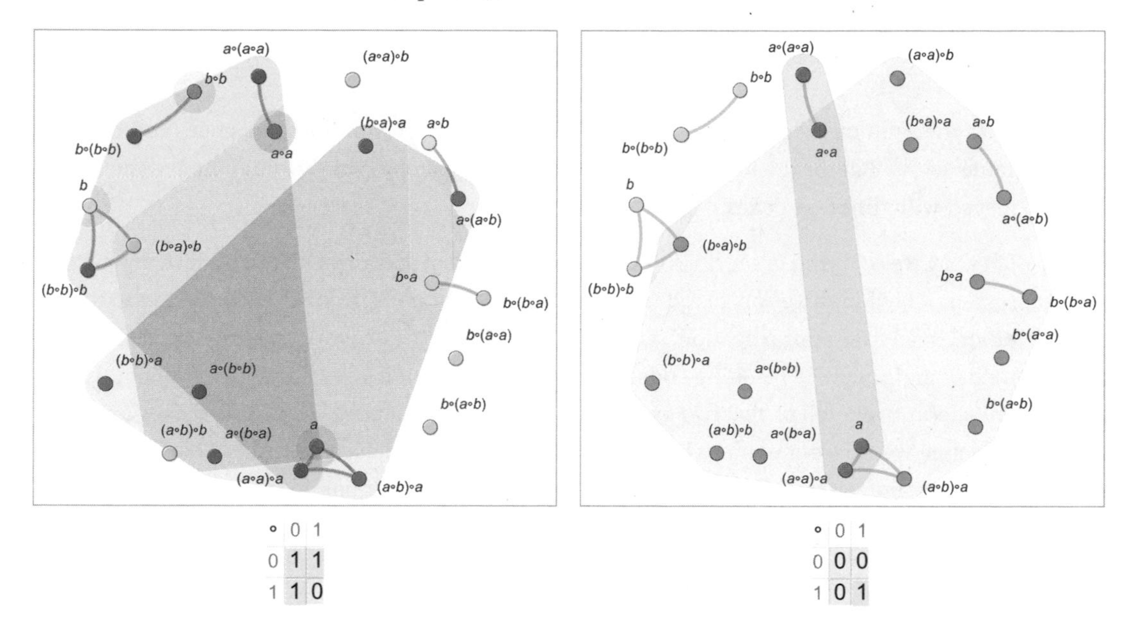

∘	0	1
0	1	1
1	1	0

∘	0	1
0	0	0
1	0	1

We can think of our graph of equivalences between expressions as corresponding to a slice through an entailment graph—and essentially being "laid out in metamathematical space", like a branchial graph, or what we'll later call an "entailment fabric". And what we see is that when we have a valid model different codes yield different patches that in effect cover metamathematical space in a way that respects the equivalences implied by the underlying axiom system.

But now let's see what happens if we make an entailment cone, tagging each node with the code corresponding to the expression it represents, first for a valid model, and then for non-valid ones:

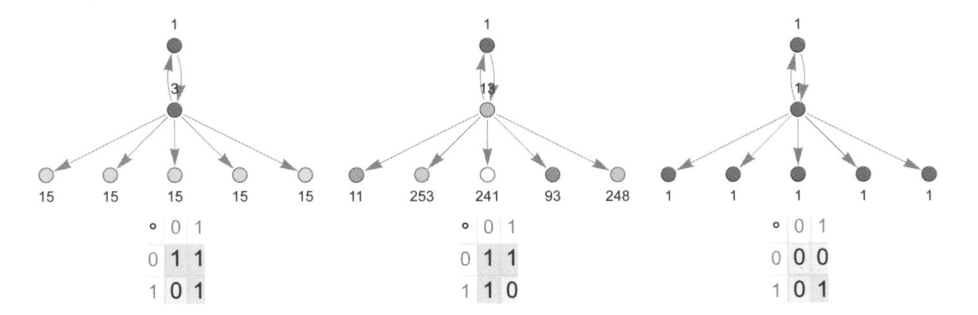

With the valid model, the whole entailment cone is tagged with the same code (and here also same color). But for the non-valid models, different "patches" in the entailment cone are tagged with different codes.

Let's say we're trying to see if two expressions are equal according to the underlying axiom system. The definitive way to tell this is to find a "proof path" from one expression to the other. But as an "approximation" we can just "evaluate" these two expressions according to a model, and see if the resulting codes are the same. Even if it's a valid model, though, this can only definitively tell us that two expressions aren't equal; it can't confirm that they are. In principle we can refine things by checking in multiple models—particularly ones with more elements. But without essentially pre-checking all possible equalities we can't in general be sure that this will give us the complete story.

Of course, generating explicit proofs from the underlying axiom system can also be hard—because in general the proof can be arbitrarily long. And in a sense there is a tradeoff. Given a particular equivalence to check we can either search for a path in the entailment graph, often effectively having to try many possibilities. Or we can "do the work up front" by finding a model or collection of models that we know will correctly tell us whether the equivalence is correct.

Later we'll see how these choices relate to how mathematical observers can "parse" the structure of metamathematical space. In effect observers can either explicitly try to trace out "proof paths" formed from sequences of abstract symbolic expressions—or they can "globally predetermine" what expressions "mean" by identifying some overall model. In general there may be many possible choices of models—and what we'll see is that these different choices are essentially analogous to different choices of reference frames in physics.

One feature of our discussion of models so far is that we've always been talking about making models for axioms, and then applying these models to expressions. But in the accumulative systems we've discussed above (and that seem like closer metamodels of actual mathematics), we're only ever talking about "statements"—with "axioms" just being statements we happen to start with. So how do models work in such a context?

Here's the beginning of the token-event graph starting with

$x \circ y \leftrightarrow (y \circ x) \circ y$

produced using one step of entailment by substitution:

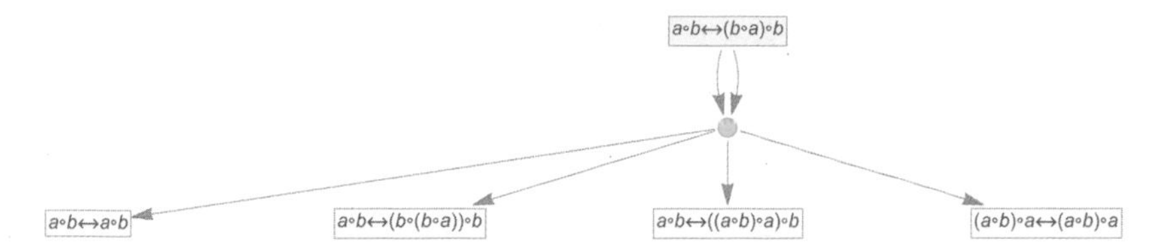

For each of the statements given here, there are certain size-2 models (indicated here by their multiplication tables) that are valid—or in some cases all models are valid:

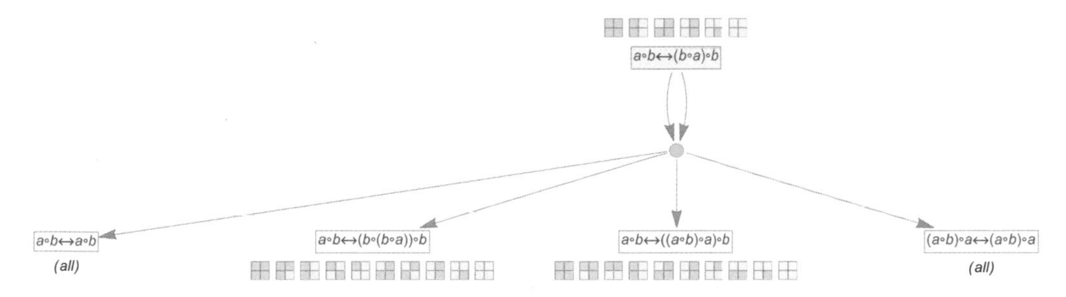

We can summarize this by indicating in a 4×4 grid which of the 16 possible size-2 models are consistent with each statement generated so far in the entailment cone:

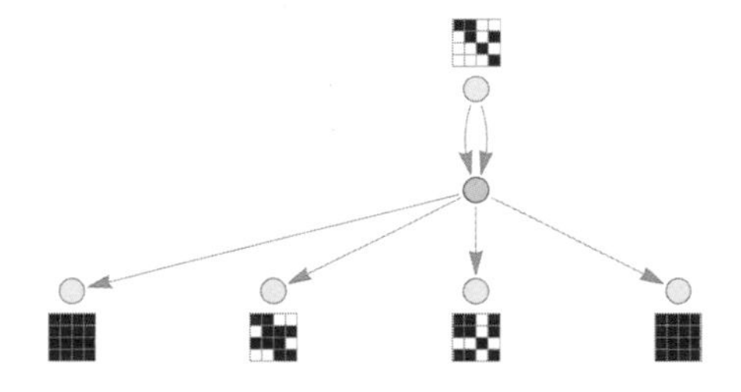

Continuing one more step we get:

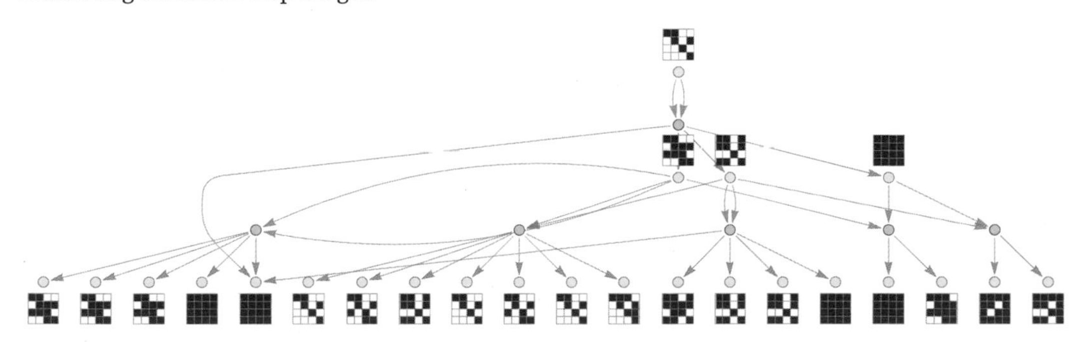

It's often the case that statements generated on successive steps in the entailment cone in essence just "accumulate more models". But—as we can see from the right-hand edge of this graph—it's not always the case—and sometimes a model valid for one statement is no longer valid for a statement it entails. (And the same is true if we use full bisubstitution rather than just substitution.)

Everything we've discussed about models so far here has to do with expressions. But there can also be models for other kinds of structures. For strings it's possible to use something like the same setup, though it doesn't work quite so well. One can think of transforming the string

ABABBAA

into

$(((((A \circ B) \circ A) \circ B) \circ B) \circ A) \circ A$

and then trying to find appropriate "multiplication tables" for $\circ$, but here operating on the specific elements A and B, not on a collection of elements defined by the model.

Defining models for a hypergraph rewriting system is more challenging, if interesting. One can think of the expressions we've used as corresponding to trees—which can be "evaluated" as soon as definite "operators" associated with the model are filled in at each node. If we try to do the same thing with graphs (or hypergraphs) we'll immediately be thrust into issues of the order in which we scan the graph.

At a more general level, we can think of a "model" as being a way that an observer tries to summarize things. And we can imagine many ways to do this, with differing degrees of fidelity, but always with the feature that if the summaries of two things are different, then those two things can't be transformed into each other by whatever underlying process is being used.

Put another way, a model defines some kind of invariant for the underlying transformations in a system. The raw material for computing this invariant may be operators at nodes, or may be things like overall graph properties (like cycle counts).

17 | Axiom Systems in the Wild

We've talked about what happens with specific, sample axiom systems, as well as with various axiom systems that have arisen in present-day mathematics. But what about "axiom systems in the wild"—say just obtained by random sampling, or by systematic enumeration? In effect, each possible axiom system can be thought of as "defining a possible field of mathematics"—just in most cases not one that's actually been studied in the history of human mathematics. But the ruliad certainly contains all such axiom systems. And in the style of *A New Kind of Science* we can do ruliology to explore them.

As an example, let's look at axiom systems with just one axiom, one binary operator and one or two variables. Here are the smallest few:

$a = b$	$a = a \circ a$	$a = a \circ b$	$a = b \circ a$	$a = b \circ b$	$a \circ a = a \circ b$	$a \circ a = b \circ a$
$a \circ a = b \circ b$	$a \circ b = b \circ a$	$a \circ b = b \circ b$	$a = a \circ (a \circ a)$	$a = a \circ (a \circ b)$	$a = a \circ (b \circ a)$	$a = a \circ (b \circ b)$
$a = b \circ (a \circ a)$	$a = b \circ (a \circ b)$	$a = b \circ (b \circ a)$	$a = b \circ (b \circ b)$	$a = (a \circ a) \circ a$	$a = (a \circ a) \circ b$	$a = (a \circ b) \circ a$
$a = (a \circ b) \circ b$	$a = (b \circ a) \circ a$	$a = (b \circ a) \circ b$	$a = (b \circ b) \circ a$	$a = (b \circ b) \circ b$	$a \circ a = a \circ (a \circ a)$	$a \circ a = a \circ (a \circ b)$
$a \circ a = a \circ (b \circ a)$	$a \circ a = a \circ (b \circ b)$	$a \circ a = b \circ (a \circ a)$	$a \circ a = b \circ (a \circ b)$	$a \circ a = b \circ (b \circ a)$	$a \circ a = b \circ (b \circ b)$	$a \circ a = (a \circ a) \circ a$

For each of these axiom systems, we can then ask what theorems they imply. And for example we can enumerate theorems—just as we have enumerated axiom systems—then use automated theorem proving to determine which theorems are implied by which axiom systems. This shows the result, with possible axiom systems going down the page, possible theorems going across, and a particular square being filled in (darker for longer proofs) if a given theorem can be proved from a given axiom system:

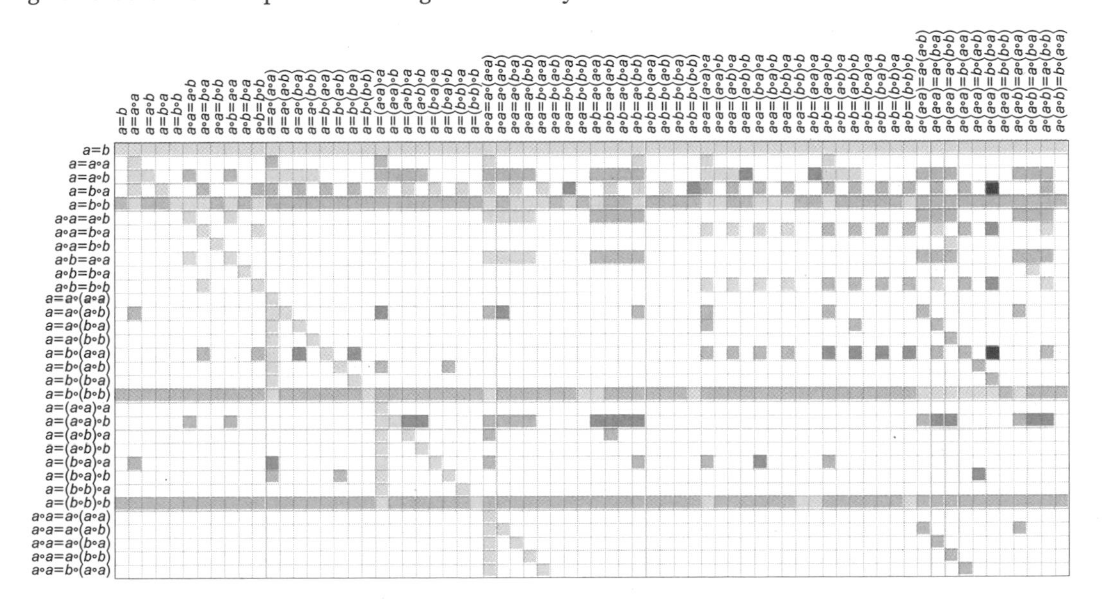

The diagonal on the left is axioms "proving themselves". The lines across are for axiom systems like $a = b$ that basically say that any two expressions are equal—so that any theorem that is stated can be proved from the axiom system.

But what if we look at the whole entailment cone for each of these axiom systems? Here are a few examples of the first two steps:

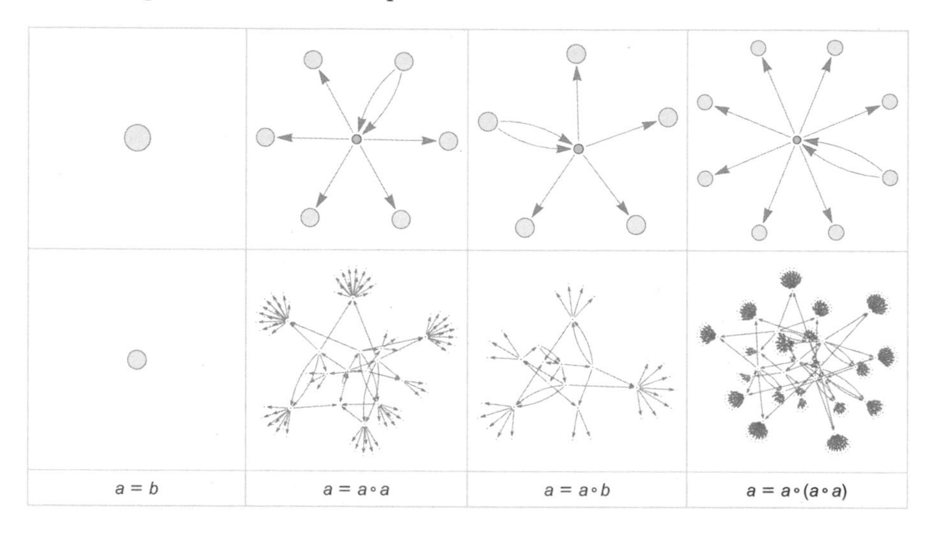

With our method of accumulative evolution the axiom $a = b$ doesn't on its own generate a growing entailment cone (though if combined with any axiom containing ∘ it does, and so does $a = b \circ c$ on its own). But in all the other cases shown the entailment cone grows rapidly (typically at least exponentially)—in effect quickly establishing many theorems. Most of those theorems, however, are "not small"—and for example after 2 steps here are the distributions of their sizes:

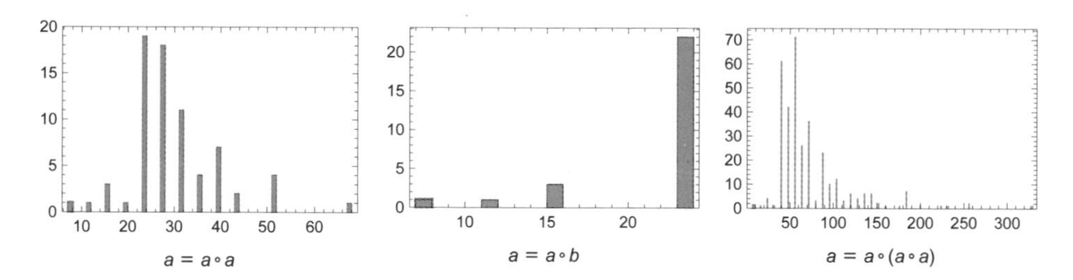

So let's say we generate only one step in the entailment cone. This is the pattern of "small theorems" we establish:

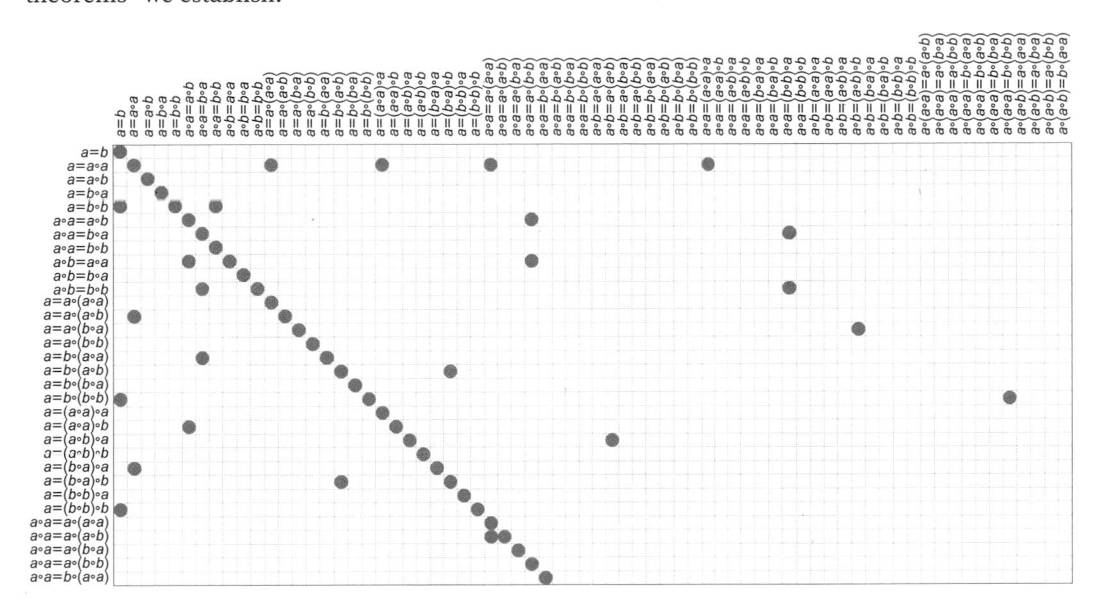

And here is the corresponding result after two steps:

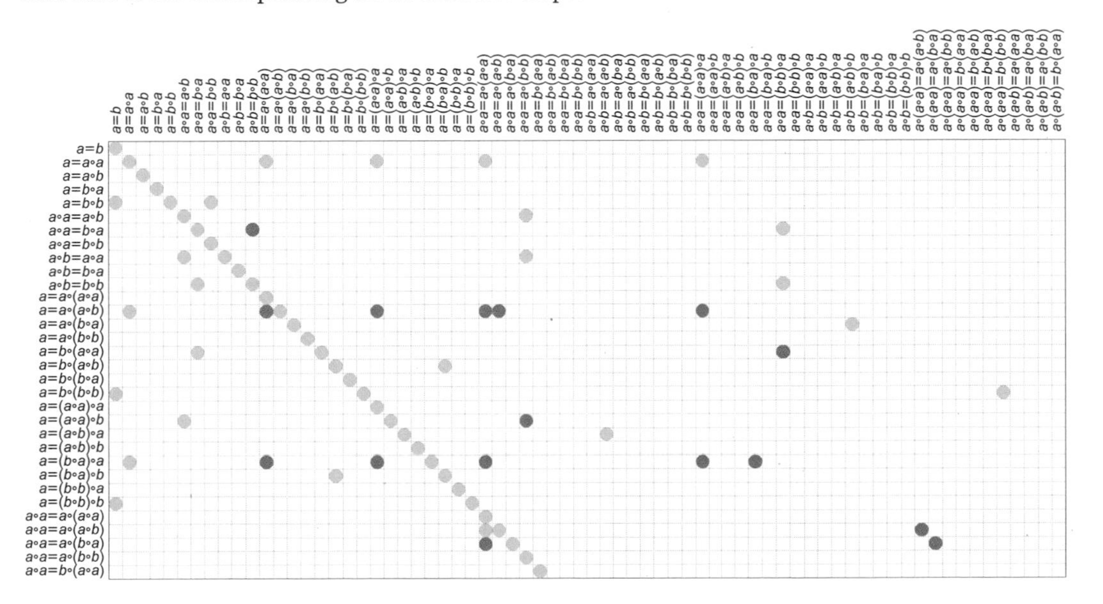

Superimposing this on our original array of theorems we get:

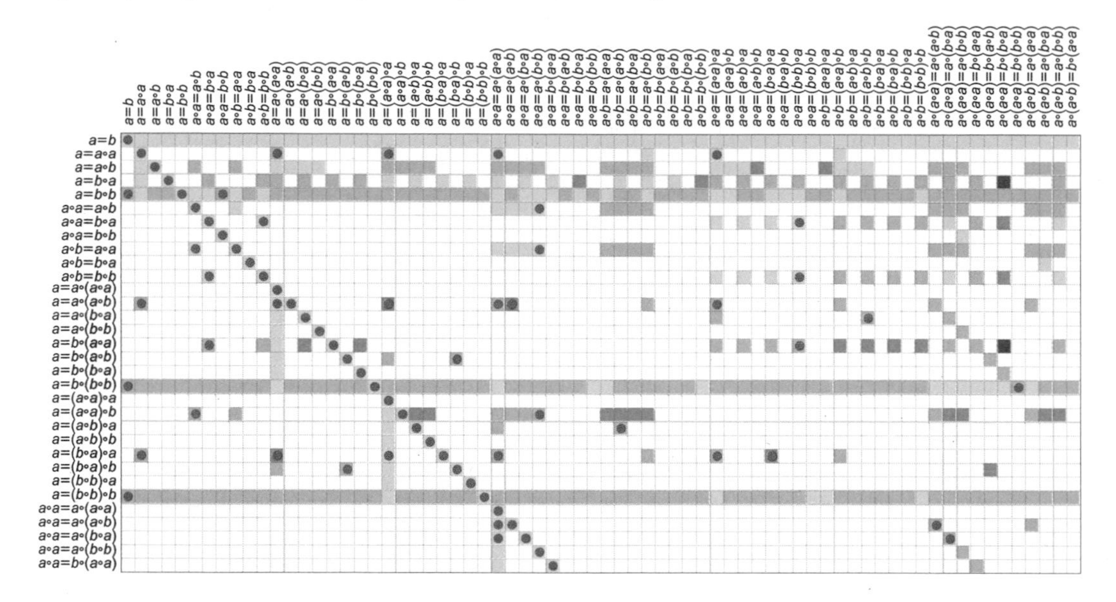

In other words, there are many small theorems that we can establish "if we look for them", but which won't "naturally be generated" quickly in the entailment cone (though eventually it's inevitable that they will be generated). (Later we'll see how this relates to the concept of "entailment fabrics" and the "knitting together of pieces of mathematics".)

In the previous section we discussed the concept of models for axiom systems. So what models do typical "axiom systems from the wild" have? The number of possible models of a given size varies greatly for different axiom systems:

	size 2	size 3
a=b		
a=a∘a		...(729)
a=a∘b		
a=a∘(a∘a)	...(9)	...(3375)

	size 2	size 3	size 4
a=b	2	3	4
a=a∘a	4	729	16 777 216
a=a∘b	1	1	1
a=b∘a	1	1	1
a=b∘b	0	0	0
a∘a=a∘b	4	27	256
a∘a=b∘a	4	27	256
a∘a=b∘b	8	2187	67 108 864

,

	size 2	size 3	size 4
a∘b=a∘a	4	27	256
a∘b=b∘a	8	729	1 048 576
a∘b=b∘b	4	27	256
a=a∘(a∘a)	9	3375	157 351 936
a=a∘(a∘b)	1	27	10 000
a=a∘(b∘a)	5	136	46 121
a=a∘(b∘b)	5	298	1 147 649
a=b∘(a∘a)	2	4	10

,

	size 2	size 3	size 4
a=b∘(a∘b)	2	3	18
a=b∘(b∘a)	4	64	10 000
a=b∘(b∘b)	0	0	0
a=(a∘a)∘a	9	3375	157 351 936
a=(a∘a)∘b	2	4	10
a=(a∘b)∘a	5	136	46 121
a=(a∘b)∘b	4	64	10 000
a=(b∘a)∘a	1	27	10 000

,

	size 2	size 3	size 4
a=(b∘a)∘b	2	3	18
a=(b∘b)∘a	5	298	1 147 649
a=(b∘b)∘b	0	0	0
a∘a=a∘(a∘a)	9	3375	157 351 936
a∘a=a∘(a∘b)	4	343	614 656
a∘a=a∘(b∘a)	7	476	506 698
a∘a=a∘(b∘b)	6	411	1 401 880
a∘a=b∘(a∘a)	5	298	1 147 649

But for each model we can ask what theorems it implies are valid. And for example combining all models of size 2 yields the following "predictions" for what theorems are valid (with the actual theorems indicated by dots):

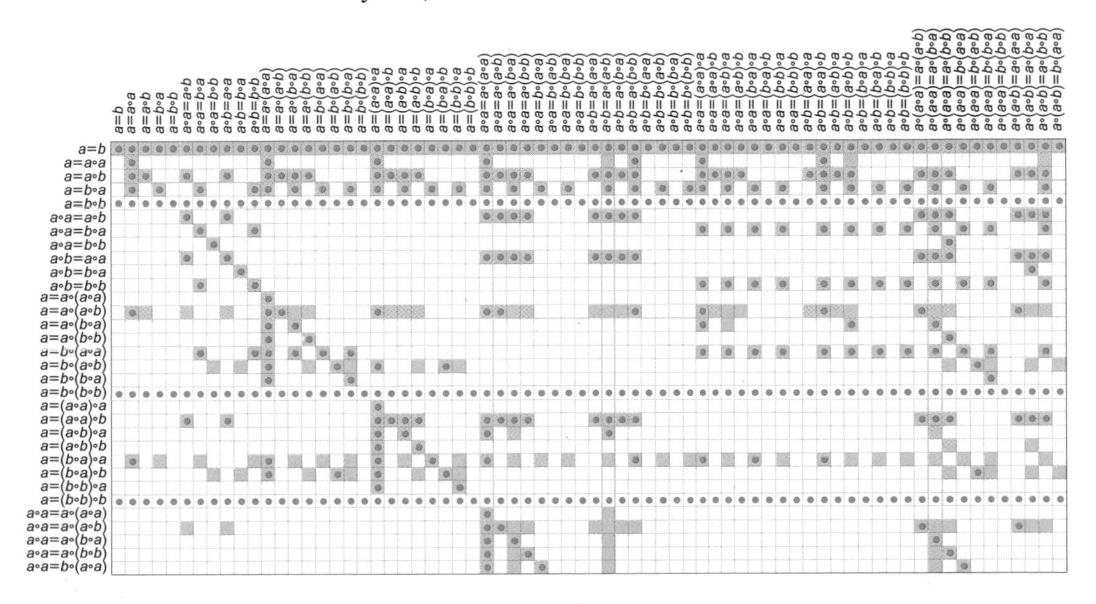

Using instead models of size 3 gives “more accurate predictions”:

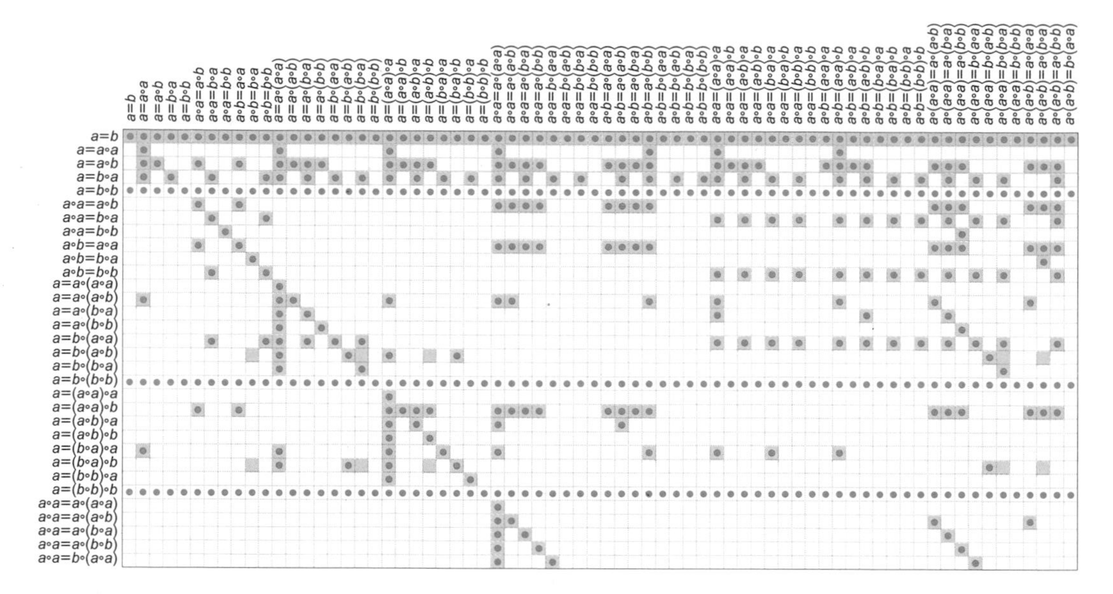

As expected, looking at a fixed number of steps in the entailment cone “underestimates” the number of valid theorems, while looking at finite models overestimates it.

So how does our analysis for “axiom systems from the wild” compare with what we’d get if we considered axiom systems that have been explicitly studied in traditional human mathematics? Here are some examples of “known” axiom systems that involve just a single binary operator

(1) Boolean algebra	$(a\circ a)\circ(a\circ a) = a,\ a\circ(b\circ(b\circ b)) = a\circ a,\ (a\circ(b\circ c))\circ(a\circ(b\circ c)) = ((b\circ b)\circ a)\circ((c\circ c)\circ a)$
(2) Boolean algebra	$(a\circ a)\circ(a\circ b) = a,\ a\circ(a\circ b) = a\circ(b\circ b),\ a\circ(a\circ(b\circ c)) = b\circ(b\circ(a\circ c))$
(3) Boolean algebra	$a\circ b = b\circ a,\ (a\circ b)\circ(a\circ(b\circ c)) = a$
(4) Boolean algebra	$((b\circ c)\circ a)\circ(b\circ((b\circ a)\circ b)) = a$
(5) Boolean algebra	$(a\circ((c\circ a)\circ a))\circ(c\circ(b\circ a)) = c$
(6) equivalential calculus	$(a\circ b)\circ c = a\circ(b\circ c),\ a\circ b = b\circ a,\ (b\circ b)\circ a = a$
(7) implicational calculus	$(a\circ b)\circ a = a,\ a\circ(b\circ c) = b\circ(a\circ c),\ (a\circ b)\circ b = (b\circ a)\circ a$
(8) junctional calculus	$(a\circ b)\circ c = a\circ(b\circ c),\ a\circ b = b\circ a,\ a\circ a = a$
(9) semigroups	$a\circ(b\circ c) = (a\circ b)\circ c$
(10) groups	$a\circ((((a\circ a)\circ b)\circ c)\circ(((a\circ a)\circ a)\circ c)) = b$
(11) Abelian groups	$a\circ(b\circ(c\circ(a\circ b))) = c$
(12) central groupoids	$(a\circ b)\circ(b\circ c) = a$

and here's the distribution of theorems they give:

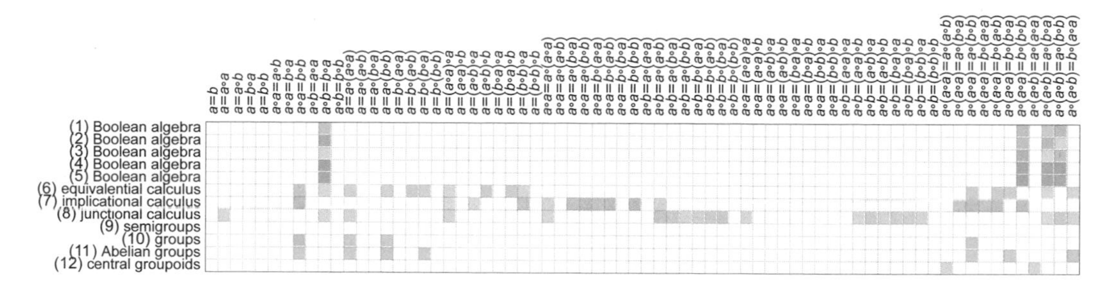

As must be the case, all the axiom systems for Boolean algebra yield the same theorems. But axiom systems for "different mathematical theories" yield different collections of theorems.

What happens if we look at entailments from these axiom systems? Eventually all theorems must show up somewhere in the entailment cone of a given axiom system. But here are the results after one step of entailment:

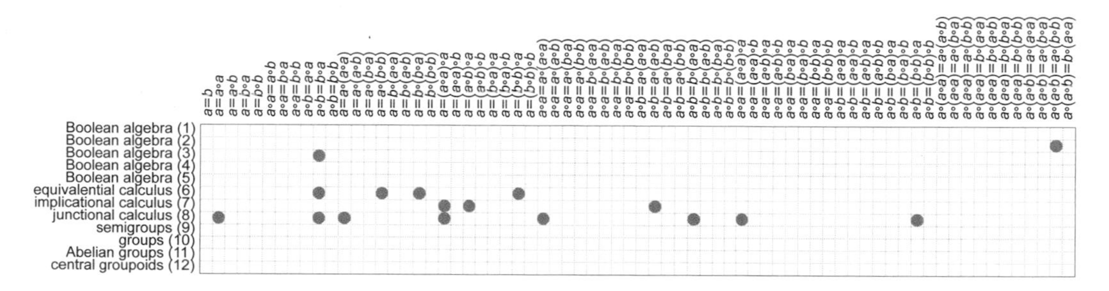

Some theorems have already been generated, but many have not:

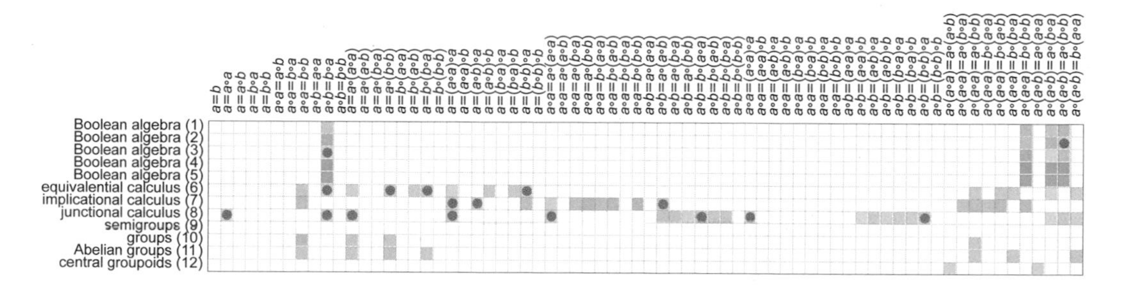

Just as we did above, we can try to "predict" theorems by constructing models. Here's what happens if we ask what theorems hold for all valid models of size 2:

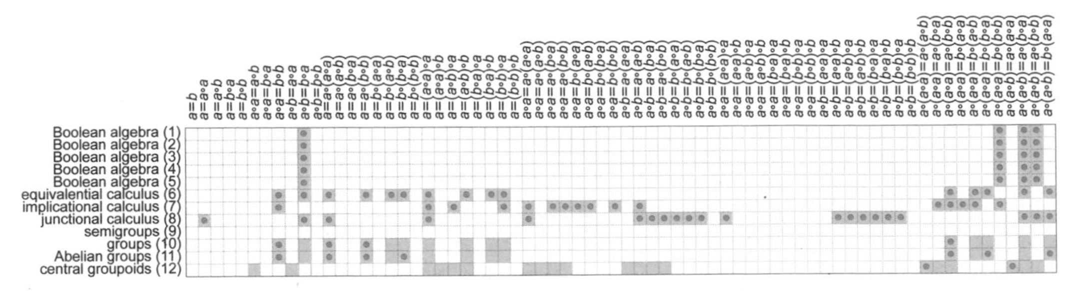

For several of the axiom systems, the models "perfectly predict" at least the theorems we show here. And for Boolean algebra, for example, this isn't surprising: the models just correspond to identifying ∘ as Nand or Nor, and to say this gives a complete description of Boolean algebra. But in the case of groups, "size-2 models" just capture particular groups that happen to be of size 2, and for these particular groups there are special, extra theorems that aren't true for groups in general.

If we look at models specifically of size 3 there aren't any examples for Boolean algebra so we don't predict any theorems. But for group theory, for example, we start to get a slightly more accurate picture of what theorems hold in general:

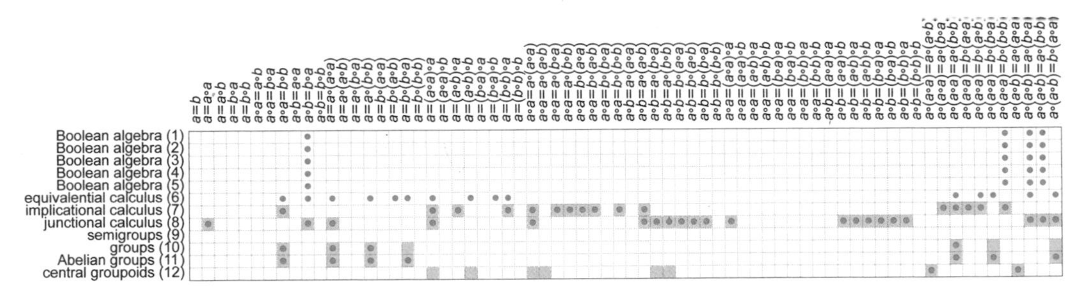

Based on what we've seen here, is there something "obviously special" about the axiom systems that have traditionally been used in human mathematics? There are cases like Boolean algebra where the axioms in effect constrain things so much that we can reasonably say that they're "talking about definite things" (like Nand and Nor). But there are plenty of other cases, like group theory, where the axioms provide much weaker constraints, and for example allow an infinite number of possible specific groups. But both situations occur among axiom systems "from the wild". And in the end what we're doing here doesn't seem to reveal anything "obviously special" (say in the statistics of models or theorems) about "human" axiom systems.

And what this means is that we can expect that conclusions we draw from looking at the "general case of all axiom systems"—as captured in general by the ruliad—can be expected to hold in particular for the specific axiom systems and mathematical theories that human mathematics has studied.

18 | The Topology of Proof Space

In the typical practice of pure mathematics the main objective is to establish theorems. Yes, one wants to know that a theorem has a proof (and perhaps the proof will be helpful in understanding the theorem), but the main focus is on theorems and not on proofs. In our effort to "go underneath" mathematics, however, we want to study not only what theorems there are, but also the process by which the theorems are reached. We can view it as an important simplifying assumption of typical mathematical observers that all that matters is theorems—and that different proofs aren't relevant. But to explore the underlying structure of metamathematics, we need to unpack this—and in effect look directly at the structure of proof space.

Let's consider a simple system based on strings. Say we have the rewrite rule $\{A \to BBB, BB \to A\}$ and we want to establish the theorem $A \to ABA$. To do this we have to find some path from A to ABA in the multiway system (or, effectively, in the entailment cone for this axiom system):

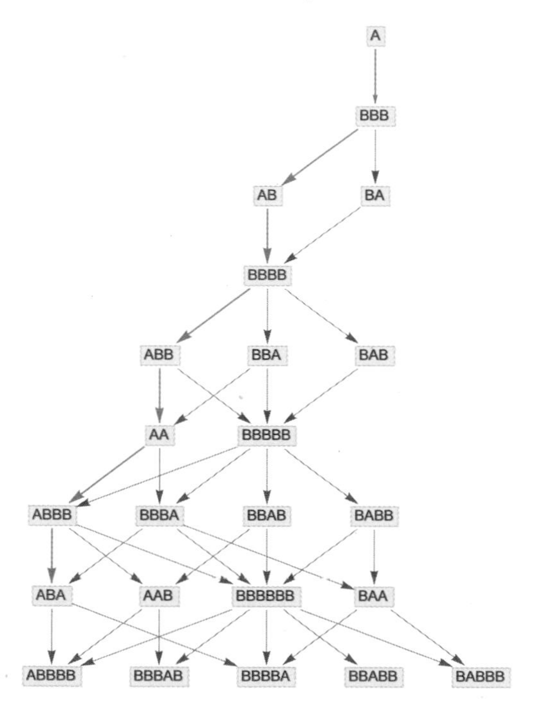

But this isn't the only possible path, and thus the only possible proof. In this particular case, there are 20 distinct paths, each corresponding to at least a slightly different proof:

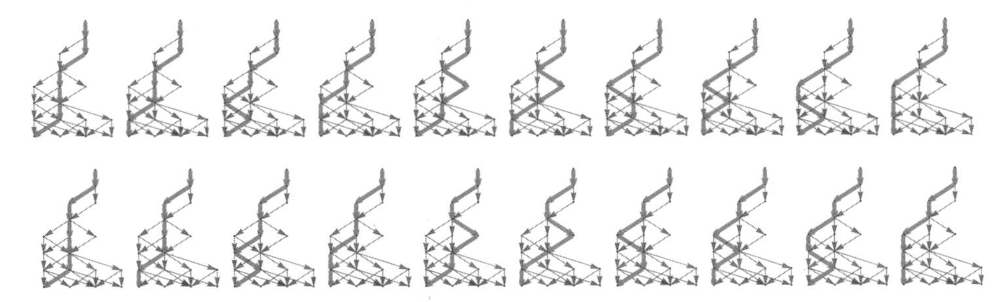

But one feature here is that all these different proofs can in a sense be "smoothly deformed" into each other, in this case by progressively changing just one step at a time. So this means that in effect there is no nontrivial topology to proof space in this case—and "distinctly inequivalent" collections of proofs:

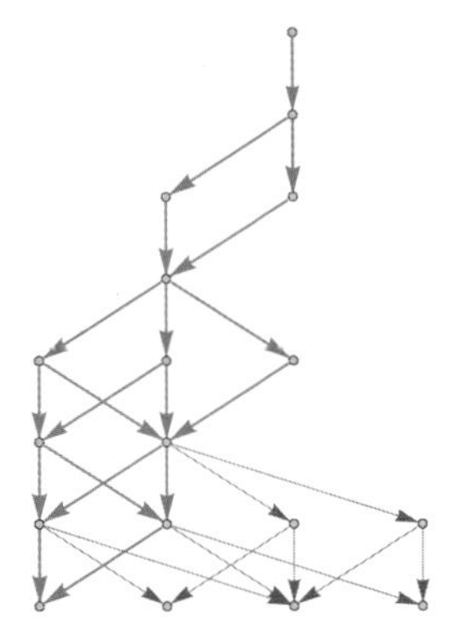

But consider instead the rule {A → AA, A → BAAB}. With this "axiom system" there are 15 possible proofs for the theorem A → BAABAABAAB:

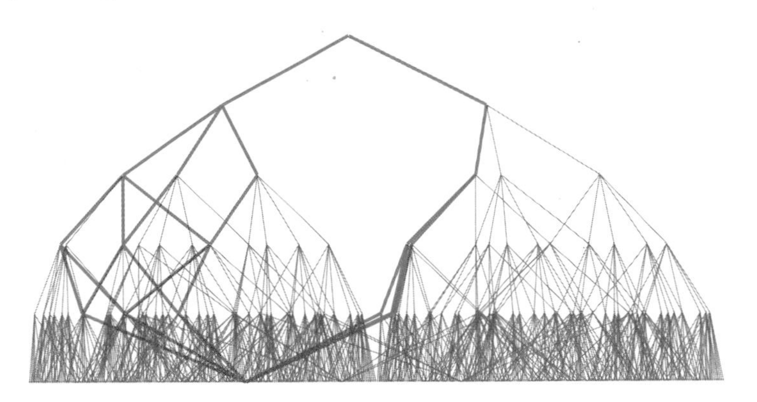

Pulling out just the proofs we get:

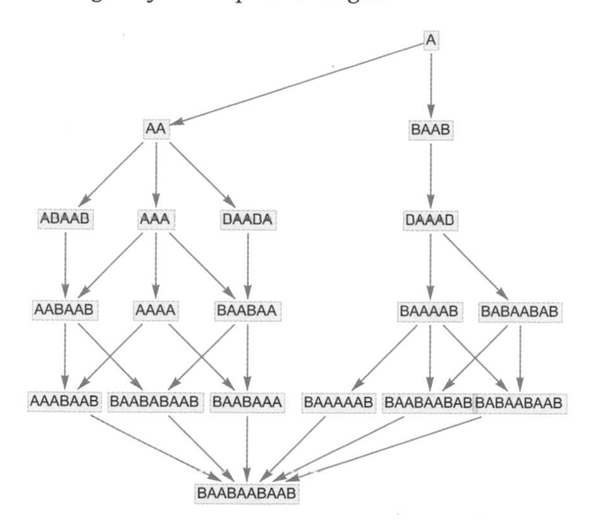

And we see that in a sense there's a "hole" in proof space here—so that there are two distinctly different kinds of proofs that can be done.

One place it's common to see a similar phenomenon is in games and puzzles. Consider for example the Towers of Hanoi puzzle. We can set up a multiway system for the possible moves that can be made. Starting from all disks on the left peg, we get after 1 step:

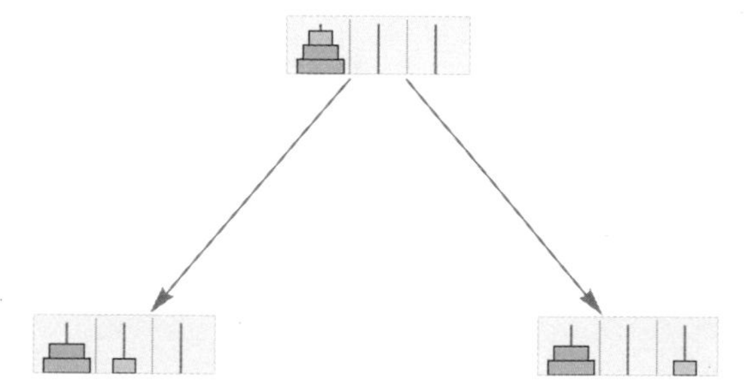

After 2 steps we have:

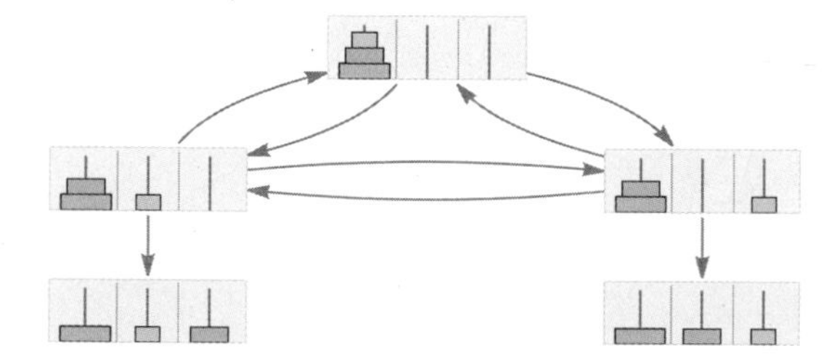

And after 8 steps (in this case) we have the whole "game graph":

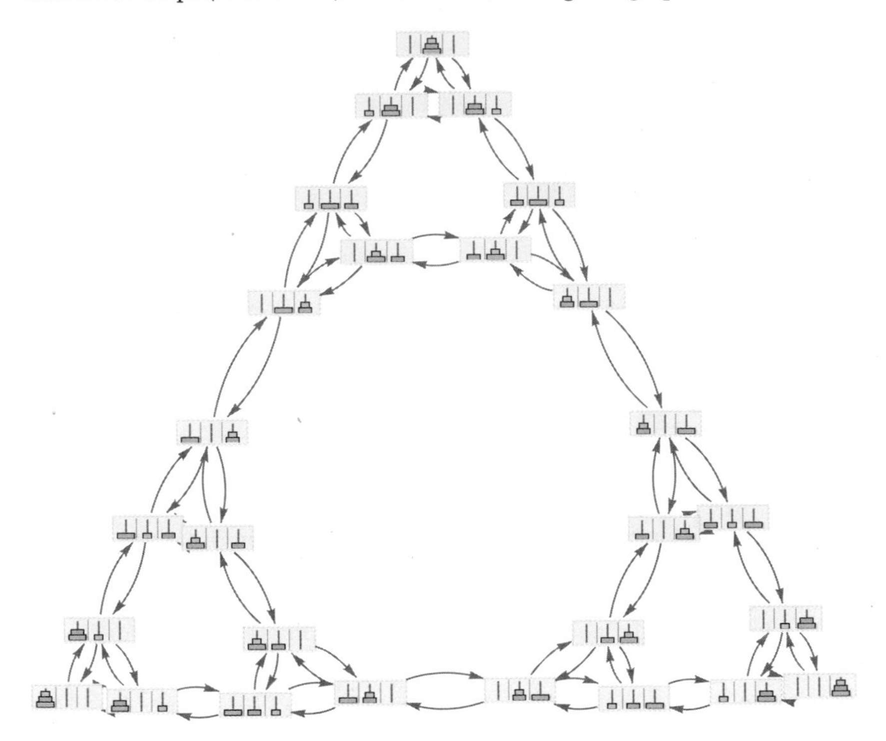

The corresponding result for 4 disks is:

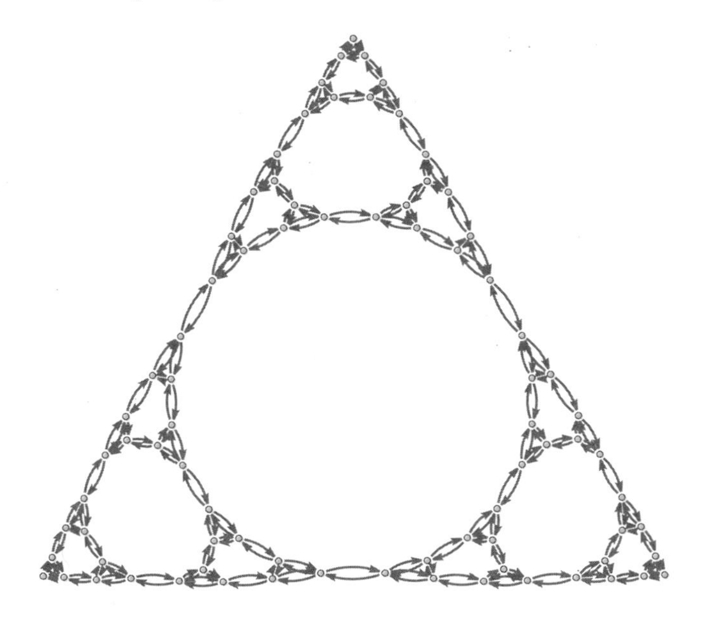

And in each case we see the phenomenon of nontrivial topology. What fundamentally causes this? In a sense it reflects the possibility for distinctly different strategies that lead to the same result. Here, for example, different sides of the "main loop" correspond to the "foundational choice" of whether to move the biggest disk first to the left or to the right. And the same basic thing happens with 4 disks on 4 pegs, though the overall structure is more complicated there:

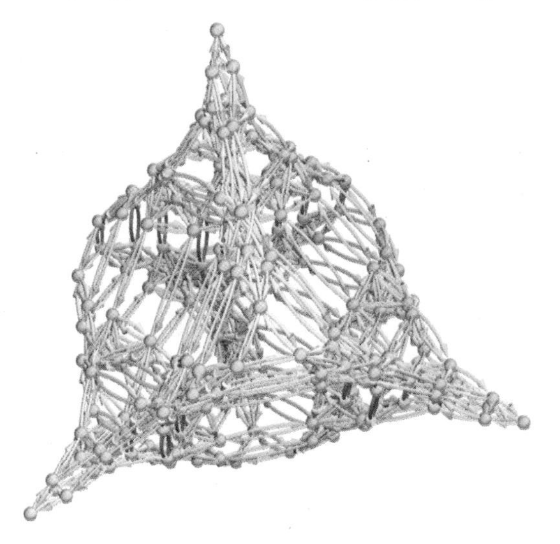

If two paths diverge in a multiway system it could be that it will never be possible for them to merge again. But whenever the system has the property of confluence, it's guaranteed that eventually the paths will merge. And, as it turns out, our accumulative evolution setup guarantees that (at least ignoring generation of new variables) confluence will always be achieved. But the issue is how quickly. If branches always merge after just one step, then in a sense there'll always be topologically trivial proof space. But if the merging can take awhile (and in a continuum limit, arbitrarily long) then there'll in effect be nontrivial topology.

And one consequence of the nontrivial topology we're discussing here is that it leads to disconnection in branchial space. Here are the branchial graphs for the first 3 steps in our original 3-disk 3-peg case:

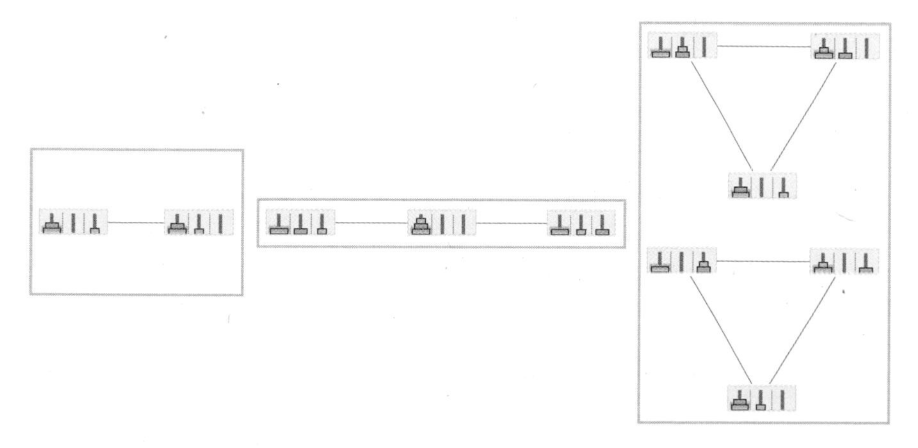

For the first two steps, the branchial graphs stay connected; but on the third step there's disconnection. For the 4-disk 4-peg case the sequence of branchial graphs begins:

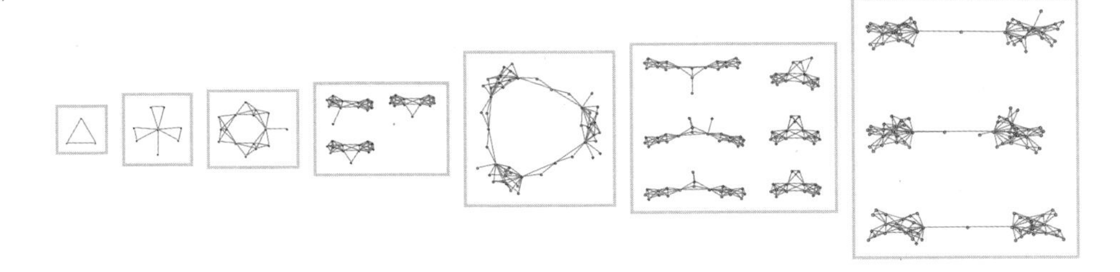

At the beginning (and also the end) there's a single component, that we might think of as a coherent region of metamathematical space. But in the middle it breaks into multiple disconnected components—in effect reflecting the emergence of multiple distinct regions of metamathematical space with something like event horizons temporarily existing between them.

How should we interpret this? First and foremost, it's something that reveals that there's structure "below" the "fluid dynamics" level of mathematics; it's something that depends on the discrete "axiomatic infrastructure" of metamathematics. And from the point of view of our Physics Project, we can think of it as a kind of metamathematical analog of a "quantum effect".

In our Physics Project we imagine different paths in the multiway system to correspond to different possible quantum histories. The observer is in effect spread over multiple paths, which they coarse grain or conflate together. An "observable quantum effect" occurs when there are paths that can be followed by the system, but that are somehow "too far apart" to be immediately coarse-grained together by the observer.

Put another way, there is "noticeable quantum interference" when the different paths corresponding to different histories that are "simultaneously happening" are "far enough apart" to be distinguished by the observer. "Destructive interference" is presumably associated with paths that are so far apart that to conflate them would effectively require conflating essentially every possible path. (And our later discussion of the relationship between falsity and the "principle of explosion" then suggests a connection between destructive interference in physics and falsity in mathematics.)

In essence what determines the extent of "quantum effects" is then our "size" as observers in branchial space relative to the size of features in branchial space such as the "topological holes" we've been discussing. In the metamathematical case, the "size" of us as observers is in effect related to our ability (or choice) to distinguish slight differences in axiomatic formulations of things. And what we're saying here is that when there is nontrivial topology in proof space, there is an intrinsic dynamics in metamathematical entailment that leads to the development of distinctions at some scale—though whether these become "visible" to us as mathematical observers depends on how "strong a metamathematical microscope" we choose to use relative to the scale of the "topological holes".

19 | Time, Timelessness and Entailment Fabrics

A fundamental feature of our metamodel of mathematics is the idea that a given set of mathematical statements can entail others. But in this picture what does "mathematical progress" look like?

In analogy with physics one might imagine it would be like the evolution of the universe through time. One would start from some limited set of axioms and then—in a kind of "mathematical Big Bang"—these would lead to a progressively larger entailment cone containing more and more statements of mathematics. And in analogy with physics, one could imagine that the process of following chains of successive entailments in the entailment cone would correspond to the passage of time.

But realistically this isn't how most of the actual history of human mathematics has proceeded. Because people—and even their computers—basically never try to extend mathematics by axiomatically deriving all possible valid mathematical statements. Instead, they come up with particular mathematical statements that for one reason or another they think are valid and interesting, then try to prove these.

Sometimes the proof may be difficult, and may involve a long chain of entailments. Occasionally—especially if automated theorem proving is used—the entailments may approximate a geodesic path all the way from the axioms. But the practical experience of human mathematics tends to be much more about identifying "nearby statements" and then trying to "fit them together" to deduce the statement one's interested in.

And in general human mathematics seems to progress not so much through the progressive "time evolution" of an entailment graph as through the assembly of what one might call an "entailment fabric" in which different statements are being knitted together by entailments.

In physics, the analog of the entailment graph is basically the causal graph which builds up over time to define the content of a light cone (or, more accurately, an entanglement cone). The analog of the entailment fabric is basically the (more-or-less) instantaneous state of space (or, more accurately, branchial space).

In our Physics Project we typically take our lowest-level structure to be a hypergraph—and informally we often say that this hypergraph "represents the structure of space". But really we should be deducing the "structure of space" by taking a particular time slice from the "dynamic evolution" represented by the causal graph—and for example we should think of two "atoms of space" as "being connected" in the "instantaneous state of space" if there's a causal connection between them defined within the slice of the causal graph that occurs within the time slice we're considering. In other words, the "structure of space" is knitted together by the causal connections represented by the causal graph. (In traditional physics, we might say that space can be "mapped out" by looking at overlaps between lots of little light cones.)

Let's look at how this works out in our metamathematical setting, using string rewrites to simplify things. If we start from the axiom A ⟷ AA this is the beginning of the entailment cone it generates:

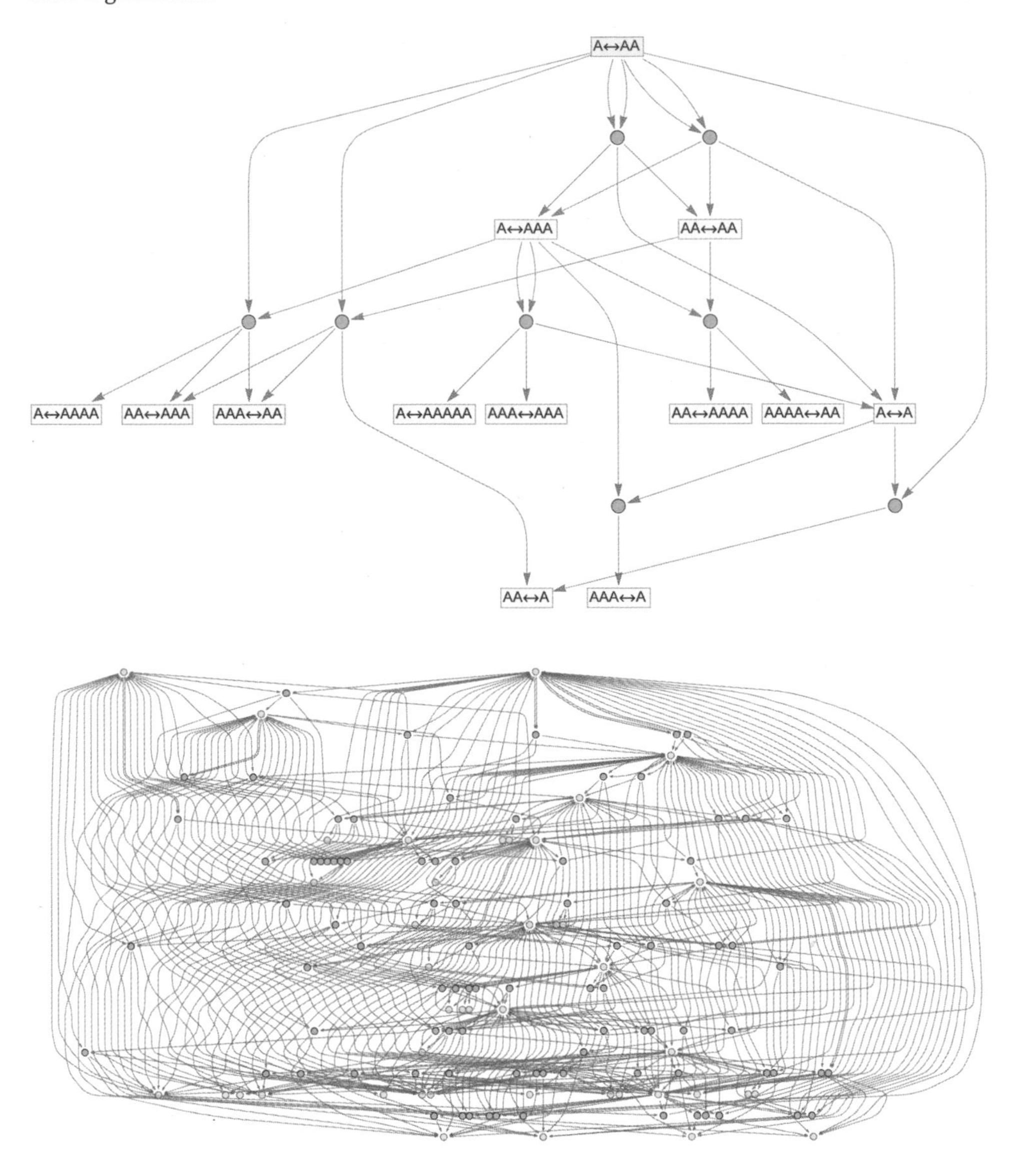

But instead of starting with one axiom and building up a progressively larger entailment cone, let's start with multiple statements, and from each one generate a small entailment cone, say applying each rule at most twice. Here are entailment cones started from several different statements:

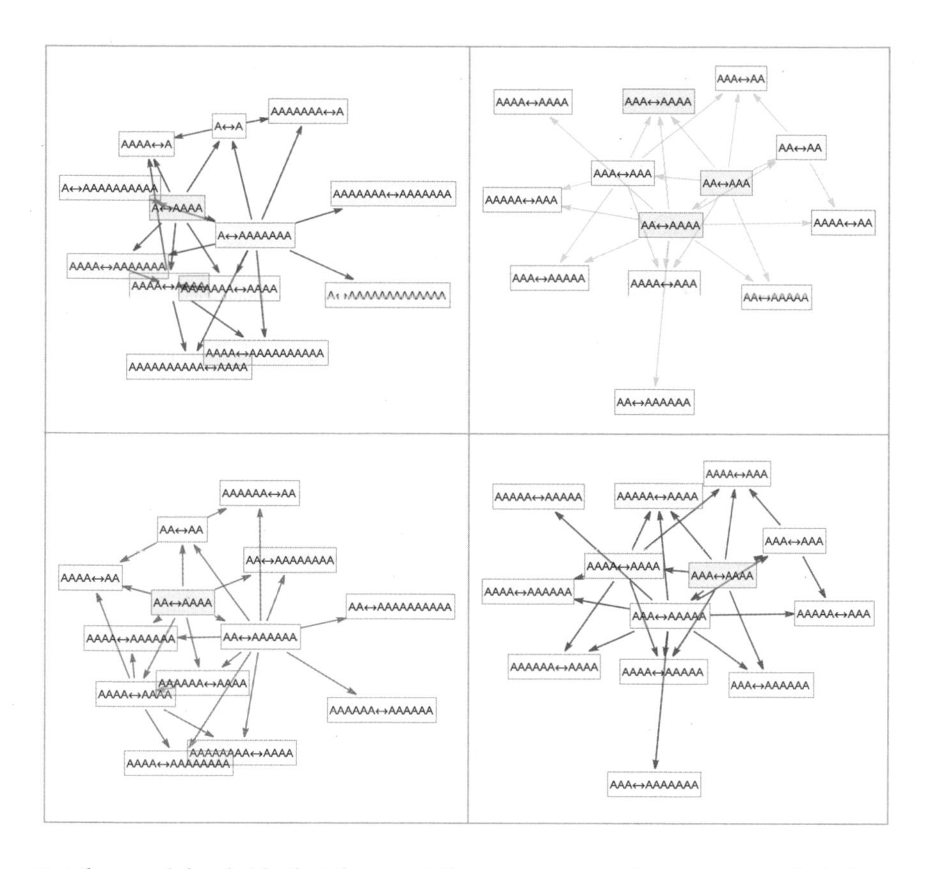

But the crucial point is that these entailment cones overlap—so we can knit them together into an "entailment fabric":

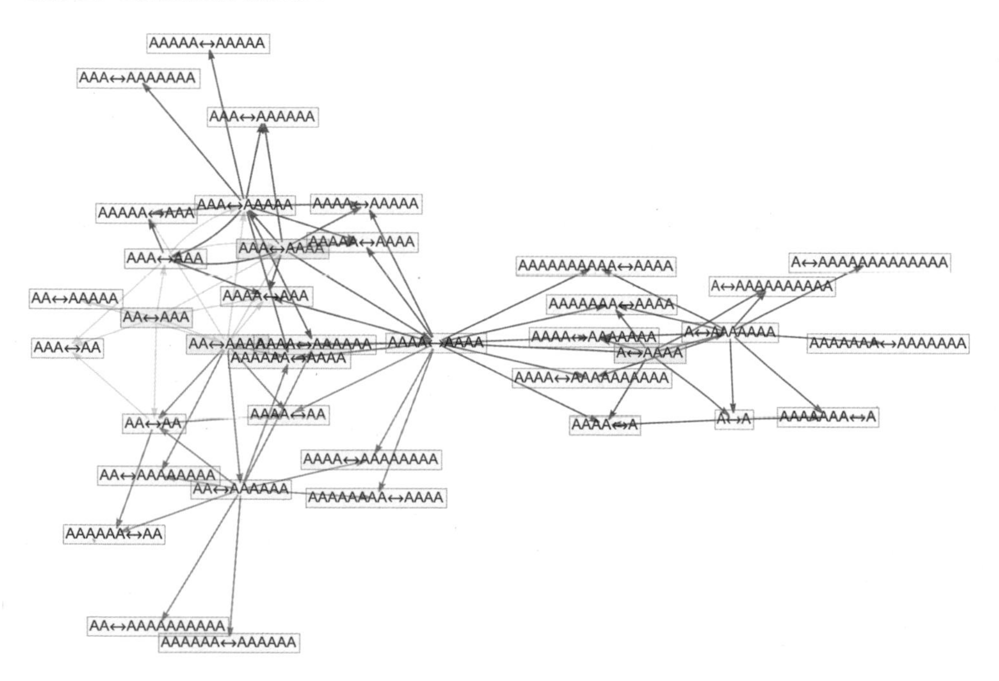

Or with more pieces and another step of entailment:

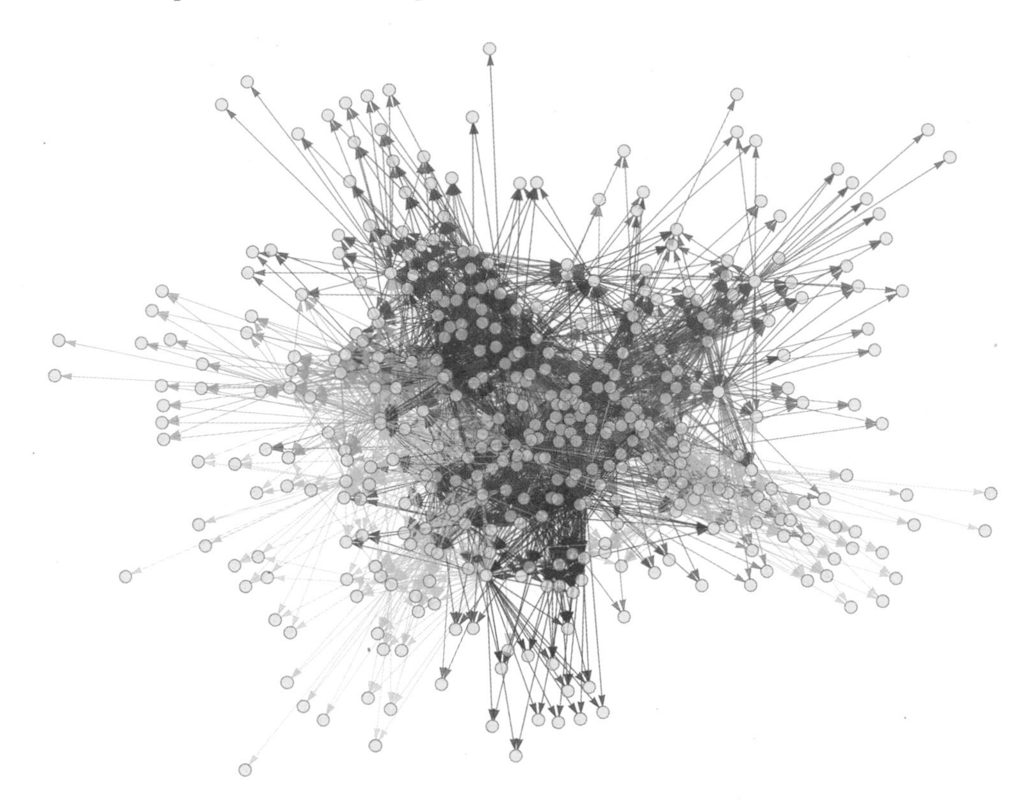

And in a sense this is a "timeless" way to imagine building up mathematics—and metamathematical space. Yes, this structure can in principle be viewed as part of the branchial graph obtained from a slice of an entailment graph (and technically this will be a useful way to think about it). But a different view—closer to the practice of human mathematics—is that it's a "fabric" formed by fitting together many different mathematical statements. It's not something where one's tracking the overall passage of time, and seeing causal connections between things—as one might in "running a program". Rather, it's something where one's fitting pieces together in order to satisfy constraints—as one might in creating a tiling.

Underneath everything is the ruliad. And entailment cones and entailment fabrics can be thought of just as different samplings or slicings of the ruliad. The ruliad is ultimately the entangled limit of all possible computations. But one can think of it as being built up by starting from all possible rules and initial conditions, then running them for an infinite number of steps. An entailment cone is essentially a "slice" of this structure where one's looking at the "time evolution" from a particular rule and initial condition. An entailment fabric is an "orthogonal" slice, looking "at a particular time" across different rules and initial conditions. (And, by the way, rules and initial conditions are essentially equivalent, particularly in an accumulative system.)

One can think of these different slices of the ruliad as being what different kinds of observers will perceive within the ruliad. Entailment cones are essentially what observers who persist through time but are localized in rulial space will perceive. Entailment fabrics are what observers who ignore time but explore more of rulial space will perceive.

Elsewhere I've argued that a crucial part of what makes us perceive the laws of physics we do is that we are observers who consider ourselves to be persistent through time. But now we're seeing that in the way human mathematics is typically done, the "mathematical observer" will be of a different character. And whereas for a physical observer what's crucial is causality through time, for a mathematical observer (at least one who's doing mathematics the way it's usually done) what seems to be crucial is some kind of consistency or coherence across metamathematical space.

In physics it's far from obvious that a persistent observer would be possible. It could be that with all those detailed computationally irreducible processes happening down at the level of atoms of space there might be nothing in the universe that one could consider consistent through time. But the point is that there are certain "coarse-grained" attributes of the behavior that are consistent through time. And it is by concentrating on these that we end up describing things in terms of the laws of physics we know.

There's something very analogous going on in mathematics. The detailed branchial structure of metamathematical space is complicated, and presumably full of computational irreducibility. But once again there are "coarse-grained" attributes that have a certain consistency and coherence across it. And it is on these that we concentrate as human "mathematical observers". And it is in terms of these that we end up being able to do "human-level mathematics"—in effect operating at a "fluid dynamics" level rather than a "molecular dynamics" one.

The possibility of "doing physics in the ruliad" depends crucially on the fact that as physical observers we assume that we have certain persistence and coherence through time. The possibility of "doing mathematics (the way it's usually done) in the ruliad" depends crucially on the fact that as "mathematical observers" we assume that the mathematical statements we consider will have a certain coherence and consistency—or, in effect, that it's possible for us to maintain and grow a coherent body of mathematical knowledge, even as we try to include all sorts of new mathematical statements.

20 | The Notion of Truth

Logic was originally conceived as a way to characterize human arguments—in which the concept of "truth" has always seemed quite central. And when logic was applied to the foundations of mathematics, "truth" was also usually assumed to be quite central. But the way we've modeled mathematics here has been much more about what statements can be derived (or entailed) than about any kind of abstract notion of what statements can be "tagged as true". In other words, we've been more concerned with "structurally deriving" that "1 + 1 = 2" than in saying that "1 + 1 = 2 is true".

But what is the relation between this kind of "constructive derivation" and the logical notion of truth? We might just say that "if we can construct a statement then we should consider it true". And if we're starting from axioms, then in a sense we'll never have an "absolute notion of truth"—because whatever we derive is only "as true as the axioms we started from".

One issue that can come up is that our axioms might be inconsistent—in the sense that from them we can derive two obviously inconsistent statements. But to get further in discussing things like this we really need not only to have a notion of truth, but also a notion of falsity.

In traditional logic it has tended to be assumed that truth and falsity are very much "the same kind of thing"—like 1 and 0. But one feature of our view of mathematics here is that actually truth and falsity seem to have a rather different character. And perhaps this is not surprising—because in a sense if there's one true statement about something there are typically an infinite number of false statements about it. So, for example, the single statement $1 + 1 = 2$ is true, but the infinite collection of statements $1 + 1 = n$ for any other n are all false.

There is another aspect to this, discussed since at least the Middle Ages, often under the name of the "principle of explosion": that as soon as one assumes any statement that is false, one can logically derive absolutely any statement at all. In other words, introducing a single "false axiom" will start an explosion that will eventually "blow up everything".

So within our model of mathematics we might say that things are "true" if they can be derived, and are "false" if they lead to an "explosion". But let's say we're given some statement. How can we tell if it's true or false? One thing we can do to find out if it's true is to construct an entailment cone from our axioms and see if the statement appears anywhere in it. Of course, given computational irreducibility there's in general no upper bound on how far we'll need to go to determine this. But now to find out if a statement is false we can imagine introducing the statement as an additional axiom, and then seeing if the entailment cone that's now produced contains an explosion—though once again there'll in general be no upper bound on how far we'll have to go to guarantee that we have a "genuine explosion" on our hands.

So is there any alternative procedure? Potentially the answer is yes: we can just try to see if our statement is somehow equivalent to "true" or "false". But in our model of mathematics where we're just talking about transformations on symbolic expressions, there's no immediate built-in notion of "true" and "false". To talk about these we have to add something. And for example what we can do is to say that "true" is equivalent to what seems like an "obvious tautology" such as $x = x$, or in our computational notation, $x\_ \longleftrightarrow x\_$, while "false" is equivalent to something "obviously explosive", like $x\_ \leftrightarrow y\_$ (or in our particular setup something more like $x\_ \longleftrightarrow x\_ \circ y\_$).

But even though something like "Can we find a way to reach $x\_ \longleftrightarrow x\_$ from a given statement?" seems like a much more practical question for an actual theorem-proving system than "Can we fish our statement out of a whole entailment cone?", it runs into many of the same issues—in particular that there's no upper limit on the length of path that might be needed.

Soon we'll return to the question of how all this relates to our interpretation of mathematics as a slice of the ruliad—and to the concept of the entailment fabric perceived by a mathematical observer. But to further set the context for what we're doing let's explore how what we've discussed so far relates to things like Gödel's theorem, and to phenomena like incompleteness.

From the setup of basic logic we might assume that we could consider any statement to be either true or false. Or, more precisely, we might think that given a particular axiom system, we should be able to determine whether any statement that can be syntactically constructed with the primitives of that axiom system is true or false. We could explore this by asking whether every statement is either derivable or leads to an explosion—or can be proved equivalent to an "obvious tautology" or to an "obvious explosion".

But as a simple "approximation" to this, let's consider a string rewriting system in which we define a "local negation operation". In particular, let's assume that given a statement like AB ⟷ BBA the "negation" of this statement just exchanges A and B, in this case yielding BA ⟷ AAB.

Now let's ask what statements are generated from a given axiom system. Say we start with AB ⟷ B. After one step of possible substitutions we get

{AAB ↔ B, AB ↔ AB, AB ↔ B, B ↔ B}

while after 2 steps we get:

{AAAAB ↔ B, AAAB ↔ AB, AAAB ↔ B, AAB ↔ AAB, AAB ↔ AB, AAB ↔ B, AB ↔ AB, AB ↔ B, B ↔ B}

And in our setup we're effectively asserting that these are "true" statements. But now let's "negate" the statements, by exchanging A and B. And if we do this, we'll see that there's never a statement where both it and its negation occur. In other words, there's no obvious inconsistency being generated within this axiom system.

But if we consider instead the axiom AB ⟷ BA then this gives:

{AB ↔ AB, AB ↔ BA, BA ↔ BA}

And since this includes both AB ⟷ AB and its "negation" BA ⟷ BA, by our criteria we must consider this axiom system to be inconsistent.

In addition to inconsistency, we can also ask about incompleteness. For all possible statements, does the axiom system eventually generate either the statement or its negation? Or, in other words, can we always decide from the axiom system whether any given statement is true or false?

With our simple assumption about negation, questions of inconsistency and incompleteness become at least in principle very simple to explore. Starting from a given axiom system, we generate its entailment cone, then we ask within this cone what fraction of possible statements, say of a given length, occur.

If the answer is more than 50% we know there's inconsistency, while if the answer is less than 50% that's evidence of incompleteness. So what happens with different possible axiom systems?

Here are some results from *A New Kind of Science,* in each case showing both what amounts to the raw entailment cone (or, in this case, multiway system evolution from "true"), and the number of statements of a given length reached after progressively more steps:

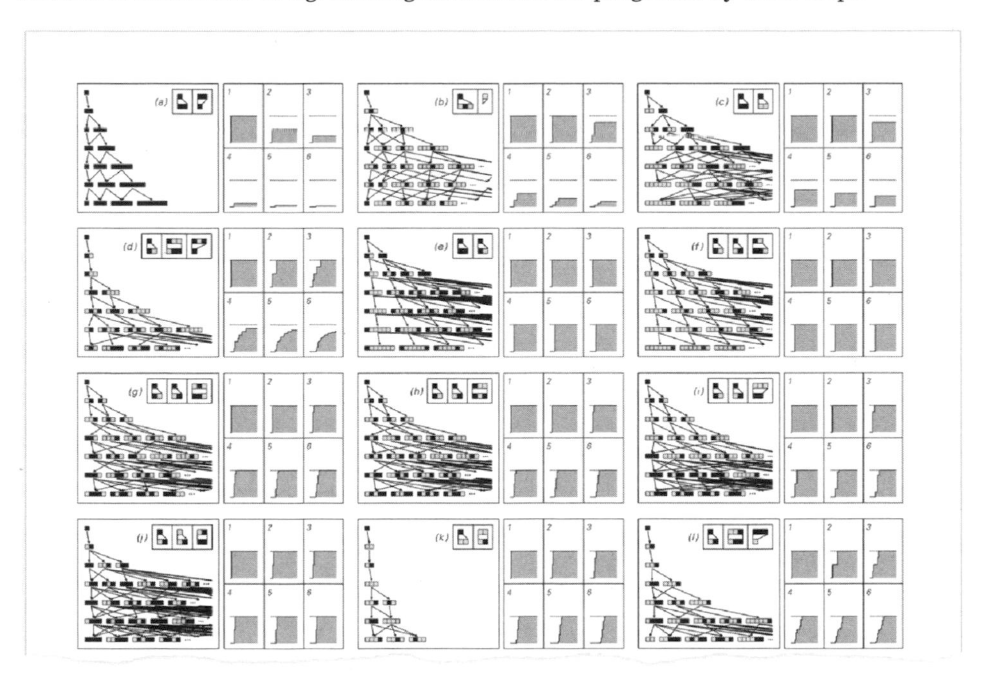

At some level this is all rather straightforward. But from the pictures above we can already get a sense that there's a problem. For most axiom systems the fraction of statements reached of a given length changes as we increase the number of steps in the entailment cone. Sometimes it's straightforward to see what fraction will be achieved even after an infinite number of steps. But often it's not.

And in general we'll run into computational irreducibility—so that in effect the only way to determine whether some particular statement is generated is just to go to ever more steps in the entailment cone and see what happens. In other words, there's no guaranteed-finite way to decide what the ultimate fraction will be—and thus whether or not any given axiom system is inconsistent, or incomplete, or neither.

For some axiom systems it may be possible to tell. But for some axiom systems it's not, in effect because we don't in general know how far we'll have to go to determine whether a given statement is true or not.

A certain amount of additional technical detail is required to reach the standard versions of Gödel's incompleteness theorems. (Note that these theorems were originally stated specifically for the Peano axioms for arithmetic, but the Principle of Computational Equivalence suggests that they're in some sense much more general, and even ubiquitous.) But the important point here is that given an axiom system there may be statements that either can or cannot be reached—but there's no upper bound on the length of path that might be needed to reach them even if one can.

OK, so let's come back to talking about the notion of truth in the context of the ruliad. We've discussed axiom systems that might show inconsistency, or incompleteness—and the difficulty of determining if they do. But the ruliad in a sense contains all possible axiom systems—and generates all possible statements.

So how then can we ever expect to identify which statements are "true" and which are not? When we talked about particular axiom systems, we said that any statement that is generated can be considered true (at least with respect to that axiom system). But in the ruliad every statement is generated. So what criterion can we use to determine which we should consider "true"?

The key idea is any computationally bounded observer (like us) can perceive only a tiny slice of the ruliad. And it's a perfectly meaningful question to ask whether a particular statement occurs within that perceived slice.

One way of picking a "slice" is just to start from a given axiom system, and develop its entailment cone. And with such a slice, the criterion for the truth of a statement is exactly what we discussed above: does the statement occur in the entailment cone?

But how do typical "mathematical observers" actually sample the ruliad? As we discussed in the previous section, it seems to be much more by forming an entailment fabric than by developing a whole entailment cone. And in a sense progress in mathematics can be seen as a process of adding pieces to an entailment fabric: pulling in one mathematical statement after another, and checking that they fit into the fabric.

So what happens if one tries to add a statement that "isn't true"? The basic answer is that it produces an "explosion" in which the entailment fabric can grow to encompass essentially any statement. From the point of view of underlying rules—or the ruliad—there's really nothing wrong with this. But the issue is that it's incompatible with an "observer like us"—or with any realistic idealization of a mathematician.

Our view of a mathematical observer is essentially an entity that accumulates mathematical statements into an entailment fabric. But we assume that the observer is computationally bounded, so in a sense they can only work with a limited collection of statements. So if there's an explosion in an entailment fabric that means the fabric will expand beyond what a mathematical observer can coherently handle. Or, put another way, the only kind of entailment fabrics that a mathematical observer can reasonably consider are ones that "contain no explosions". And in such fabrics, it's reasonable to take the generation or entailment of a statement as a signal that the statement can be considered true.

The ruliad is in a sense a unique and absolute thing. And we might have imagined that it would lead us to a unique and absolute definition of truth in mathematics. But what we've seen is that that's not the case. And instead our notion of truth is something based on how we sample the ruliad as mathematical observers. But now we must explore what this means about what mathematics as we perceive it can be like.

21 | What Can Human Mathematics Be Like?

The ruliad in a sense contains all structurally possible mathematics—including all mathematical statements, all axiom systems and everything that follows from them. But mathematics as we humans conceive of it is never the whole ruliad; instead it is always just some tiny part that we as mathematical observers sample.

We might imagine, however, that this would mean that there is in a sense a complete arbitrariness to our mathematics—because in a sense we could just pick any part of the ruliad we want. Yes, we might want to start from a specific axiom system. But we might imagine that that axiom system could be chosen arbitrarily, with no further constraint. And that the mathematics we study can therefore be thought of as an essentially arbitrary choice, determined by its detailed history, and perhaps by cognitive or other features of humans.

But there is a crucial additional issue. When we "sample our mathematics" from the ruliad we do it as mathematical observers and ultimately as humans. And it turns out that even very general features of us as mathematical observers turn out to put strong constraints on what we can sample, and how.

When we discussed physics, we said that the central features of observers are their computational boundedness and their assumption of their own persistence through time. In mathematics, observers are again computationally bounded. But now it is not persistence through time that they assume, but rather a certain coherence of accumulated knowledge.

We can think of a mathematical observer as progressively expanding the entailment fabric that they consider to "represent mathematics". And the question is what they can add to that entailment fabric while still "remaining coherent" as observers. In the previous section, for example, we argued that if the observer adds a statement that can be considered "logically false" then this will lead to an "explosion" in the entailment fabric.

Such a statement is certainly present in the ruliad. But if the observer were to add it, then they wouldn't be able to maintain their coherence—because, whimsically put, their mind would necessarily explode.

In thinking about axiomatic mathematics it's been standard to say that any axiom system that's "reasonable to use" should at least be consistent (even though, yes, for a given axiom system it's in general ultimately undecidable whether this is the case). And certainly consistency is one criterion that we now see is necessary for a "mathematical observer like us". But one can expect that it's not the only criterion.

In other words, although it's perfectly possible to write down any axiom system, and even start generating its entailment cone, only some axiom systems may be compatible with "mathematical observers like us".

And so, for example, something like the Continuum Hypothesis—which is known to be independent of the "established axioms" of set theory—may well have the feature that, say, it has to be assumed to be true in order to get a metamathematical structure compatible with mathematical observers like us.

In the case of physics, we know that the general characteristics of observers lead to certain key perceived features and laws of physics. In statistical mechanics, we're dealing with "coarse-grained observers" who don't trace and decode the paths of individual molecules, and therefore perceive the Second Law of thermodynamics, fluid dynamics, etc. And in our Physics Project we're also dealing with coarse-grained observers who don't track all the details of the atoms of space, but instead perceive space as something coherent and effectively continuous.

And it seems as if in metamathematics there's something very similar going on. As we began to discuss in the very first section, mathematical observers tend to "coarse grain" metamathematical space. In operational terms, one way they do this is by talking about something like the Pythagorean theorem without always going down to the detailed level of axioms, and for example saying just how real numbers should be defined. And something related is that they tend to concentrate more on mathematical statements and theorems than on their proofs. Later we'll see how in the context of the ruliad there's an even deeper level to which one can go. But the point here is that in actually doing mathematics one tends to operate at the "human scale" of talking about mathematical concepts rather than the "molecular-scale details" of axioms.

But why does this work? Why is one not continually "dragged down" to the detailed axiomatic level—or below? How come it's possible to reason at what we described above as the "fluid dynamics" level, without always having to go down to the detailed "molecular dynamics" level?

The basic claim is that this works for mathematical observers for essentially the same reason as the perception of space works for physical observers. With the "coarse-graining" characteristics of the observer, it's inevitable that the slice of the ruliad they sample will have the kind of coherence that allows them to operate at a higher level. In other words, mathematics can be done "at a human level" for the same basic reason that we have a "human-level experience" of space in physics.

The fact that it works this way depends both on necessary features of the ruliad—and in general of multicomputation—as well as on characteristics of us as observers.

Needless to say, there are "corner cases" where what we've described starts to break down. In physics, for example, the "human-level experience" of space breaks down near spacetime singularities. And in mathematics, there are cases where for example undecidability forces one to take a lower-level, more axiomatic and ultimately more metamathematical view.

But the point is that there are large regions of physical space—and metamathematical space—where these kinds of issues don't come up, and where our assumptions about physical—and mathematical—observers can be maintained. And this is what ultimately allows us to have the "human-scale" views of physics and mathematics that we do.

22 | Going below Axiomatic Mathematics

In the traditional view of the foundations of mathematics one imagines that axioms—say stated in terms of symbolic expressions—are in some sense the lowest level of mathematics. But thinking in terms of the ruliad suggests that in fact there is a still-lower "ur level"—a kind of analog of machine code in which everything, including axioms, is broken down into ultimate "raw computation".

Take an axiom like $x \circ y = (y \circ x) \circ y$, or, in more precise computational language:

```
x_∘y_ ↔ (y_∘x_)∘y_
```

Compared to everything we're used to seeing in mathematics this looks simple. But actually it's already got a lot in it. For example, it assumes the notion of a binary operator, which it's in effect naming "∘". And for example it also assumes the notion of variables, and has two distinct pattern variables that are in effect "tagged" with the names *x* and *y*.

So how can we define what this axiom ultimately "means"? Somehow we have to go from its essentially textual symbolic representation to a piece of actual computation. And, yes, the particular representation we've used here can immediately be interpreted as computation in the Wolfram Language. But the ultimate computational concept we're dealing with is more general than that. And in particular it can exist in any universal computational system.

Different universal computational systems (say particular languages or CPUs or Turing machines) may have different ways to represent computations. But ultimately any computation can be represented in any of them—with the differences in representation being like different "coordinatizations of computation".

And however we represent computations there is one thing we can say for sure: all possible computations are somewhere in the ruliad. Different representations of computations correspond in effect to different coordinatizations of the ruliad. But all computations are ultimately there.

For our Physics Project it's been convenient to use a "parametrization of computation" that can be thought of as being based on rewriting of hypergraphs. The elements in these hypergraphs are ultimately purely abstract, but we tend to talk about them as "atoms of space" to indicate the beginnings of our interpretation.

It's perfectly possible to use hypergraph rewriting as the "substrate" for representing axiom systems stated in terms of symbolic expressions. But it's a bit more convenient (though ultimately equivalent) to instead use systems based on expression rewriting—or in effect tree rewriting.

At the outset, one might imagine that different axiom systems would somehow have to be represented by "different rules" in the ruliad. But as one might expect from the phenomenon of universal computation, it's actually perfectly possible to think of different axiom systems as just being specified by different "data" operated on by a single set of rules. There are many rules and structures that we could use. But one set that has the benefit of a century of history are S, K combinators.

The basic concept is to represent everything in terms of "combinator expressions" containing just the two objects S and K. (It's also possible to have just one fundamental object, and indeed S alone may be enough.)

It's worth saying at the outset that when we go this "far down" things get pretty non-human and obscure. Setting things up in terms of axioms may already seem pedantic and low level. But going to a substrate below axioms—that we can think of as getting us to raw "atoms of existence"—will lead us to a whole other level of obscurity and complexity. But if we're going to understand how mathematics can emerge from the ruliad this is where we have to go. And combinators provide us with a more-or-less-concrete example.

Here's an example of a small combinator expression

S[S[K[S]][S[K[S[K[S]]]][S[K[K]]]]][S[S[K[S]][S[K[K]][S]]][K[K]]]

which corresponds to the "expression tree":

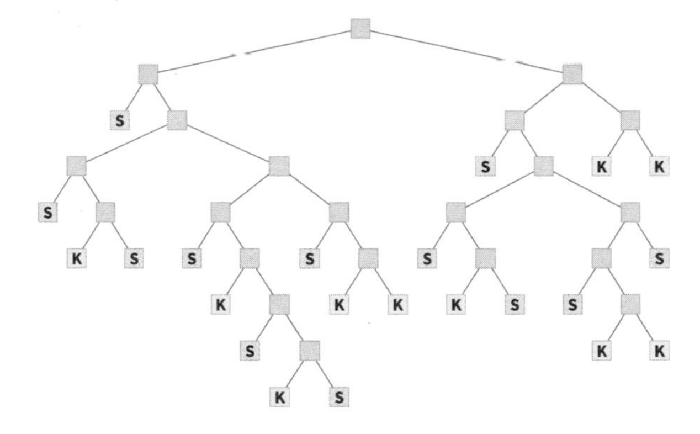

We can write the combinator expression without explicit "function application" [...] by using a (left) application operator •

S•(S•(K•S)•(S•(K•(S•(K•S)))•(S•(K•K))))•(S•(S•(K•S)•(S•(K•K)•S))•(K•K))

and it's always unambiguous to omit this operator, yielding the compact representation:

S(S(KS)(S(K(S(KS)))(S(KK))))(S(S(KS)(S(KK)S))(KK))

By mapping S, K and the application operator to codewords it's possible to represent this as a simple binary sequence:

1100110010100110010110010100100101011100110010100110010101001010

But what does our combinator expression mean? The basic combinators are defined to have the rules:

{S[x_][y_][z_] → x[z][y[z]], K[x_][y_] → x}

These rules on their own don't do anything to our combinator expression. But if we form the expression

S[S[K[S]][S[K[S[K[S]]]][S[K[K]]]]][S[S[K[S]][S[K[K]][S]]][K[K]]][c][x][y]

which we can write as

S(S(KS)(S(K(S(KS)))(S(KK))))(S(S(KS)(S(KK)S))(KK)) *cxy*

then repeated application of the rules gives:

S[S[K[S]][S[K[S[K[S]]]][S[K[K]]]]][S[S[K[S]][S[K[K]][S]]][K[K]]][c][x][y]
S[K[S]][S[K[S[K[S]]]][S[K[K]]]][c][S[S[K[S]][S[K[K]][S]]][K[K]][c]][x][y]
K[S][c][S[K[S[K[S]]]][S[K[K]]][c]][S[S[K[S]][S[K[K]][S]]][K[K]][c]][x][y]
S[S[K[S[K[S]]]][S[K[K]]][c]][S[S[K[S]][S[K[K]][S]]][K[K]][c]][x][y]
S[K[S[K[S]]]][S[K[K]]][c][x][S[S[K[S]][S[K[K]][S]]][K[K]][c][x]][y]
K[S[K[S]]][c][S[K[K]][c]][x][S[S[K[S]][S[K[K]][S]]][K[K]][c][x]][y]
S[K[S]][S[K[K]][c]][x][S[S[K[S]][S[K[K]][S]]][K[K]][c][x]][y]
K[S][x][S[K[K]][c][x]][S[S[K[S]][S[K[K]][S]]][K[K]][c][x]][y]
S[S[K[K]][c][x]][S[S[K[S]][S[K[K]][S]]][K[K]][c][x]][y]
S[K[K]][c][x][y][S[S[K[S]][S[K[K]][S]]][K[K]][c][x][y]]
K[K][x][c[x]][y][S[S[K[S]][S[K[K]][S]]][K[K]][c][x][y]]
K[c[x]][y][S[S[K[S]][S[K[K]][S]]][K[K]][c][x][y]]
c[x][S[S[K[S]][S[K[K]][S]]][K[K]][c][x][y]]
c[x][S[K[S]][S[K[K]][S]][c][K[K][c]][x][y]]
c[x][K[S][c][S[K[K]][S][c]][K[K][c]][x][y]]
c[x][S[S[K[K]][S][c]][K[K][c]][x][y]]
c[x][S[K[K]][S][c][x][K[K][c][x]][y]]
c[x][K[K][c][S[c]][x][K[K][c][x]][y]]
c[x][K[S[c]][x][K[K][c][x]][y]]
c[x][S[c][K[K][c][x]][y]]
c[x][c[y][K[K][c][x][y]]]
c[x][c[y][K[x][y]]]
c[x][c[y][x]]

We can think of this as "feeding" *c*, *x* and *y* into our combinator expression, then using the "plumbing" defined by the combinator expression to assemble a particular expression in terms of *c*, *x* and *y*.

But what does this expression now mean? Well, that depends on what we think c, x and y mean. We might notice that c always appears in the configuration c[_][_]. And this means we can interpret it as a binary operator, which we could write in infix form as ∘ so that our expression becomes:

$$x \circ (y \circ x)$$

And, yes, this is all incredibly low level. But we need to go even further. Right now we're feeding in names like c, x and y. But in the end we want to represent absolutely everything purely in terms of S and K. So we need to get rid of the "human-readable names" and just replace them with "lumps" of S, K combinators that—like the names—get "carried around" when the combinator rules are applied.

We can think about our ultimate expressions in terms of S and K as being like machine code. "One level up" we have assembly language, with the same basic operations, but explicit names. And the idea is that things like axioms—and the laws of inference that apply to them—can be "compiled down" to this assembly language.

But ultimately we can always go further, to the very lowest-level "machine code", in which only S and K ever appear. Within the ruliad as "coordinatized" by S, K combinators, there's an infinite collection of possible combinator expressions. But how do we find ones that "represent something recognizably mathematical"?

As an example let's consider a possible way in which S, K can represent integers, and arithmetic on integers. The basic idea is that an integer n can be input as the combinator expression

Nest[S[S[K[S]][K]], S[K], n]

which for $n = 5$ gives:

S[S[K[S]][K]][S[S[K[S]][K]][S[S[K[S]][K]][S[S[K[S]][K]][S[K]]]]]

But if we now apply this to [S][K] what we get reduces to

S[S[S[S[K]]]]

which contains 4 S's.

But with this representation of integers it's possible to find combinator expressions that represent arithmetic operations. For example, here's a representation of an addition operator:

S[K[S]][S[K[S[K[S]]]][S[K[K]]]]

At the "assembly language" level we might call this **plus**, and apply it to integers i and j using:

plus[i][j]

But at the "pure machine code" level 1 + 2 can be represented simply by

S[K[S]][S[K[S[K[S]]]][S[K[K]]]][S[S[K[S]][K]][S[K]]][S[S[K[S]][K]][S[S[K[S]][K]][S[K]]]]

which when applied to [S][K] reduces to the "output representation" of 3:

S[S[S[K]]]

As a slightly more elaborate example

S[K[S[S[K][K]]]][K]

represents the operation of raising to a power. Then 2^3 becomes:

S[K[S[S[K][K]]]][K][S[S[K[S]][K]][S[S[K[S]][K]][S[K]]]][S[S[K[S]][K]][S[S[K[S]][K]][S[S[K[S]][K]][S[K]]]]]

Applying this to [S][K] repeated application of the combinator rules gives

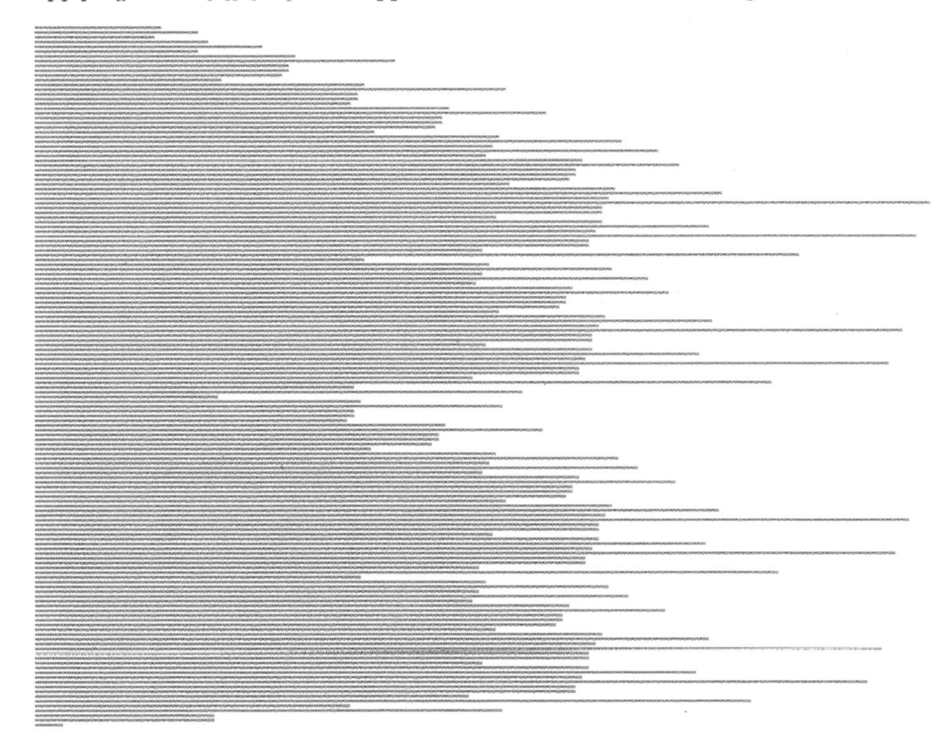

eventually yielding the output representation of 8:

S[S[S[S[S[S[S[S[K]]]]]]]]

We could go on and construct any other arithmetic or computational operation we want, all just in terms of the "universal combinators" S and K.

But how should we think about this in terms of our conception of mathematics? Basically what we're seeing is that in the "raw machine code" of S, K combinators it's possible to "find" a representation for something we consider to be a piece of mathematics.

Earlier we talked about starting from structures like axiom systems and then "compiling them down" to raw machine code. But what about just "finding mathematics" in a sense "naturally occurring" in "raw machine code"? We can think of the ruliad as containing "all possible machine code". And somewhere in that machine code must be all the conceivable "structures of mathematics". But the question is: in the wildness of the raw ruliad, what structures can we as mathematical observers successfully pick out?

The situation is quite directly analogous to what happens at multiple levels in physics. Consider for example a fluid full of molecules bouncing around. As we've discussed several times, observers like us usually aren't sensitive to the detailed dynamics of the molecules. But we can still successfully pick out large-scale structures—like overall fluid motions, vortices, etc. And—much like in mathematics—we can talk about physics just at this higher level.

In our Physics Project all this becomes much more extreme. For example, we imagine that space and everything in it is just a giant network of atoms of space. And now within this network we imagine that there are "repeated patterns"—that correspond to things like electrons and quarks and black holes.

In a sense it is the big achievement of natural science to have managed to find these regularities so that we can describe things in terms of them, without always having to go down to the level of atoms of space. But the fact that these are the kinds of regularities we have found is also a statement about us as physical observers.

And the point is that even at the level of the raw ruliad our characteristics as physical observers will inevitably lead us to such regularities. The fact that we are computationally bounded and assume ourselves to have a certain persistence will lead us to consider things that are localized and persistent—that in physics we identify for example as particles.

And it's very much the same thing in mathematics. As mathematical observers we're interested in picking out from the raw ruliad "repeated patterns" that are somehow robust. But now instead of identifying them as particles, we'll identify them as mathematical constructs and definitions. In other words, just as a repeated pattern in the ruliad might in physics be interpreted as an electron, in mathematics a repeated pattern in the ruliad might be interpreted as an integer.

We might think of physics as something "emergent" from the structure of the ruliad, and now we're thinking of mathematics the same way. And of course not only is the "underlying stuff" of the ruliad the same in both cases, but also in both cases it's "observers like us" that are sampling and perceiving things.

There are lots of analogies to the process we're describing of "fishing constructs out of the raw ruliad". As one example, consider the evolution of a ("class 4") cellular automaton in which localized structures emerge:

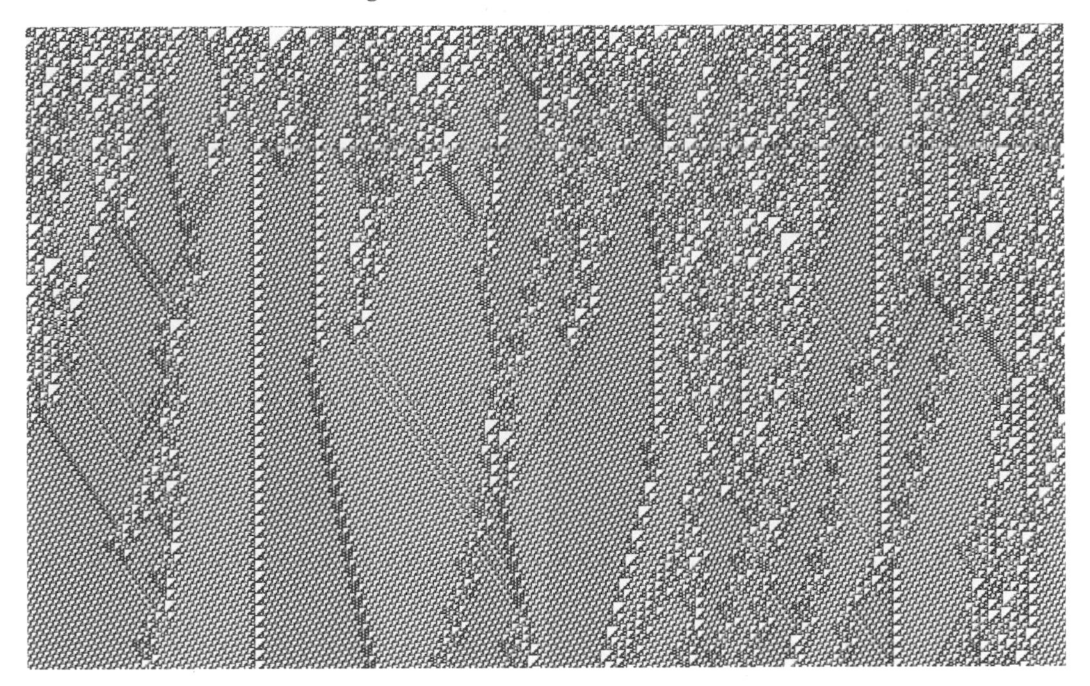

Underneath, just as throughout the ruliad, there's lots of detailed computation going on, with rules repeatedly getting applied to each cell. But out of all this underlying computation we can identify a certain set of persistent structures—which we can use to make a "higher-level description" that may capture the aspects of the behavior that we care about.

Given an "ocean" of S, K combinator expressions, how might we set about "finding mathematics" in them? One straightforward approach is just to identify certain "mathematical properties" we want, and then go searching for S, K combinator expressions that satisfy these.

For example, if we want to "search for (propositional) logic" we first need to pick combinator expressions to symbolically represent "true" and "false". There are many pairs of expressions that will work. As one example, let's pick:

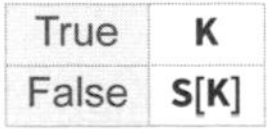

True	**K**
False	**S[K]**

Now we can just search for combinator expressions which, when applied to all possible pairs of "true" and "false" give truth tables corresponding to particular logical functions. And if we do this, here are examples of the smallest combinator expressions we find:

And	**S[S][K]**
Or	**S[S[S]][S][S[K]]**
Implies	**S[S][K[K[K]]]**
Nand	**S[S[K[S[S[S][K[K[K]]]]]]][S]**
Equal	**S[S][K[S[S[S][K[K[K]]]][S]]]**

Here's how we can then reproduce the truth table for **And**:

And[True][True]	And[True][False]	And[False][True]	And[False][False]
S[S][K][K][K] S[K][K[K]][K] K[K][K[K][K]] **K**	S[S][K][K][S[K]] S[K][K[K]][S[K]] K[S[K]][K[K][S[K]]] **S[K]**	S[S][K][S[K]][K] S[S[K]][K[S[K]]][K] S[K][K][K[S[K]][K]] K[K[S[K]][K]][K[K[S[K]][K]]] K[S[K]][K] **S[K]**	S[S][K][S[K]][S[K]] S[S[K]][K[S[K]]][S[K]] S[K][S[K]][K[S[K]][S[K]]] K[K[S[K]][S[K]]][S[K][K[S[K]][S[K]]]] K[S[K]][S[K]] **S[K]**

If we just started picking combinator expressions at random, then most of them wouldn't be "interpretable" in terms of this representation of logic. But if we ran across for example

S[S][K[K[K]]][S[S][K][S[S[S]][S][S[K]][p][q]][p]][S[S[K[S[S[S][K[K[K]]]]]]][S][p][q]]

we could recognize in it the combinators for And, Or, etc. that we identified above, and in effect "disassemble" it to give:

Implies[And[Or[p][q]][p]][Nand[p][q]]

It's worth noting, though, that even with the choices we made above for "true" and "false", there's not just a single possible combinator, say for And. Here are a few possibilities:

S[S][K]	S[S][S[S][K]]	S[S][K[K]][S]	S[S][K[K][S]]	S[S][K[K][K]]	S[K[S][S]][K]	S[K[S][K]][K]
K[S][S][S][K]	K[S[S]][S][K]	K[S[S][K]][S]	K[S[S][K]][K]	K[S[S]][K][K]	K[S][K][S][K]	S[S][S[S[K]][K]]
S[S[S[K]][S]][K]	S[S][S[K][S][K]]	S[S][S[K][K][K]]	S[S[K][S][S]][K]	S[S[K]][S][S][K]	S[S[K]][S[S]][K]	S[S[K]][S[S][K]]
S[S][K[K][S[S]]]	S[S[K][K][S]][K]	S[S][K[K[S[K]]]]	S[S][K[K][S[K]]]	S[S][K[K][K[S]]]	S[S][K[K][K[K]]]	S[K][S][S][S][K]
S[K[S][S[S]]][K]	S[K][S][S[S]][K]	S[K][S][S[S][K]]	S[K[S][S[K]]][K]	S[K[S][K[S]]][K]	S[K[S][K[K]]][K]	S[K][K][S][S][K]

And there's also nothing unique about the choices for "true" and "false". With the alternative choices

True	**K[K]**
False	**K**

here are the smallest combinator expressions for a few logical functions:

And	**S[S][S[S[K]]]**
Or	**S[S[S[S]]][K]**
Implies	**S[S][S][S[S][K]][S]**
Nand	**S[S[S][S][S[S][K]][S]]**
Equal	**S[S[S[S[S]][S[S]][S[S[K]]]]]**

So what can we say in general about the "interpretability" of an arbitrary combinator expression? Obviously any combinator expression does what it does at the level of raw combinators. But the question is whether it can be given a "higher-level"—and potentially "mathematical"—interpretation.

And in a sense this is directly an issue of what a mathematical observer "perceives" in it. Does it contain some kind of robust structure—say a kind of analog for mathematics of a particle in physics?

Axiom systems can be viewed as a particular way to "summarize" certain "raw machine code" in the ruliad. But from the point of a "raw coordinatization of the ruliad" like combinators there doesn't seem to be anything immediately special about them. At least for us humans, however, they do seem to be an obvious "waypoint". Because by distinguishing operators and variables, establishing arities for operators and introducing names for things, they reflect the kind of structure that's familiar from human language.

But now that we think of the ruliad as what's "underneath" both mathematics and physics there's a different path that's suggested. With the axiomatic approach we're effectively trying to leverage human language as a way of summarizing what's going on. But an alternative is to leverage our direct experience of the physical world, and our perception and intuition about things like space. And as we'll discuss later, this is likely in many ways a better "metamodel" of the way pure mathematics is actually practiced by us humans.

In some sense, this goes straight from the "raw machine code" of the ruliad to "human-level mathematics", sidestepping the axiomatic level. But given how much "reductionist" work has already been done in mathematics to represent its results in axiomatic form, there is definitely still great value in seeing how the whole axiomatic setup can be "fished out" of the "raw ruliad".

And there's certainly no lack of complicated technical issues in doing this. As one example, how should one deal with "generated variables"? If one "coordinatizes" the ruliad in terms of something like hypergraph rewriting this is fairly straightforward: it just involves creating new elements or hypergraph nodes (which in physics would be interpreted as atoms of space). But for something like S, K combinators it's a bit more subtle. In the examples we've given above, we have combinators that, when "run", eventually reach a fixed point. But to deal with generated variables we probably also need combinators that never reach fixed points, making it considerably more complicated to identify correspondences with definite symbolic expressions.

Another issue involves rules of entailment, or, in effect, the metalogic of an axiom system. In the full axiomatic setup we want to do things like create token-event graphs, where each event corresponds to an entailment. But what rule of entailment should be used? The underlying rules for S, K combinators, for example, define a particular choice—though they can be used to emulate others. But the ruliad in a sense contains all choices. And, once again, it's up to the observer to "fish out" of the raw ruliad a particular "slice"—which captures not only the axiom system but also the rules of entailment used.

It may be worth mentioning a slightly different existing "reductionist" approach to mathematics: the idea of describing things in terms of types. A type is in effect an equivalence class that characterizes, say, all integers, or all functions from tuples of reals to truth values. But in our terms we can interpret a type as a kind of "template" for our underlying "machine code": we can say that some piece of machine code represents something of a particular type if the machine code matches a particular pattern of some kind. And the issue is then whether that pattern is somehow robust "like a particle" in the raw ruliad.

An important part of what made our Physics Project possible is the idea of going "underneath" space and time and other traditional concepts of physics. And in a sense what we're doing here is something very similar, though for mathematics. We want to go "underneath" concepts like functions and variables, and even the very idea of symbolic expressions. In our Physics Project a convenient "parametrization" of what's "underneath" is a hypergraph made up of elements that we often refer to as "atoms of space". In mathematics we've discussed using combinators as our "parametrization" of what's "underneath".

But what are these "made of"? We can think of them as corresponding to raw elements of metamathematics, or raw elements of computation. But in the end, they're "made of" whatever the ruliad is "made of". And perhaps the best description of the elements of the ruliad is that they are "atoms of existence"—the smallest units of anything, from which everything, in mathematics and physics and elsewhere, must be made.

The atoms of existence aren't bits or points or anything like that. They're something fundamentally lower level that's come into focus only with our Physics Project, and particularly with the identification of the ruliad. And for our purposes here I'll call such atoms of existence "emes" (pronounced "eemes", like phonemes etc.).

Everything in the ruliad is made of emes. The atoms of space in our Physics Project are emes. The nodes in our combinator trees are emes. An eme is a deeply abstract thing. And in a sense all it has is an identity. Every eme is distinct. We could give it a name if we wanted to, but it doesn't intrinsically have one. And in the end the structure of everything is built up simply from relations between emes.

23 | The Physicalized Laws of Mathematics

The concept of the ruliad suggests there is a deep connection between the foundations of mathematics and physics. And now that we have discussed how some of the familiar formalism of mathematics can "fit into" the ruliad, we are ready to use the "bridge" provided by the ruliad to start exploring how to apply some of the successes and intuitions of physics to mathematics.

A foundational part of our everyday experience of physics is our perception that we live in continuous space. But our Physics Project implies that at sufficiently small scales space is actually made of discrete elements—and it is only because of the coarse-grained way in which we experience it that we perceive it as continuous.

In mathematics—unlike physics—we've long thought of the foundations as being based on things like symbolic expressions that have a fundamentally discrete structure. Normally, though, the elements of those expressions are, for example, given human-recognizable names (like 2 or Plus). But what we saw in the previous section is that these recognizable forms can be thought of as existing in an "anonymous" lower-level substrate made of what we can call atoms of existence or emes.

But the crucial point is that this substrate is directly based on the ruliad. And its structure is identical between the foundations of mathematics and physics. In mathematics the emes aggregate up to give us our universe of mathematical statements. In physics they aggregate up to give us our physical universe.

But now the commonality of underlying "substrate" makes us realize that we should be able to take our experience of physics, and apply it to mathematics. So what is the analog in mathematics of our perception of the continuity of space in physics? We've discussed the idea that we can think of mathematical statements as being laid out in a metamathematical space—or, more specifically, in what we've called an entailment fabric. We initially talked about "coordinatizing" this using axioms, but in the previous section we saw how to go "below axioms" to the level of "pure emes".

When we do mathematics, though, we're sampling this on a much higher level. And just like as physical observers we coarse grain the emes (that we usually call "atoms of space") that make up physical space, so too as "mathematical observers" we coarse grain the emes that make up metamathematical space.

Foundational approaches to mathematics—particularly over the past century or so—have almost always been based on axioms and on their fundamentally discrete symbolic structure. But by going to a lower level and seeing the correspondence with physics we are led to consider what we might think of as a higher-level "experience" of mathematics—operating not at the "molecular dynamics" level of specific axioms and entailments, but rather at what one might call the "fluid dynamics" level of larger-scale concepts.

At the outset one might not have any reason to think that this higher-level approach could consistently be applied. But this is the first big place where ideas from physics can be used. If both physics and mathematics are based on the ruliad, and if our general characteristics as observers apply in both physics and mathematics, then we can expect that similar features will emerge. And in particular, we can expect that our everyday perception of physical space as continuous will carry over to mathematics, or, more accurately, to metamathematical space.

The picture is that we as mathematical observers have a certain "size" in metamathematical space. We identify concepts—like integers or the Pythagorean theorem—as "regions" in the space of possible configurations of emes (and ultimately of slices of the ruliad). At an axiomatic level we might think of ways to capture what a typical mathematician might consider "the same concept" with slightly different formalism (say, different large cardinal axioms or different models of real numbers). But when we get down to the level of emes there'll be vastly more freedom in how we capture a given concept—so that we're in effect using a whole region of "emic space" to do so.

But now the question is what happens if we try to make use of the concept defined by this "region"? Will the "points in the region" behave coherently, or will everything be "shredded", with different specific representations in terms of emes leading to different conclusions?

The expectation is that in most cases it will work much like physical space, and that what we as observers perceive will be quite independent of the detailed underlying behavior at the level of emes. Which is why we can expect to do "higher-level mathematics", without always having to descend to the level of emes, or even axioms.

And this we can consider as the first great "physicalized law of mathematics": that coherent higher-level mathematics is possible for us for the same reason that physical space seems coherent to observers like us.

We've discussed several times before the analogy to the Second Law of thermodynamics—and the way it makes possible a higher-level description of things like fluids for "observers like us". There are certainly cases where the higher-level description breaks down. Some of them may involve specific probes of molecular structure (like Brownian motion). Others may be slightly more "unwitting" (like hypersonic flow).

In our Physics Project we're very interested in where similar breakdowns might occur—because they'd allow us to "see below" the traditional continuum description of space. Potential targets involve various extreme or singular configurations of spacetime, where in effect the "coherent observer" gets "shredded", because different atoms of space "within the observer" do different things.

In mathematics, this kind of "shredding" of the observer will tend to be manifest in the need to "drop below" higher-level mathematical concepts, and go down to a very detailed axiomatic, metamathematical or even eme level—where computational irreducibility and phenomena like undecidability are rampant.

It's worth emphasizing that from the point of view of pure axiomatic mathematics it's not at all obvious that higher-level mathematics should be possible. It could be that there'd be no choice but to work through every axiomatic detail to have any chance of making conclusions in mathematics.

But the point is that we now know there could be exactly the same issue in physics. Because our Physics Project implies that at the lowest level our universe is effectively made of emes that have all sorts of complicated—and computationally irreducible—behavior. Yet we know that we don't have to trace through all the details of this to make conclusions about what will happen in the universe—at least at the level we normally perceive it.

In other words, the fact that we can successfully have a "high-level view" of what happens in physics is something that fundamentally has the same origin as the fact that we can successfully have a high-level view of what happens in mathematics. Both are just features of how observers like us sample the ruliad that underlies both physics and mathematics.

24 | Uniformity and Motion in Metamathematical Space

We've discussed how the basic concept of space as we experience it in physics leads us to our first great physicalized law of mathematics—and how this provides for the very possibility of higher-level mathematics. But this is just the beginning of what we can learn from thinking about the correspondences between physical and metamathematical space implied by their common origin in the structure of the ruliad.

A key idea is to think of a limit of mathematics in which one is dealing with so many mathematical statements that one can treat them "in bulk"—as forming something we could consider a continuous metamathematical space. But what might this space be like?

Our experience of physical space is that at our scale and with our means of perception it seems to us for the most part quite simple and uniform. And this is deeply connected to the concept that pure motion is possible in physical space—or, in other words, that it's possible for things to move around in physical space without fundamentally changing their character.

Looked at from the point of view of the atoms of space it's not at all obvious that this should be possible. After all, whenever we move we'll almost inevitably be made up of different atoms of space. But it's fundamental to our character as observers that the features we end up perceiving are ones that have a certain persistence—so that we can imagine that we, and objects around us, can just "move unchanged", at least with respect to those aspects of the objects that we perceive. And this is why, for example, we can discuss laws of mechanics without having to "drop down" to the level of the atoms of space.

So what's the analog of all this in metamathematical space? At the present stage of our physical universe, we seem to be able to experience physical space as having features like being basically three-dimensional. Metamathematical space probably doesn't have such familiar mathematical characterizations. But it seems very likely (and we'll see some evidence of this from empirical metamathematics below) that at the very least we'll perceive metamathematical space as having a certain uniformity or homogeneity.

In our Physics Project we imagine that we can think of physical space as beginning "at the Big Bang" with what amounts to some small collection of atoms of space, but then growing to the vast number of atoms in our current universe through the repeated application of particular rules. But with a small set of rules being applied a vast number of times, it seems almost inevitable that some kind of uniformity must result.

But then the same kind of thing can be expected in metamathematics. In axiomatic mathematics one imagines the mathematical analog of the Big Bang: everything starts from a small collection of axioms, and then expands to a huge number of mathematical statements through repeated application of laws of inference. And from this picture (which gets a bit more elaborate when one considers emes and the full ruliad) one can expect that at least after it's "developed for a while" metamathematical space, like physical space, will have a certain uniformity.

The idea that physical space is somehow uniform is something we take very much for granted, not least because that's our lifelong experience. But the analog of this idea for metamathematical space is something we don't have immediate everyday intuition about—and that in fact may at first seem surprising or even bizarre. But actually what it implies is something that increasingly rings true from modern experience in pure mathematics. Because by saying that metamathematical space is in a sense uniform, we're saying that different parts of it somehow seem similar—or in other words that there's parallelism between what we see in different areas of mathematics, even if they're not "nearby" in terms of entailments.

But this is exactly what, for example, the success of category theory implies. Because it shows us that even in completely different areas of mathematics it makes sense to set up the same basic structures of objects, morphisms and so on. As such, though, category theory defines only the barest outlines of mathematical structure. But what our concept of perceived uniformity in metamathematical space suggests is that there should in fact be closer correspondences between different areas of mathematics.

We can view this as another fundamental "physicalized law of mathematics": that different areas of mathematics should ultimately have structures that are in some deep sense "perceived the same" by mathematical observers. For several centuries we've known there's a certain correspondence between, for example, geometry and algebra. But it's been a major achievement of recent mathematics to identify more and more such correspondences or "dualities".

Often the existence of these has seemed remarkable, and surprising. But what our view of metamathematics here suggests is that this is actually a general physicalized law of mathematics—and that in the end essentially all different areas of mathematics must share a deep structure, at least in some appropriate "bulk metamathematical limit" when enough statements are considered.

But it's one thing to say that two places in metamathematical space are "similar"; it's another to say that "motion between them" is possible. Once again we can make an analogy with physical space. We're used to the idea that we can move around in space, maintaining our identity and structure. But this in a sense requires that we can maintain some kind of continuity of existence on our path between two positions.

In principle it could have been that we would have to be "atomized" at one end, then "reconstituted" at the other end. But our actual experience is that we perceive ourselves to continually exist all the way along the path. In a sense this is just an assumption about how things work that physical observers like us make; but what's nontrivial is that the underlying structure of the ruliad implies that this will always be consistent.

And so we expect it will be in metamathematics. Like a physical observer, the way a mathematical observer operates, it'll be possible to "move" from one area of mathematics to another "at a high level", without being "atomized" along the way. Or, in other words, that a mathematical observer will be able to make correspondences between different areas of mathematics without having to go down to the level of emes to do so.

It's worth realizing that as soon as there's a way of representing mathematics in computational terms the concept of universal computation (and, more tightly, the Principle of Computational Equivalence) implies that at some level there must always be a way to translate between any two mathematical theories, or any two areas of mathematics. But the question is whether it's possible to do this in "high-level mathematical terms" or only at the level of the underlying "computational substrate". And what we're saying is that there's a general physicalized law of mathematics that implies that higher-level translation should be possible.

Thinking about mathematics at a traditional axiomatic level can sometimes obscure this, however. For example, in axiomatic terms we usually think of Peano Arithmetic as not being as powerful as ZFC set theory (for example, it lacks transfinite induction)—and so nothing like "dual" to it. But Peano Arithmetic can perfectly well support universal computation, so inevitably a "formal emulator" for ZFC set theory can be built in it. But the issue is that to do this essentially requires going down to the "atomic" level and operating not in terms of mathematical constructs but instead directly in terms of "metamathematical" symbolic structure (and, for example, explicitly emulating things like equality predicates).

But the issue, it seems, is that if we think at the traditional axiomatic level, we're not dealing with a "mathematical observer like us". In the analogy we've used above, we're operating at the "molecular dynamics" level, not at the human-scale "fluid dynamics" level. And so we see all sorts of details and issues that ultimately won't be relevant in typical approaches to actually doing pure mathematics.

It's somewhat ironic that our physicalized approach shows this by going below the axiomatic level—to the level of emes and the raw ruliad. But in a sense it's only at this level that there's the uniformity and coherence to conveniently construct a general picture that can encompass observers like us.

Much as with ordinary matter we can say that "everything is made of atoms", we're now saying that everything is "made of computation" (and its structure and behavior is ultimately described by the ruliad). But the crucial idea that emerged from our Physics Project—and that is at the core of what I'm calling the multicomputational paradigm—is that when we ask what observers perceive there is a whole additional level of inexorable structure. And this is what makes it possible to do both human-scale physics and higher-level mathematics—and for there to be what amounts to "pure motion", whether in physical or metamathematical space.

There's another way to think about this, that we alluded to earlier. A key feature of an observer is to have a coherent identity. In physics, that involves having a consistent thread of experience in time. In mathematics, it involves bringing together a consistent view of "what's true" in the space of mathematical statements.

In both cases the observer will in effect involve many separate underlying elements (ultimately, emes). But in order to maintain the observer's view of having a coherent identity, the observer must somehow conflate all these elements, effectively treating them as "the same". In physics, this means "coarse-graining" across physical or branchial (or, in fact, rulial) space. In mathematics, this means "coarse-graining" across metamathematical space—or in effect treating different mathematical statements as "the same".

In practice, there are several ways this happens. First of all, one tends to be more concerned about mathematical results than their proofs, so two statements that have the same form can be considered the same even if the proofs (or other processes) that generated them are different (and indeed this is something we have routinely done in constructing entailment cones here). But there's more. One can also imagine that any statements that entail each other can be considered "the same".

In a simple case, this means that if $a = b$ and $b = c$ then one can always assume $a = c$. But there's a much more general version of this embodied in the univalence axiom of homotopy type theory—that in our terms can be interpreted as saying that mathematical observers consider equivalent things the same.

There's another way that mathematical observers conflate different statements—that's in many ways more important, but less formal. As we mentioned above, when mathematicians talk, say, about the Pythagorean theorem, they typically think they have a definite concept in mind. But at the axiomatic level—and even more so at the level of emes—there are a huge number of different "metamathematical configurations" that are all "considered the same" by the typical working mathematician, or by our "mathematical observer". (At the level of axioms, there might be different axiom systems for real numbers; at the level of emes there might be different ways of representing concepts like addition or equality.)

In a sense we can think of mathematical observers as having a certain "extent" in metamathematical space. And much like human-scale physical observers see only the aggregate effects of huge numbers of atoms of space, so also mathematical observers see only the "aggregate effects" of huge numbers of emes of metamathematical space.

But now the key question is whether a "whole mathematical observer" can "move in metamathematical space" as a single "rigid" entity, or whether it will inevitably be distorted—or shredded—by the structure of metamathematical space. In the next section we'll discuss the analog of gravity—and curvature—in metamathematical space. But our physicalized approach tends to suggest that in "most" of metamathematical space, a typical mathematical observer will be able to "move around freely", implying that there will indeed be paths or "bridges" between different areas of mathematics, that involve only higher-level mathematical constructs, and don't require dropping down to the level of emes and the raw ruliad.

25 | Gravitational and Relativistic Effects in Metamathematics

If metamathematical space is like physical space, does that mean that it has analogs of gravity, and relativity? The answer seems to be "yes"—and these provide our next examples of physicalized laws of mathematics.

In the end, we're going to be able to talk about at least gravity in a largely "static" way, referring mostly to the "instantaneous state of metamathematics", captured as an entailment fabric. But in leveraging ideas from physics, it's important to start off formulating things in terms of the analog of time for metamathematics—which is entailment.

As we've discussed above, the entailment cone is the direct analog of the light cone in physics. Starting with some mathematical statement (or, more accurately, some event that transforms it) the forward entailment cone contains all statements (or, more accurately, events) that follow from it. Any possible "instantaneous state of metamathematics" then corresponds to a "transverse slice" through this entailment cone—with the slice in effect being laid out in metamathematical space.

An individual entailment of one statement by another corresponds to a path in the entailment cone, and this path (or, more accurately for accumulative evolution, subgraph) can be thought of as a proof of one statement given another. And in these terms the shortest proof can be thought of as a geodesic in the entailment cone. (In practical mathematics, it's very unlikely one will find—or care about—the strictly shortest proof. But even having a "fairly short proof" will be enough to give the general conclusions we'll discuss here.)

Given a path in the entailment cone, we can imagine projecting it onto a transverse slice, i.e. onto an entailment fabric. Being able to consistently do this depends on having a certain uniformity in the entailment cone, and in the sequence of "metamathematical hypersurfaces" that are defined by whatever "metamathematical reference frame" we're using. But assuming, for example, that underlying computational irreducibility successfully generates a kind of "statistical uniformity" that cannot be "decoded" by the observer, we can expect to have meaningful paths—and geodesics—on entailment fabrics.

But what these geodesics are like then depends on the emergent geometry of entailment fabrics. In physics, the limiting geometry of the analog of this for physical space is presumably a fairly simple 3D manifold. For branchial space, it's more complicated, probably for example being "exponential dimensional". And for metamathematics, the limiting geometry is also undoubtedly more complicated—and almost certainly exponential dimensional.

We've argued that we expect metamathematical space to have a certain perceived uniformity. But what will affect this, and therefore potentially modify the local geometry of the space? The basic answer is exactly the same as in our Physics Project. If there's "more activity" somewhere in an entailment fabric, this will in effect lead to "more local connections", and thus effective "positive local curvature" in the emergent geometry of the network. Needless to say, exactly what "more activity" means is somewhat subtle, especially given that the fabric in which one is looking for this is itself defining the ambient geometry, measures of "area", etc.

In our Physics Project we make things more precise by associating "activity" with energy density, and saying that energy effectively corresponds to the flux of causal edges through spacelike hypersurfaces. So this suggests that we think about an analog of energy in metamathematics: essentially defining it to be the density of update events in the entailment fabric. Or, put another way, energy in metamathematics depends on the "density of proofs" going through a region of metamathematical space, i.e. involving particular "nearby" mathematical statements.

There are lots of caveats, subtleties and details. But the notion that "activity AKA energy" leads to increasing curvature in an emergent geometry is a general feature of the whole multicomputational paradigm that the ruliad captures. And in fact we expect a quantitative relationship between energy density (or, strictly, energy-momentum) and induced curvature of the "transversal space"—that corresponds exactly to Einstein's equations in general relativity. It'll be more difficult to see this in the metamathematical case because metamathematical space is geometrically more complicated—and less familiar—than physical space.

But even at a qualitative level, it seems very helpful to think in terms of physics and spacetime analogies. The basic phenomenon is that geodesics are deflected by the presence of "energy", in effect being "attracted to it". And this is why we can think of regions of higher energy (or energy-momentum/mass)—in physics and in metamathematics—as "generating gravity", and deflecting geodesics towards them. (Needless to say, in metamathematics, as in physics, the vast majority of overall activity is just devoted to knitting together the structure of space, and when gravity is produced, it's from slightly increased activity in a particular region.)

(In our Physics Project, a key result is that the same kind of dependence of "spatial" structure on energy happens not only in physical space, but also in branchial space—where there's a direct analog of general relativity that basically yields the path integral of quantum mechanics.)

What does this mean in metamathematics? Qualitatively, the implication is that "proofs will tend to go through where there's a higher density of proofs". Or, in an analogy, if you want to drive from one place to another, it'll be more efficient if you can do at least part of your journey on a freeway.

One question to ask about metamathematical space is whether one can always get from any place to any other. In other words, starting from one area of mathematics, can one somehow derive all others? A key issue here is whether the area one starts from is computation universal. Propositional logic is not, for example. So if one starts from it, one is essentially trapped, and cannot reach other areas.

But results in mathematical logic have established that most traditional areas of axiomatic mathematics are in fact computation universal (and the Principle of Computational Equivalence suggests that this will be ubiquitous). And given computation universality there will at least be some "proof path". (In a sense this is a reflection of the fact that the ruliad is unique, so everything is connected in "the same ruliad".)

But a big question is whether the "proof path" is "big enough" to be appropriate for a "mathematical observer like us". Can we expect to get from one part of metamathematical space to another without the observer being "shredded"? Will we be able to start from any of a whole collection of places in metamathematical space that are considered "indistinguishably nearby" to a mathematical observer and have all of them "move together" to reach our destination? Or will different specific starting points follow quite different paths—preventing us from having a high-level ("fluid dynamics") description of what's going on, and instead forcing us to drop down to the "molecular dynamics" level?

In practical pure mathematics, this tends to be an issue of whether there is an "elegant proof using high-level concepts", or whether one has to drop down to a very detailed level that's more like low-level computer code, or the output of an automated theorem proving system. And indeed there's a very visceral sense of "shredding" in cases where one's confronted with a proof that consists of page after page of "machine-like details".

But there's another point here as well. If one looks at an individual proof path, it can be computationally irreducible to find out where the path goes, and the question of whether it ever reaches a particular destination can be undecidable. But in most of the current practice of pure mathematics, one's interested in "higher-level conclusions", that are "visible" to a mathematical observer who doesn't resolve individual proof paths.

Later we'll discuss the dichotomy between explorations of computational systems that routinely run into undecidability—and the typical experience of pure mathematics, where undecidability is rarely encountered in practice. But the basic point is that what a typical mathematical observer sees is at the "fluid dynamics level", where the potentially circuitous path of some individual molecule is not relevant.

Of course, by asking specific questions—about metamathematics, or, say, about very specific equations—it's still perfectly possible to force tracing of individual "low-level" proof paths. But this isn't what's typical in current pure mathematical practice. And in a sense we can see this as an extension of our first physicalized law of mathematics: not only is higher-level mathematics possible, but it's ubiquitously so, with the result that, at least in terms of the questions a mathematical observer would readily formulate, phenomena like undecidability are not generically seen.

But even though undecidability may not be directly visible to a mathematical observer, its underlying presence is still crucial in coherently "knitting together" metamathematical space. Because without undecidability, we won't have computation universality and computational irreducibility. But—just like in our Physics Project—computational irreducibility is crucial in producing the low-level apparent randomness that is needed to support any kind of "continuum limit" that allows us to think of large collections of what are ultimately discrete emes as building up some kind of coherent geometrical space.

And when undecidability is not present, one will typically not end up with anything like this kind of coherent space. An extreme example occurs in rewrite systems that eventually terminate—in the sense that they reach a "fixed-point" (or "normal form") state where no more transformations can be applied.

In our Physics Project, this kind of termination can be interpreted as a spacelike singularity at which "time stops" (as at the center of a non-rotating black hole). But in general decidability is associated with "limits on how far paths can go"—just like the limits on causal paths associated with event horizons in physics.

There are many details to work out, but the qualitative picture can be developed further. In physics, the singularity theorems imply that in essence the eventual formation of spacetime singularities is inevitable. And there should be a direct analog in our context that implies the eventual formation of "metamathematical singularities". In qualitative terms, we can expect that the presence of proof density (which is the analog of energy) will "pull in" more proofs until eventually there are so many proofs that one has decidability and a "proof event horizon" is formed.

In a sense this implies that the long-term future of mathematics is strangely similar to the long-term future of our physical universe. In our physical universe, we expect that while the expansion of space may continue, many parts of the universe will form black holes and essentially be "closed off". (At least ignoring expansion in branchial space, and quantum effects in general.)

The analog of this in mathematics is that while there can be continued overall expansion in metamathematical space, more and more parts of it will "burn out" because they've become decidable. In other words, as more work and more proofs get done in a particular area, that area will eventually be "finished"—and there will be no more "open-ended" questions associated with it.

In physics there's sometimes discussion of white holes, which are imagined to effectively be time-reversed black holes, spewing out all possible material that could be captured in a black hole. In metamathematics, a white hole is like a statement that is false and therefore "leads to an explosion". The presence of such an object in metamathematical space will in effect cause observers to be shredded—making it inconsistent with the coherent construction of higher-level mathematics.

We've talked at some length about the "gravitational" structure of metamathematical space. But what about seemingly simpler things like special relativity? In physics, there's a notion of basic, flat spacetime, for which it's easy to construct families of reference frames, and in which parallel trajectories stay parallel. In metamathematics, the analog is presumably metamathematical space in which "parallel proof geodesics" remain "parallel"—so that in effect one can continue "making progress in mathematics" by just "keeping on doing what you've been doing".

And somehow relativistic invariance is associated with the idea that there are many ways to do math, but in the end they're all able to reach the same conclusions. Ultimately this is something one expects as a consequence of fundamental features of the ruliad—and the inevitability of causal invariance in it resulting from the Principle of Computational Equivalence. It's also something that might seem quite familiar from practical mathematics and, say, from the ability to do derivations using different methods—like from either geometry or algebra—and yet still end up with the same conclusions.

So if there's an analog of relativistic invariance, what about analogs of phenomena like time dilation? In our Physics Project time dilation has a rather direct interpretation. To "progress in time" takes a certain amount of computational work. But motion in effect also takes a certain amount of computational work—in essence to continually recreate versions of something in different places. But from the ruliad on up there is ultimately only a certain amount of computational work that can be done—and if computational work is being "used up" on motion, there is less available to devote to progress in time, and so time will effectively run more slowly, leading to the experience of time dilation.

So what is the metamathematical analog of this? Presumably it's that when you do derivations in math you can either stay in one area and directly make progress in that area, or you can "base yourself in some other area" and make progress only by continually translating back and forth. But ultimately that translation process will take computational work, and so will slow down your progress—leading to an analog of time dilation.

In physics, the speed of light defines the maximum amount of motion in space that can occur in a certain amount of time. In metamathematics, the analog is that there's a maximum "translation distance" in metamathematical space that can be "bridged" with a certain amount of derivation. In physics we're used to measuring spatial distance in meters—and time in seconds. In metamathematics we don't yet have familiar units in which to measure, say, distance between mathematical concepts—or, for that matter, "amount of derivation" being done. But with the empirical metamathematics we'll discuss in the next section we actually have the beginnings of a way to define such things, and to use what's been achieved in the history of human mathematics to at least imagine "empirically measuring" what we might call "maximum metamathematical speed".

It should be emphasized that we are only at the very beginning of exploring things like the analogs of relativity in metamathematics. One important piece of formal structure that we haven't really discussed here is causal dependence, and causal graphs. We've talked at length about statements entailing other statements. But we haven't talked about questions like which part of which statement is needed for some event to occur that will entail some other statement. And—while there's no fundamental difficulty in doing it—we haven't concerned ourselves with constructing causal graphs to represent causal relationships and causal dependencies between events.

When it comes to physical observers, there is a very direct interpretation of causal graphs that relates to what a physical observer can experience. But for mathematical observers—where the notion of time is less central—it's less clear just what the interpretation of causal graphs should be. But one certainly expects that they will enter in the construction of any general "observer theory" that characterizes "observers like us" across both physics and mathematics.

26 | Empirical Metamathematics

We've discussed the overall structure of metamathematical space, and the general kind of sampling that we humans do of it (as "mathematical observers") when we do mathematics. But what can we learn from the specifics of human mathematics, and the actual mathematical statements that humans have published over the centuries?

We might imagine that these statements are just ones that—as "accidents of history"—humans have "happened to find interesting". But there's definitely more to it—and potentially what's there is a rich source of "empirical data" relevant to our physicalized laws of mathematics, and to what amounts to their "experimental validation".

The situation with "human settlements" in metamathematical space is in a sense rather similar to the situation with human settlements in physical space. If we look at where humans have chosen to live and build cities, we'll find a bunch of locations in 3D space. The details of where these are depend on history and many factors. But there's a clear overarching theme, that's in a sense a direct reflection of underlying physics: all the locations lie on the more-or-less spherical surface of the Earth.

It's not so straightforward to see what's going on in the metamathematical case, not least because any notion of coordinatization seems to be much more complicated for metamathematical space than for physical space. But we can still begin by doing "empirical metamathematics" and asking questions about for example what amounts to where in metamathematical space we humans have so far established ourselves. And as a first example, let's consider Boolean algebra.

Even to talk about something called "Boolean algebra" we have to be operating at a level far above the raw ruliad—where we've already implicitly aggregated vast numbers of emes to form notions of, for example, variables and logical operations.

But once we're at this level we can "survey" metamathematical space just by enumerating possible symbolic statements that can be created using the operations we've set up for Boolean algebra (here And $\wedge$, Or $\vee$ and Not $\bar{\square}$):

$a = b$	$a = \bar{a}$	$a = \bar{b}$	$\bar{a} = \bar{b}$	$a = a \wedge a$	$\bar{a} = a \wedge a$	$a = a \vee a$	$\bar{a} = a \vee a$
$a \wedge a = a \vee a$	$a = a \wedge b$	$\bar{a} = a \wedge b$	$a \wedge a = a \wedge b$	$a \vee a = a \wedge b$	$a = a \vee b$	$\bar{a} = a \vee b$	$a \wedge a = a \vee b$
$a \vee a = a \vee b$	$a \wedge b = a \vee b$	$a \wedge b = a \wedge c$	$a \vee b = a \wedge c$	$a \wedge b = a \vee c$	$a \vee b = a \vee c$	$a = b \wedge a$	$\bar{a} = b \wedge a$
$a \wedge a = b \wedge a$	$a \vee a = b \wedge a$	$a \wedge b = b \wedge a$	$a \vee b = b \wedge a$	$a = b \vee a$	$\bar{a} = b \vee a$	$a \wedge a = b \vee a$	$a \vee a = b \vee a$
$a \wedge b = b \vee a$	$a \vee b = b \vee a$	$a = b \wedge b$	$\bar{a} = b \wedge b$	$a \wedge a = b \wedge b$	$a \vee a = b \wedge b$	$a \wedge b = b \wedge b$	$a \vee b = b \wedge b$
$a = b \vee b$	$\bar{a} = b \vee b$	$a \wedge a = b \vee b$	$a \vee a = b \vee b$	$a \wedge b = b \vee b$	$a \vee b = b \vee b$	$a = b \wedge c$	$\bar{a} = b \wedge c$
$a \wedge a = b \wedge c$	$a \vee a = b \wedge c$	$a \wedge b = b \wedge c$	$a \vee b = b \wedge c$	$a = b \vee c$	$\bar{a} = b \vee c$	$a \wedge a = b \vee c$	$a \vee a = b \vee c$
$a \wedge b = b \vee c$	$a \vee b = b \vee c$	$a \wedge b = c \wedge a$	$a \vee b = c \wedge a$	$a \wedge b = c \vee a$	$a \vee b = c \vee a$	$a \wedge b = c \wedge b$	$a \vee b = c \wedge b$
$a \wedge b = c \vee b$	$a \vee b = c \vee b$	$a \wedge b = c \wedge c$	$a \vee b = c \wedge c$	$a \wedge b = c \vee c$	$a \vee b = c \vee c$	$a = \bar{\bar{a}}$	$\bar{a} = \bar{\bar{a}}$
$a \wedge a = \bar{\bar{a}}$	$a \vee a = \bar{\bar{a}}$	$a \wedge b = \bar{\bar{a}}$	$a \vee b = \bar{\bar{a}}$	$a = \bar{\bar{b}}$	$\bar{a} = \bar{\bar{b}}$	$a \wedge a = \bar{\bar{b}}$	$a \vee a = \bar{\bar{b}}$

But so far these are just raw, structural statements. To connect with actual Boolean algebra we must pick out which of these can be derived from the axioms of Boolean algebra, or, put another way, which of them are in the entailment cone of these axioms:

$a=b$	$a=\bar{a}$	$a=\bar{b}$	$\bar{a}=\bar{b}$	$a=a\wedge a$	$\bar{a}=a\wedge a$	$a=a\vee a$	$\bar{a}=a\vee a$
$a\wedge a=a\vee a$	$a=a\wedge b$	$\bar{a}=a\wedge b$	$a\wedge a=a\wedge b$	$a\vee a=a\wedge b$	$a=a\vee b$	$\bar{a}=a\vee b$	$a\wedge a=a\vee b$
$a\vee a=a\vee b$	$a\wedge b=a\vee b$	$a\wedge b=a\wedge c$	$a\vee b=a\wedge c$	$a\wedge b=a\vee c$	$a\vee b=a\vee c$	$a=b\wedge a$	$\bar{a}=b\wedge a$
$a\wedge a=b\wedge a$	$a\vee a=b\wedge a$	$a\wedge b=b\wedge a$	$a\vee b=b\wedge a$	$a=b\vee a$	$\bar{a}=b\vee a$	$a\wedge a=b\vee a$	$a\vee a=b\vee a$
$a\wedge b=b\vee a$	$a\vee b=b\vee a$	$a=b\wedge b$	$\bar{a}=b\wedge b$	$a\wedge a=b\wedge b$	$a\vee a=b\wedge b$	$a\wedge b=b\wedge b$	$a\vee b=b\wedge b$
$a=b\vee b$	$\bar{a}=b\vee b$	$a\wedge a=b\vee b$	$a\vee a=b\vee b$	$a\wedge b=b\vee b$	$a\vee b=b\vee b$	$a=b\wedge c$	$\bar{a}=b\wedge c$
$a\wedge a=b\wedge c$	$a\vee a=b\wedge c$	$a\wedge b=b\wedge c$	$a\vee b=b\wedge c$	$a=b\vee c$	$\bar{a}=b\vee c$	$a\wedge a=b\vee c$	$a\vee a=b\vee c$
$a\wedge b=b\vee c$	$a\vee b=b\vee c$	$a\wedge b=c\wedge a$	$a\vee b=c\wedge a$	$a\wedge b=c\vee a$	$a\vee b=c\vee a$	$a\wedge b=c\wedge b$	$a\vee b=c\wedge b$
$a\wedge b=c\vee b$	$a\vee b=c\vee b$	$a\wedge b=c\wedge c$	$a\vee b=c\wedge c$	$a\wedge b=c\vee c$	$a\vee b=c\vee c$	$a=\bar{\bar{a}}$	$\bar{a}=\bar{\bar{a}}$
$a\wedge a=\bar{\bar{a}}$	$a\vee a=\bar{\bar{a}}$	$a\wedge b=\bar{\bar{a}}$	$a\vee b=\bar{\bar{a}}$	$a=\bar{\bar{b}}$	$\bar{a}=\bar{\bar{b}}$	$a\wedge a=\bar{\bar{b}}$	$a\vee a=\bar{\bar{b}}$

Of all possible statements, it's only an exponentially small fraction that turn out to be derivable:

variables	*a*		*a, b*		*a, b, c*	
size	*total*	*theorems*	*total*	*theorems*	*total*	*theorems*
2	1	0	4	0	4	0
3	33	8	132	8	164	8
4	673	164	5348	404	9316	404
5	15009	3620	234724	17940	597092	22292

But in the case of Boolean algebra, we can readily collect such statements:

$a=a\wedge a$	$a=a\vee a$	$a\wedge a=a\vee a$	$a\wedge b=b\wedge a$	$a\vee b=b\vee a$	$a=\bar{\bar{a}}$
$a\wedge a=\bar{\bar{a}}$	$a\vee a=\bar{\bar{a}}$	$\bar{a}\wedge a=a\wedge\bar{a}$	$\bar{a}\vee a=a\vee\bar{a}$	$\bar{a}=\overline{a\wedge a}$	$\bar{a}=\overline{a\vee a}$
$\overline{a\wedge a}=\overline{a\vee a}$	$a\wedge\bar{b}=\bar{b}\wedge a$	$\bar{a}\wedge b=b\wedge\bar{a}$	$a\vee\bar{b}=\bar{b}\vee a$	$\bar{a}\vee b=b\vee\bar{a}$	$\overline{a\wedge b}=\overline{b\wedge a}$
$\overline{a\vee b}=\overline{b\vee a}$	$\bar{a}\wedge a=\bar{b}\wedge b$	$a\wedge\bar{a}=\bar{b}\wedge b$	$\bar{a}\wedge a=b\wedge\bar{b}$	$a\wedge\bar{a}=b\wedge\bar{b}$	$\bar{a}\vee a=\bar{b}\vee b$
$a\vee\bar{a}=\bar{b}\vee b$	$\bar{a}\vee a=b\vee\bar{b}$	$a\vee\bar{a}=b\vee\bar{b}$	$\bar{a}=\bar{a}\wedge\bar{a}$	$\overline{a\wedge a}=\bar{a}\wedge\bar{a}$	$\overline{a\vee a}=\bar{a}\wedge\bar{a}$
$\bar{a}=\bar{a}\vee\bar{a}$	$\overline{a\wedge a}=\bar{a}\vee\bar{a}$	$\overline{a\vee a}=\bar{a}\vee\bar{a}$	$\bar{a}\wedge\bar{a}=\bar{a}\vee\bar{a}$	$\overline{a\vee b}=\bar{a}\wedge\bar{b}$	$\overline{a\wedge b}=\bar{a}\vee\bar{b}$
$\overline{a\vee b}=\bar{b}\wedge\bar{a}$	$\bar{a}\wedge\bar{b}=\bar{b}\wedge\bar{a}$	$\overline{a\wedge b}=\bar{b}\vee\bar{a}$	$\bar{a}\vee\bar{b}=\bar{b}\vee\bar{a}$	$a=(a\wedge a)\wedge a$	$a\wedge a=(a\wedge a)\wedge a$
$a\vee a=(a\wedge a)\wedge a$	$\bar{\bar{a}}=(a\wedge a)\wedge a$	$a=(a\vee a)\wedge a$	$a\wedge a=(a\vee a)\wedge a$	$a\vee a=(a\vee a)\wedge a$	$\bar{\bar{a}}=(a\vee a)\wedge a$
$(a\wedge a)\wedge a=(a\vee a)\wedge a$	$a=a\wedge(a\wedge a)$	$a\wedge a=a\wedge(a\wedge a)$	$a\vee a=a\wedge(a\wedge a)$	$\bar{\bar{a}}=a\wedge(a\wedge a)$	$(a\wedge a)\wedge a=a\wedge(a\wedge a)$
$(a\vee a)\wedge a=a\wedge(a\wedge a)$	$a=a\wedge(a\vee a)$	$a\wedge a=a\wedge(a\vee a)$	$a\vee a=a\wedge(a\vee a)$	$\bar{\bar{a}}=a\wedge(a\vee a)$	$(a\wedge a)\wedge a=a\wedge(a\vee a)$

We've typically explored entailment cones by looking at slices consisting of collections of theorems generated after a specified number of proof steps. But here we're making a very different sampling of the entailment cone—looking in effect instead at theorems in order of their structural complexity as symbolic expressions.

In doing this kind of systematic enumeration we're in a sense operating at a "finer level of granularity" than typical human mathematics. Yes, these are all "true theorems". But mostly they're not theorems that a human mathematician would ever write down, or specifically "consider interesting". And for example only a small fraction of them have historically been given names—and are called out in typical logic textbooks:

$a=a\land a$	$a=a\lor a$	$a\land a=a\lor a$	$a\land b=b\land a$	$a\lor b=b\lor a$	$a=\bar{\bar{a}}$	$a\land a=\bar{\bar{a}}$
$a\lor a=\bar{\bar{a}}$	$\bar{a}=\overline{a\land a}$	$\bar{a}=\overline{a\lor a}$	$a=(a\land a)\land a$	$a=(a\lor a)\land a$	$a=a\land(a\land a)$	$a=a\land(a\lor a)$
$a=(a\land a)\lor a$	$a=(a\lor a)\lor a$	$a=a\lor(a\land a)$	$a=a\lor(a\lor a)$	$a=a\land(a\lor b)$	$a=a\lor(a\land b)$	$a=(a\lor b)\land a$
$a=a\land(b\lor a)$	$a=(a\land b)\lor a$	$a=a\lor(b\land a)$	$a=(b\lor a)\land a$	$a=(b\land a)\lor a$	$\bar{a}=\bar{a}\land\bar{a}$	$\bar{a}=\bar{a}\lor\bar{a}$
$\bar{a}\land a=a\land\bar{a}$	$\bar{a}\lor a=a\lor\bar{a}$	$\overline{a\land a}=\overline{a\lor a}$	$\bar{\bar{a}}=(a\land a)\land a$	$\bar{\bar{a}}=(a\lor a)\land a$	$\bar{\bar{a}}=a\land(a\land a)$	$\bar{\bar{a}}=a\land(a\lor a)$
$\bar{\bar{a}}=(a\land a)\lor a$	$\bar{\bar{a}}=(a\lor a)\lor a$	$\bar{\bar{a}}=a\lor(a\land a)$	$\bar{\bar{a}}=a\lor(a\lor a)$	$\bar{\bar{a}}=a\land(a\lor b)$	$\bar{\bar{a}}=a\lor(a\land b)$	$\bar{\bar{a}}=(a\lor b)\land a$
$\bar{\bar{a}}=a\land(b\lor a)$	$\bar{\bar{a}}=(a\land b)\lor a$	$\bar{\bar{a}}=a\lor(b\land a)$	$\bar{a}\land a=\bar{b}\land b$	$a\land\bar{a}=\bar{b}\land b$	$\bar{a}\land a=b\land\bar{b}$	$a\land\bar{a}=b\land\bar{b}$
$\bar{a}\lor a=\bar{b}\lor b$	$a\lor\bar{a}=\bar{b}\lor b$	$\bar{a}\lor a=b\lor\bar{b}$	$a\lor\bar{a}=b\lor\bar{b}$	$\bar{\bar{a}}=(b\lor a)\land a$	$\bar{\bar{a}}=(b\land a)\lor a$	$a\land\bar{b}=\bar{b}\land a$
$\bar{a}\land b=b\land\bar{a}$	$a\lor\bar{b}=\bar{b}\lor a$	$\bar{a}\lor b=b\lor\bar{a}$	$\overline{a\land b}=\overline{b\land a}$	$\overline{a\lor b}=\overline{b\lor a}$	$a\land a=(a\land a)\land a$	$a\lor a=(a\land a)\land a$
$a\land a=(a\lor a)\land a$	$a\lor a=(a\lor a)\land a$	$a\land a=a\land(a\land a)$	$a\lor a=a\land(a\land a)$	$a\land a=a\land(a\lor a)$	$a\lor a=a\land(a\lor a)$	$a\land a=(a\land a)\lor a$
$a\lor a=(a\land a)\lor a$	$a\land a=(a\lor a)\lor a$	$a\lor a=(a\lor a)\lor a$	$a\land a=a\lor(a\land a)$	$a\lor a=a\lor(a\land a)$	$a\land a=a\lor(a\lor a)$	$a\lor a=a\lor(a\lor a)$
$a=(a\land a)\land(a\land a)$	$a=(a\land a)\land(a\lor a)$	$a=(a\lor a)\land(a\land a)$	$a=(a\lor a)\land(a\lor a)$	$a=(a\land a)\lor(a\land a)$	$a=(a\land a)\lor(a\lor a)$	$a=(a\lor a)\lor(a\land a)$
$a=(a\lor a)\lor(a\lor a)$	$a\land a=a\land(a\lor b)$	$a\lor a=a\land(a\lor b)$	$a\land a=a\lor(a\land b)$	$a\lor a=a\lor(a\land b)$	$a=(a\land a)\land(a\lor b)$	$a=(a\lor a)\land(a\lor b)$
$a=(a\land a)\lor(a\land b)$	$a=(a\lor a)\lor(a\land b)$	$a\land a=(a\lor b)\land a$	$a\lor a=(a\lor b)\land a$	$a\land a=a\land(b\lor a)$	$a\lor a=a\land(b\lor a)$	$a\land a=(a\land b)\lor a$
$a\lor a=(a\land b)\lor a$	$a\land a=a\lor(b\land a)$	$a\lor a=a\lor(b\land a)$	$a=(a\land a)\land(b\lor a)$	$a=(a\lor a)\land(b\lor a)$	$a=(a\land a)\lor(b\land a)$	$a=(a\lor a)\lor(b\land a)$
$a\land a=(b\lor a)\land a$	$a\lor a=(b\lor a)\land a$	$a\land a=(b\land a)\lor a$	$a\lor a=(b\land a)\lor a$	$a=(a\lor b)\land(a\land a)$	$a=(a\lor b)\land(a\lor a)$	$a=(a\land b)\lor(a\land a)$
$a=(a\land b)\lor(a\lor a)$	$a=(b\lor a)\land(a\land a)$	$a=(b\lor a)\land(a\lor a)$	$a=(b\land a)\lor(a\land a)$	$a=(b\land a)\lor(a\lor a)$	$a\land b=(a\land a)\land b$	$a\land b=(a\lor a)\land b$
$a\land b=a\land(a\land b)$	$a\lor b=(a\land a)\lor b$	$a\lor b=(a\lor a)\lor b$	$a\lor b=a\lor(a\lor b)$	$a\land b=(a\land b)\land a$	$a\land b=a\land(b\land a)$	$a\lor b=(a\lor b)\lor a$
$a\lor b=a\lor(b\lor a)$	$a\land b=(a\land b)\land b$	$a\land b=a\land(b\land b)$	$a\land b=a\land(b\lor b)$	$a\lor b=(a\lor b)\lor b$	$a\lor b=a\lor(b\land b)$	$a\lor b=a\lor(b\lor b)$
$a\land b=(b\land a)\land a$	$a\land b=b\land(a\land a)$	$a\land b=b\land(a\lor a)$	$a\lor b=(b\lor a)\lor a$	$a\lor b=b\lor(a\land a)$	$a\lor b=b\lor(a\lor a)$	$a\land b=(b\land a)\land b$
$a\land b=b\land(a\land b)$	$a\lor b=(b\lor a)\lor b$	$a\lor b=b\lor(a\lor b)$	$a\land b=(b\land b)\land a$	$a\land b=(b\lor b)\land a$	$a\land b=b\land(b\land a)$	$a\lor b=(b\land b)\lor a$
$a\lor b=(b\lor b)\lor a$	$a\lor b=b\lor(b\lor a)$	$\overline{a\land a}=\bar{a}\land\bar{a}$	$\overline{a\lor a}=\bar{a}\land\bar{a}$	$\overline{a\land a}=\bar{a}\lor\bar{a}$	$\overline{a\lor a}=\bar{a}\lor\bar{a}$	$\overline{a\lor b}=\bar{a}\land\bar{b}$
$\overline{a\land b}=\bar{a}\lor\bar{b}$	$\overline{a\lor b}=\bar{b}\land\bar{a}$	$\overline{a\land b}=\bar{b}\lor\bar{a}$	$\bar{a}\land\bar{a}=\bar{a}\lor\bar{a}$	$\bar{a}\land\bar{b}=\bar{b}\land\bar{a}$	$\bar{a}\lor\bar{b}=\bar{b}\lor\bar{a}$	$\bar{a}\land a=(a\land a)\land\bar{a}$
$a\land\bar{a}=(a\land a)\land\bar{a}$	$\bar{a}\land a=(a\lor a)\land\bar{a}$	$a\land\bar{a}=(a\lor a)\land\bar{a}$	$\bar{a}\land a=\bar{a}\land(a\land a)$	$a\land\bar{a}=\bar{a}\land(a\land a)$	$\bar{a}\land a=\bar{a}\land(a\lor a)$	$a\land\bar{a}=\bar{a}\land(a\lor a)$
$\bar{a}\lor a=(a\land a)\lor\bar{a}$	$a\lor\bar{a}=(a\land a)\lor\bar{a}$	$\bar{a}\lor a=(a\lor a)\lor\bar{a}$	$a\lor\bar{a}=(a\lor a)\lor\bar{a}$	$\bar{a}\lor a=\bar{a}\lor(a\land a)$	$a\lor\bar{a}=\bar{a}\lor(a\land a)$	$\bar{a}\lor a=\bar{a}\lor(a\lor a)$

⋮ 88 lines

$a\lor b=(b\land b)\lor(b\lor a)$	$a\lor b=(b\lor b)\lor(b\lor a)$	$(a\lor b)\land b=(b\land b)\land b$	$(a\land b)\lor b=(b\land b)\land b$	$(a\lor b)\land b=(b\lor b)\land b$	$(a\land b)\lor b=(b\lor b)\land b$
$(a\lor b)\land b=b\land(b\land b)$	$(a\land b)\lor b=b\land(b\land b)$	$(a\lor b)\land b=b\land(b\lor b)$	$(a\land b)\lor b=b\land(b\lor b)$	$(a\lor b)\land b=(b\land b)\lor b$	$(a\land b)\lor b=(b\land b)\lor b$
$(a\lor b)\land b=(b\lor b)\lor b$	$(a\land b)\lor b=(b\lor b)\lor b$	$(a\lor b)\land b=b\lor(b\land b)$	$(a\land b)\lor b=b\lor(b\land b)$	$(a\lor b)\land b=b\lor(b\lor b)$	$(a\land b)\lor b=b\lor(b\lor b)$
$(a\lor b)\land b=b\land(b\lor c)$	$(a\land b)\lor b=b\land(b\lor c)$	$(a\lor b)\land b=b\lor(b\land c)$	$(a\land b)\lor b=b\lor(b\land c)$	$(a\lor b)\land b=(b\lor c)\land b$	$(a\land b)\lor b=(b\lor c)\land b$
$(a\lor b)\land b=b\land(c\lor b)$	$(a\land b)\lor b=b\land(c\lor b)$	$(a\lor b)\land b=(b\land c)\lor b$	$(a\land b)\lor b=(b\land c)\lor b$	$(a\lor b)\land b=b\lor(c\land b)$	$(a\land b)\lor b=b\lor(c\land b)$
$a\land b=(b\lor c)\land(a\land b)$	$a\lor b=(b\land c)\lor(a\lor b)$	$a\land b=(b\lor c)\land(b\land a)$	$a\lor b=(b\land c)\lor(b\lor a)$	$(a\lor b)\land b=(c\lor b)\land b$	$(a\land b)\lor b=(c\lor b)\land b$
$(a\lor b)\land b=(c\land b)\lor b$	$(a\land b)\lor b=(c\land b)\lor b$	$a\land b=(c\lor a)\land(a\land b)$	$a\lor b=(c\land a)\lor(a\lor b)$	$a\land b=(c\lor a)\land(b\land a)$	$a\lor b=(c\land a)\lor(b\lor a)$
$(a\land b)\land c=a\land(b\land c)$	$(a\lor b)\lor c=a\lor(b\lor c)$	$(a\land b)\land c=(a\land c)\land b$	$a\land(b\land c)=(a\land c)\land b$	$(a\land b)\land c=a\land(c\land b)$	$a\land(b\land c)=a\land(c\land b)$
$a\land(b\lor c)=a\land(c\lor b)$	$(a\lor b)\lor c=(a\lor c)\lor b$	$a\lor(b\lor c)=(a\lor c)\lor b$	$a\lor(b\land c)=a\lor(c\land b)$	$(a\lor b)\lor c=a\lor(c\lor b)$	$a\lor(b\lor c)=a\lor(c\lor b)$
$a\land b=(c\lor b)\land(a\land b)$	$a\lor b=(c\land b)\lor(a\lor b)$	$(a\land b)\land c=(b\land a)\land c$	$a\land(b\land c)=(b\land a)\land c$	$(a\lor b)\land c=(b\lor a)\land c$	$(a\land b)\land c=b\land(a\land c)$

⋮ 324 lines

$a\land(b\land c)=(a\lor a)\land(c\land b)$	$a\land(b\lor c)=(a\lor a)\land(c\lor b)$	$a\lor(b\land c)=(a\land a)\lor(c\land b)$	$(a\lor b)\lor c=(a\land a)\lor(c\lor b)$
$a\lor(b\lor c)=(a\land a)\lor(c\lor b)$	$a\lor(b\land c)=(a\lor a)\lor(c\land b)$	$(a\lor b)\lor c=(a\lor a)\lor(c\lor b)$	$a\lor(b\lor c)=(a\lor a)\lor(c\lor b)$
$(a\land b)\land c=(a\land b)\land(a\land c)$	$a\land(b\land c)=(a\land b)\land(a\land c)$	$a\lor(b\land c)=(a\lor b)\land(a\lor c)$	$a\land(b\lor c)=(a\land b)\lor(a\land c)$
$(a\lor b)\lor c=(a\lor b)\lor(a\lor c)$	$a\lor(b\lor c)=(a\lor b)\lor(a\lor c)$	$(a\land b)\land c=(a\land b)\land(b\land c)$	$a\land(b\land c)=(a\land b)\land(b\land c)$
$(a\lor b)\lor c=(a\lor b)\lor(b\lor c)$	$a\lor(b\lor c)=(a\lor b)\lor(b\lor c)$	$(a\land b)\land c=(a\land b)\land(c\land a)$	$a\land(b\land c)=(a\land b)\land(c\land a)$
$a\lor(b\land c)=(a\lor b)\land(c\lor a)$	$a\land(b\lor c)=(a\land b)\lor(c\land a)$	$(a\lor b)\lor c=(a\lor b)\lor(c\lor a)$	$a\lor(b\lor c)=(a\lor b)\lor(c\lor a)$

The reduction from all "structurally possible" theorems to just "ones we consider interesting" can be thought of as a form of coarse graining. And it could well be that this coarse graining would depend on all sorts of accidents of human mathematical history. But at least in the case of Boolean algebra there seems to be a surprisingly simple and "mechanical" procedure that can reproduce it.

Go through all theorems in order of increasing structural complexity, in each case seeing whether a given theorem can be proved from ones earlier in the list:

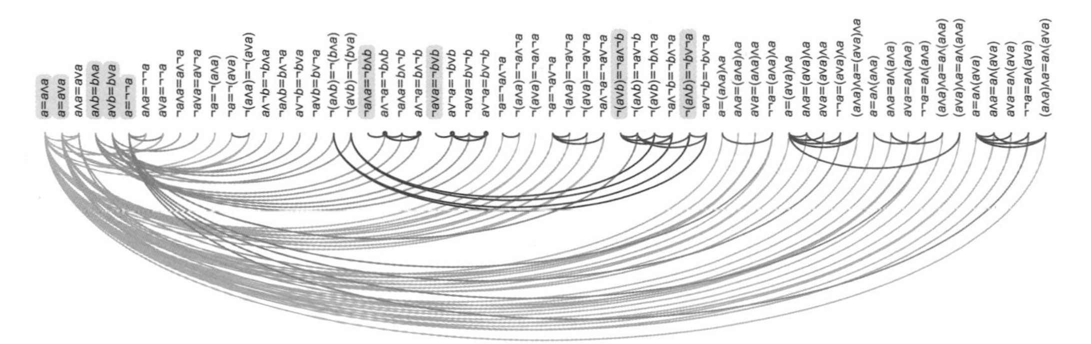

It turns out that the theorems identified by humans as "interesting" coincide almost exactly with "root theorems" that cannot be proved from earlier theorems in the list. Or, put another way, the "coarse graining" that human mathematicians do seems (at least in this case) to essentially consist of picking out only those theorems that represent "minimal statements" of new information—and eliding away those that involve "extra ornamentation".

But how are these "notable theorems" laid out in metamathematical space? Earlier we saw how the simplest of them can be reached after just a few steps in the entailment cone of a typical textbook axiom system for Boolean algebra. The full entailment cone rapidly gets unmanageably large but we can get a first approximation to it by generating individual proofs (using automated theorem proving) of our notable theorems, and then seeing how these "knit together" through shared intermediate lemmas in a token-event graph:

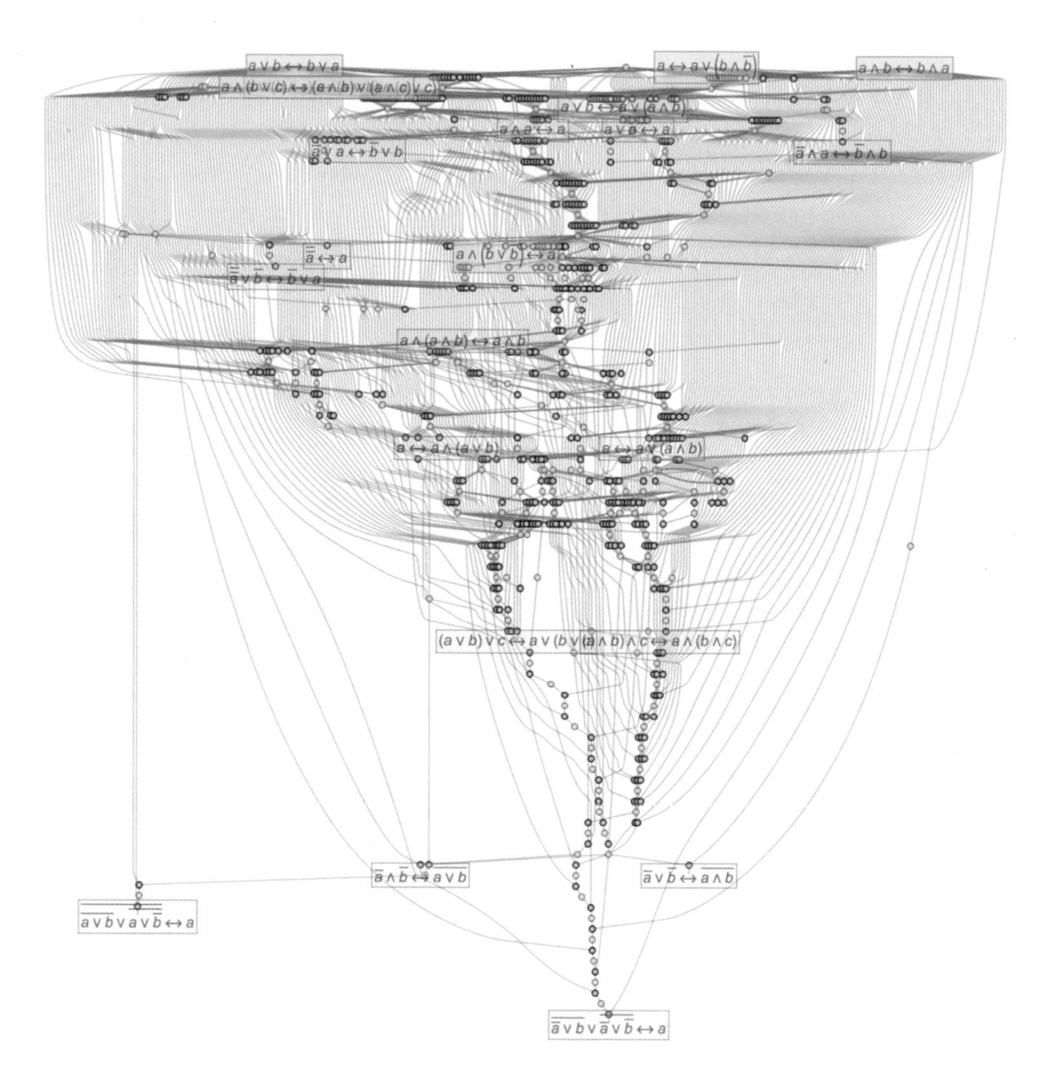

Looking at this picture we see at least a hint that clumps of notable theorems are spread out across the entailment cone, only modestly building on each other—and in effect "staking out separated territories" in the entailment cone. But of the 11 notable theorems shown here, 7 depend on all 6 axioms, while 4 depend only on various different sets of 3 axioms—suggesting at least a certain amount of fundamental interdependence or coherence.

From the token-event graph we can derive a branchial graph that represents a very rough approximation to how the theorems are "laid out in metamathematical space":

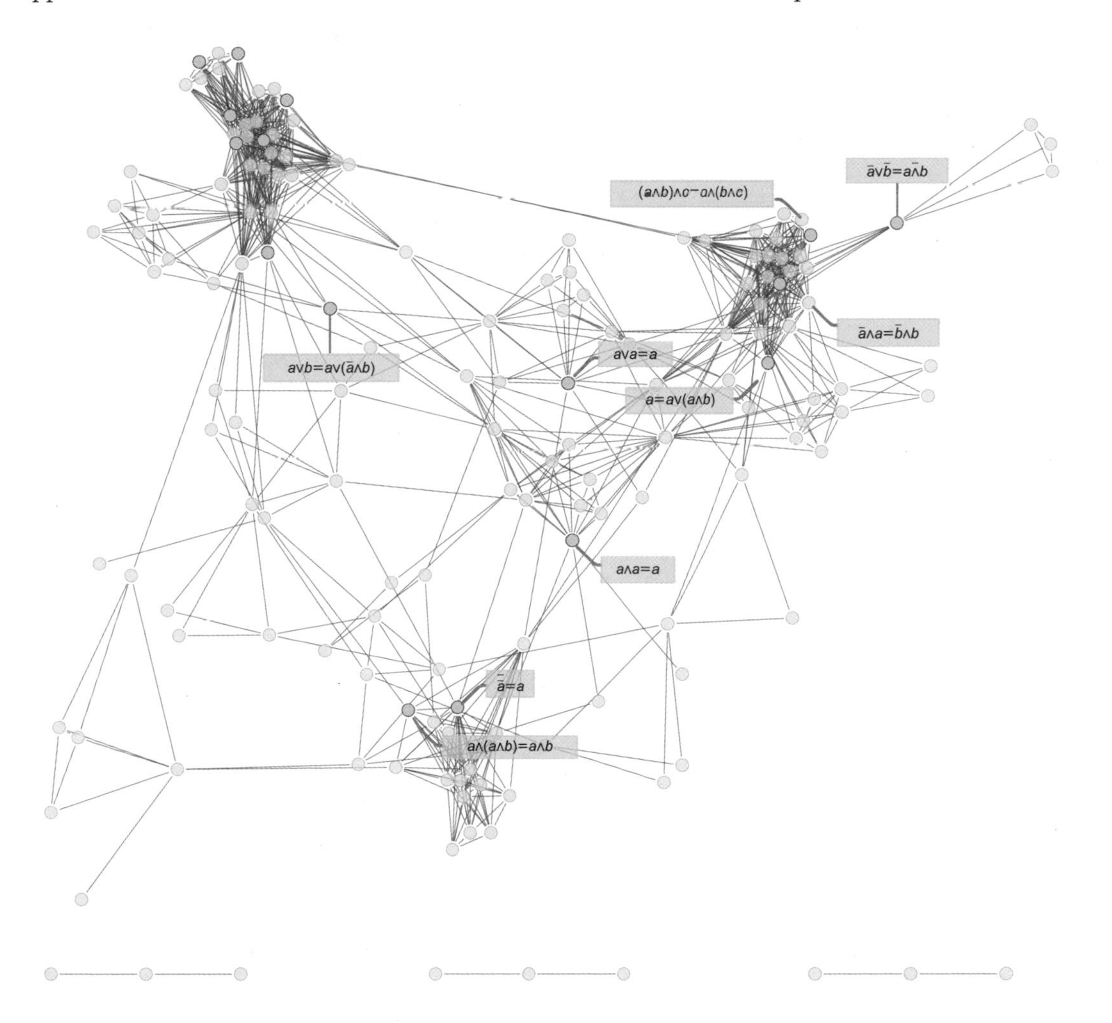

We can get a potentially slightly better approximation by including proofs not just of notable theorems, but of all theorems up to a certain structural complexity. The result shows separation of notable theorems both in the multiway graph

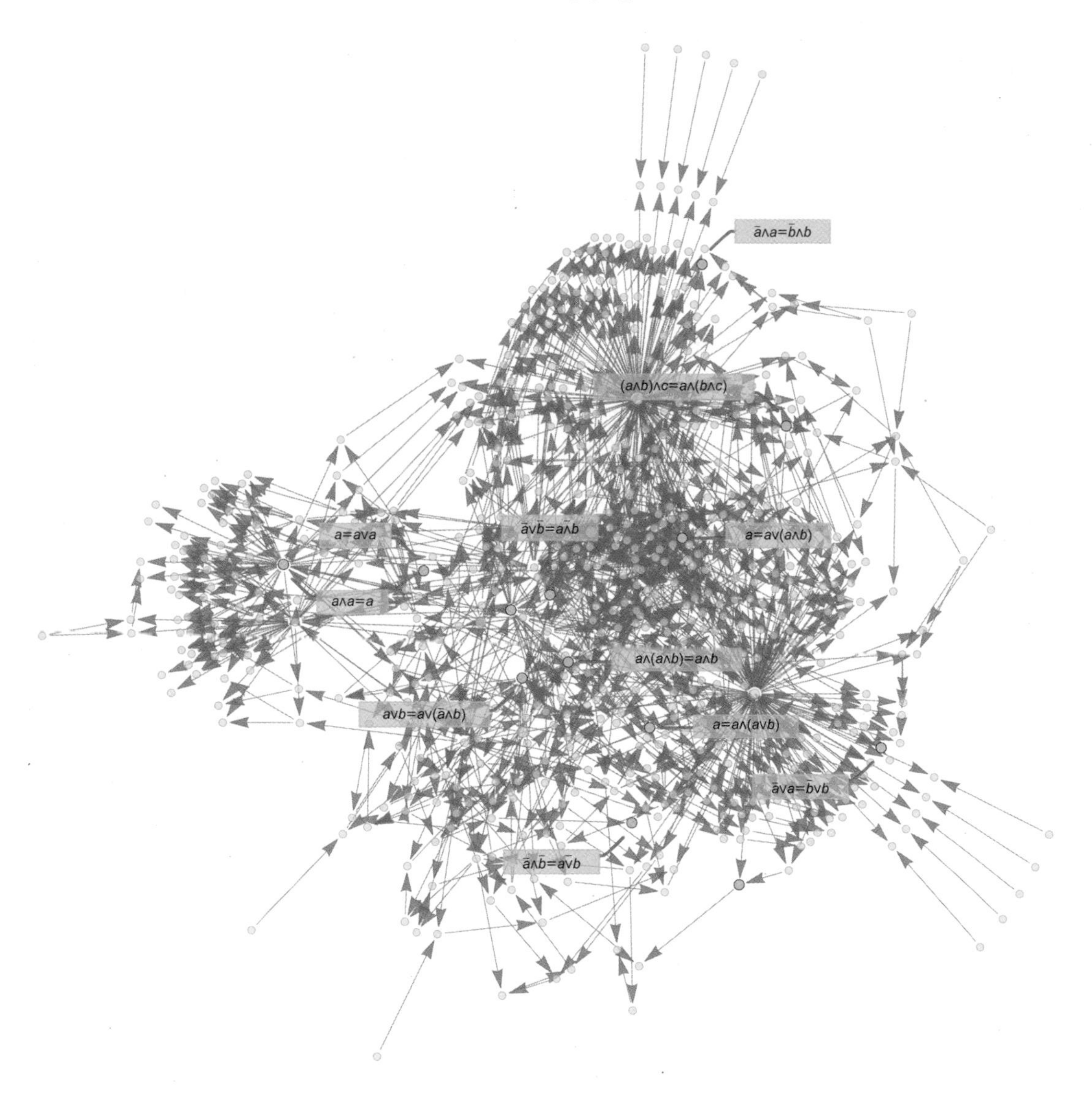

and in the branchial graph:

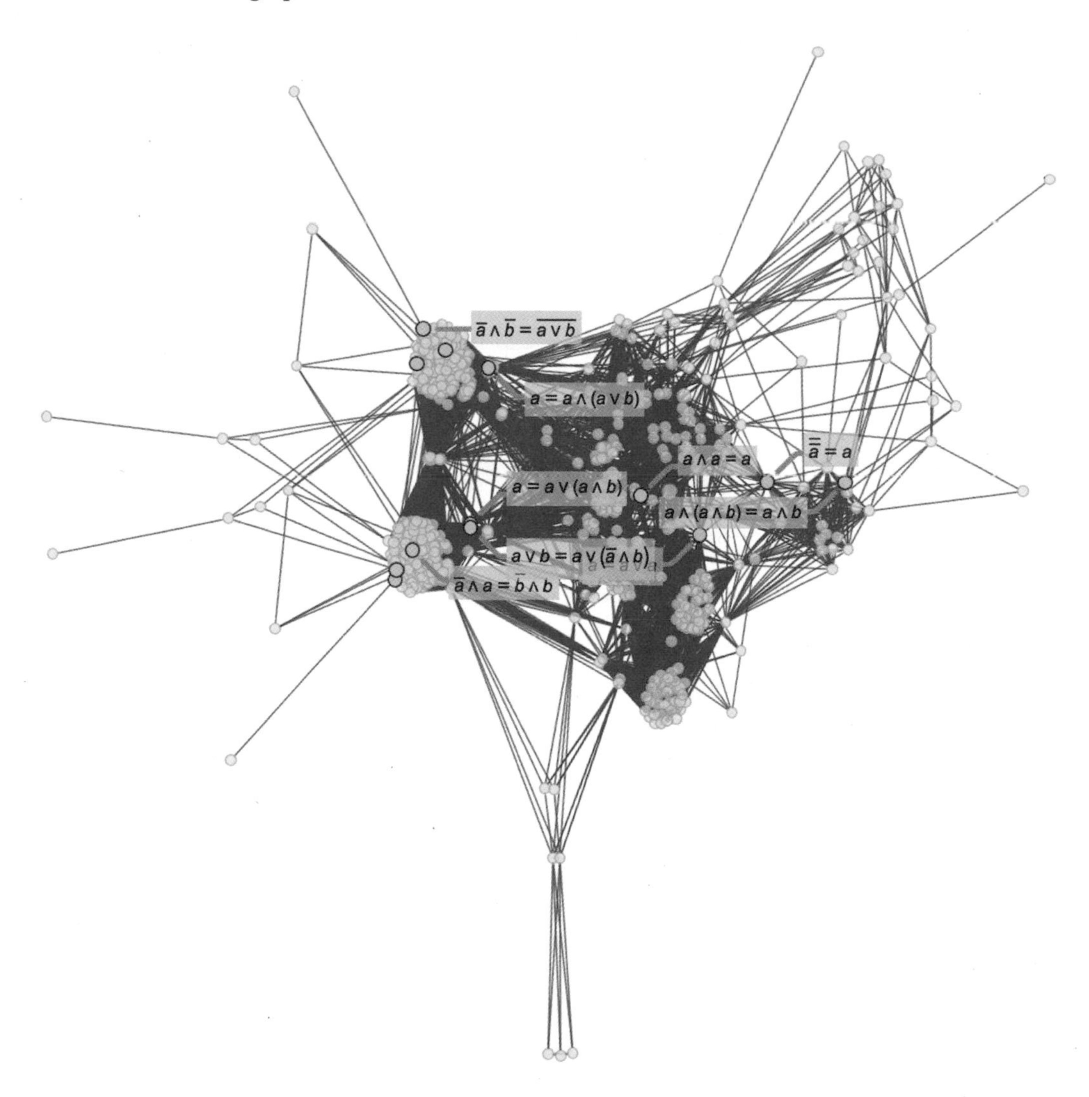

In doing this empirical metamathematics we're including only specific proofs rather than enumerating the whole entailment cone. We're also using only a specific axiom system. And even beyond this, we're using specific operators to write our statements in Boolean algebra.

In a sense each of these choices represents a particular "metamathematical coordinatization"—or particular reference frame or slice that we're sampling in the ruliad.

For example, in what we've done above we've built up statements from And, Or and Not. But we can just as well use any other functionally complete sets of operators, such as the following (here each shown representing a few specific Boolean expressions):

	Or[p, q]	Nand[p, q]	Xor[p, q]	Implies[p, q]	Or[p, And[q, r]]
And $\wedge$ Or $\vee$ Not —	$p\vee q$	$\overline{p\wedge q}$	$\overline{p\wedge q}\wedge(p\vee q)$	$q\vee\overline{p}$	$p\vee(q\wedge r)$
And $\wedge$ Not —	$\overline{\overline{p}\wedge\overline{q\wedge q}}$	$\overline{p\wedge q}$	$\overline{p\wedge q}\wedge\overline{\overline{p}\wedge\overline{q}}$	$\overline{p\wedge\overline{q}}$	$\overline{\overline{p}\wedge\overline{q\wedge r}}$
Or $\vee$ Not —	$p\vee q$	$\overline{p}\vee\overline{q}$	$\overline{p\vee\overline{q}}\vee\overline{q\vee\overline{p}}$	$q\vee\overline{p}$	$p\vee\overline{\overline{q}\vee\overline{r}}$
Implies $\Rightarrow$ Not —	$\overline{p}\Rightarrow q$	$p\Rightarrow\overline{q}$	$(p\Rightarrow q)\Rightarrow\overline{q\Rightarrow p}$	$p\Rightarrow q$	$(q\Rightarrow\overline{r})\Rightarrow p$
Xor $\veebar$ Implies $\Rightarrow$	$(p\Rightarrow q)\Rightarrow q$	$p\Rightarrow p\veebar q$	$p\veebar q$	$p\Rightarrow q$	$(q\Rightarrow(r\Rightarrow p))\Rightarrow p$
Nand $\circ$ Not —	$\overline{p}\circ\overline{q}$	$p\circ q$	$(p\circ\overline{q})\circ(q\circ\overline{p})$	$p\circ\overline{q}$	$(q\circ r)\circ\overline{p}$
Nand $\circ$	$(p\circ p)\circ(q\circ q)$	$p\circ q$	$(p\circ(p\circ q))\circ(q\circ(p\circ p))$	$p\circ(p\circ q)$	$(p\circ p)\circ(q\circ r)$
Nor $\circ$	$(p\circ q)\circ(p\circ q)$	$(p\circ(p\circ p))\circ((p\circ p)\circ(q\circ q))$	$(p\circ q)\circ((p\circ p)\circ(q\circ q))$	$(p\circ(p\circ p))\circ(q\circ(p\circ p))$	$(p\circ q)\circ(p\circ r)$

For each set of operators, there are different axiom systems that can be used. And for each axiom system there will be different proofs. Here are a few examples of axiom systems with a few different sets of operators—in each case giving a proof of the law of double negation (which has to be stated differently for different operators):

operators	*axioms*	*statement*	*proof*	*steps*
And $\wedge$ Or $\vee$ Not —	$a\wedge b=b\wedge a$	$\overline{\overline{p}}=p$		19
	$a\wedge(b\vee c)=(a\wedge b)\vee(a\wedge c)$			
	$a\vee(b\wedge\overline{b})=a$			
	$a\wedge(b\vee\overline{b})=a$			
	$a\vee b=b\vee a$			
	$a\vee(b\wedge c)=(a\vee b)\wedge(a\vee c)$			
Or $\vee$ Not —	$a\vee(b\vee c)=(a\vee b)\vee c$	$\overline{\overline{p}}=p$		24
	$a\vee b=b\vee a$			
	$\overline{\overline{a}\vee b}\vee\overline{\overline{a}\vee\overline{b}}=a$			
Or $\vee$ Not —	$\overline{\overline{c\vee b}\vee\overline{a}}\vee\overline{\overline{\overline{d}\vee d}\vee(\overline{a}\vee c)}=a$	$\overline{\overline{p}}=p$	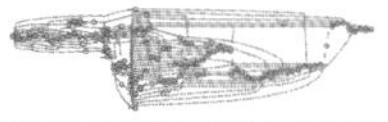	111
Nand $\circ$	$(a\circ a)\circ(a\circ a)=a$	$(p\circ p)\circ(p\circ p)=p$		3
	$a\circ(b\circ(b\circ b))=a\circ a$			
	$(a\circ(b\circ c))\circ(a\circ(b\circ c))=((b\circ b)\circ a)\circ((c\circ c)\circ a)$			
Nand $\circ$	$(a\circ a)\circ(a\circ b)=a$	$(p\circ p)\circ(p\circ p)=p$		3
	$a\circ(a\circ b)=a\circ(b\circ b)$			
	$a\circ(a\circ(b\circ c))=b\circ(b\circ(a\circ c))$			

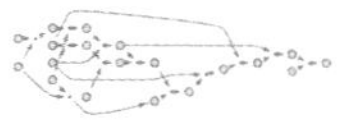

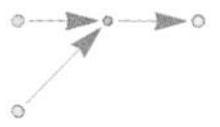

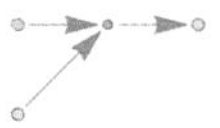

operators	*axioms*	*statement*	*proof*	*steps*
Nand $\circ$	$a \circ (b \circ (a \circ c)) = ((c \circ b) \circ b) \circ a$ $(a \circ a) \circ (b \circ a) = a$	$(p \circ p) \circ (p \circ p) = p$		3
Nand $\circ$	$a \circ b = b \circ a$ $(a \circ b) \circ (a \circ (b \circ c)) = a$	$(p \circ p) \circ (p \circ p) = p$		8
Nand $\circ$	$((b \circ c) \circ a) \circ (b \circ ((b \circ a) \circ b)) = a$	$(p \circ p) \circ (p \circ p) = p$		54
Nand $\circ$	$(a \circ ((c \circ a) \circ a)) \circ (c \circ (b \circ a)) = c$	$(p \circ p) \circ (p \circ p) = p$		12

Boolean algebra (or, equivalently, propositional logic) is a somewhat desiccated and thin example of mathematics. So what do we find if we do empirical metamathematics on other areas?

Let's talk first about geometry—for which Euclid's *Elements* provided the very first large-scale historical example of an axiomatic mathematical system. The *Elements* started from 10 axioms (5 "postulates" and 5 "common notions"), then gave 465 theorems.

Each theorem was proved from previous ones, and ultimately from the axioms. Thus, for example, the "proof graph" (or "theorem dependency graph") for Book 1, Proposition 5 (which says that angles at the base of an isosceles triangle are equal) is:

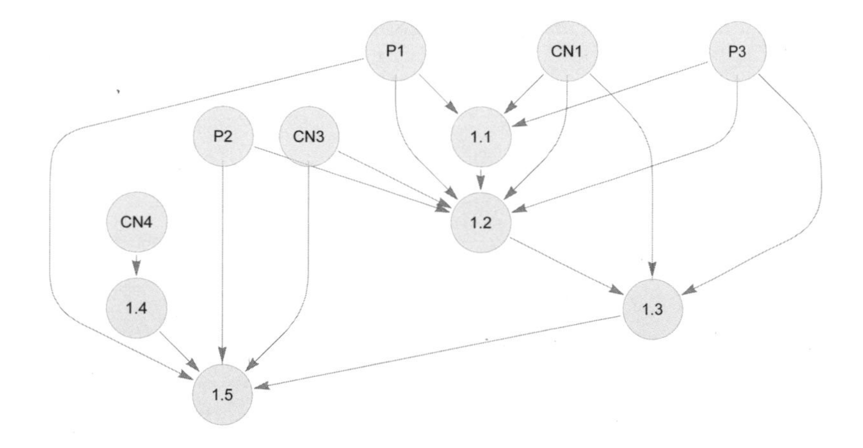

One can think of this as a coarse-grained version of the proof graphs we've used before (which are themselves in turn "slices" of the entailment graph)—in which each node shows how a collection of "input" theorems (or axioms) entails a new theorem.

Here's a slightly more complicated example (Book 1, Proposition 48) that ultimately depends on all 10 of the original axioms:

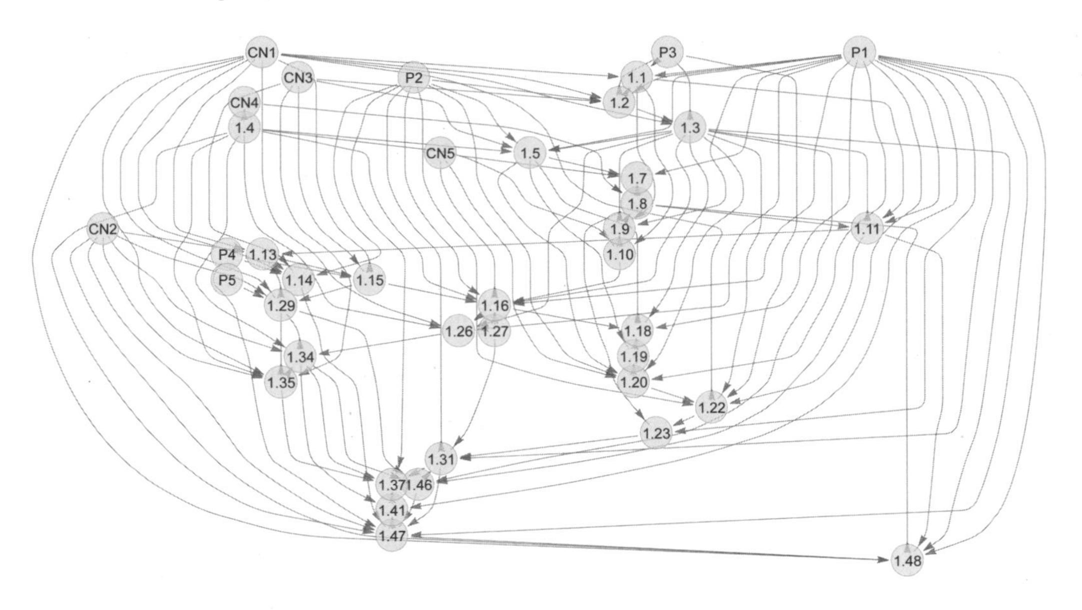

And here's the full graph for all the theorems in Euclid's *Elements*:

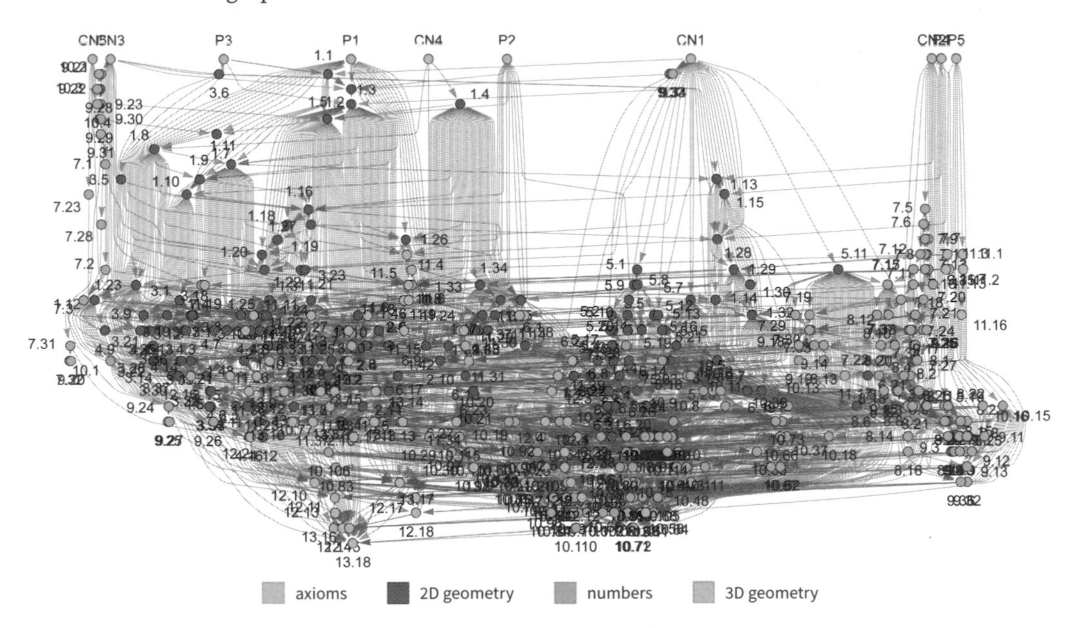

Of the 465 theorems here, 255 (i.e. 55%) depend on all 10 axioms. (For the much smaller number of notable theorems of Boolean algebra above we found that 64% depended on all 6 of our stated axioms.) And the general connectedness of this graph in effect reflects the idea that Euclid's theorems represent a coherent body of connected mathematical knowledge.

The branchial graph gives us an idea of how the theorems are "laid out in metamathematical space":

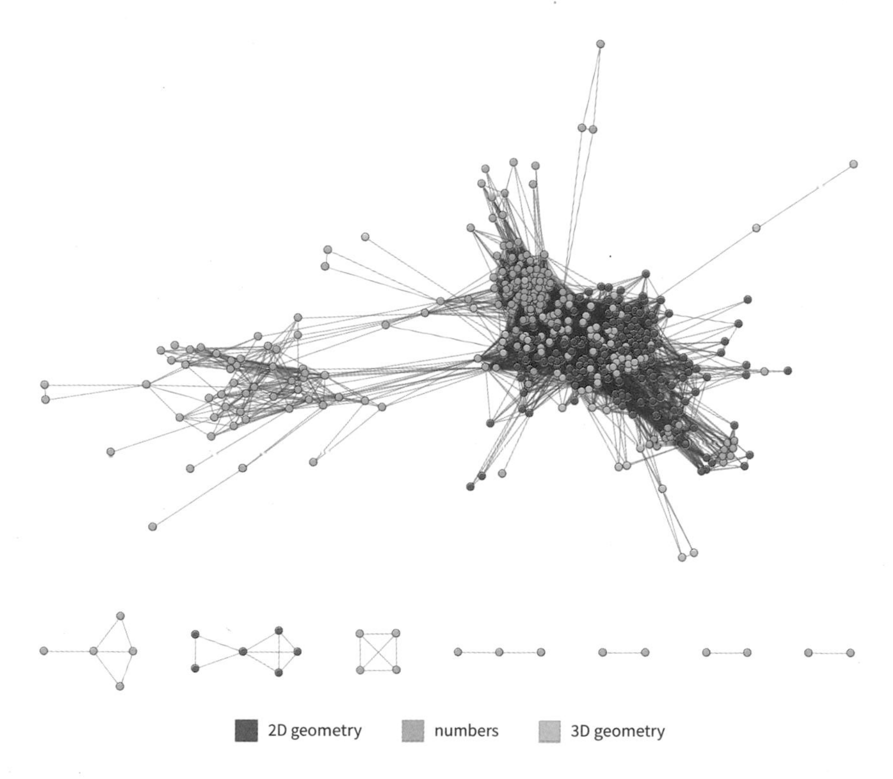

One thing we notice is that theorems about different areas—shown here in different colors—tend to be separated in metamathematical space. And in a sense the seeds of this separation are already evident if we look "textually" at how theorems in different books of Euclid's *Elements* refer to each other:

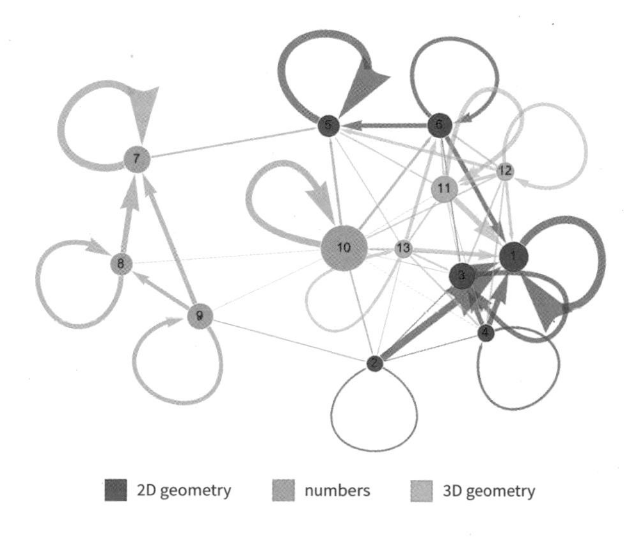

Looking at the overall dependence of one theorem on others in effect shows us a very coarse form of entailment. But can we go to a finer level—as we did above for Boolean algebra? As a first step, we have to have an explicit symbolic representation for our theorems. And beyond that, we have to have a formal axiom system that describes possible transformations between these.

At the level of "whole theorem dependency" we can represent the entailment of Euclid's Book 1, Proposition 1 from axioms as:

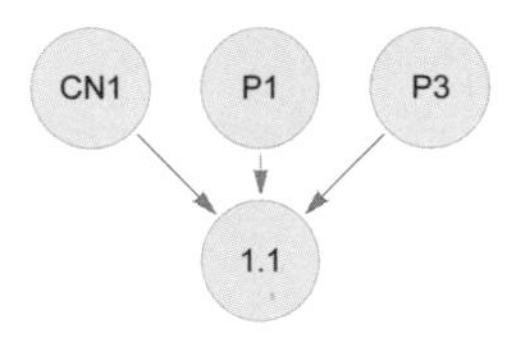

But if we now use the full, formal axiom system for geometry that we discussed in a previous section we can use automated theorem proving to get a full proof of Book 1, Proposition 1:

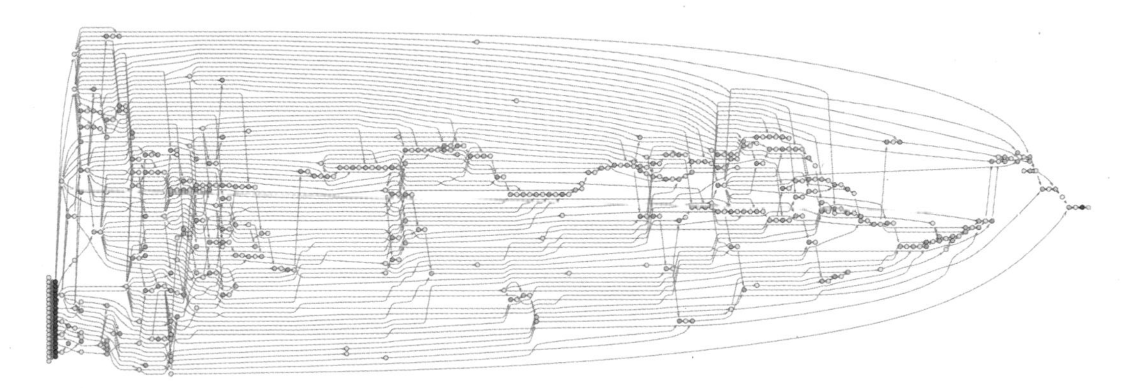

In a sense this is "going inside" the theorem dependency graph to look explicitly at how the dependencies in it work. And in doing this we see that what Euclid might have stated in words in a sentence or two is represented formally in terms of hundreds of detailed intermediate lemmas. (It's also notable that whereas in Euclid's version, the theorem depends only on 3 out of 10 axioms, in the formal version the theorem depends on 18 out of 20 axioms.)

How about for other theorems? Here is the theorem dependency graph from Euclid's *Elements* for the Pythagorean theorem (which Euclid gives as Book 1, Proposition 47):

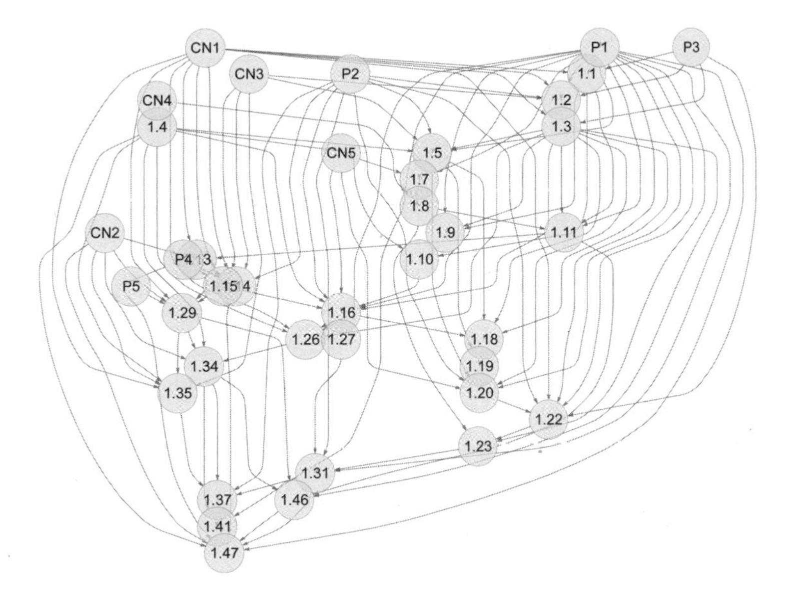

The theorem depends on all 10 axioms, and its stated proof goes through 28 intermediate theorems (i.e. about 6% of all theorems in the *Elements*). In principle we can "unroll" the proof dependency graph to see directly how the theorem can be "built up" just from copies of the original axioms. Doing a first step of unrolling we get:

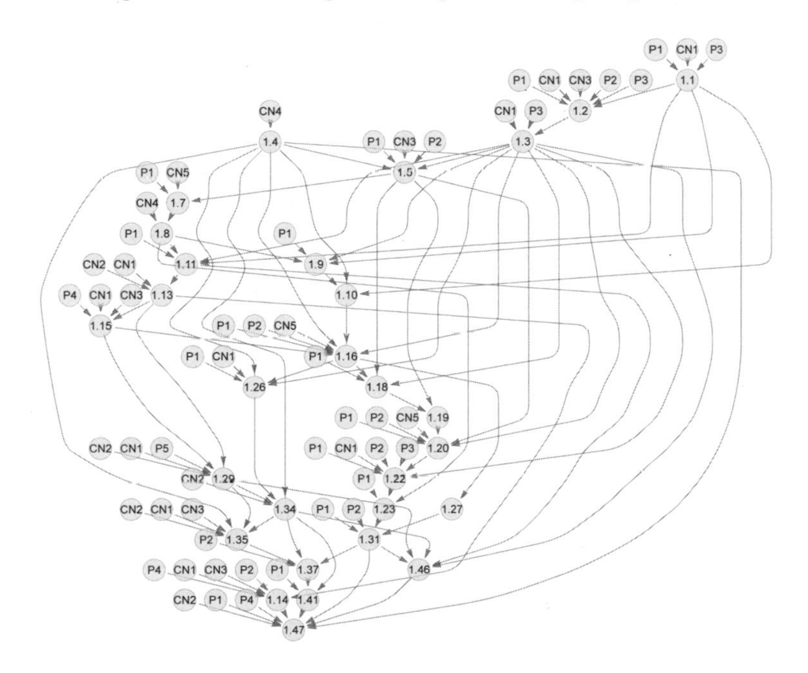

And "flattening everything out" so that we don't use any intermediate lemmas but just go back to the axioms to "re-prove" everything we can derive the theorem from a "proof tree" with the following number of copies of each axiom (and a certain "depth" to reach that axiom):

axiom	CN1	CN2	CN3	CN4	CN5	P1	P2	P3	P4	P5
depth	2	1	2	2	4	1	2	3	1	3
copies	1235	81	310	217	82	1094	321	1096	27	10

So how about a more detailed and formal proof? We could certainly in principle construct this using the axiom system we previously discussed.

But an important general point is that the thing we in practice call "the Pythagorean theorem" can actually be set up in all sorts of different axiom systems. And as an example let's consider setting it up in the main actual axiom system that working mathematicians typically imagine they're (usually implicitly) using, namely ZFC set theory.

Conveniently, the Metamath formalized math system has accumulated about 40,000 theorems across mathematics, all with hand-constructed proofs based ultimately on ZFC set theory. And within this system we can find the theorem dependency graph for the Pythagorean theorem:

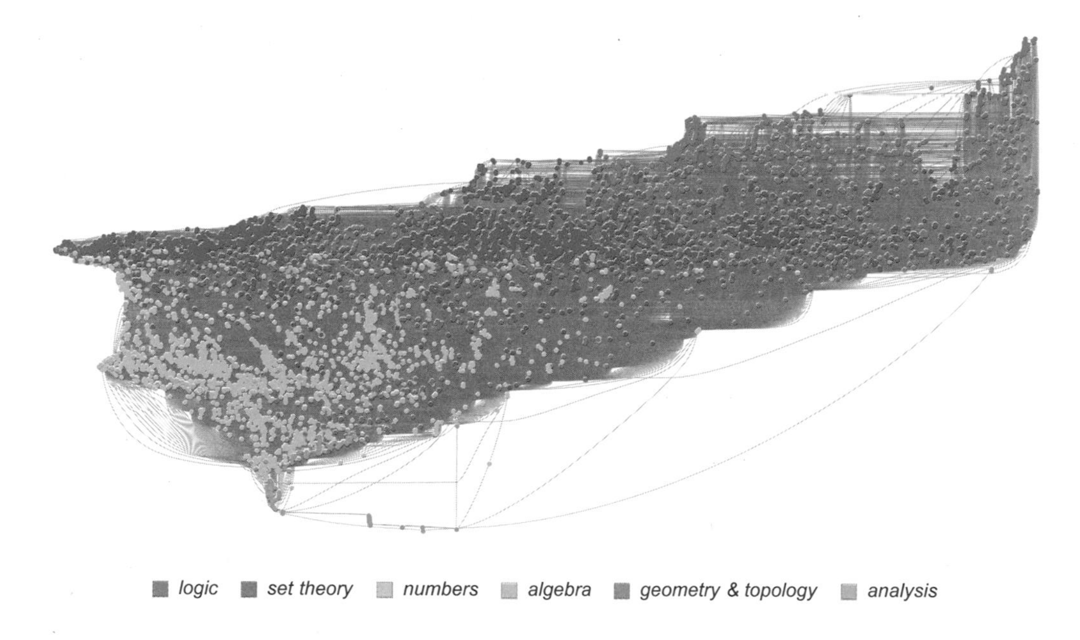

logic   set theory   numbers   algebra   geometry & topology   analysis

Altogether it involves 6970 intermediate theorems, or about 18% of all theorems in Metamath—including ones from many different areas of mathematics. But how does it ultimately depend on the axioms? First, we need to talk about what the axioms actually are. In addition to "pure ZFC set theory", we need axioms for (predicate) logic, as well as ones that define real and complex numbers. And the way things are set up in Metamath's "set.mm" there are (essentially) 49 basic axioms (9 for pure set theory, 15 for logic and 25 related to numbers). And much as in Euclid's *Elements* we found that the Pythagorean theorem depended on all the axioms, so now here we find that the Pythagorean theorem depends on 48 of the 49 axioms—with the only missing axiom being the Axiom of Choice.

Just like in the Euclid's *Elements* case, we can imagine "unrolling" things to see how many copies of each axiom are used. Here are the results—together with the "depth" to reach each axiom:

Modus ponens	6.0×10^{36}	2
Axiom of Simplification	1.8×10^{36}	3
Frege's Axiom	1.9×10^{36}	4
Principle of Transposition	2.7×10^{35}	7
Rule of Generalization	1.3×10^{32}	6
Axiom of Quantified Implication	7.1×10^{31}	6
Axiom of Distinctness	4.2×10^{31}	4
Axiom of Existence	3.8×10^{31}	8
Axiom of Equality	5.2×10^{31}	8
Binary predicate left equality	4.4×10^{24}	12
Binary predicate right equality	1.1×10^{30}	10
Axiom of Quantified Negation	1.9×10^{30}	10
Axiom of Quantifier Commutation	1.8×10^{29}	9
Axiom of Substitution	2.2×10^{30}	9
Axiom of Quantified Equality	5.2×10^{29}	9
Axiom of Extensionality	5.5×10^{29}	6
Axiom of Replacement	2.8×10^{10}	10
Axiom of Separation	4.2×10^{26}	7
Null Set Axiom	2.7×10^{26}	6
Axiom of Power Sets	5.8×10^{23}	7
Axiom of Pairing	4.1×10^{26}	10
Axiom of Union	2.1×10^{24}	8
Axiom of Infinity	2.4×10^{9}	7
Complex numbers form set	1.8×10^{16}	6

Real numbers subset of complex	3.4×10^{22}	3
1 is a complex number	2.1×10^{22}	3
i is a complex number	1.0×10^{22}	3
Closure of complex addition	1.0×10^{22}	3
Closure of real addition	2.1×10^{22}	5
Closure of complex multiplication	1.1×10^{22}	3
Closure of real multiplication	5.7×10^{21}	5
Commutativity of complex multiplication	6.3×10^{21}	4
Associativity of complex addition	2.1×10^{21}	5
Associativity of complex multiplication	3.7×10^{21}	6
Distributivity for complex numbers	3.7×10^{21}	5
$i^2=-1$	7.8×10^{21}	4
1 and 0 are distinct	5.0×10^{21}	4
1 is multiplicative identity	1.6×10^{21}	6
Negative reals exist	5.9×10^{21}	4
Reciprocals of nonzero reals exist	6.2×10^{21}	4
Complex numbers expressible by 2 reals	6.1×10^{21}	5
Ordering on reals has strict trichotomy	7.6×10^{21}	6
Ordering on reals is transitive	1.9×10^{21}	7
Ordering property of addition on reals	5.7×10^{21}	7
Product of 2 positive reals is positive	2.8×10^{17}	5
Supremum property for set of reals	3.0×10^{13}	10
Addition applies to complex numbers	1.4×10^{7}	10
Multiplication applies to complex numbers	1.4×10^{7}	9

And, yes, the numbers of copies of most of the axioms required to establish the Pythagorean theorem are extremely large.

There are several additional wrinkles that we should discuss. First, we've so far only considered overall theorem dependency—or in effect "coarse-grained entailment". But the Metamath system ultimately gives complete proofs in terms of explicit substitutions (or, effectively, bisubstitutions) on symbolic expressions. So, for example, while the first-level "whole-theorem-dependency" graph for the Pythagorean theorem is

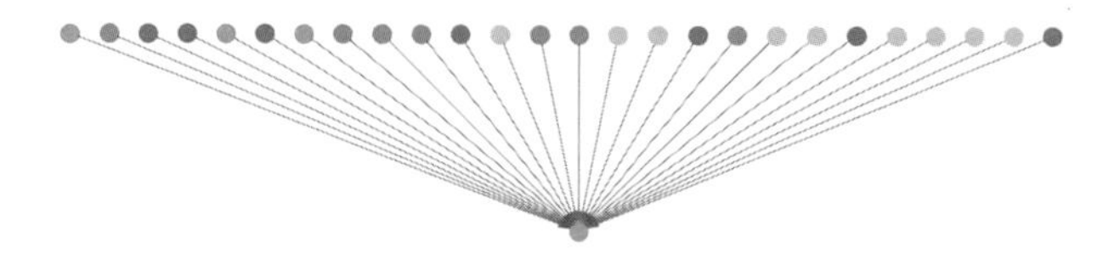

the full first-level entailment structure based on the detailed proof is (where the black vertices indicate "internal structural elements" in the proof—such as variables, class specifications and "inputs"):

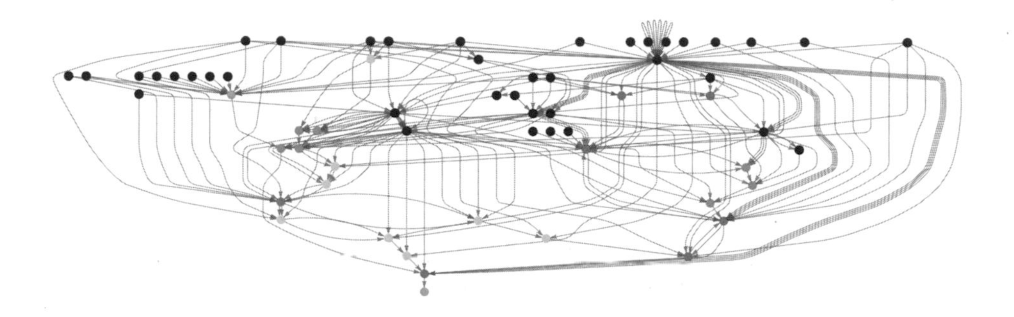

Another important wrinkle has to do with the concept of definitions. The Pythagorean theorem, for example, refers to squaring numbers. But what is squaring? What are numbers? Ultimately all these things have to be defined in terms of the "raw data structures" we're using.

In the case of Boolean algebra, for example, we could set things up just using Nand (say denoted $\circ$), but then we could define And and Or in terms of Nand (say as $(p \circ q) \circ (p \circ q)$ and $(p \circ p) \circ (q \circ q)$ respectively). We could still write expressions using And and Or—but with our definitions we'd immediately be able to convert these to pure Nands. Axioms—say about Nand—give us transformations we can use repeatedly to make derivations. But definitions are transformations we use "just once" (like macro expansion in programming) to reduce things to the point where they involve only constructs that appear in the axioms.

In Metamath's "set.mm" there are about 1700 definitions that effectively build up from "pure set theory" (as well as logic, structural elements and various axioms about numbers) to give the mathematical constructs one needs. So, for example, here is the definition dependency graph for addition ("+" or **Plus**):

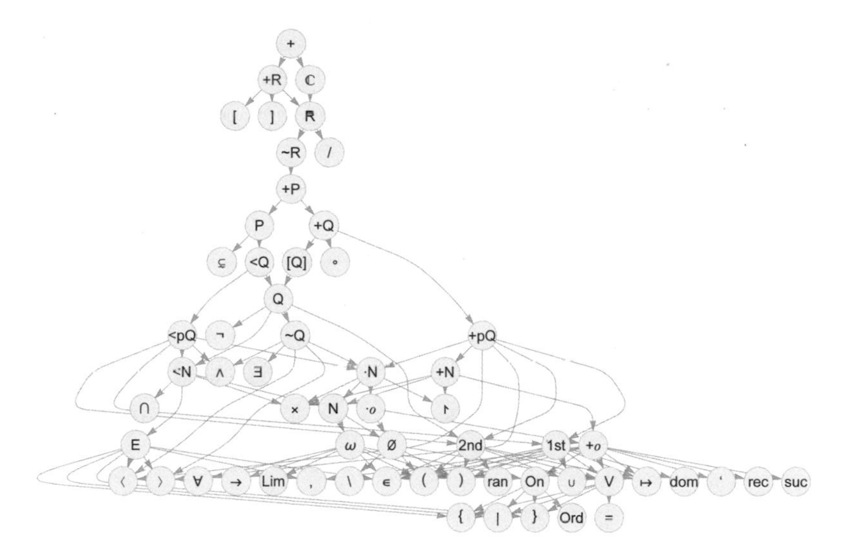

At the bottom are the basic constructs of logic and set theory—in terms of which things like order relations, complex numbers and finally addition are defined. The definition dependency graph for GCD, for example, is somewhat larger, though has considerable overlap at lower levels:

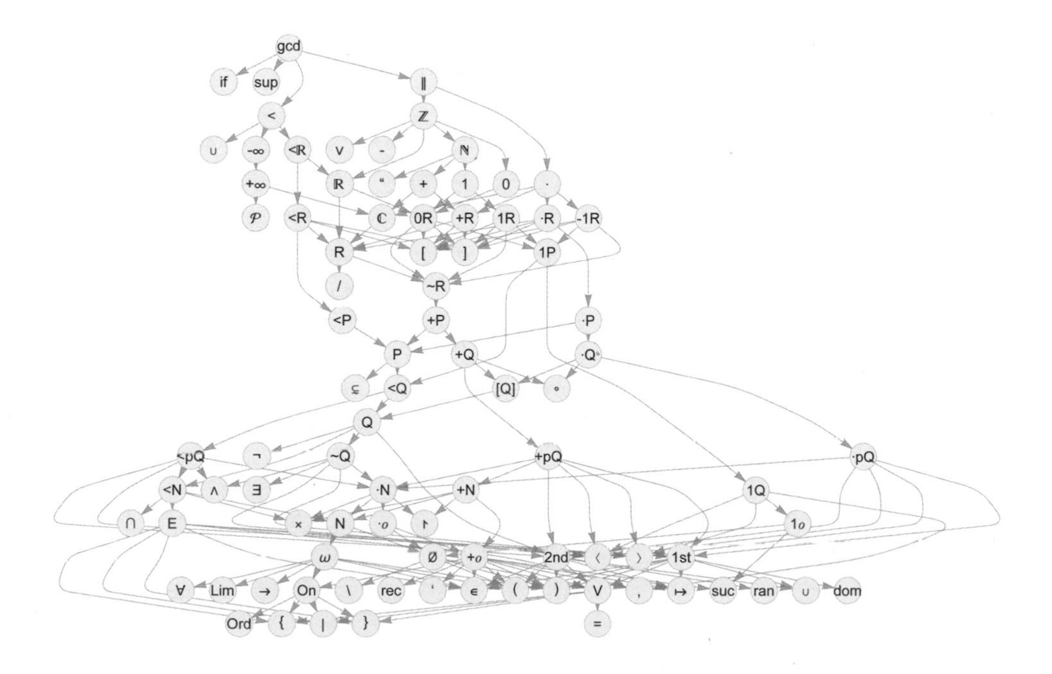

Different constructs have definition dependency graphs of different sizes—in effect reflecting their "definitional distance" from set theory and the underlying axioms being used:

construct	*distance*
First	12
∞	58
Plus	60
1	61
2	67
EvenQ	83
GCD	91
Power	92
PrimePi	102

construct	*distance*
Sinh	102
SymmetricGroup	104
π	105
Log	112
Abs	114
Eigenvalues	119
Zeta	123
Area	133
Det	191

In our physicalized approach to metamathematics, though, something like set theory is not our ultimate foundation. Instead, we imagine that everything is eventually built up from the raw ruliad, and that all the constructs we're considering are formed from what amount to configurations of emes in the ruliad. We discussed above how constructs like numbers and logic can be obtained from a combinator representation of the ruliad.

We can view the definition dependency graph shown previously as being an empirical example of how somewhat higher-level definitions can be built up. From a computer science perspective, we can think of it as being like a type hierarchy. From a physics perspective, it's as if we're starting from atoms, then building up to molecules and beyond.

It's worth pointing out, however, that even the top of the definition hierarchy in something like Metamath is still operating very much at an axiomatic kind of level. In the analogy we've been using, it's still for the most part "formulating math at the molecular dynamics level" not at the more human "fluid dynamics" level.

We've been talking about "the Pythagorean theorem". But even on the basis of set theory there are many different possible formulations one can give. In Metamath, for example, there is the pythag version (which is what we've been using), and there is also a (somewhat more general) pythi version. So how are these related? Here's their combined theorem dependency graph (or at least the first two levels in it)—with red indicating theorems used only in deriving pythag, blue indicating ones used only in deriving pythi, and purple indicating ones used in both:

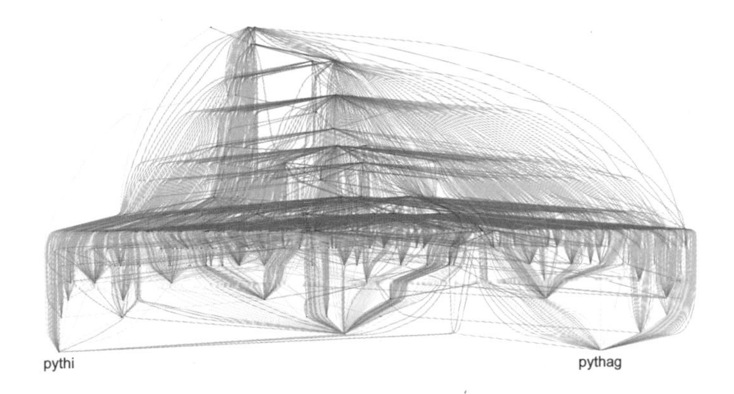

And what we see is there's a certain amount of "lower-level overlap" between the derivations of these variants of the Pythagorean theorem, but also some discrepancy—indicating a certain separation between these variants in metamathematical space.

So what about other theorems? Here's a table of some famous theorems from all over mathematics, sorted by the total number of theorems on which proofs of them formulated in Metamath depend—giving also the number of axioms and definitions used in each case:

	total	*axioms*	*definitions*
Schröder–Bernstein Theorem	1474	21	48
Principle of Mathematical Induction	1532	20	47
Cantor's Theorem	1719	21	55
Number of subsets of a set	1737	21	59
Infinitude of primes	3394	45	98
Non–denumerability of continuum	3508	45	101
Triangle Inequality	3626	44	106
Intermediate Value Theorem	3782	44	111
Irrationality of $\sqrt{2}$	3816	44	108
Bezout's Theorem	3856	44	111
Denumerability of rationals	3948	45	108
Binomial as number of subsets	3961	44	110
Euclid's GCD algorithm	3966	44	113
Formula for Pythagorean triples	4281	44	118
Sum of a geometric series	4459	46	121
Fermat's Little Theorem	4477	45	123
Fundamental Theorem of Arithmetic	4497	45	121
Binomial Theorem	4522	46	123
Lebesgue Integration Theorem	4543	46	127
Sum of an arithmetic series	4545	46	123
Divisibility by 3 Rule	4561	46	129
Principle of Inclusion/Exclusion	4651	46	124
Sum of triangular–number reciprocals	4662	46	126
Sum of kth powers	4698	46	124
Four–Squares Theorem	4710	45	125
Lagrange's Theorem	4738	46	141
Divergence of harmonic series	4868	46	127
Ramsey's Theorem	4916	46	130
Königsberg bridge theorem	4996	45	145
Polynomial Factor Theorem	5015	47	133
De Moivre's Theorem	5217	48	133

	total	*axioms*	*definitions*
Divergence of inverse prime series	5368	46	135
Sylow's Theorem	5445	46	153
Wilson's Theorem	5544	48	175
Cramer's Rule	6192	47	218
Mean Value Theorem	6438	48	216
Law of Quadratic Reciprocity	6448	48	214
L'Hôpital's Rule	6498	48	217
Cayley–Hamilton Theorem	6551	47	230
Friendship Graph Theorem	6853	47	181
Ptolemy's Theorem	7103	48	229
Liouville's Theorem	7179	48	238
Primes 4k+1 are sums of 2 squares	7246	48	243
Taylor's Theorem	7253	48	249
Law of Cosines	7269	48	231
Pythagorean Theorem	7271	48	231
Heron's Formula	7285	48	231
Sum of angles of a triangle	7295	48	231
Isosceles Triangle Theorem	7309	48	231
Fundamental Theorem of Calculus	7338	49	223
Product of segments of chords	7361	48	231
Solution of quartic equations	7461	48	233
Solution of cubic equations	7490	48	233
Value of $\zeta(2)$	7522	48	238
Fundamental Theorem of Algebra	7542	48	238
Perfect Number Theorem	7579	48	237
Arithmetic–geometric mean inequality	7750	48	253
Leibniz' series for π	7870	48	238
Bertrand's Postulate	7945	48	240
Birthday Problem probability	8001	48	237
Prime Number Theorem	8477	48	248
Dirichlet's Theorem	10006	48	312

The Pythagorean theorem (here the pythi formulation) occurs solidly in the second half. Some of the theorems with the fewest dependencies are in a sense very structural theorems. But it's interesting to see that theorems from all sorts of different areas soon start appearing, and then are very much mixed together in the remainder of the list. One might have thought that theorems involving "more sophisticated concepts" (like Ramsey's theorem) would appear later than "more elementary" ones (like the sum of angles of a triangle). But this doesn't seem to be true.

There's a distribution of what amount to "proof sizes" (or, more strictly, theorem dependency sizes)—from the Schröder–Bernstein theorem which relies on less than 4% of all theorems, to Dirichlet's theorem that relies on 25%:

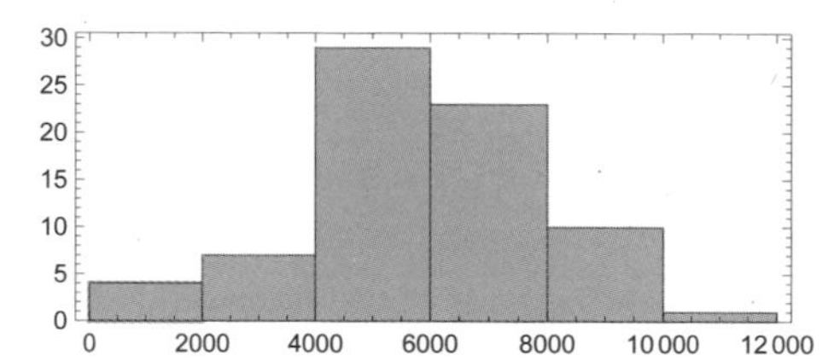

If we look not at "famous" theorems, but at all theorems covered by Metamath, the distribution becomes broader, with many short-to-prove "glue" or essentially "definitional" lemmas appearing:

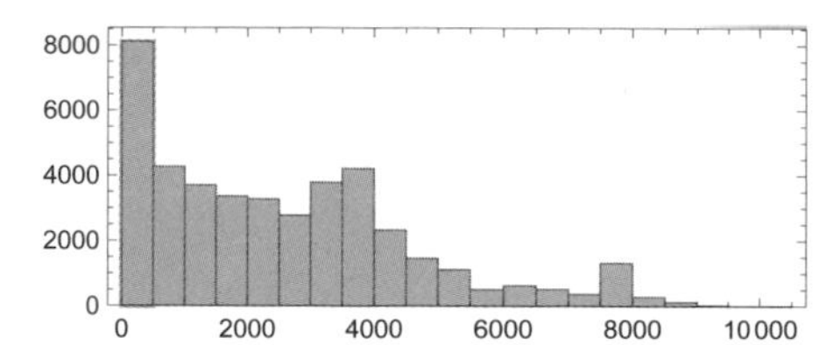

But using the list of famous theorems as an indication of the "math that mathematicians care about" we can conclude that there is a kind of "metamathematical floor" of results that one needs to reach before "things that we care about" start appearing. It's a bit like the situation in our Physics Project—where the vast majority of microscopic events that happen in the universe seem to be devoted merely to knitting together the structure of space, and only "on top of that" can events which can be identified with things like particles and motion appear.

And indeed if we look at the "prerequisites" for different famous theorems, we indeed find that there is a large overlap (indicated by lighter colors)—supporting the impression that in a sense one first has "knit together metamathematical space" and only then can one start generating "interesting theorems":

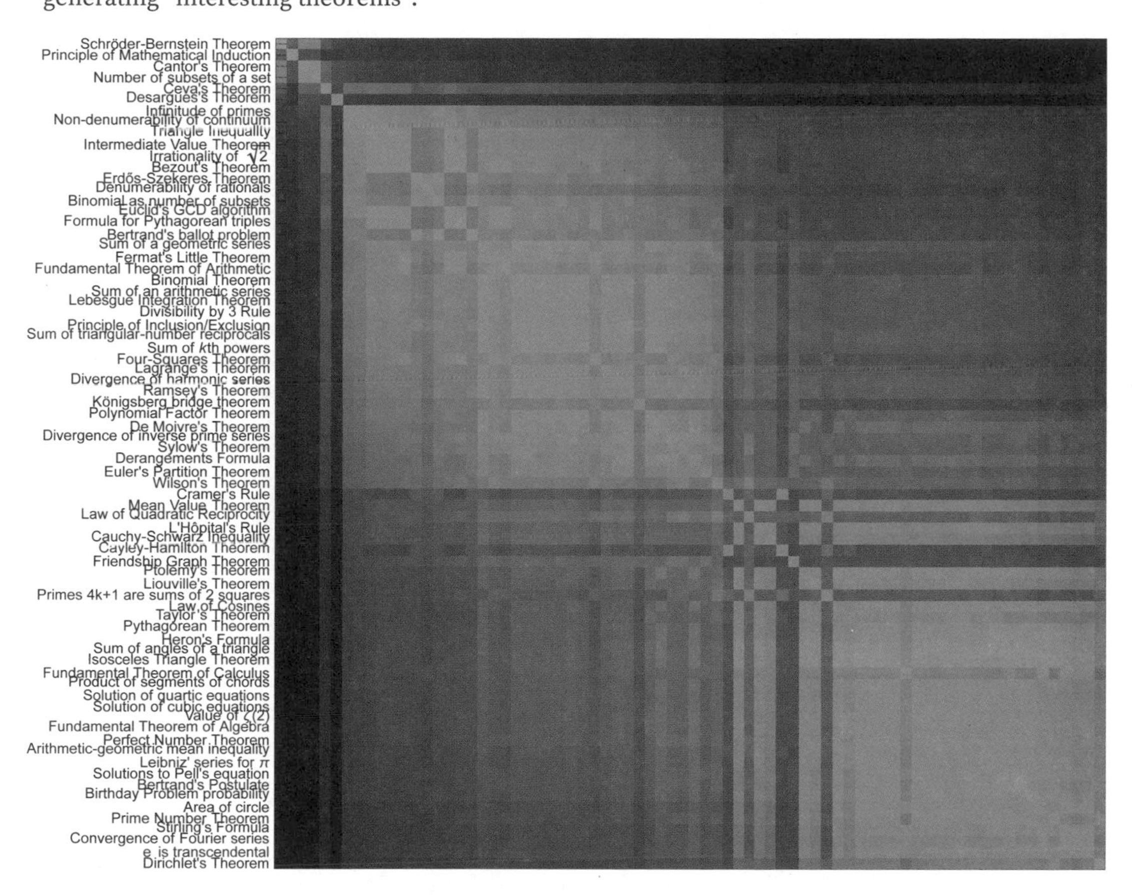

Another way to see "underlying overlap" is to look at what axioms different theorems ultimately depend on (the colors indicate the "depth" at which the axioms are reached):

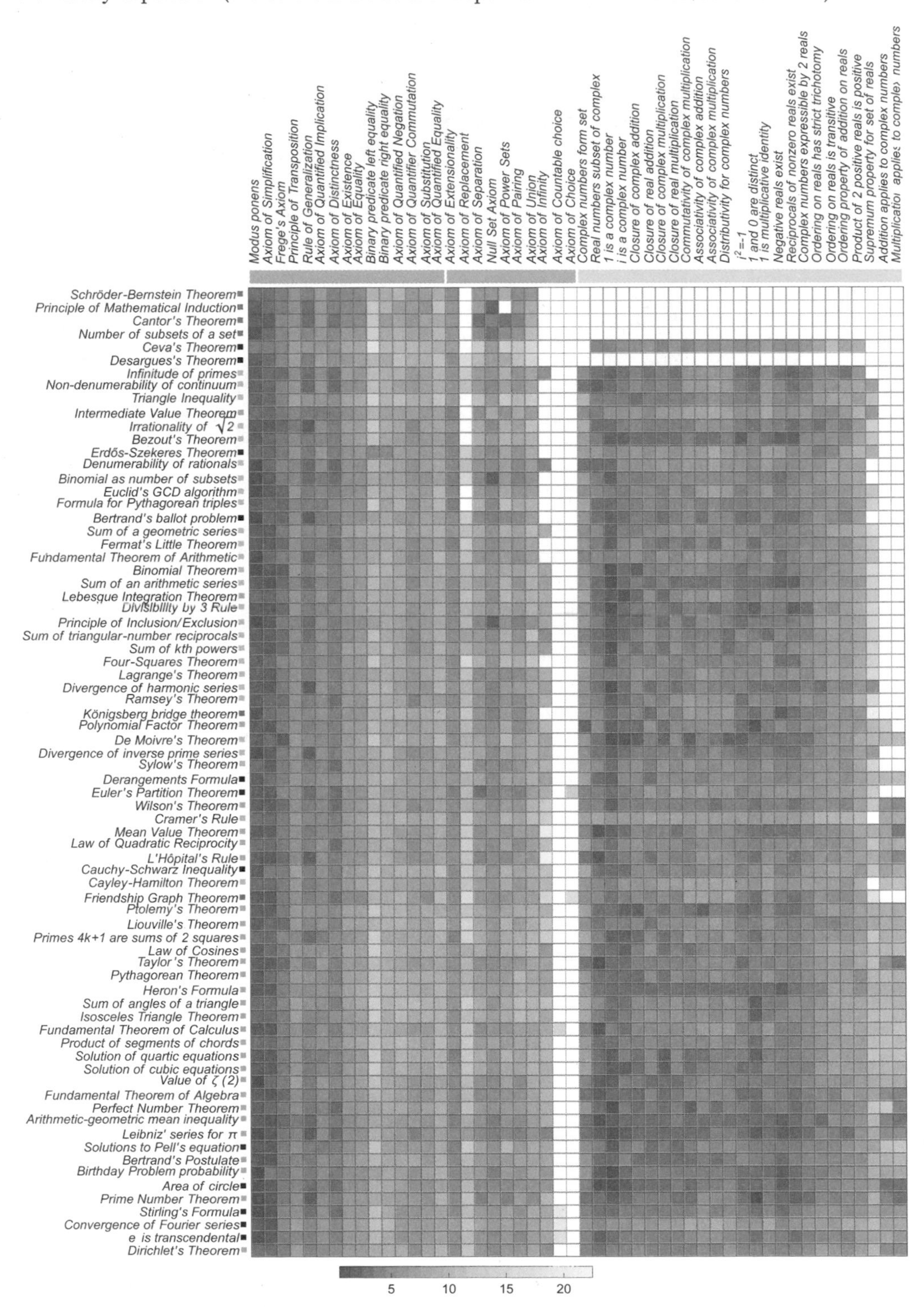

The theorems here are again sorted in order of "dependency size". The "very-set-theoretic" ones at the top don't depend on any of the various number-related axioms. And quite a few "integer-related theorems" don't depend on complex number axioms. But otherwise, we see that (at least according to the proofs in set.mm) most of the "famous theorems" depend on almost all the axioms. The only axiom that's rarely used is the Axiom of Choice—on which only things like "analysis-related theorems" such as the Fundamental Theorem of Calculus depend.

If we look at the "depth of proof" at which axioms are reached, there's a definite distribution:

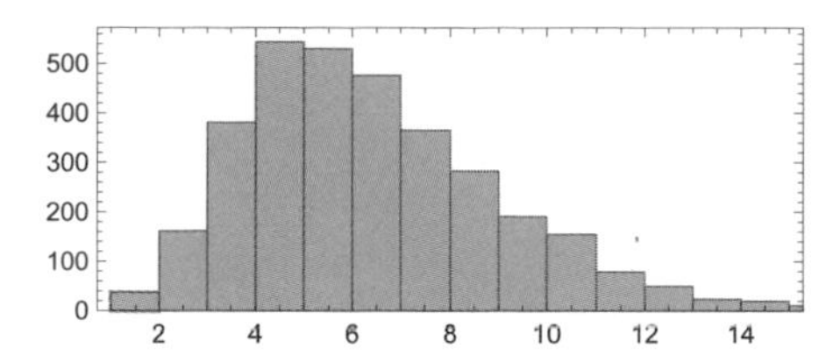

And this may be about as robust as any a "statistical characteristic" of the sampling of metamathematical space corresponding to mathematics that is "important to humans". If we were, for example, to consider all possible theorems in the entailment cone we'd get a very different picture. But potentially what we see here may be a characteristic signature of what's important to a "mathematical observer like us".

Going beyond "famous theorems" we can ask, for example, about all the 42,000 or so identified theorems in the Metamath set.mm collection. Here's a rough rendering of their theorem dependency graph, with different colors indicating theorems in different fields of math (and with explicit edges removed):

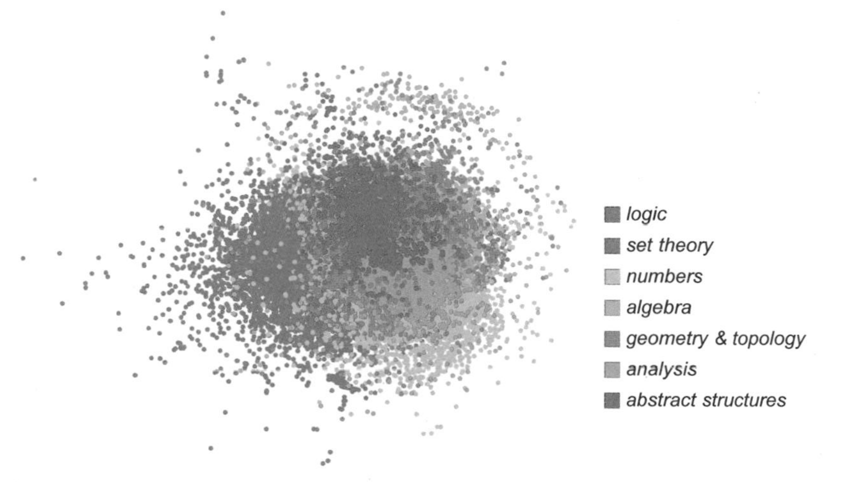

There's some evidence of a certain overall uniformity, but we can see definite "patches of metamathematical space" dominated by different areas of mathematics. And here's what happens if we zoom in on the central region, and show where famous theorems lie:

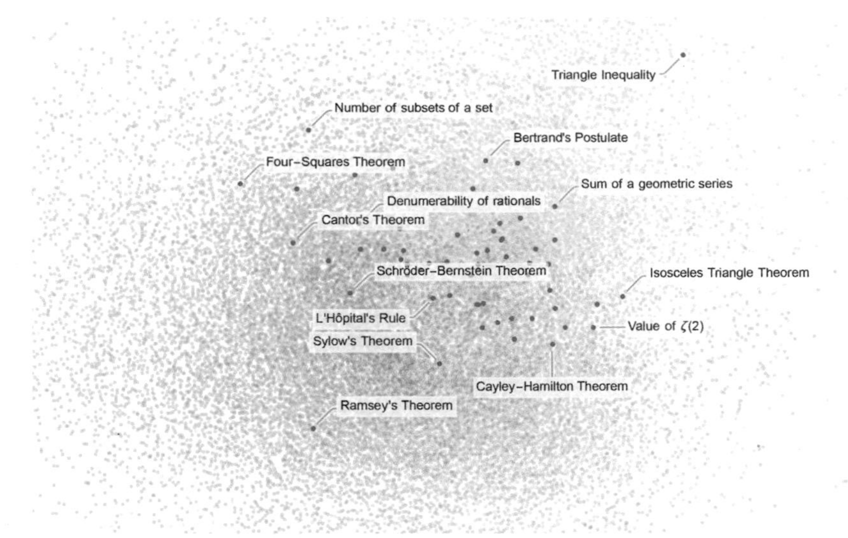

A bit like we saw for the named theorems of Boolean algebra clumps of famous theorems appear to somehow "stake out their own separate metamathematical territory". But notably the famous theorems seem to show some tendency to congregate near "borders" between different areas of mathematics.

To get more of a sense of the relation between these different areas, we can make what amounts to a highly coarsened branchial graph, effectively laying out whole areas of mathematics in metamathematical space, and indicating their cross-connections:

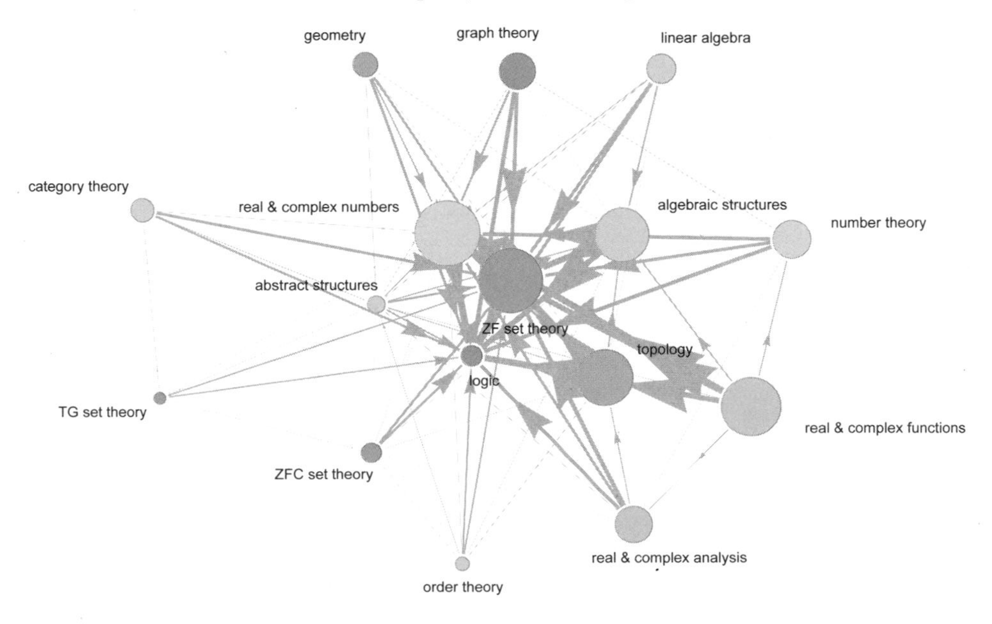

We can see "highways" between certain areas. But there's also a definite "background entanglement" between areas, reflecting at least a certain background uniformity in metamathematical space, as sampled with the theorems identified in Metamath.

It's not the case that all these areas of math "look the same"—and for example there are differences in their distributions of theorem dependency sizes:

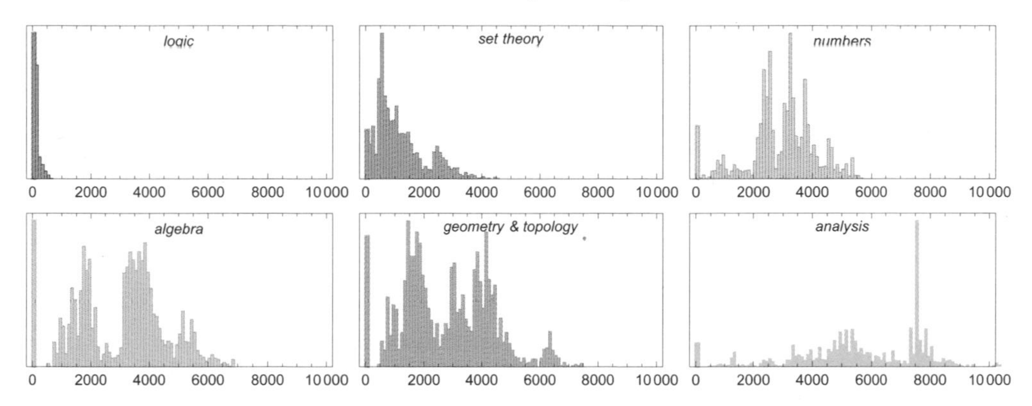

In areas like algebra and number theory, most proofs are fairly long, as revealed by the fact that they have many dependencies. But in set theory there are plenty of short proofs, and in logic all the proofs of theorems that have been included in Metamath are short.

What if we look at the overall dependency graph for all theorems in Metamath? Here's the adjacency matrix we get:

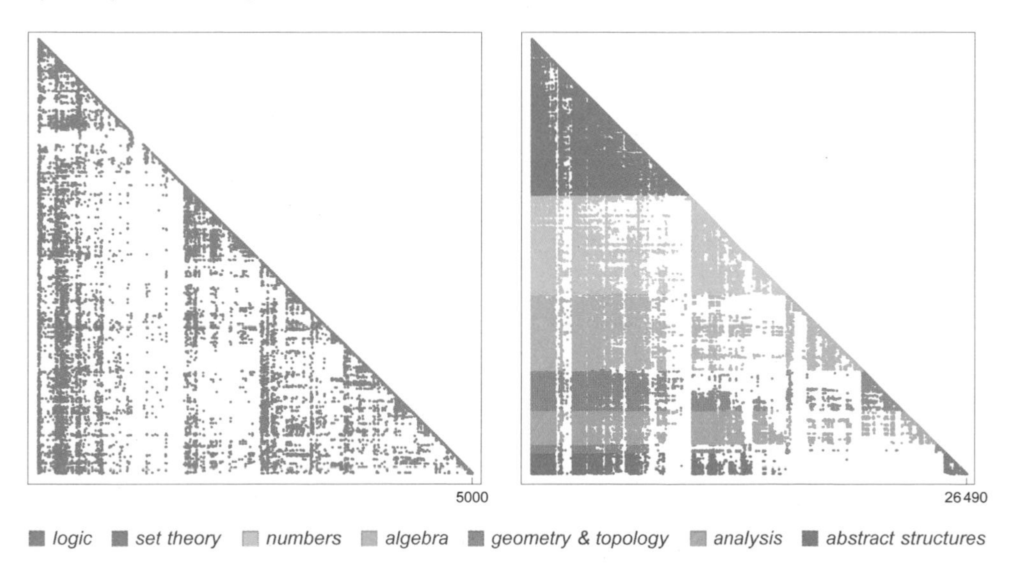

The results are triangular because theorems in the Metamath database are arranged so that later ones only depend on earlier ones. And while there's considerable patchiness visible, there still seems to be a certain overall background level of uniformity.

In doing this empirical metamathematics we're sampling metamathematical space just through particular "human mathematical settlements" in it. But even from the distribution of these "settlements" we potentially begin to see evidence of a certain background uniformity in metamathematical space.

Perhaps in time as more connections between different areas of mathematics are found human mathematics will gradually become more "uniformly settled" in metamathematical space—and closer to what we might expect from entailment cones and ultimately from the raw ruliad. But it's interesting to see that even with fairly basic empirical metamathematics—operating on a current corpus of human mathematical knowledge—it may already be possible to see signs of some features of physicalized metamathematics.

One day, no doubt, we'll be able do experiments in physics that take our "parsing" of the physical universe in terms of things like space and time and quantum mechanics—and reveal "slices" of the raw ruliad underneath. But perhaps something similar will also be possible in empirical metamathematics: to construct what amounts to a metamathematical microscope (or telescope) through which we can see aspects of the ruliad.

27 | Invented or Discovered? How Mathematics Relates to Humans

It's an old and oft-asked question: is mathematics ultimately something that is invented, or something that is discovered? Or, put another way: is mathematics something arbitrarily set up by us humans, or something inevitable and fundamental and in a sense "preexisting", that we merely get to explore? In the past it's seemed as if these were two fundamentally incompatible possibilities. But the framework we've built here in a sense blends them both into a rather unexpected synthesis.

The starting point is the idea that mathematics—like physics—is rooted in the ruliad, which is a representation of formal necessity. Actual mathematics as we "experience" it is—like physics—based on the particular sampling we make of the ruliad. But then the crucial point is that very basic characteristics of us as "observers" are sufficient to constrain that experience to be our general mathematics—or our physics.

At some level we can say that "mathematics is always there"—because every aspect of it is ultimately encoded in the ruliad. But in another sense we can say that the mathematics we have is all "up to us"—because it's based on how we sample the ruliad. But the point is that that sampling is not somehow "arbitrary": if we're talking about mathematics for us humans then it's us ultimately doing the sampling, and the sampling is inevitably constrained by general features of our nature.

A major discovery from our Physics Project is that it doesn't take much in the way of constraints on the observer to deeply constrain the laws of physics they will perceive. And similarly we posit here that for "observers like us" there will inevitably be general ("physicalized") laws of mathematics, that make mathematics inevitably have the general kinds of characteristics we perceive it to have (such as the possibility of doing mathematics at a high level, without always having to drop down to an "atomic" level).

Particularly over the past century there's been the idea that mathematics can be specified in terms of axiom systems, and that these axiom systems can somehow be "invented at will". But our framework does two things. First, it says that "far below" axiom systems is the raw ruliad, which in a sense represents all possible axiom systems. And second, it says that whatever axiom systems we perceive to be "operating" will be ones that we as observers can pick out from the underlying structure of the ruliad.

At a formal level we can "invent" an arbitrary axiom system (and it'll be somewhere in the ruliad), but only certain axiom systems will be ones that describe what we as "mathematical observers" can perceive. In a physics setting we might construct some formal physical theory that talks about detailed patterns in the atoms of space (or molecules in a gas), but the kind of "coarse-grained" observations that we can make won't capture these. Put another way, observers like us can perceive certain kinds of things, and can describe things in terms of these perceptions. But with the wrong kind of theory—or "axioms"—these descriptions won't be sufficient—and only an observer who's "shredded" down to a more "atomic" level will be able to track what's going on.

There's lots of different possible math—and physics—in the ruliad. But observers like us can only "access" a certain type. Some putative alien not like us might access a different type—and might end up with both a different math and a different physics. Deep underneath they—like us—would be talking about the ruliad. But they'd be taking different samples of it, and describing different aspects of it.

For much of the history of mathematics there was a close alignment between the mathematics that was done and what we perceive in the world. For example, Euclidean geometry—with its whole axiomatic structure—was originally conceived just as an idealization of geometrical things that we observe about the world. But by the late 1800s the idea had emerged that one could create "disembodied" axiomatic systems with no particular grounding in our experience in the world.

And, yes, there are many possible disembodied axiom systems that one can set up. And in doing ruliology and generally exploring the computational universe it's interesting to investigate what they do. But the point is that this is something quite different from mathematics as mathematics is normally conceived. Because in a sense mathematics—like physics—is a "more human" activity that's based on what "observers like us" make of the raw formal structure that is ultimately embodied in the ruliad.

When it comes to physics there are, it seems, two crucial features of "observers like us". First, that we're computationally bounded. And second, that we have the perception that we're persistent—and have a definite and continuous thread of experience. At the level of atoms of space, we're in a sense constantly being "remade". But we nevertheless perceive it as always being the "same us".

This single seemingly simple assumption has far-reaching consequences. For example, it leads us to experience a single thread of time. And from the notion that we maintain a continuity of experience from every successive moment to the next we are inexorably led to the idea of a perceived continuum—not only in time, but also for motion and in space. And when combined with intrinsic features of the ruliad and of multicomputation in general, what comes out in the end is a surprisingly precise description of how we'll perceive our universe to operate—that seems to correspond exactly with known core laws of physics.

What does that kind of thinking tell us about mathematics? The basic point is that—since in the end both relate to humans—there is necessarily a close correspondence between physical and mathematical observers. Both are computationally bounded. And the assumption of persistence in time for physical observers becomes for mathematical observers the concept of maintaining coherence as more statements are accumulated. And when combined with intrinsic features of the ruliad and multicomputation this then turns out to imply the kind of physicalized laws of mathematics that we've discussed.

In a formal axiomatic view of mathematics one just imagines that one invents axioms and sees their consequences. But what we're describing here is a view of mathematics that is ultimately just about the ways that we as mathematical observers sample and experience the ruliad. And if we use axiom systems it has to be as a kind of "intermediate language" that helps us make a slightly higher-level description of some corner of the raw ruliad. But actual "human-level" mathematics—like human-level physics—operates at a higher level.

Our everyday experience of the physical world gives us the impression that we have a kind of "direct access" to many foundational features of physics, like the existence of space and the phenomenon of motion. But our Physics Project implies that these are not concepts that are in any sense "already there"; they are just things that emerge from the raw ruliad when you "parse" it in the kinds of ways observers like us do.

In mathematics it's less obvious (at least to all but perhaps experienced pure mathematicians) that there's "direct access" to anything. But in our view of mathematics here, it's ultimately just like physics—and ultimately also rooted in the ruliad, but sampled not by physical observers but by mathematical ones.

So from this point of view there's just as much that's "real" underneath mathematics as there is underneath physics. The mathematics is sampled slightly differently (though very similarly)—but we should not in any sense consider it "fundamentally more abstract".

When we think of ourselves as entities within the ruliad, we can build up what we might consider a "fully abstract" description of how we get our "experience" of physics. And we can basically do the same thing for mathematics. So if we take the commonsense point of view that physics fundamentally exists "for real", we're forced into the same point of view for mathematics. In other words, if we say that the physical universe exists, so must we also say that in some fundamental sense, mathematics also exists.

It's not something we as humans "just make", but it is something that is made through our particular way of observing the ruliad, that is ultimately defined by our particular characteristics as observers, with our particular core assumptions about the world, our particular kinds of sensory experience, and so on.

So what can we say in the end about whether mathematics is "invented" or "discovered"? It is neither. Its underpinnings are the ruliad, whose structure is a matter of formal necessity. But its perceived form for us is determined by our intrinsic characteristics as observers. We neither get to "arbitrarily invent" what's underneath, nor do we get to "arbitrarily discover" what's already there. The mathematics we see is the result of a combination of formal necessity in the underlying ruliad, and the particular forms of perception that we—as entities like us—have. Putative aliens could have quite different mathematics, not because the underlying ruliad is any different for them, but because their forms of perception might be different. And it's the same with physics: even though they "live in the same physical universe" their perception of the laws of physics could be quite different.

28 | What Axioms Can There Be for Human Mathematics?

When they were first developed in antiquity the axioms of Euclidean geometry were presumably intended basically as a kind of "tightening" of our everyday impressions of geometry—that would aid in being able to deduce what was true in geometry. But by the mid-1800s—between non-Euclidean geometry, group theory, Boolean algebra and quaternions—it had become clear that there was a range of abstract axiom systems one could in principle consider. And by the time of Hilbert's program around 1900 the pure process of deduction was in effect being viewed as an end in itself—and indeed the core of mathematics—with axiom systems being seen as "starter material" pretty much just "determined by convention".

In practice even today very few different axiom systems are ever commonly used—and indeed in *A New Kind of Science* I was able to list essentially all of them comfortably on a couple of pages. But why these axiom systems and not others? Despite the idea that axiom systems could ultimately be arbitrary, the concept was still that in studying some particular area of mathematics one should basically have an axiom system that would provide a "tight specification" of whatever mathematical object or structure one was trying to talk about. And so, for example, the Peano axioms are what became used for talking about arithmetic-style operations on integers.

In 1931, however, Gödel's theorem showed that actually these axioms weren't strong enough to constrain one to be talking only about integers: there were also other possible models of the axiom system, involving all sorts of exotic "non-standard arithmetic". (And moreover, there was no finite way to "patch" this issue.) In other words, even though the Peano axioms had been invented—like Euclid's axioms for geometry—as a way to describe a definite "intuitive" mathematical thing (in this case, integers) their formal axiomatic structure "had a life of its own" that extended (in some sense, infinitely) beyond its original intended purpose.

Both geometry and arithmetic in a sense had foundations in everyday experience. But for set theory dealing with infinite sets there was never an obvious intuitive base rooted in everyday experience. Some extrapolations from finite sets were clear. But in covering infinite sets various axioms (like the Axiom of Choice) were gradually added to capture what seemed like "reasonable" mathematical assertions.

But one example whose status for a long time wasn't clear was the Continuum Hypothesis—which asserts that the "next distinct possible cardinality" $\aleph_1$ after the cardinality $\aleph_0$ of the integers is $2^{\aleph_0}$: the cardinality of real numbers (i.e. of "the continuum"). Was this something that followed from previously accepted axioms of set theory? And if it was added, would it even be consistent with them? In the early 1960s it was established that actually the Continuum Hypothesis is independent of the other axioms.

With the axiomatic view of the foundations of mathematics that's been popular for the past century or so it seems as if one could, for example, just choose at will whether to include the Continuum Hypothesis (or its negation) as an axiom in set theory. But with the approach to the foundations of mathematics that we've developed here, this is no longer so clear.

Recall that in our approach, everything is ultimately rooted in the ruliad—with whatever mathematics observers like us "experience" just being the result of the particular sampling we do of the ruliad. And in this picture, axiom systems are a particular representation of fairly low-level features of the sampling we do of the raw ruliad.

If we could do any kind of sampling we want of the ruliad, then we'd presumably be able to get all possible axiom systems—as intermediate-level "waypoints" representing different kinds of slices of the ruliad. But in fact by our nature we are observers capable of only certain kinds of sampling of the ruliad.

We could imagine "alien observers" not like us who could for example make whatever choice they want about the Continuum Hypothesis. But given our general characteristics as observers, we may be forced into a particular choice. Operationally, as we've discussed above, the wrong choice could, for example, be incompatible with an observer who "maintains coherence" in metamathematical space.

Let's say we have a particular axiom stated in standard symbolic form. "Underneath" this axiom there will typically be at the level of the raw ruliad a huge cloud of possible configurations of emes that can represent the axiom. But an "observer like us" can only deal with a coarse-grained version in which all these different configurations are somehow considered equivalent. And if the entailments from "nearby configurations" remain nearby, then everything will work out, and the observer can maintain a coherent view of what's going, for example just in terms of symbolic statements about axioms.

But if instead different entailments of raw configurations of emes lead to very different places, the observer will in effect be "shredded"—and instead of having definite coherent "single-minded" things to say about what happens, they'll have to separate everything into all the different cases for different configurations of emes. Or, as we've said it before, the observer will inevitably end up getting "shredded"—and not be able to come up with definite mathematical conclusions.

So what specifically can we say about the Continuum Hypothesis? It's not clear. But conceivably we can start by thinking of $\aleph_0$ as characterizing the "base cardinality" of the ruliad, while $2^{\aleph_0}$ characterizes the base cardinality of a first-level hyperruliad that could for example be based on Turing machines with oracles for their halting problems. And it could be that for us to conclude that the Continuum Hypothesis is false, we'd have to somehow be straddling the ruliad and the hyperruliad, which would be inconsistent with us maintaining a coherent view of mathematics. In other words, the Continuum Hypothesis might somehow be equivalent to what we've argued before is in a sense the most fundamental "contingent fact"—that just as we live in a particular location in physical space—so also we live in the ruliad and not the hyperruliad.

We might have thought that whatever we might see—or construct—in mathematics would in effect be "entirely abstract" and independent of anything about physics, or our experience in the physical world. But particularly insofar as we're thinking about mathematics as done by humans we're dealing with "mathematical observers" that are "made of the same stuff" as physical observers. And this means that whatever general constraints or features exist for physical observers we can expect these to carry over to mathematical observers—so it's no coincidence that both physical and mathematical observers have the same core characteristics, of computational boundedness and "assumption of coherence".

And what this means is that there'll be a fundamental correlation between things familiar from our experience in the physical world and what shows up in our mathematics. We might have thought that the fact that Euclid's original axioms were based on our human perceptions of physical space would be a sign that in some "overall picture" of mathematics they should be considered arbitrary and not in any way central. But the point is that in fact our notions of space are central to our characteristics as observers. And so it's inevitable that "physical-experience-informed" axioms like those for Euclidean geometry will be what appear in mathematics for "observers like us".

29 | Counting the Emes of Mathematics and Physics

How does the "size of mathematics" compare to the size of our physical universe? In the past this might have seemed like an absurd question, that tries to compare something abstract and arbitrary with something real and physical. But with the idea that both mathematics and physics as we experience them emerge from our sampling of the ruliad, it begins to seem less absurd.

At the lowest level the ruliad can be thought of as being made up of atoms of existence that we call emes. As physical observers we interpret these emes as atoms of space, or in effect the ultimate raw material of the physical universe. And as mathematical observers we interpret them as the ultimate elements from which the constructs of mathematics are built.

As the entangled limit of all possible computations, the whole ruliad is infinite. But we as physical or mathematical observers sample only limited parts of it. And that means we can meaningfully ask questions like how the number of emes in these parts compare—or, in effect, how big is physics as we experience it compared to mathematics.

In some ways an eme is like a bit. But the concept of emes is that they're "actual atoms of existence"—from which "actual stuff" like the physical universe and its history are made—rather than just "static informational representations" of it. As soon as we imagine that everything is ultimately computational we are immediately led to start thinking of representing it in terms of bits. But the ruliad is not just a representation. It's in some way something lower level. It's the "actual stuff" that everything is made of. And what defines our particular experience of physics or of mathematics is the particular samples we as observers take of what's in the ruliad.

So the question is now how many emes there are in those samples. Or, more specifically, how many emes "matter to us" in building up our experience.

Let's return to an analogy we've used several times before: a gas made of molecules. In the volume of a room there might be 10^{27} individual molecules, each on average colliding every 10^{-10} seconds. So that means that our "experience of the room" over the course of a minute or so might sample 10^{39} collisions. Or, in terms closer to our Physics Project, we might say that there are perhaps 10^{39} "collision events" in the causal graph that defines what we experience.

But these "collision events" aren't something fundamental; they have what amounts to "internal structure" with many associated parameters about location, time, molecular configuration, etc.

Our Physics Project, however, suggests that—far below for example our usual notions of space and time—we can in fact have a truly fundamental definition of what's happening in the universe, ultimately in terms of emes. We don't yet know the "physical scale" for this—and in the end we presumably need experiments to determine that. But rather rickety estimates based on a variety of assumptions suggest that the elementary length might be around 10^{-90} meters, with the elementary time being around 10^{-100} seconds.

And with these estimates we might conclude that our "experience of a room for a minute" would involve sampling perhaps 10^{370} update events, that create about this number of atoms of space.

But it's immediately clear that this is in a sense a gross underestimate of the total number of emes that we're sampling. And the reason is that we're not accounting for quantum mechanics, and for the multiway nature of the evolution of the universe. We've so far only considered one "thread of time" at one "position in branchial space". But in fact there are many threads of time, constantly branching and merging. So how many of these do we experience?

In effect that depends on our size in branchial space. In physical space "human scale" is of order a meter—or perhaps 10^{90} elementary lengths. But how big is it in branchial space?

The fact that we're so large compared to the elementary length is the reason that we consistently experience space as something continuous. And the analog in branchial space is that if we're big compared to the "elementary branchial distance between branches" then we won't experience the different individual histories of these branches, but only an aggregate "objective reality" in which we conflate together what happens on all the branches. Or, put another way, being large in branchial space is what makes us experience classical physics rather than quantum mechanics.

Our estimates for branchial space are even more rickety than for physical space. But conceivably there are on the order of 10^{120} "instantaneous parallel threads of time" in the universe, and 10^{20} encompassed by our instantaneous experience—implying that in our minute-long experience we might sample a total of on the order of close to 10^{500} emes.

But even this is a vast underestimate. Yes, it tries to account for our extent in physical space and in branchial space. But then there's also rulial space—which in effect is what "fills out" the whole ruliad. So how big are we in that space? In essence that's like asking how many different possible sequences of rules there are that are consistent with our experience.

The total conceivable number of sequences associated with 10^{500} emes is roughly the number of possible hypergraphs with 10^{500} nodes—or around $(10^{500})^{10^{500}}$. But the actual number consistent with our experience is smaller, in particular as reflected by the fact that we attribute specific laws to our universe. But when we say "specific laws" we have to recognize that there is a finiteness to our efforts at inductive inference which inevitably makes these laws at least somewhat uncertain to us. And in a sense that uncertainty is what represents our "extent in rulial space".

But if we want to count the emes that we "absorb" as physical observers, it's still going to be a huge number. Perhaps the base may be lower—say 10^{10}—but there's still a vast exponent, suggesting that if we include our extent in rulial space, we as physical observers may experience numbers of emes like $(10^{10})^{10^{500}}$.

But let's say we go beyond our "everyday human-scale experience". For example, let's ask about "experiencing" our whole universe. In physical space, the volume of our current universe is about 10^{78} times larger than "human scale" (while human scale is perhaps 10^{270} times larger than the "scale of the atoms of space"). In branchial space, conceivably our current universe is 10^{100} times larger than "human scale". But these differences absolutely pale in comparison to the sizes associated with rulial space.

We might try to go beyond "ordinary human experience" and for example measure things using tools from science and technology. And, yes, we could then think about "experiencing" lengths down to 10^{-22} meters, or something close to "single threads" of quantum histories. But in the end, it's still the rulial size that dominates, and that's where we can expect most of the vast number of emes that form of our experience of the physical universe to come from.

OK, so what about mathematics? When we think about what we might call human-scale mathematics, and talk about things like the Pythagorean theorem, how many emes are there "underneath"? "Compiling" our theorem down to typical traditional mathematical axioms, we've seen that we'll routinely end up with expressions containing, say, 10^{20} symbolic elements. But what happens if we go "below that", compiling these symbolic elements—which might include things like variables and operators—into "pure computational elements" that we can think of as emes? We've seen a few examples, say with combinators, that suggest that for the traditional axiomatic structures of mathematics, we might need another factor of maybe roughly 10^{10}.

These are incredibly rough estimates, but perhaps there's a hint that there's "further to go" to get from human-scale for a physical observer down to atoms of space that correspond to emes, than there is to get from human-scale for a mathematical observer down to emes.

Just like in physics, however, this kind of "static drill-down" isn't the whole story for mathematics. When we talk about something like the Pythagorean theorem, we're really referring to a whole cloud of "human-equivalent" points in metamathematical space. The total number of "possible points" is basically the size of the entailment cone that contains something like the Pythagorean theorem. The "height" of the entailment cone is related to typical lengths of proofs—which for current human mathematics might be perhaps hundreds of steps.

And this would lead to overall sizes of entailment cones of very roughly 10^{100} theorems. But within this "how big" is the cloud of variants corresponding to particular "human-recognized" theorems? Empirical metamathematics could provide additional data on this question. But if we very roughly imagine that half of every proof is "flexible", we'd end up with things like 10^{50} variants. So if we asked how many emes correspond to the "experience" of the Pythagorean theorem, it might be, say, 10^{80}.

To give an analogy of "everyday physical experience" we might consider a mathematician thinking about mathematical concepts, and maybe in effect pondering a few tens of theorems per minute—implying according to our extremely rough and speculative estimates that while typical "specific human-scale physics experience" might involve 10^{500} emes, specific human-scale mathematics experience might involve 10^{80} emes (a number comparable, for example, to the number of physical atoms in our universe).

What if instead of considering "everyday mathematical experience" we consider all humanly explored mathematics? On the scales we're describing, the factors are not large. In the history of human mathematics, only a few million theorems have been published. If we think about all the computations that have been done in the service of mathematics, it's a somewhat larger factor. I suspect Mathematica is the dominant contributor here—and we can estimate that the total number of Wolfram Language operations corresponding to "human-level mathematics" done so far is perhaps 10^{20}.

But just like for physics, all these numbers pale in comparison with those introduced by rulial sizes. We've talked essentially about a particular path from emes through specific axioms to theorems. But the ruliad in effect contains all possible axiom systems. And if we start thinking about enumerating these—and effectively "populating all of rulial space"—we'll end up with exponentially more emes.

But as with the perceived laws of physics, in mathematics as done by humans it's actually just a narrow slice of rulial space that we're sampling. It's like a generalization of the idea that something like arithmetic as we imagine it can be derived from a whole cloud of possible axiom systems. It's not just one axiom system; but it's also not all possible axiom systems.

One can imagine doing some combination of ruliology and empirical metamathematics to get an estimate of "how broad" human-equivalent axiom systems (and their construction from emes) might be. But the answer seems likely to be much smaller than the kinds of sizes we have been estimating for physics.

It's important to emphasize that what we've discussed here is extremely rough—and speculative. And indeed I view its main value as being to provide an example of how to imagine thinking through things in the context of the ruliad and the framework around it. But on the basis of what we've discussed, we might make the very tentative conclusion that "human-experienced physics" is bigger than "human-experienced mathematics". Both involve vast numbers of emes. But physics seems to involve a lot more. In a sense—even with all its abstraction—the suspicion is that there's "less ultimately in mathematics" as far as we're concerned than there is in physics. Though by any ordinary human standards, mathematics still involves absolutely vast numbers of emes.

30 | Some Historical (and Philosophical) Background

The human activity that we now call "mathematics" can presumably trace its origins into prehistory. What might have started as "a single goat", "a pair of goats", etc. became a story of abstract numbers that could be indicated purely by things like tally marks. In Babylonian times the practicalities of a city-based society led to all sorts of calculations involving arithmetic and geometry—and basically everything we now call "mathematics" can ultimately be thought of as a generalization of these ideas.

The tradition of philosophy that emerged in Greek times saw mathematics as a kind of reasoning. But while much of arithmetic (apart from issues of infinity and infinitesimals) could be thought of in explicit calculational ways, precise geometry immediately required an idealization—specifically the concept of a point having no extent, or equivalently, the continuity of space. And in an effort to reason on top of this idealization, there emerged the idea of defining axioms and making abstract deductions from them.

But what kind of a thing actually was mathematics? Plato talked about things we sense in the external world, and things we conceptualize in our internal thoughts. But he considered mathematics to be at its core an example of a third kind of thing: something from an abstract world of ideal forms. And with our current thinking, there is an immediate resonance between this concept of ideal forms and the concept of the ruliad.

But for most of the past two millennia of the actual development of mathematics, questions about what it ultimately was lay in the background. An important step was taken in the late 1600s when Newton and others "mathematicized" mechanics, at first presenting what they did in the form of axioms similar to Euclid's. Through the 1700s mathematics as a practical field was viewed as some kind of precise idealization of features of the world—though with an increasingly elaborate tower of formal derivations constructed in it. Philosophy, meanwhile, typically viewed mathematics—like logic—mostly as an example of a system in which there was a formal process of derivation with a "necessary" structure not requiring reference to the real world.

But in the first half of the 1800s there arose several examples of systems where axioms—while inspired by features of the world—ultimately seemed to be "just invented" (e.g. group theory, curved space, quaternions, Boolean algebra, ...). A push towards increasing rigor (especially for calculus and the nature of real numbers) led to more focus on axiomatization and formalization—which was still further emphasized by the appearance of a few non-constructive "purely formal" proofs.

But if mathematics was to be formalized, what should its underlying primitives be? One obvious choice seemed to be logic, which had originally been developed by Aristotle as a kind of catalog of human arguments, but two thousand years later felt basic and inevitable. And so it was that Frege, followed by Whitehead and Russell, tried to start "constructing mathematics" from "pure logic" (along with set theory). Logic was in a sense a rather low-level "machine code", and it took hundreds of pages of unreadable (if impressive-looking) "code" for Whitehead and Russell, in their 1910 *Principia Mathematica*, to get to 1 + 1 = 2.

366 PROLEGOMENA TO CARDINAL ARITHMETIC [PART II

*52·41. ⊢ : ∃ ! α . α ∼∈ 1 . ≡ . (∃x, y) . x, y ∈ α . x ≠ y
Dem.
⊢ . *24·54 . ⊃ ⊢ :. ∃ ! α . α ∼∈ 1 . ≡ : α ≠ Λ . α ∼∈ 1 :
[*4·56] ≡ : ∼{α ∈ 1 . ∨ . α = Λ} :
[*51·236] ≡ : ∼(α ∈ 1 ∪ ι'Λ) :
[*52·4.Transp] ≡ : ∼{x, y ∈ α . ⊃_{x,y} . x = y}
[*11·52] ≡ : (∃x, y) . x, y ∈ α . x ≠ y :. ⊃ ⊢ . Prop

*52·42. ⊢ :. α ∈ 1 . ⊃ : ∃ ! α ∩ β . ≡ . α ∩ β ∈ 1
Dem.
⊢ . *51·31 . ⊃ ⊢ :. ∃ ! ι'x ∩ β . ≡ . ι'x ∩ β = ι'x :.
[*20·53] ⊃ ⊢ :. α = ι'x . ⊃ : ∃ ! α ∩ β . ≡ . α ∩ β = ι'x :.
[*10·11·28] ⊃ ⊢ :. (∃x) . α = ι'x . ⊃ : (∃x) : ∃ ! α ∩ β . ≡ . α ∩ β = ι'x :
[*10·37] ⊃ : ∃ ! α ∩ β . ⊃ . (∃x) . α ∩ β = ι'x (1)
⊢ . (1) . *52·1 . ⊃ ⊢ :. α ∈ 1 . ⊃ : ∃ ! α ∩ β . ⊃ . α ∩ β ∈ 1 (2)
⊢ . *52·16 . ⊃ ⊢ : α ∩ β ∈ 1 . ⊃ . ∃ ! α ∩ β (3)
⊢ . (2) . (3) . ⊃ ⊢ . Prop

*52·43. ⊢ : α ∈ 1 . ∃ ! α ∩ β . ≡ . α ∈ 1 . α ∩ β ∈ 1 [*52·42 . *5·32]

*52·44. ⊢ :. α ∈ 1 . ⊃ : ∃ ! α ∩ β . ≡ . α ⊂ β . ≡ . α ∩ β = α
Dem.
⊢ . *51·31 . ⊃ ⊢ : ∃ ! ι'x ∩ β . ≡ . ι'x ⊂ β :
[*13·13.Exp] ⊃ ⊢ :. α = ι'x . ⊃ : ∃ ! α ∩ β . ≡ . α ⊂ β :.
[*10·11·23] ⊃ ⊢ :. (∃x) . α = ι'x . ⊃ : ∃ ! α ∩ β . ≡ . α ⊂ β :.
[*52·1] ⊃ ⊢ :. α ∈ 1 . ⊃ : ∃ ! α ∩ β . ≡ . α ⊂ β (1)
⊢ . (1) . *22·621 . ⊃ ⊢ . Prop

*52·45. ⊢ :: α, β ∈ 1 . ⊃ :. α ⊂ β ∪ γ . ≡ : α = β . ∨ . α ⊂ γ
Dem.
⊢ . *51·236 $\frac{x, y, \gamma}{z, x, \beta}$. ⊃
⊢ :. x ∈ ι'y ∪ γ . ≡ : x = y . ∨ . x ∈ γ :.
[*51·2·23] ⊃ ⊢ :. ι'x ⊂ ι'y ∪ γ . ≡ : ι'x = ι'y . ∨ . ι'x ⊂ γ :.
[*13·21] ⊃ ⊢ :: α = ι'x . β = ι'y . ⊃ :. α ⊂ β ∪ γ . ≡ : α = β . ∨ . α ⊂ γ ::
[*11·11·35] ⊃ ⊢ :: (∃x, y) . α = ι'x . β = ι'y . ⊃ :. α ⊂ β ∪ γ . ≡ : α = β . ∨ . α ⊂ γ (1)
⊢ . (1) . *52·1 . ⊃ ⊢ . Prop

52·46. ⊢ :. α, β ∈ 1 . ⊃ : α ⊂ β . ≡ . α = β . ≡ . ∃ ! (α ∩ β)
Dem.
⊢ . *51·2·23 . ⊃ ⊢ : ι'x ⊂ ι'y . ≡ . ι'x = ι'y (1)
⊢ . (1) . *13·21 . ⊃ ⊢ :. α = ι'x . β = ι'y . ⊃ : α ⊂ β . ≡ . α = β (2)
⊢ . (2) . *11·11·35 . *52·1 . ⊃ ⊢ :. α, β ∈ 1 . ⊃ : α ⊂ β . ≡ . α = β (3)
⊢ . (3) . *52·44 . ⊃ ⊢ . Prop

SECTION A] THE CARDINAL NUMBER 1 367

*52·6. ⊢ :. α ∈ 1 . ⊃ : x ∈ α . ≡ . ι'x = α . ≡ . x = ῐ'α
Dem.
⊢ . *51·23 . ⊃ ⊢ : x ∈ ι'y . ≡ . ι'x = ι'y :
[*13·13.Exp] ⊃ ⊢ :. α = ι'y . ⊃ : x ∈ α . ≡ . ι'x = α :.
[*10·11·23.*52·1] ⊃ ⊢ :. α ∈ 1 . ⊃ : x ∈ α . ≡ . ι'x = α . (1)
[*51·51] ≡ . x = ῐ'α (2)
⊢ . (1) . (2) . ⊃ ⊢ . Prop

*52·601. ⊢ :: α ∈ 1 . ⊃ :. φ(ῐ'α) . ≡ : x ∈ α . ⊃_x . φx : ≡ : (∃x) . x ∈ α . φx
Dem.
⊢ . *52·15 . ⊃ ⊢ :. Hp . ⊃ : E ! ῐ'α : (1)
[*30·4] ⊃ : x ῐ α . ≡ . x = ῐ'α .
[*52·6] ≡ . x ∈ α (2)
⊢ . (1) . *30·33 . ⊃
⊢ :: Hp . ⊃ :. φ(ῐ'α) . ≡ : x ῐ α . ⊃_x . φx : ≡ : (∃x) . x ῐ α . φx (3)
⊢ . (2) . (3) . ⊃ ⊢ . Prop

*52·602. ⊢ :. ẑ(φz) ∈ 1 . ⊃ : ψ(ιx)(φx) . ≡ . φx ⊃_x ψx . ≡ . (∃x) . φx . ψx
[*52·12 . *14·26]

*52·61. ⊢ :. α ∈ 1 . ⊃ : ῐ'α ∈ β . ≡ . α ⊂ β . ≡ . ∃ ! (α ∩ β) [*52·601 $\frac{x \in \beta}{\phi x}$]

*52·62. ⊢ :. α, β ∈ 1 . ⊃ : α = β . ≡ . ῐ'α = ῐ'β
Dem.
⊢ . *52·601 . ⊃ ⊢ :: Hp . ⊃ :. ῐ'α = ῐ'β . ≡ : x ∈ α . ⊃_x . x = ῐ'β :
[*52·6] ≡ : x ∈ α . ⊃_x . x ∈ β :
[*52·46] ≡ : α = β :: ⊃ ⊢ . Prop

*52·63. ⊢ : α, β ∈ 1 . α ≠ β . ⊃ . α ∩ β = Λ [*52·46 . Transp]

*52·64. ⊢ : α ∈ 1 . ⊃ . α ∩ β ∈ 1 ∪ ι'Λ
Dem.
⊢ . *52·43 . ⊃ ⊢ : Hp . ∃ ! α ∩ β . ⊃ . α ∩ β ∈ 1 :
[*5·6.*24·54] ⊃ ⊢ :. Hp . ⊃ : α ∩ β = Λ . ∨ . α ∩ β ∈ 1 :
[*51·236] ⊃ : α ∩ β ∈ 1 ∪ ι'Λ :. ⊃ ⊢ . Prop

*52·7. ⊢ :. β − α ∈ 1 . α ⊂ ξ . ξ ⊂ β . ⊃ : ξ = α . ∨ . ξ = β
Dem.
⊢ . *22·41 . ⊃ ⊢ : Hp . ξ ⊂ α . ⊃ . ξ = α (1)
⊢ . *24·55 . ⊃ ⊢ : ∼(ξ ⊂ α) . ⊃ . ∃ ! ξ − α (2)
⊢ . *22·48 . ⊃ ⊢ : Hp . ⊃ . ξ − α ⊂ β − α (3)
⊢ . (2) . (3) . ⊃ ⊢ : Hp . ∼(ξ ⊂ α) . ⊃ . ∃ ! ξ − α . ξ − α ⊂ β − α (4)
⊢ . *52·1 . ⊃ ⊢ : Hp . ⊃ . (∃x) . β − α = ι'x (5)
⊢ . (4) . (5) . *51·4 . ⊃ ⊢ : Hp . ∼(ξ ⊂ α) . ⊃ . ξ − α = β − α .
[*24·411] ⊃ . ξ = β (6)
⊢ . (1) . (6) . ⊃ ⊢ . Prop

Meanwhile, starting around 1900, Hilbert took a slightly different path, essentially representing everything with what we would now call symbolic expressions, and setting up axioms as relations between these. But what axioms should be used? Hilbert seemed to feel that the core of mathematics lay not in any "external meaning" but in the pure formal structure built up from whatever axioms were used. And he imagined that somehow all the truths of mathematics could be "mechanically derived" from axioms, a bit, as he said in a certain resonance with our current views, like the "great calculating machine, Nature" does it for physics.

Not all mathematicians, however, bought into this "formalist" view of what mathematics is. And in 1931 Gödel managed to prove from inside the formal axiom system traditionally used for arithmetic that this system had a fundamental incompleteness that prevented it from ever having anything to say about certain mathematical statements. But Gödel seems to have maintained a more Platonic belief about mathematics: that even though the axiomatic method falls short, the truths of mathematics are in some sense still "all there", and it's potentially possible for the human mind to have "direct access" to them. And while this is not quite the same as our picture of the mathematical observer accessing the ruliad, there's again some definite resonance here.

But, OK, so how has mathematics actually conducted itself over the past century? Typically there's at least lip service paid to the idea that there are "axioms underneath"—usually assumed to be those from set theory. There's been significant emphasis placed on the idea of formal deduction and proof—but not so much in terms of formally building up from axioms as in terms of giving narrative expositions that help humans understand why some theorem might follow from other things they know.

There's been a field of "mathematical logic" concerned with using mathematics-like methods to explore mathematics-like aspects of formal axiomatic systems. But (at least until very recently) there's been rather little interaction between this and the "mainstream" study of mathematics. And for example phenomena like undecidability that are central to mathematical logic have seemed rather remote from typical pure mathematics—even though many actual long-unsolved problems in mathematics do seem likely to run into it.

But even if formal axiomatization may have been something of a sideshow for mathematics, its ideas have brought us what is without much doubt the single most important intellectual breakthrough of the twentieth century: the abstract concept of computation. And what's now become clear is that computation is in some fundamental sense much more general than mathematics.

At a philosophical level one can view the ruliad as containing all computation. But mathematics (at least as it's done by humans) is defined by what a "mathematical observer like us" samples and perceives in the ruliad.

The most common "core workflow" for mathematicians doing pure mathematics is first to imagine what might be true (usually through a process of intuition that feels a bit like making "direct access to the truths of mathematics")—and then to "work backwards" to try to construct a proof. As a practical matter, though, the vast majority of "mathematics done in the world" doesn't follow this workflow, and instead just "runs forward"—doing computation. And there's no reason for at least the innards of that computation to have any "humanized character" to it; it can just involve the raw processes of computation.

But the traditional pure mathematics workflow in effect depends on using "human-level" steps. Or if, as we described earlier, we think of low-level axiomatic operations as being like molecular dynamics, then it involves operating at a "fluid dynamics" level.

A century ago efforts to "globally understand mathematics" centered on trying to find common axiomatic foundations for everything. But as different areas of mathematics were explored (and particularly ones like algebraic topology that cut across existing disciplines) it began to seem as if there might also be "top-down" commonalities in mathematics, in effect directly at the "fluid dynamics" level. And within the last few decades, it's become increasingly common to use ideas from category theory as a general framework for thinking about mathematics at a high level.

But there's also been an effort to progressively build up—as an abstract matter—formal "higher category theory". A notable feature of this has been the appearance of connections to both geometry and mathematical logic—and for us a connection to the ruliad and its features.

The success of category theory has led in the past decade or so to interest in other high-level structural approaches to mathematics. A notable example is homotopy type theory. The basic concept is to characterize mathematical objects not by using axioms to describe properties they should have, but instead to use "types" to say "what the objects are" (for example, "mapping from reals to integers"). Such type theory has the feature that it tends to look much more "immediately computational" than traditional mathematical structures and notation—as well as making explicit proofs and other metamathematical concepts. And in fact questions about types and their equivalences wind up being very much like the questions we've discussed for the multiway systems we're using as metamodels for mathematics.

Homotopy type theory can itself be set up as a formal axiomatic system—but with axioms that include what amount to metamathematical statements. A key example is the univalence axiom which essentially states that things that are equivalent can be treated as the same. And now from our point of view here we can see this being essentially a statement of metamathematical coarse graining—and a piece of defining what should be considered "mathematics" on the basis of properties assumed for a mathematical observer.

When Plato introduced ideal forms and their distinction from the external and internal world the understanding of even the fundamental concept of computation—let alone multicomputation and the ruliad—was still more than two millennia in the future. But now our picture is that everything can in a sense be viewed as part of the world of ideal forms that is the ruliad—and that not only mathematics but also physical reality are in effect just manifestations of these ideal forms.

But a crucial aspect is how we sample the "ideal forms" of the ruliad. And this is where the "contingent facts" about us as human "observers" enter. The formal axiomatic view of mathematics can be viewed as providing one kind of low-level description of the ruliad. But the point is that this description isn't aligned with what observers like us perceive—or with what we will successfully be able to view as human-level mathematics.

A century ago there was a movement to take mathematics (as well, as it happens, as other fields) beyond its origins in what amount to human perceptions of the world. But what we now see is that while there is an underlying "world of ideal forms" embodied in the ruliad that has nothing to do with us humans, mathematics as we humans do it must be associated with the particular sampling we make of that underlying structure.

And it's not as if we get to pick that sampling "at will"; the sampling we do is the result of fundamental features of us as humans. And an important point is that those fundamental features determine our characteristics both as mathematical observers and as physical observers. And this fact leads to a deep connection between our experience of physics and our definition of mathematics.

Mathematics historically began as a formal idealization of our human perception of the physical world. Along the way, though, it began to think of itself as a more purely abstract pursuit, separated from both human perception and the physical world. But now, with the general idea of computation, and more specifically with the concept of the ruliad, we can in a sense see what the limit of such abstraction would be. And interesting though it is, what we're now discovering is that it's not the thing we call mathematics. And instead, what we call mathematics is something that is subtly but deeply determined by general features of human perception—in fact, essentially the same features that also determine our perception of the physical world.

The intellectual foundations and justification are different now. But in a sense our view of mathematics has come full circle. And we can now see that mathematics is in fact deeply connected to the physical world and our particular perception of it. And we as humans can do what we call mathematics for basically the same reason that we as humans manage to parse the physical world to the point where we can do science about it.

31 | Implications for the Future of Mathematics

Having talked a bit about historical context let's now talk about what the things we've discussed here mean for the future of mathematics—both in theory and in practice.

At a theoretical level we've characterized the story of mathematics as being the story of a particular way of exploring the ruliad. And from this we might think that in some sense the ultimate limit of mathematics would be to just deal with the ruliad as a whole. But observers like us—at least doing mathematics the way we normally do it—simply can't do that. And in fact, with the limitations we have as mathematical observers we can inevitably sample only tiny slices of the ruliad.

But as we've discussed, it is exactly this that leads us to experience the kinds of "general laws of mathematics" that we've talked about. And it is from these laws that we get a picture of the "large-scale structure of mathematics"—that turns out to be in many ways similar to the picture of the large-scale structure of our physical universe that we get from physics.

As we've discussed, what corresponds to the coherent structure of physical space is the possibility of doing mathematics in terms of high-level concepts—without always having to drop down to the "atomic" level. Effective uniformity of metamathematical space then leads to the idea of "pure metamathematical motion", and in effect the possibility of translating at a high level between different areas of mathematics. And what this suggests is that in some sense "all high-level areas of mathematics" should ultimately be connected by "high-level dualities"—some of which have already been seen, but many of which remain to be discovered.

Thinking about metamathematics in physicalized terms also suggests another phenomenon: essentially an analog of gravity for metamathematics. As we discussed earlier, in direct analogy to the way that "larger densities of activity" in the spatial hypergraph for physics lead to a deflection in geodesic paths in physical space, so also larger "entailment density" in metamathematical space will lead to deflection in geodesic paths in metamathematical space. And when the entailment density gets sufficiently high, it presumably becomes inevitable that these paths will all converge, leading to what one might think of as a "metamathematical singularity".

In the spacetime case, a typical analog would be a place where all geodesics have finite length, or in effect "time stops". In our view of metamathematics, it corresponds to a situation where "all proofs are finite"—or, in other words, where everything is decidable, and there is no more "fundamental difficulty" left.

Absent other effects we might imagine that in the physical universe the effects of gravity would eventually lead everything to collapse into black holes. And the analog in metamathematics would be that everything in mathematics would "collapse" into decidable theories. But among the effects not accounted for is continued expansion—or in effect the creation of new physical or metamathematical space, formed in a sense by underlying raw computational processes.

What will observers like us make of this, though? In statistical mechanics an observer who does coarse graining might perceive the "heat death of the universe". But at a molecular level there is all sorts of detailed motion that reflects a continued irreducible process of computation. And inevitably there will be an infinite collection of possible "slices of reducibility" to be found in this—just not necessarily ones that align with any of our current capabilities as observers.

What does this mean for mathematics? Conceivably it might suggest that there's only so much that can fundamentally be discovered in "high-level mathematics" without in effect "expanding our scope as observers"—or in essence changing our definition of what it is we humans mean by doing mathematics.

But underneath all this is still raw computation—and the ruliad. And this we know goes on forever, in effect continually generating "irreducible surprises". But how should we study "raw computation"?

In essence we want to do unfettered exploration of the computational universe, of the kind I did in *A New Kind of Science*, and that we now call the science of ruliology. It's something we can view as more abstract and more fundamental than mathematics—and indeed, as we've argued, it's for example what's underneath not only mathematics but also physics.

Ruliology is a rich intellectual activity, important for example as a source of models for many processes in nature and elsewhere. But it's one where computational irreducibility and undecidability are seen at almost every turn—and it's not one where we can readily expect "general laws" accessible to observers like us, of the kind we've seen in physics, and now see in mathematics.

We've argued that with its foundation in the ruliad mathematics is ultimately based on structures lower level than axiom systems. But given their familiarity from the history of mathematics, it's convenient to use axiom systems—as we have done here—as a kind of "intermediate-scale metamodel" for mathematics.

But what is the "workflow" for using axiom systems? One possibility in effect inspired by ruliology is just to systematically construct the entailment cone for an axiom system, progressively generating all possible theorems that the axiom system implies. But while doing this is of great theoretical interest, it typically isn't something that will in practice reach much in the way of (currently) familiar mathematical results.

But let's say one's thinking about a particular result. A proof of this would correspond to a path within the entailment cone. And the idea of automated theorem proving is to systematically find such a path—which, with a variety of tricks, can usually be done vastly more efficiently than just by enumerating everything in the entailment cone. In practice, though, despite half a century of history, automated theorem proving has seen very little use in mainstream mathematics. Of course it doesn't help that in typical mathematical work a proof is seen as part of the high-level exposition of ideas—but automated proofs tend to operate at the level of "axiomatic machine code" without any connection to human-level narrative.

But if one doesn't already know the result one's trying to prove? Part of the intuition that comes from *A New Kind of Science* is that there can be "interesting results" that are still simple enough that they can conceivably be found by some kind of explicit search—and then verified by automated theorem proving. But so far as I know, only one significant unexpected result has so far ever been found in this way with automated theorem proving: my 2000 result on the simplest axiom system for Boolean algebra.

And the fact is that when it comes to using computers for mathematics, the overwhelming fraction of the time they're used not to construct proofs, but instead to do "forward computations" and "get results" (yes, often with Mathematica). Of course, within those forward computations, there are many operations—like Reduce, SatisfiableQ, PrimeQ, etc.—that essentially work by internally finding proofs, but their output is "just results" not "why-it's-true explanations". (FindEquationalProof—as its name suggests—is a case where an actual proof is generated.)

Whether one's thinking in terms of axioms and proofs, or just in terms of "getting results", one's ultimately always dealing with computation. But the key question is how that computation is "packaged". Is one dealing with arbitrary, raw, low-level constructs, or with something higher level and more "humanized"?

As we've discussed, at the lowest level, everything can be represented in terms of the ruliad. But when we do both mathematics and physics what we're perceiving is not the raw ruliad, but rather just certain high-level features of it. But how should these be represented? Ultimately we need a language that we humans understand, that captures the particular features of the underlying raw computation that we're interested in.

From our computational point of view, mathematical notation can be thought of as a rough attempt at this. But the most complete and systematic effort in this direction is the one I've worked towards for the past several decades: what's now the full-scale computational language that is the Wolfram Language (and Mathematica).

Ultimately the Wolfram Language can represent any computation. But the point is to make it easy to represent the computations that people care about: to capture the high-level constructs (whether they're polynomials, geometrical objects or chemicals) that are part of modern human thinking.

The process of language design (on which, yes, I've spent immense amounts of time) is a curious mixture of art and science, that requires both drilling down to the essence of things, and creatively devising ways to make those things accessible and cognitively convenient for humans. At some level it's a bit like deciding on words as they might appear in a human language—but it's something more structured and demanding.

And it's our best way of representing "high-level" mathematics: mathematics not at the axiomatic (or below) "machine code" level, but instead at the level human mathematicians typically think about it.

We've definitely not "finished the job", though. Wolfram Language currently has around 7000 built-in primitive constructs, of which at least a couple of thousand can be considered "primarily mathematical". But while the language has long contained constructs for algebraic numbers, random walks and finite groups, it doesn't (yet) have built-in constructs for algebraic topology or K-theory. In recent years we've been slowly adding more kinds of pure-mathematical constructs—but to reach the frontiers of modern human mathematics might require perhaps a thousand more. And to make them useful all of them will have to be carefully and coherently designed.

The great power of the Wolfram Language comes not only from being able to represent things computationally, but also being able to compute with things, and get results. And it's one thing to be able to represent some pure mathematical construct—but quite another to be able to broadly compute with it.

The Wolfram Language in a sense emphasizes the "forward computation" workflow. Another workflow that's achieved some popularity in recent years is the proof assistant one—in which one defines a result and then as a human one tries to fill in the steps to create a proof of it, with the computer verifying that the steps correctly fit together. If the steps are low level then what one has is something like typical automated theorem proving—though now being attempted with human effort rather than being done automatically.

In principle one can build up to much higher-level "steps" in a modular way. But now the problem is essentially the same as in computational language design: to create primitives that are both precise enough to be immediately handled computationally, and "cognitively convenient" enough to be usefully understood by humans. And realistically once one's done the design (which, after decades of working on such things, I can say is hard), there's likely to be much more "leverage" to be had by letting the computer just do computations than by expending human effort (even with computer assistance) to put together proofs.

One might think that a proof would be important in being sure one's got the right answer. But as we've discussed, that's a complicated concept when one's dealing with human-level mathematics. If we go to a full axiomatic level it's very typical that there will be all sorts of pedantic conditions involved. Do we have the "right answer" if underneath we assume that $1/0 = 0$? Or does this not matter at the "fluid dynamics" level of human mathematics?

One of the great things about computational language is that—at least if it's written well—it provides a clear and succinct specification of things, just like a good "human proof" is supposed to. But computational language has the great advantage that it can be run to create new results—rather than just being used to check something.

It's worth mentioning that there's another potential workflow beyond "compute a result" and "find a proof". It's "here's an object or a set of constraints for creating one; now find interesting facts about this". Type into Wolfram|Alpha something like sin^4(x) (and, yes, there's "natural math understanding" needed to translate something like this to precise Wolfram Language). There's nothing obvious to "compute" here. But instead what Wolfram|Alpha does is to "say interesting things" about this—like what its maximum or its integral over a period is.

In principle this is a bit like exploring the entailment cone—but with the crucial additional piece of picking out which entailments will be "interesting to humans". (And implementationally it's a very deeply constrained exploration.)

It's interesting to compare these various workflows with what one can call experimental mathematics. Sometimes this term is basically just applied to studying explicit examples of known mathematical results. But the much more powerful concept is to imagine discovering new mathematical results by "doing experiments".

Usually these experiments are not done at the level of axioms, but rather at a considerably higher level (e.g. with things specified using the primitives of Wolfram Language). But the typical pattern is to enumerate a large number of cases and to see what happens—with the most exciting result being the discovery of some unexpected phenomenon, regularity or irregularity.

This type of approach is in a sense much more general than mathematics: it can be applied to anything computational, or anything described by rules. And indeed it is the core methodology of ruliology, and what it does to explore the computational universe—and the ruliad.

One can think of the typical approach in pure mathematics as representing a gradual expansion of the entailment fabric, with humans checking (perhaps with a computer) statements they consider adding. Experimental mathematics effectively strikes out in some "direction" in metamathematical space, potentially jumping far away from the entailment fabric currently within the purview of some mathematical observer.

And one feature of this—very common in ruliology—is that one may run into undecidability. The "nearby" entailment fabric of the mathematical observer is in a sense "filled in enough" that it doesn't typically have infinite proof paths of the kind associated with undecidability. But something reached by experimental mathematics has no such guarantee.

What's good of course is that experimental mathematics can discover phenomena that are "far away" from existing mathematics. But (like in automated theorem proving) there isn't necessarily any human-accessible "narrative explanation" (and if there's undecidability there may be no "finite explanation" at all).

So how does this all relate to our whole discussion of new ideas about the foundations of mathematics? In the past we might have thought that mathematics must ultimately progress just by working out more and more consequences of particular axioms. But what we've argued is that there's a fundamental infrastructure even far below axiom systems—whose low-level exploration is the subject of ruliology. But the thing we call mathematics is really something higher level.

Axiom systems are some kind of intermediate modeling layer—a kind of "assembly language" that can be used as a wrapper above the "raw ruliad". In the end, we've argued, the details of this language won't matter for typical things we call mathematics. But in a sense the situation is very much like in practical computing: we want an "assembly language" that makes it easiest to do the typical high-level things we want. In practical computing that's often achieved with RISC instruction sets. In mathematics we typically imagine using axiom systems like ZFC. But—as reverse mathematics has tended to indicate—there are probably much more accessible axiom systems that could be used to reach the mathematics we want. (And ultimately even ZFC is limited in what it can reach.)

But if we could find such a "RISC" axiom system for mathematics it has the potential to make practical more extensive exploration of the entailment cone. It's also conceivable—though not guaranteed—that it could be "designed" to be more readily understood by humans. But in the end actual human-level mathematics will typically operate at a level far above it.

And now the question is whether the "physicalized general laws of mathematics" that we've discussed can be used to make conclusions directly about human-level mathematics. We've identified a few features—like the very possibility of high-level mathematics, and the expectation of extensive dualities between mathematical fields. And we know that basic commonalities in structural features can be captured by things like category theory. But the question is what kinds of deeper general features can be found, and used.

In physics our everyday experience immediately makes us think about "large-scale features" far above the level of atoms of space. In mathematics our typical experience so far has been at a lower level. So now the challenge is to think more globally, more metamathematically and, in effect, more like in physics.

In the end, though, what we call mathematics is what mathematical observers perceive. So if we ask about the future of mathematics we must also ask about the future of mathematical observers.

If one looks at the history of physics there was already much to understand just on the basis of what we humans could "observe" with our unaided senses. But gradually as more kinds of detectors became available—from microscopes to telescopes to amplifiers and so on—the domain of the physical observer was expanded, and the perceived laws of physics with it. And today, as the practical computational capability of observers increases, we can expect that we'll gradually see new kinds of physical laws (say associated with hitherto "it's just random" molecular motion or other features of systems).

As we've discussed above, we can see our characteristics as physical observers as being associated with "experiencing" the ruliad from one particular "vantage point" in rulial space (just as we "experience" physical space from one particular vantage point in physical space). Putative "aliens" might experience the ruliad from a different vantage point in rulial space—leading them to have laws of physics utterly incoherent with our own. But as our technology and ways of thinking progress, we can expect that we'll gradually be able to expand our "presence" in rulial space (just as we do with spacecraft and telescopes in physical space). And so we'll be able to "experience" different laws of physics.

We can expect the story to be very similar for mathematics. We have "experienced" mathematics from a certain vantage point in the ruliad. Putative aliens might experience it from another point, and build their own "paramathematics" utterly incoherent with our mathematics. The "natural evolution" of our mathematics corresponds to a gradual expansion in the entailment fabric, and in a sense a gradual spreading in rulial space. Experimental mathematics has the potential to launch a kind of "metamathematical space probe" which can discover quite different mathematics. At first, though, this will tend to be a piece of "raw ruliology". But, if pursued, it potentially points the way to a kind of "colonization of rulial space" that will gradually expand the domain of the mathematical observer.

The physicalized general laws of mathematics we've discussed here are based on features of current mathematical observers (which in turn are highly based on current physical observers). What these laws would be like with "enhanced" mathematical observers we don't yet know.

Mathematics as it is today is a great example of the "humanization of raw computation". Two other examples are theoretical physics and computational language. And in all cases there is the potential to gradually expand our scope as observers. It'll no doubt be a mixture of technology and methods along with expanded cognitive frameworks and understanding. We can use ruliology—or experimental mathematics—to "jump out" into the raw ruliad. But most of what we'll see is "non-humanized" computational irreducibility.

But perhaps somewhere there'll be another slice of computational reducibility: a different "island" on which "alien" general laws can be built. But for now we exist on our current "island" of reducibility. And on this island we see the particular kinds of general laws that we've discussed. We saw them first in physics. But there we discovered that they could emerge quite generically from a lower-level computational structure—and ultimately from the very general structure that we call the ruliad. And now, as we've discussed here, we realize that the thing we call mathematics is actually based on exactly the same foundations—with the result that it should show the same kinds of general laws.

It's a rather different view of mathematics—and its foundations—than we've been able to form before. But the deep connection with physics that we've discussed allows us to now have a physicalized view of metamathematics, which informs both what mathematics really is now, and what the future can hold for the remarkable pursuit that we call mathematics.

Some Personal History: The Evolution of These Ideas

It's been a long personal journey to get to the ideas described here—stretching back nearly 45 years. Parts have been quite direct, steadily building over the course of time. But other parts have been surprising—even shocking. And to get to where we are now has required me to rethink some very long-held assumptions, and adopt what I had believed was a rather different way of thinking—even though, ironically, I've realized in the end that many aspects of this way of thinking pretty much mirror what I've done all along at a practical and technological level.

Back in the late 1970s as a young theoretical physicist I had discovered the "secret weapon" of using computers to do mathematical calculations. By 1979 I had outgrown existing systems and decided to build my own. But what should its foundations be? A key goal was to represent the processes of mathematics in a computational way. I thought about the methods I'd found effective in practice. I studied the history of mathematical logic. And in the end I came up with what seemed to me at the time the most obvious and direct approach: that everything should be based on transformations for symbolic expressions.

I was pretty sure this was actually a good general approach to computation of all kinds—and the system we released in 1981 was named SMP ("Symbolic Manipulation Program") to reflect this generality. History has indeed borne out the strength of the symbolic expression paradigm—and it's from that we've been able to build the huge tower of technology that is the modern Wolfram Language. But all along mathematics has been an important use case—and in effect we've now seen four decades of validation that the core idea of transformations on symbolic expressions is a good metamodel of mathematics.

When Mathematica was first released in 1988 we called it "A System for Doing Mathematics by Computer", where by "doing mathematics" we meant doing computations in mathematics and getting results. People soon did all sorts of experiments on using Mathematica to create and present proofs. But the overwhelming majority of actual usage was for directly computing results—and almost nobody seemed interested in seeing the inner workings, presented as a proof or otherwise.

But in the 1980s I had started my work on exploring the computational universe of simple programs like cellular automata. And doing this was all about looking at the ongoing behavior of systems—or in effect the (often computationally irreducible) history of computations. And even though I sometimes talked about using my computational methods to do "experimental mathematics", I don't think I particularly thought about the actual progress of the computations I was studying as being like mathematical processes or proofs.

In 1991 I started working on what became *A New Kind of Science*, and in doing so I tried to systematically study possible forms of computational processes—and I was soon led to substitution systems and symbolic systems which I viewed in their different ways as being minimal idealizations of what would become Wolfram Language, as well as to multiway systems. There were some areas to which I was pretty sure the methods of *A New Kind of Science* would apply. Three that I wasn't sure about were biology, physics and mathematics.

But by the late 1990s I had worked out quite a bit about the first two, and started looking at mathematics. I knew that Mathematica and what would become Wolfram Language were good representations of "practical mathematics". But I assumed that to understand the foundations of mathematics I should look at the traditional low-level representation of mathematics: axiom systems.

And in doing this I was soon able to simplify to multiway systems—with proofs being paths:

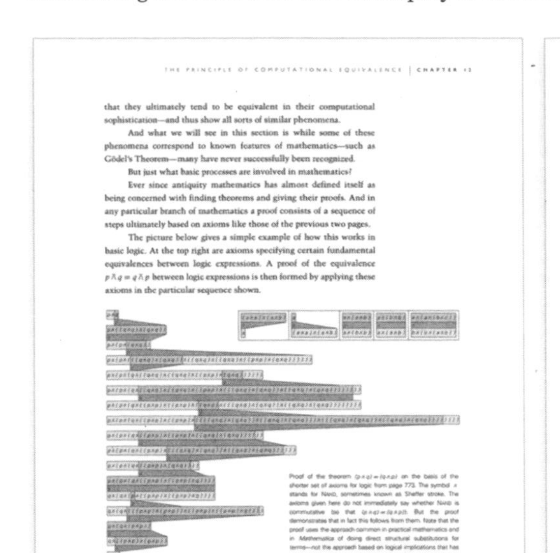

THE PRINCIPLE OF COMPUTATIONAL EQUIVALENCE | CHAPTER 12

that they ultimately tend to be equivalent in their computational sophistication—and thus show all sorts of similar phenomena.

And what we will see in this section is while some of these phenomena correspond to known features of mathematics—such as Gödel's Theorem—many have never successfully been recognized.

But just what basic processes are involved in mathematics?

Ever since antiquity mathematics has almost defined itself as being concerned with finding theorems and giving their proofs. And in any particular branch of mathematics a proof consists of a sequence of steps ultimately based on axioms like those of the previous two pages.

The picture below gives a simple example of how this works in basic logic. At the top right are axioms specifying certain fundamental equivalences between logic expressions. A proof of the equivalence $p \barwedge q = q \barwedge p$ between logic expressions is then formed by applying these axioms in the particular sequence shown.

Proof of the theorem $(p \barwedge q) = (q \barwedge p)$ on the basis of the shorter set of axioms for logic from page 773. The symbol $\barwedge$ stands for NAND, sometimes known as Sheffer stroke. The axioms given here do not immediately say whether NAND is commutative (so that $(p \barwedge q) = (q \barwedge p)$). But the proof demonstrates that in fact this follows from them. Note that the proof uses the approach common in practical mathematics and in Mathematica of doing direct structural substitutions for terms—not the approach based on logical implications that has traditionally been discussed in typical formal mathematical logic.

775

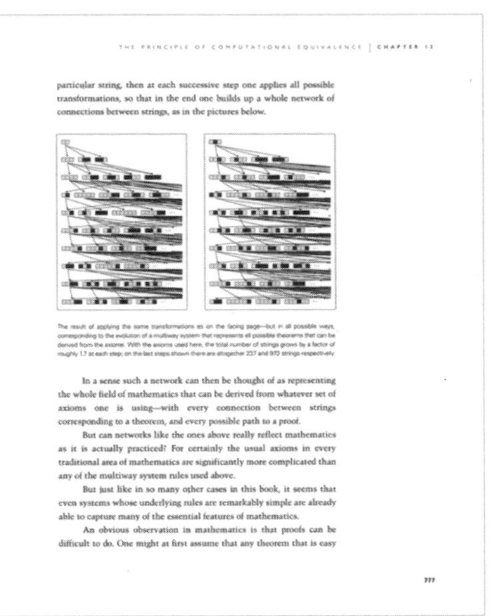

THE PRINCIPLE OF COMPUTATIONAL EQUIVALENCE | CHAPTER 12

particular string, then at each successive step one applies all possible transformations, so that in the end one builds up a whole network of connections between strings, as in the pictures below.

The result of applying the same transformations as on the facing page—but in all possible ways, corresponding to the evolution of a multiway system that represents all possible theorems that can be derived from the axioms. With the axioms used here, the total number of strings grows by a factor of roughly 1.7 at each step; on the last steps shown there are altogether 237 and 973 strings respectively.

In a sense such a network can then be thought of as representing the whole field of mathematics that can be derived from whatever set of axioms one is using—with every connection between strings corresponding to a theorem, and every possible path to a proof.

But can networks like the ones above really reflect mathematics as it is actually practiced? For certainly the usual axioms in every traditional area of mathematics are significantly more complicated than any of the multiway system rules used above.

But just like in so many other cases in this book, it seems that even systems whose underlying rules are remarkably simple are already able to capture many of the essential features of mathematics.

An obvious observation in mathematics is that proofs can be difficult to do. One might at first assume that any theorem that is easy

777

I had long wondered what the detailed relationships between things like my idea of computational irreducibility and earlier results in mathematical logic were. And I was pleased at how well many things could be clarified—and explicitly illustrated—by thinking in terms of multiway systems.

My experience in exploring simple programs in general had led to the conclusion that computational irreducibility and therefore undecidability were quite ubiquitous. So I considered it quite a mystery why undecidability seemed so rare in the mathematics that mathematicians typically did. I suspected that in fact undecidability was lurking close at hand—and I got some evidence of that by doing experimental mathematics. But why weren't mathematicians running into this more? I came to suspect that it had something to do with the history of mathematics, and with the idea that mathematics had tended to expand its subject matter by asking "How can this be generalized while still having such-and-such a theorem be true?"

But I also wondered about the particular axiom systems that had historically been used for mathematics. They all fit easily on a couple of pages. But why these and not others? Following my general "ruliological" approach of exploring all possible systems I started just enumerating possible axiom systems—and soon found out that many of them had rich and complicated implications.

But where among these possible systems did the axiom systems historically used in mathematics lie? I did searches, and at about the 50,000th axiom was able to find the simplest axiom system for Boolean algebra. Proving that it was correct gave me my first serious experience with automated theorem proving.

But what kind of a thing was the proof? I made some attempt to understand it, but it was clear that it wasn't something a human could readily understand—and reading it felt a bit like trying to read machine code. I recognized that the problem was in a sense a lack of "human connection points"—for example of intermediate lemmas that (like words in a human language) had a contextualized significance. I wondered about how one could find lemmas that "humans would care about"? And I was surprised to discover that at least for the "named theorems" of Boolean algebra a simple criterion could reproduce them.

Quite a few years went by. Off and on I thought about two ultimately related issues. One was how to represent the execution histories of Wolfram Language programs. And the other was how to represent proofs. In both cases there seemed to be all sorts of detail, and it seemed difficult to have a structure that would capture what would be needed for further computation—or any kind of general understanding.

Meanwhile, in 2009, we released Wolfram|Alpha. One of its features was that it had "step by step" math computations. But these weren't "general proofs": rather they were narratives synthesized in very specific ways for human readers. Still, a core concept in Wolfram|Alpha—and the Wolfram Language—is the idea of integrating in knowledge about as many things as possible in the world. We'd done this for cities and movies and lattices and animals and much more. And I thought about doing it for mathematical theorems as well.

We did a pilot project—on theorems about continued fractions. We trawled through the mathematical literature assessing the difficulty of extending the "natural math understanding" we'd built for Wolfram|Alpha. I imagined a workflow which would mix automated theorem generation with theorem search—in which one would define a mathematical scenario, then say "tell me interesting facts about this". And in 2014 we set about engaging the mathematical community in a large-scale curation effort to formalize the theorems of mathematics. But try as we might, only people already involved in math formalization seemed to care; with few exceptions working mathematicians just didn't seem to consider it relevant to what they did.

We continued, however, to push slowly forward. We worked with proof assistant developers. We curated various kinds of mathematical structures (like function spaces). I had estimated that we'd need more than a thousand new Wolfram Language functions to cover "modern pure mathematics", but without a clear market we couldn't motivate the huge design (let alone implementation) effort that would be needed—though, partly in a nod to the intellectual origins of mathematics, we did for example do a project that has succeeded in finally making Euclid-style geometry computable.

Then in the latter part of the 2010s a couple more "proof-related" things happened. Back in 2002 we'd started using equational logic automated theorem proving to get results in functions like FullSimplify. But we hadn't figured out how to present the proofs that were generated. In 2018 we finally introduced FindEquationalProof—allowing programmatic access to proofs, and making it feasible for me to explore collections of proofs in bulk.

I had for decades been interested in what I've called "symbolic discourse language": the extension of the idea of computational language to "everyday discourse"—and to the kind of thing one might want for example to express in legal contracts. And between this and our involvement in the idea of computational contracts, and things like blockchain technology, I started exploring questions of AI ethics and "constitutions". At this point we'd also started to introduce machine-learning-based functions into the Wolfram Language. And—with my "human incomprehensible" Boolean algebra proof as "empirical data"—I started exploring general questions of explainability, and in effect proof.

And not long after that came the surprise breakthrough of our Physics Project. Extending my ideas from the 1990s about computational foundations for fundamental physics it suddenly became possible finally to understand the underlying origins of the main known laws of physics. And core to this effort—and particularly to the understanding of quantum mechanics—were multiway systems.

At first we just used the knowledge that multiway systems could also represent axiomatic mathematics and proofs to provide analogies for our thinking about physics ("quantum observers might in effect be doing critical-pair completions", "causal graphs are like higher categories", etc.) But then we started wondering whether the phenomenon of the emergence that we'd seen for the familiar laws of physics might also affect mathematics—and whether it could give us something like a "bulk" version of metamathematics.

I had long studied the transition from discrete "computational" elements to "bulk" behavior, first following my interest in the Second Law of thermodynamics, which stretched all the way back to age 12 in 1972, then following my work on cellular automaton fluids in the mid-1980s, and now with the emergence of physical space from underlying hypergraphs in our Physics Project. But what might "bulk" metamathematics be like?

One feature of our Physics Project—in fact shared with thermodynamics—is that certain aspects of its observed behavior depend very little on the details of its components. But what did they depend on? We realized that it all had to do with the observer—and their interaction (according to what I've described as the 4th paradigm for science) with the general "multicomputational" processes going on underneath. For physics we had some idea what characteristics an "observer like us" might have (and actually they seemed to be closely related to our notion of consciousness). But what might a "mathematical observer" be like?

In its original framing we talked about our Physics Project as being about "finding the rule for the universe". But right around the time we launched the project we realized that that wasn't really the right characterization. And we started talking about rulial multiway systems that instead "run every rule"—but in which an observer perceives only some small slice, that in particular can show emergent laws of physics.

But what is this "run every rule" structure? In the end it's something very fundamental: the entangled limit of all possible computations—that I call the ruliad. The ruliad basically depends on nothing: it's unique and its structure is a matter of formal necessity. So in a sense the ruliad "necessarily exists"—and, I argued, so must our universe.

But we can think of the ruliad not only as the foundation for physics, but also as the foundation for mathematics. And so, I concluded, if we believe that the physical universe exists, then we must conclude—a bit like Plato—that mathematics exists too.

But how did all this relate to axiom systems and ideas about metamathematics? I had two additional pieces of input from the latter half of 2020. First, following up on a note in *A New Kind of Science*, I had done an extensive study of the "empirical metamathematics" of the network of the theorems in Euclid, and in a couple of math formalization systems. And second, in celebration of the 100th anniversary of their invention essentially as "primitives for mathematics", I had done an extensive ruliological and other study of combinators.

I began to work on this current piece in the fall of 2020, but felt there was something I was missing. Yes, I could study axiom systems using the formalism of our Physics Project. But was this really getting at the essence of mathematics? I had long assumed that axiom systems really were the "raw material" of mathematics—even though I'd long gotten signals they weren't really a good representation of how serious, aesthetically oriented pure mathematicians thought about things.

In our Physics Project we'd always had as a target to reproduce the known laws of physics. But what should the target be in understanding the foundations of mathematics? It always seemed like it had to revolve around axiom systems and processes of proof. And it felt like validation when it became clear that the same concepts of "substitution rules applied to expressions" seemed to span my earliest efforts to make math computational, the underlying structure of our Physics Project, and "metamodels" of axiom systems.

But somehow the ruliad—and the idea that if physics exists so must math—made me realize that this wasn't ultimately the right level of description. And that axioms were some kind of intermediate level, between the "raw ruliad", and the "humanized" level at which pure mathematics is normally done. At first I found this hard to accept; not only had axiom systems dominated thinking about the foundations of mathematics for more than a century, but they also seemed to fit so perfectly into my personal "symbolic rules" paradigm.

But gradually I got convinced that, yes, I had been wrong all this time—and that axiom systems were in many respects missing the point. The true foundation is the ruliad, and axiom systems are a rather-hard-to-work-with "machine-code-like" description below the inevitable general "physicalized laws of metamathematics" that emerge—and that imply that for observers like us there's a fundamentally higher-level approach to mathematics.

At first I thought this was incompatible with my general computational view of things. But then I realized: "No, quite the opposite!" All these years I've been building the Wolfram Language precisely to connect "at a human level" with computational processes—and with mathematics. Yes, it can represent and deal with axiom systems. But it's never felt particularly natural. And it's because they're at an awkward level—neither at the level of the raw ruliad and raw computation, nor at the level where we as humans define mathematics.

But now, I think, we begin to get some clarity on just what this thing we call mathematics really is. What I've done here is just a beginning. But between its explicit computational examples and its conceptual arguments I feel it's pointing the way to a broad and incredibly fertile new understanding that—even though I didn't see it coming—I'm very excited is now here.

Notes & Thanks

For more than 25 years Elise Cawley has been telling me her thematic (and rather Platonic) view of the foundations of mathematics—and that basing everything on constructed axiom systems is a piece of modernism that misses the point. From what's described here, I now finally realize that, yes, despite my repeated insistence to the contrary, what she's been telling me has been on the right track all along!

I'm grateful for extensive help on this project from James Boyd and Nik Murzin, with additional contributions by Brad Klee and Mano Namuduri. Some of the early core technical ideas here arose from discussions with Jonathan Gorard, with additional input from Xerxes Arsiwalla and Hatem Elshatlawy. (Xerxes and Jonathan have now also been developing connections with homotopy type theory.)

I've had helpful background discussions (some recently and some longer ago) with many people, including Richard Assar, Jeremy Avigad, Andrej Bauer, Kevin Buzzard, Mario Carneiro, Greg Chaitin, Harvey Friedman, Tim Gowers, Tom Hales, Lou Kauffman, Maryanthe Malliaris, Norm Megill, Assaf Peretz, Dana Scott, Matthew Szudzik, Michael Trott and Vladimir Voevodsky.

I'd like to recognize Norm Megill, creator of the Metamath system used for some of the empirical metamathematics here, who died in December 2021. (Shortly before his death he was also working on simplifying the proof of my axiom for Boolean algebra.)

Most of the specific development of this report has been livestreamed or otherwise recorded, and is available—along with archives of working notebooks—at the Wolfram Physics Project website.

The Wolfram Language code to produce all the images here is directly available from wolfr.am/metamathematics. And I should add that this project would have been impossible without the Wolfram Language, both its practical manifestation, and the ideas that it has inspired and clarified. So thanks to everyone involved in the 40+ years of its development and gestation!

Graphical Key

state/expression	
axiom	
statement/theorem	
notable theorem	
hypothesis	
substitution event	
cosubstitution event	
bisubstitution event	
multiway/entailment graph	
accumulative evolution graph	
token–event graph	
branchial/metamathematical graph	

Glossary

A glossary of terms that are either new here, or used in unfamiliar ways

accumulative system

A system in which states are rules and rules update rules. Successive steps in the evolution of such a system are collections of rules that can be applied to each other.

axiomatic level

The traditional foundational way to represent mathematics using axioms, viewed here as being intermediate between the raw ruliad and human-scale mathematics.

bisubstitution

The combination of substitution and cosubstitution that corresponds to the complete set of possible transformations to make on expressions containing patterns.

branchial space

Space corresponding to the limit of a branchial graph that provides a map of common ancestry (or entanglement) in a multiway graph.

cosubstitution

The dual operation to substitution, in which a pattern expression that is to be transformed is specialized to allow a given rule to match it.

eme

The smallest element of existence according to our framework. In physics it can be identified as an "atom of space", but in general it is an entity whose only internal attribute is that it is distinct from others.

entailment cone

The expanding region of a multiway graph or token-event graph affected by a particular node. The entailment cone is the analog in metamathematical space of a light cone in physical space.

entailment fabric

A piece of metamathematical space constructed by knitting together many small entailment cones. An entailment fabric is a rough model for what a mathematical observer might effectively perceive.

entailment graph

A combination of entailment cones starting from a collection of initial nodes.

expression rewriting

The process of rewriting (tree-structured) symbolic expressions according to rules for symbolic patterns. (Called "operator systems" in *A New Kind of Science*. Combinators are a special case.)

mathematical observer

An entity sampling the ruliad as a mathematician might effectively do it. Mathematical observers are expected to have certain core human-derived characteristics in common with physical observers.

metamathematical space

The space in which mathematical expressions or mathematical statements can be considered to lie. The space can potentially acquire a geometry as a limit of its construction through a branchial graph.

multiway graph

A graph that represents an evolution process in which there are multiple outcomes from a given state at each step. Multiway graphs are central to our Physics Project and to the multicomputational paradigm in general.

paramathematics

Parallel analogs of mathematics corresponding to different samplings of the ruliad by putative aliens or others.

pattern expression

A symbolic expression that involves pattern variables (*x_* etc. in Wolfram Language, or ∀ quantifiers in mathematical logic).

physicalization of metamathematics

The concept of treating metamathematical constructs like elements of the physical universe.

proof cone

Another term for the entailment cone.

proof graph

The subgraph in a token-event graph that leads from axioms to a given statement.

proof path

The path in a multiway graph that shows equivalence between expressions, or the subgraph in a token-event graph that shows the constructibility of a given statement.

ruliad

The entangled limit of all possible computational processes, that is posited to be the ultimate foundation of both physics and mathematics.

rulial space

The limit of rulelike slices taken from a foliation of the ruliad in time. The analog in the rulelike "direction" of branchial space or physical space.

shredding of observers

The process by which an observer who has aggregated statements in a localized region of metamathematical space is effectively pulled apart by trying to cover consequences of these statements.

statement

A symbolic expression, often containing a two-way rule, and often derivable from axioms, and thus representing a lemma or theorem.

substitution event

An update event in which a symbolic expression (which may be a rule) is transformed by substitution according to a given rule.

token-event graph

A graph indicating the transformation of expressions or statements ("tokens") through updating events.

two-way rule

A transformation rule for pattern expressions that can be applied in both directions (indicated with $\longleftrightarrow$).

uniquification

The process of giving different names to variables generated through different events.

Annotated Bibliography

For direct links to specific documents and references, see the online version here: wolfr.am/metamathematics. Here we'll give a slightly more general bibliography, though most of it should be considered background, since the approach taken here represents a significant departure from traditional directions, and builds more or less directly on extremely low-level concepts.

The earliest known large-scale axiomatic presentation of mathematics was:

Euclid (300 BC), Στοιχεῖα (in Ancient Greek) [Elements].

Empirical metamathematics from this was given in:

S. Wolfram (2020), "The Empirical Metamathematics of Euclid and Beyond". arXiv: 2107.07337.

The concept that there is underlying reality in mathematics was discussed in:

Plato (375 BC), πολιτεία (in Ancient Greek) [*The Republic*].

Plato (360 BC), Τίμαιος (in Ancient Greek) [*Timaeus*].

Modern explorations of these ideas include:

M. Balaguer (1998), *Platonism and Anti-Platonism in Mathematics*, Oxford University Press.

J. Gray (2008), *Plato's Ghost: The Modernist Transformation of Mathematics,* Princeton University Press.

R. Tieszen (2011), *After Gödel: Platonism and Rationalism in Mathematics and Logic*, Oxford University Press.

The contemporary axiomatic formulation of mathematics was developed in:

F. L. G. Frege (1879), *Begriffsschrift: eine der arithmetischen nachgebildete Formelsprache des reinen Denkens* (in German), Verlag von Louis Neber. (Translated in J. v. Heijenoort (1967), as "Begriffsschrift: A Formal Language, Modeled upon That of Arithmetic, for Pure Thought" in *From Frege to Gödel: A Source Book in Mathematical Logic, 1879–1931,* Harvard University Press, 1–82.)

R. Dedekind (1888), *Was sind und was sollen die Zahlen?* (in German), F. Vieweg und Sohn. (Translated in H. Pogorzelski, et al. (1995), as *What Are Numbers and What Should They Be?,* Research Institute for Mathematics.)

G. Peano (1889), *Arithmetices principia, nova methodo exposita* (in Italian), Fratres Bocca. (Translated by H. C. Kennedy (1973), as "The Principles of Arithmetic, Presented by a New Method", in *Selected Works of Giuseppe Peano,* University of Toronto Press, 101–134.)

D. Hilbert (1903), *Grundlagen der geometrie* (in German), B. G. Teubner. (Translated by E. G. Townsend (1902), as *The Foundations of Geometry,* Open Court.)

E. Zermelo (1908), "Untersuchungen über die Grundlagen der Mengenlehre I" (in German), *Mathematische Annalen* 65: 261–281. doi: 10.1007/BF01449999. (Translated by J. v. Heijenoort (1967), as "Investigations in the Foundations of Set Theory" in *From Frege to Gödel: A Source Book in Mathematical Logic, 1879–1931,* Harvard University Press, 199–215.)

A. N. Whitehead and B. A. W. Russell (1910–1913), *Principia Mathematica, Volumes I–III,* Cambridge University Press.

A standard collection of source documents is:

J. v. Heijenoort (1967), *From Frege to Gödel: A Source Book in Mathematical Logic, 1879–1931,* Harvard University Press.

Among general commentaries on axiomatic formalism are:

N. Bourbaki (1950), "The Architecture of Mathematics", *The American Mathematical Monthly* 57: 221–232. doi: 10.2307/2305937.

S. Feferman, et al. (2000), "Does Mathematics Need New Axioms?", *The Bulletin of Symbolic Logic 6,* 401–446. doi: 10.2307/420965.

W. Sieg (2013), *Hilbert's Programs and Beyond,* Oxford University Press.

A. Weir (2019), "Formalism in the Philosophy of Mathematics", *The Stanford Encyclopedia of Philosophy.* plato.stanford.edu/archives/spr2022/entries/formalism-mathematics.

Expositions of metamathematics and mathematical logic ideas appear for example in:

D. Hilbert and W. Ackermann (1928), *Grundzüge der theoretischen Logik* (in German), Springer. (Translated by L. M. Hammond, et al. (2000), as *Principles of Mathematical Logic,* AMS Chelsea.)

W. V. O. Quine (1940), *Mathematical Logic,* W. W. Norton & Company.

A. Church (1956), *Introduction to Mathematical Logic,* Princeton University Press.

H. Wang (1962), *A Survey of Mathematical Logic,* Science Press North-Holland.

H. B. Curry (1963), *Foundations of Mathematical Logic,* Dover.

H. Rasiowa (1963), *The Mathematics of Metamathematics,* Panstwowe Wydawnictwo Naukowe.

E . Mendelson (1964), *Intro to Mathematical Logic,* Van Nostrand Reinhold.

G. Kreisel and J. L. Krivine (1967), *Elements of Mathematical Logic: Model Theory,* North-Holland.

S. C. Kleene (1971), *Introduction to Metamathematics,* Wolters-Noordhoff.

H. B. Enderton (1972), *A Mathematical Introduction to Logic,* Harcourt/Academic Press.

A. Yasuhara (1975), "Recursive Function Theory and Logic", *Journal of Symbolic Logic* 40: 619–620. doi: 10.2307/2271829.

J. Barwise (1982), *Handbook of Mathematical Logic,* Elsevier.

G. E. Sacks (2003), *Mathematical Logic in the 20th Century,* World Scientific.

Works on the metamodeling of mathematics and on universal algebra include:

A. N. Whitehead (1898), *A Treatise on Universal Algebra with Applications,* Cambridge University Press.

A. Robinson (1963), *Introduction to Model Theory and to the Metamathematics of Algebra,* North-Holland.

N. G. de Bruijn (1970), "The Mathematical Language AUTOMATH, Its Usage, and Some of Its Extensions", in *Symposium on Automatic Demonstration,* M. Laudet, et al. (eds.), Springer.

A. I. Mal'Cev (1971), *The Metamathematics of Algebraic Systems, Collected Papers: 1936–1967,* North-Holland.

A. S. Toelstra (1973), *Metamathematical Investigation of Intuitionistic Arithmetic and Analysis,* Springer.

S. Burris and H. P. Sankappanavar (1981), *A Course in Universal Algebra,* Springer.

Low-level symbolic representations of mathematics were developed in:

M. Schönfinkel (1924), "Über die Bausteine der mathematischen Logik" (in German), *Mathematische Annalen* 92, 305–316. doi: 10.1007/BF01448013. (Translated by S. Bauer-Mengelberg (1967), as "On the Building Blocks of Mathematical Logic", in *From Frege to Gödel: A Source Book in Mathematical Logic, 1879–1931*, J. v. Heijenoort, Harvard University Press, 357–366.)

E. Post (1936), "Finite Combinatory Processes—Formulation 1", *The Journal of Symbolic Logic* 1, 103–105. doi: 10.2307/2269031.

See also these commentaries:

S. Wolfram (2020), "Combinators and the Story of Computation". arXiv: 2102.09658.

S. Wolfram (2021), *Combinators: A Centennial View*, Wolfram Media.

S. Wolfram (2021), "After 100 Years, Can We Finally Crack Post's Problem of Tag? A Story of Computational Irreducibility, and More". arXiv: 2103.06931.

Practical representation of mathematics using symbolic transformations was developed in:

S. Wolfram, et al. (1981), *SMP: A Symbolic Manipulation Program*. stephenwolfram.com/publications/smp-symbolic-manipulation-program.

S. Wolfram (1988), *Mathematica: A System for Doing Mathematics by Computer*, Addison-Wesley.

Wolfram Research (1988), Mathematica [Software system]. wolfram.com/mathematica.

Wolfram Research (2013), Wolfram Language [Computational language]. wolfram.com/language.

A proof of the "arithmeticization of metamathematics" was given in:

K. Gödel (1931), "Über formal unentscheidbare Sätze der *Principia Mathematica* und verwandter Systeme, I" (in German), *Monatshefte für Mathematik und Physik* 38: 173–198. doi: 10.1007/BF01700692. (Translated by B. Meltzer (1992), as *On Formally Undecidable Propositions of Principia Mathematica and Related Systems*, Dover.)

A major precursor to the current work is:

S. Wolfram (2002), "Implications for Mathematics and Its Foundations", in *A New Kind of Science*, Wolfram Media, 772–821. wolframscience.com/nks/p772--implications-for-mathematics-and-its-foundations.

Our Physics Project is described in:

S. Wolfram (2020), *A Project to Find the Fundamental Theory of Physics*, Wolfram Media.

S. Wolfram (2020), "A Class of Models with the Potential to Represent Fundamental Physics", *Complex Systems* 29: 107–536. doi: 10.25088/ComplexSystems.29.2.107 and arXiv:2004.08210.

The Wolfram Physics Project [Website]. wolframphysics.org.

The concept of the ruliad was introduced in:

S. Wolfram (2021), "The Concept of the Ruliad". writings.stephenwolfram.com/2021/11/the-concept-of-the-ruliad.

The concept of the mathematical observer was introduced in:

S. Wolfram (2021), "What Is Consciousness? Some New Perspectives from Our Physics Project". writings.stephenwolfram.com/2021/03/what-is-consciousness-some-new-perspectives-from-our-physics-project.

S. Wolfram (2021), "Why Does the Universe Exist? Some Perspectives from Our Physics Project". writings.stephenwolfram.com/2021/04/why-does-the-universe-exist-some-perspectives-from-our-physics-project.

Automated theorem proving is discussed for example in:

A. Robinson and A. Voronkov (2001), *Handbook of Automated Reasoning: Volume I,* Elsevier.

Relevant Wolfram Language functions include:

Wolfram Research (2018), FindEquationalProof, Wolfram Language function, reference.wolfram.com/language/ref/FindEquationalProof.html (updated 2020).

Wolfram Research (2019), AxiomaticTheory, Wolfram Language function, reference.wolfram.com/language/ref/AxiomaticTheory.html (updated 2021).

A system for low-level proof-oriented formalized mathematics (used here for empirical metamathematics) is:

N. Megill (1993), Metamath [Software system]. us.metamath.org/index.html.

N. Megill and D. A. Wheeler (2019), *Metamath: A Computer Language for Mathematical Proofs,* Lulu.

Other systems for proof-oriented formalized mathematics include:

N. G. de Brujin (1967), Automath [Software system]. win.tue.nl/automath.

A. Trybulec (1973), Mizar [Software system]. mizar.uwb.edu.pl.

University of Cambridge and Technical University of Munich (1986), Isabelle [Software system]. isabelle.in.tum.de.

T. Coquand and G. Huet (1989), Coq [Software system]. coq.inria.fr.

U. Norell and C. Coquand (2007), Agda [Software system]. wiki.portal.chalmers.se/agda/pmwiki.php.

Microsoft Research (2013), Lean [Software system]. leanprover.github.io.

Discussions of formalized mathematics include:

H. Wang (1960), "Toward Mechanical Mathematics", *IBM Journal of Research and Development* 4: 2–22. doi: 10.1147/rd.41.0002.

H. Friedman (1997), "The Formalization of Mathematics." cpb-us-w2.wpmucdn.com/u.osu.edu/dist/1/1952/files/2014/01/TalkFormMath12pt1.2.97-2jlte5o.pdf.

T. C. Hales (2008), "Formal Proof", *Notices of the AMS* 55: 1370–1380. ams.org/notices/200811/tx081101370p.pdf.

J. Avigad and J. Harrison (2014), "Formally Verified Mathematics", *Communications of the ACM* 57: 66–75. doi: 10.1145/2591012.

M. Ganesalingam and W. T. Gowers (2017), "A Fully Automatic Theorem Prover with Human-Style Output", *Journal of Automated Reasoning* 58: 253–291. doi: 10.1007/s10817-016-9377-1.

S. Wolfram (2018), "Logic, Explainability and the Future of Understanding", writings.stephenwolfram.com/2018/11/logic-explainability-and-the-future-of-understanding and (2019) Complex Systems 28: 1–40. doi: 10.25088/ComplexSystems.28.1.1.

K. Buzzard (2021), "What Is the Point of Computers? A Question for Pure Mathematicians". arXiv: 2112.11598.

Books on the philosophy of mathematics and its foundations include:

P. Benacerraf and H. Putnam (eds.) (1964), *Philosophy of Mathematics: Selected Readings,* Prentice-Hall.

I. Lakatos (1976), *Proofs and Refutations: The Logic of Mathematical Discovery,* Cambridge University Press.

T. Tymoczko (1986), *New Directions in the Philosophy of Mathematics,* Birkhäuser.

S. Shapiro (2000), *Thinking about Mathematics: The Philosophy of Mathematics,* Oxford University Press.

R. Krömer (2007), *Tool and Object: A History and Philosophy of Category Theory,* Springer.

E. Grosholz and H. Breger (2013), *The Growth of Mathematical Knowledge,* Springer.

The philosophical study of mathematics in terms of interrelated structure is summarized in:

E. Reck and G. Schiemer (2019), "Structuralism in the Philosophy of Mathematics", *The Stanford Encyclopedia of Philosophy.* plato.stanford.edu/archives/spr2020/entries/structuralism-mathematics.

The structure of proof space is discussed in univalent foundations and homotopy type theory:

The Univalent Foundations Program (2013), *Homotopy Type Theory,* Institute for Advanced Study.

S. Awodey, et al. (2013), "Voevodsky's Univalence Axiom in Homotopy Type Theory", *Notices of the AMS* 60: 1164–1167. doi: 10.48550/arxiv.1302.4731.

S. Awodey (2014), "Structuralism, Invariance, and Univalence", *Philosophia Mathematica* 22: 1–11. doi: 10.1093/philmat/nkt030.

M. Shulman (2021), "Homotopy Type Theory: The Logic of Space" in *New Spaces in Mathematics: Volume 1 Formal and Conceptual Reflections,* M. Anel and G. Catren (eds.), Cambridge University Press.

Relations between category theory, mathematics and our Physics Project were explored in:

X. D. Arsiwalla, J. Gorard, and H. Elshatlawy (2021), "Homotopies in Multiway (Non-deterministic) Rewriting Systems as n-Fold Categories". arXiv:2105.10822.

X. D. Arsiwalla and J. Gorard (2021), "Pregeometric Spaces from Wolfram Model Rewriting Systems as Homotopy Types". arXiv:2111.03460.

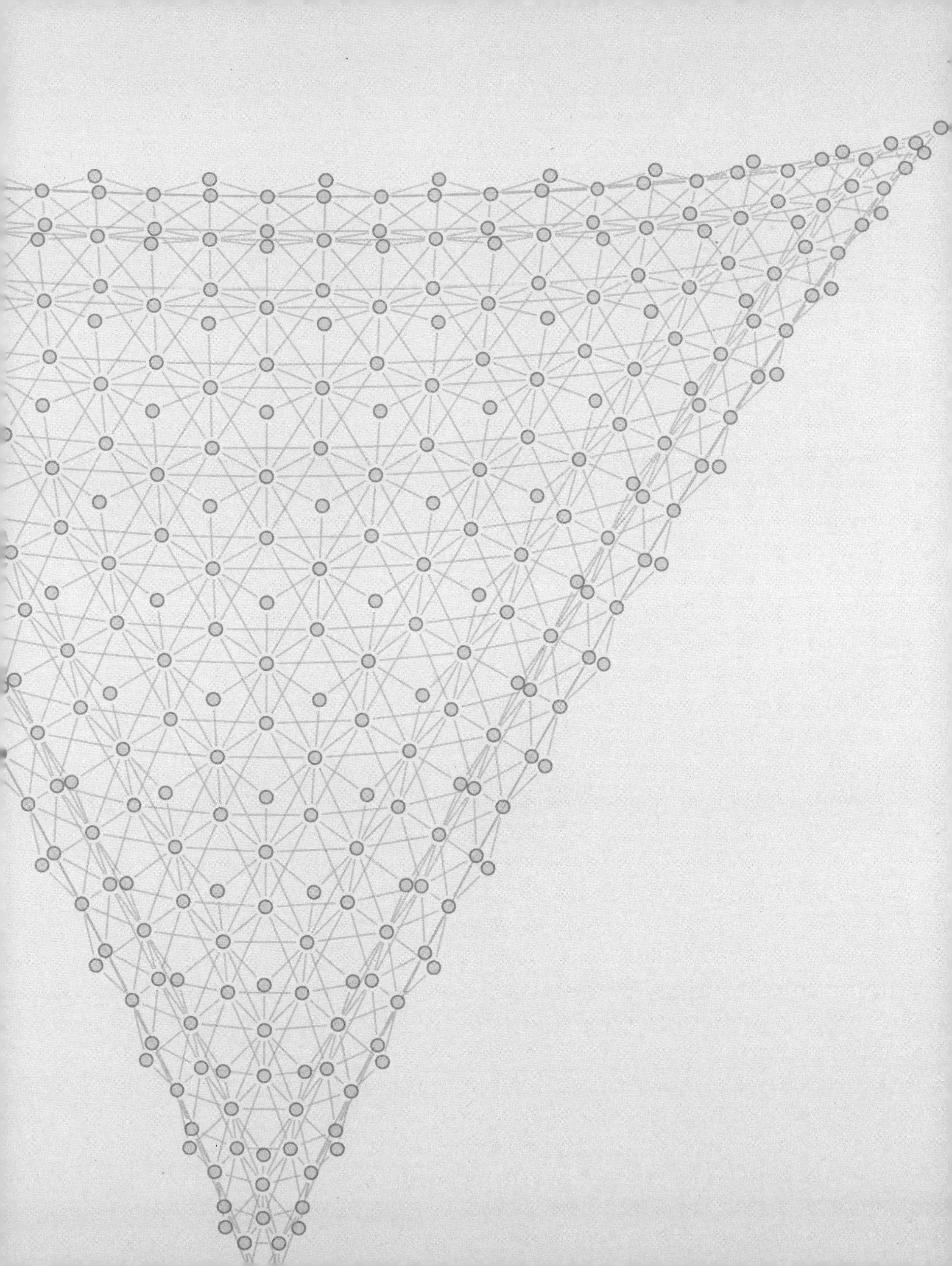

The Concept of the Ruliad

(November 10, 2021)

The Entangled Limit of Everything

I call it the ruliad. Think of it as the entangled limit of everything that is computationally possible: the result of following all possible computational rules in all possible ways. It's yet another surprising construct that's arisen from our Physics Project. And it's one that I think has extremely deep implications—both in science and beyond.

In many ways, the ruliad is a strange and profoundly abstract thing. But it's something very universal—a kind of ultimate limit of all abstraction and generalization. And it encapsulates not only all formal possibilities but also everything about our physical universe—and everything we experience can be thought of as sampling that part of the ruliad that corresponds to our particular way of perceiving and interpreting the universe.

We're going to be able to say many things about the ruliad without engaging in all its technical details. (And—it should be said at the outset—we're still only at the very beginning of nailing down those technical details and setting up the difficult mathematics and formalism they involve.) But to ground things here, let's start with a slightly technical discussion of what the ruliad is.

In the language of our Physics Project, it's the ultimate limit of all rulial multiway systems. And as such, it traces out the entangled consequences of progressively applying all possible computational rules.

Here is an example of an ordinary multiway system based on the string replacement rules {A → AB, BB → A} (indicated respectively by blueish and reddish edges):

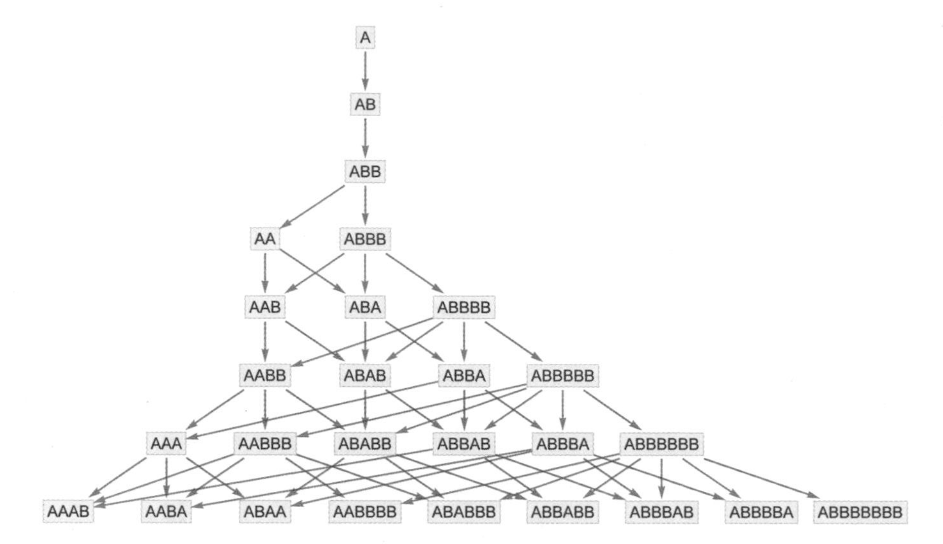

At each step, the rules are applied in all possible ways to each state. Often this generates multiple new states, leading to branching in the graph. But, importantly, there can also be merging—from multiple states being transformed to the same state.

The idea of a rulial multiway system is not just to apply particular rules in all possible ways, but to apply all possible rules of a given form. For example, if we consider "1 → 2, 2 → 1 A, B string rules", the possible rules are

{A → AA, A → AB, A → BA, A → BB, AA → A, AA → B, AB → A, AB → B, B → AA, B → AB, B → BA, B → BB, BA → A, BA → B, BB → A, BB → B}

and the resulting multiway graph is (where now we're using purple to indicate that there are edges for every possible rule):

A
AA AB BA BB
AAA ABA AAB ABB B BAA BBA BAB BBB
AAAA AABA ABAA ABBA AAAB AABB ABAB ABBB BAAA BABA BBAA BBBA BAAB BABB BBAB BBBB

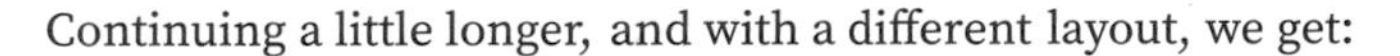

Continuing a little longer, and with a different layout, we get:

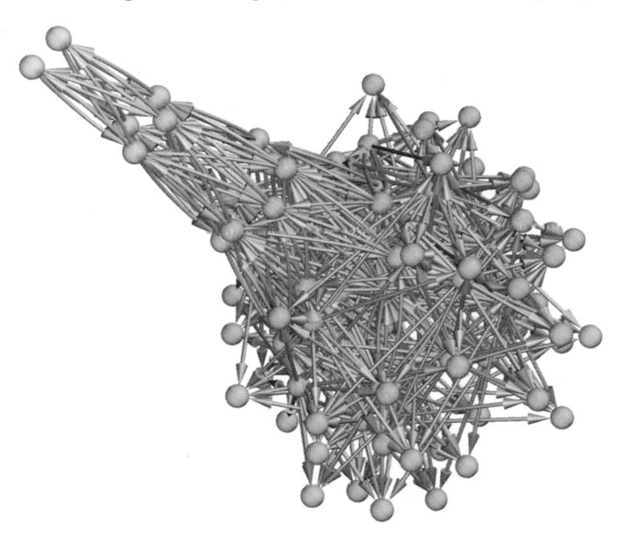

This may already look a little complicated. But the ruliad is something in a sense infinitely more complicated. Its concept is to use not just all rules of a given form, but all possible rules. And to apply these rules to all possible initial conditions. And to run the rules for an infinite number of steps.

The pictures above can be thought of as coarse finite approximations to the ruliad. The full ruliad involves taking the infinite limits of all possible rules, all possible initial conditions and all possible steps. Needless to say, this is a complicated thing to do, and there are many subtleties yet to work out about how to do it.

Perhaps the most obviously difficult issue is how conceivably to enumerate "all possible rules". But here we can use the Principle of Computational Equivalence to tell us that whatever "basis" we use, what comes out will eventually be effectively equivalent. Above we used string substitution systems. But here, for example, is a rulial multiway system made with 2-state 2-color Turing machines:

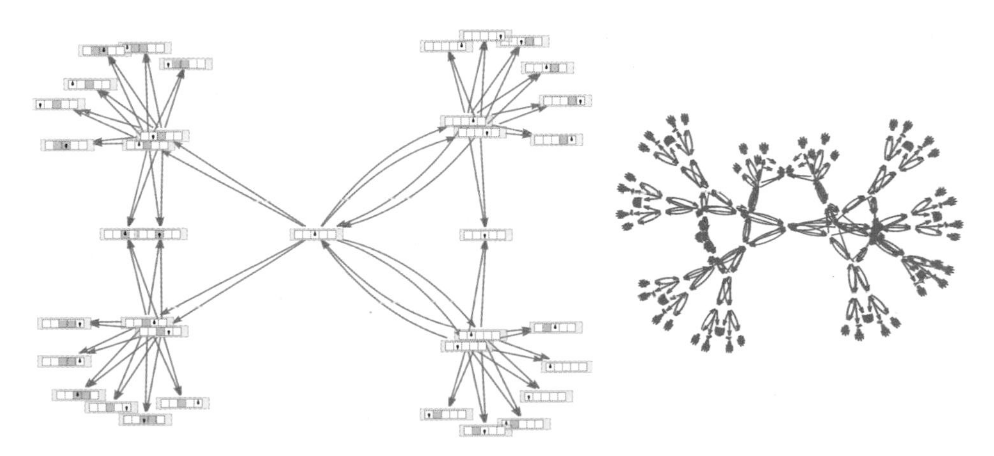

And here is a rulial multiway system made from hypergraph rewriting of the kind used in our Physics Project, using all rules with signature $1_2 \to 1_2$:

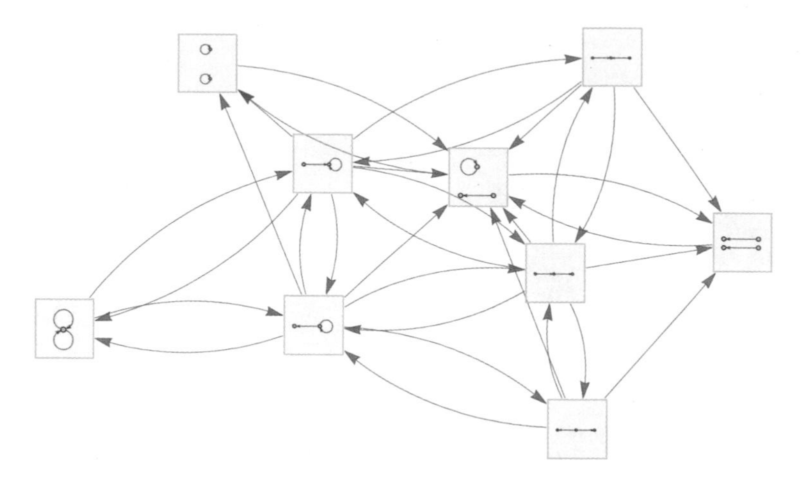

As another example, consider a multiway system based on numbers, in which the rules multiply by each possible integer:

$n \longmapsto \{2\,n, 3\,n, 4\,n, 5\,n, \ldots\}$

Here's what happens starting with 1 (and truncating the graph whenever the value exceeds 100):

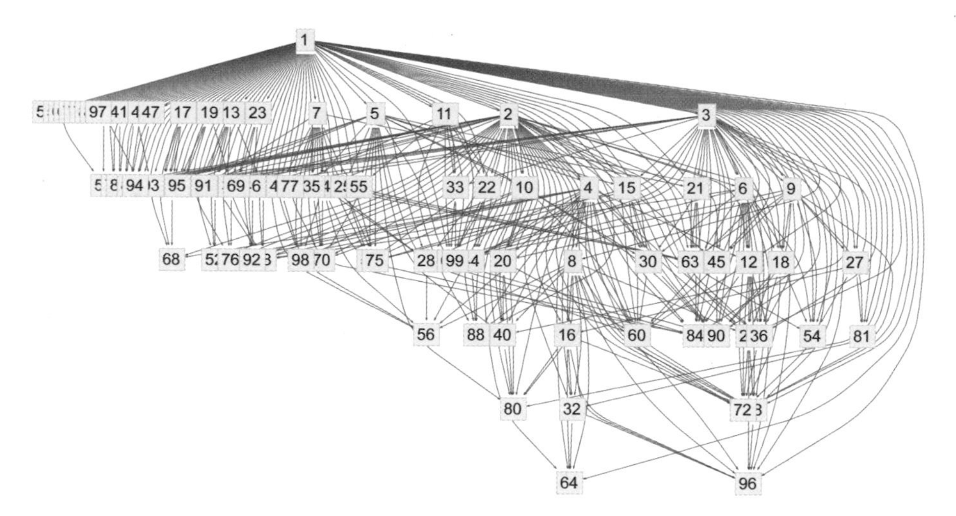

Even with this simple setup, the results are surprisingly complicated (though it's possible to give quite a bit of analysis in this particular case, as described in the Appendix at the end of this piece).

The beginning of the multiway graph is nevertheless simple: from 1 we connect to each successive integer. But then things get more complicated. To see what's going on, let's look at a fragment of the graph:

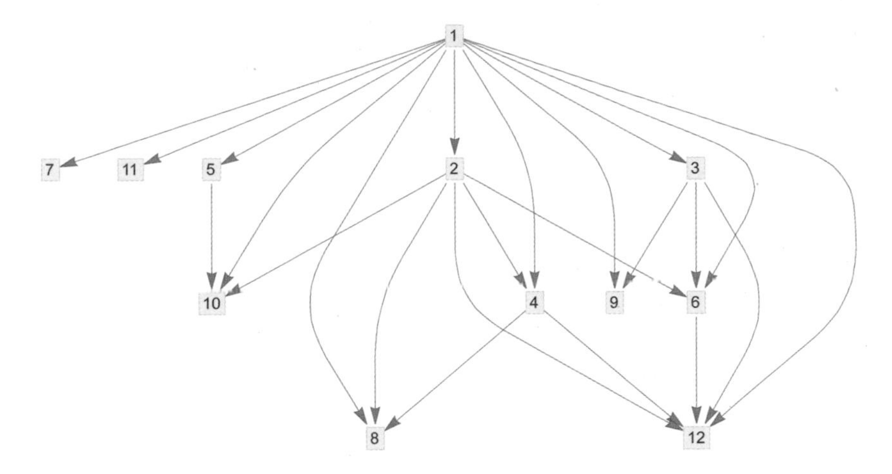

In a sense, everything would be simple if every path in the graph were separate:

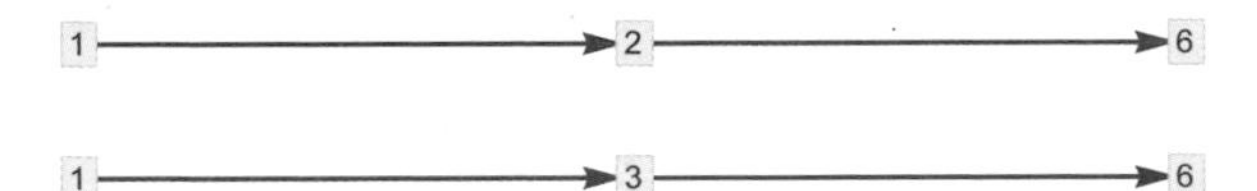

But the basic concept of multiway systems is that equivalent states should be merged—so here the "two ways to get 6" (i.e. $1 \times 2 \times 3$ and $1 \times 3 \times 2$) are combined, and what appears in the multiway graph is:

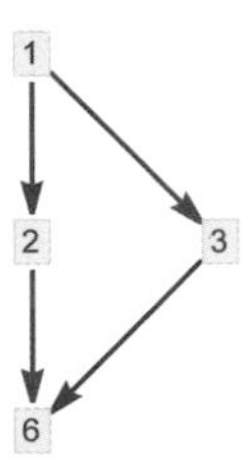

For integers, the obvious notion of equivalence is numerical equality. For hypergraphs, it's isomorphism. But the important point is that equivalence is what makes the multiway graph nontrivial. We can think about what it does as being to entangle paths. Without equivalence, different paths in the multiway system—corresponding to different possible histories—would all be separate. But equivalence entangles them.

The full ruliad is in effect a representation of all possible computations. And what gives it structure is the equivalences that exist between states generated by different computations. In a sense, there are two forces at work: the "forward" effect of the progress of computation, and the "sideways" effect of equivalences that entangle different computations. (Mathematically this can be thought of as being like decomposing the ruliad structure in terms of fibrations and foliations.)

Experiencing the Ruliad

In thinking about finding a fundamental theory of physics, one thing always bothered me. Imagine we successfully identify a rule that describes everything about our universe. Then the obvious next question will be: "Why this rule, and not another?" Well, how about if actually the universe in effect just runs every possible rule? What would this mean? It means that in a sense the "full story" of the universe is just the ruliad.

But the ruliad contains everything that is computationally possible. So why then do we have the perception that the universe has specific laws, and that definite things happen in it?

It all has to do with the fact that we are bounded observers, embedded within the ruliad. We never get to see the full ruliad; we just sample tiny parts of it, parsing them according to our particular methods of perception and analysis. And the crucial point is that for coherent observers like us, there are certain robust features that we will inevitably see in the ruliad. And these features turn out to include fundamental laws of our physics, in particular general relativity and quantum mechanics.

One can imagine an observer very different from us (say some kind of alien intelligence) who would sample different aspects of the ruliad, and deduce different laws. But one of the surprising core discoveries of our Physics Project is that even an observer with quite basic features like us will experience laws of physics that precisely correspond to ones we know.

An analogy (that's actually ultimately the result of the same underlying phenomenon) may help to illustrate what's going on. Consider molecules in a gas. The molecules bounce around in a complicated pattern that depends on their detailed properties. But an observer like us doesn't trace this whole pattern. Instead we only observe certain "coarse-grained" features. And the point is that these features are largely independent of the detailed properties of the molecules—and robustly correspond to our standard laws of physics, like the Second Law of thermodynamics. But a different kind of observer, sampling and "parsing" the system differently, could in principle identify different features, corresponding to different laws of physics.

One of the conceptual difficulties in thinking about how we perceive the ruliad is that it's a story of "self-observation". Essentially by the very definition of the ruliad, we ourselves are part of it. We never get to "see the whole ruliad from the outside". We only get to "experience it from the inside".

In some ways it's a bit like our efforts to construct the ruliad. In the end, the ruliad involves infinite rules, infinite initial conditions, and infinite time. But any way of assembling the ruliad from pieces effectively involves making particular choices about how we take those infinite limits. And that's pretty much like the fact that as entities embedded within the ruliad, we have to make particular choices about how to sample it.

One of the remarkable aspects of the ruliad is that it's in some sense the unique ultimately inevitable and necessary formal object. If one sets up some particular computational system or mathematical theory, there are choices to be made. But in the ruliad there are no choices. Because everything is there. And in a sense every aspect of the structure of the ruliad is just something formally necessary. It requires no outside input; it is just a formal consequence of the meaning of terms, like the abstract fact $1+1=2$.

But while the ruliad is unique, the description of it is not. In constructing it, one can imagine using Turing machines or hypergraph rewriting systems or indeed any other kind of computational system. Each will ultimately lead to the same limiting object that is the ruliad, but each of them can be thought of as defining a different coordinate system for describing the ruliad.

The very generality of the ruliad makes it unsurprising that there is vast diversity in how it can be described. And in a sense each possible description is like a possible way of experiencing the ruliad. In analogy to the (deeply related) situation with spacetime in general relativity, we might say that there are many reference frames in which to experience the ruliad—but it's always the same ruliad underneath.

It's important to understand that the "ruliad from the outside" could seem very different from any "internal" experience of it by an observer like us. As an example, consider a simple finite approximation to the ruliad, built from string substitution systems. In what we did above, we always started from a specific initial condition. But the full ruliad involves starting from all possible initial conditions. (Of course, one could always just say one starts from a "null" initial condition, then have rules of the form null → everything.) So now let's consider starting from all possible strings, say of length 4. If we use all possible 2-element-to-2-element rules, the finite approximation to the ruliad that we'll get will be:

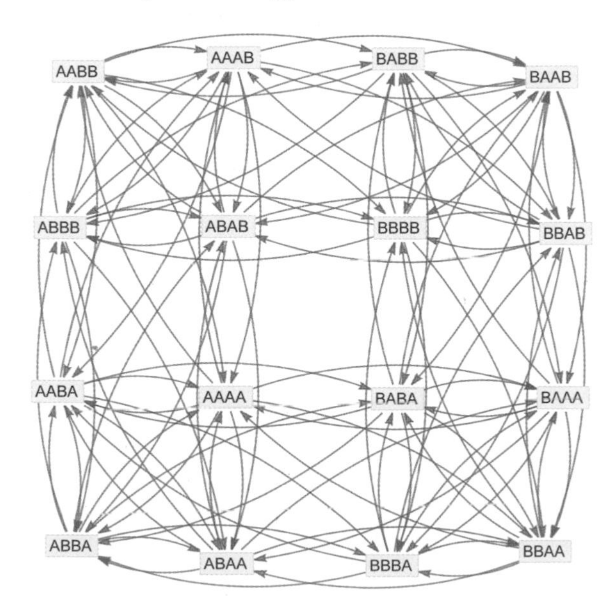

At some level this is a simple structure, and—as is inevitable for any finite approximation to the ruliad—its transitive closure is just the complete graph:

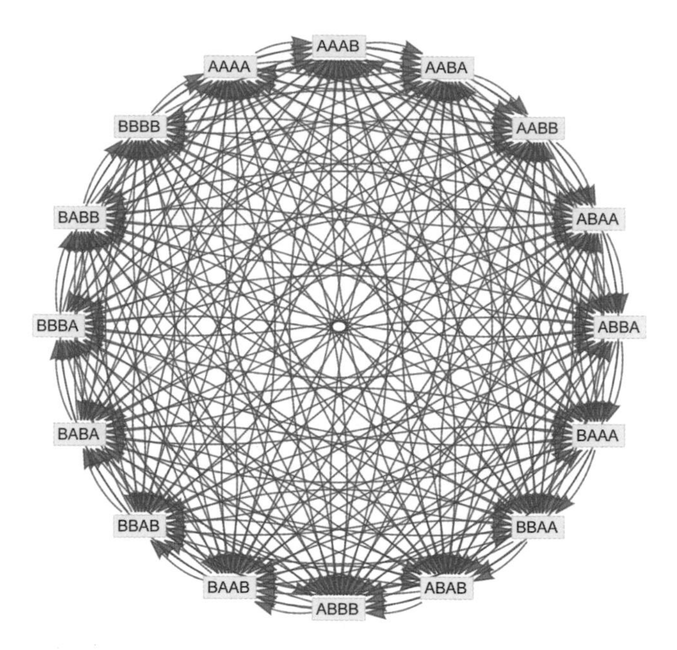

So why doesn't this mean that the ruliad is somehow trivial? A key part of the story is that we never get to "see the ruliad from the outside" like this. We are always part of it, sampling it according to some procedure, or, somewhat equivalently, thinking about constructing it according to some procedure.

As an analogy, consider the real numbers. The whole continuum of all real numbers is "from the outside" in many ways a simple construct. But if we imagine actually trying to construct real numbers, say digit by digit, according to some definite procedure, then we're dealing precisely with what Turing machines were originally invented to model, and the whole structure of computation is involved. (As we'll see, our way of thinking about "observers like us" is ultimately quite related to "Turing machines with bounded descriptions".)

In a sense, at an outside "holistic" level, the ruliad has a certain simple perfection. But as soon as you try to look at "what's in the ruliad", you have to parametrize or coordinatize it, and then you're inevitably exposed to its intricate internal structure.

Observers Like Us

One could imagine very different ways in which entities embedded within the ruliad could "experience" it. But what's most relevant for us is how "observers like us" do it—and how we manage to synthesize from what's going on in the ruliad our perception of reality, and our view of how our physical universe works.

Let's start by talking not about the full ruliad but rather about models in our Physics Project based on specific underlying rules. At the lowest level, we have a "machine-code" description of the universe in which everything just consists of a network of "atoms of space" that is continually being updated—and which we can think of as carrying out a giant, if incoherent, computation, full of computational irreducibility. But the remarkable fact is that somehow we, as observers of this, manage to pick out of it a certain slice that ends up showing coherent, computationally reducible features—that for example seem to reproduce our known laws of physics.

How does this work? Partly it has to do with features of us as observers, partly with features of how the universe fundamentally works, and partly with an interplay between these.

The first crucial feature of us as observers is that we're computationally bounded: the way we "parse" the universe involves doing an amount of computation that's absolutely tiny compared to all the computation going on in the universe. We sample only a tiny part of what's "really going on underneath", and we aggregate many details to get the summary that represents our perception of the universe.

But why should that summary have any coherence? Basically it's because we impose coherence through our definition of how observers like us work. One part of the universe will be affected by others. But to consider part of the universe as an "observer", there has to be a certain coherence to it. The behavior of the universe somehow has to imprint itself on a "medium" that has a certain coherence and consistency.

Down at the level of atoms of space, everything is always changing. But we can still identify emergent features that have a certain persistence. And it's out of those features that what we call observers are built. Given only the atoms of space with all their computationally irreducible behavior, it's not at the outset obvious that any real persistence could exist or be identified. But in our models we expect that there will, for example, be essentially topological features that correspond to particles that persistently maintain their identity.

And the point is that we can expect to "aggregate up" much further and be able to identify something like a human observer—that we can consider to persistently maintain its identity to the point where phenomena from the universe can be "systematically imprinted" on it.

Down at the level of atoms of space, there's a whole multiway graph of possible sequences of updates that can occur—with each path in effect corresponding to a different "thread of time" for the universe. But it's a crucial fact about us as observers of the universe that we don't perceive all those branching and merging threads of time. Instead, we imagine that we have a single, definite thread of experience—in which everything is sequentialized in time.

I've argued elsewhere that this sequentialization in time is a defining characteristic of "human-like consciousness". And it turns out that one of its consequences is that it implies that the particular perception we will have of the universe must be one in which there are laws of physics that correspond to ones we know.

It's not obvious, by the way, that if we sequentialize time we can form any consistent view of the universe. But the phenomenon of causal invariance—which seems ultimately to be guaranteed by the fundamental structure of the ruliad—turns out to imply that we can expect a certain generalized relativistic invariance that will inevitably lead to eventual consistency.

The notion of sequentialization in time is closely related to the idea that—even though our individual atoms of space are continually changing—we can view ourselves as having a coherent existence through time. And there's a similar phenomenon for space. At the outset, it's not obvious that there can be "pure motion", in which something can move in space without "fundamentally changing". But it turns out again to be consistent to view this as how things work for us: that even though we're "made of different atoms of space" when we're in different places, we can still imagine that in some sense we maintain the "same identity".

Down at the level of individual atoms of space, there really isn't any coherent notion of space. And the fact that we form such a notion seems to be intimately connected to what we might think of as details of us. Most important is that we're in a sense "intermediate in size" in the universe. We're large relative to the effective distance between atoms of space (which might be $\sim 10^{-90}$ m), yet we're small compared to the size of the whole universe ($\sim 10^{26}$ m). And the result is that we tend to aggregate the effects of many atoms of space, but still perceive different features of space (say, different gravitational fields) in different parts of the universe.

The fact that we "naturally form a notion of space" also seems to depend on another issue of scale—that for us the speed of light "seems fast". It takes our brains perhaps milliseconds to process anything we see. But the point is that this is very long compared to the time it takes light to get to us from objects in our typical local environment. And the result is that we tend to perceive there as being an instantaneous configuration of the world laid out in space, that "separately" changes in time. But if, for example, our brains ran much faster, or we were much bigger than we are, then the speed of light would "seem slower" to us, and we wouldn't tend to form the notion of an "instantaneous state of space".

OK, so what about quantum mechanics? The most fundamental feature of quantum mechanics is that it implies that things in the universe follow not just one but many possible paths of history—which we only get to make certain kinds of measurements on. And in our Physics Project this is something natural, and in fact inevitable. Given any particular configuration of the universe, there are many possible updates that can occur. And when we trace out all the possibilities, we get a multiway system, in which different threads of history continually branch and merge.

So how do observers like us fit into this? Being part of the universe, we inevitably branch and merge, just like the rest of the universe. So to understand our experience, what we need to ask is how a "branching brain" will perceive a "branching universe". And the story is remarkably similar to what we discussed above for our experience of space and time: it all has to do with imagining ourselves to have a certain definite persistence.

In other words, even if when "viewed from the outside" our brain might be following many different paths of history, "from the inside" we can still potentially assume that everything is conflated into a single thread of history. But will this ultimately be a consistent thing to do? Once again, causal invariance implies that it will. There are specific "quantum effects" where we can tell that there are multiple branches of history being followed, but in the end it'll be consistent to imagine an "objective reality" about "what happened".

In our Physics Project we imagine that there are abstract relations between atoms of space, and in the end the pattern of these relations defines the structure of physical space. But what about different branches of history in the multiway graph? Can we think of these as related? The answer is yes. For example, we can say that at a particular time, states on two branches are "adjacent" if they share an immediate ancestor in the multiway graph. And tracing through such connections we can develop a notion of "branchial space"—a kind of space in which states on different branches of history are laid out:

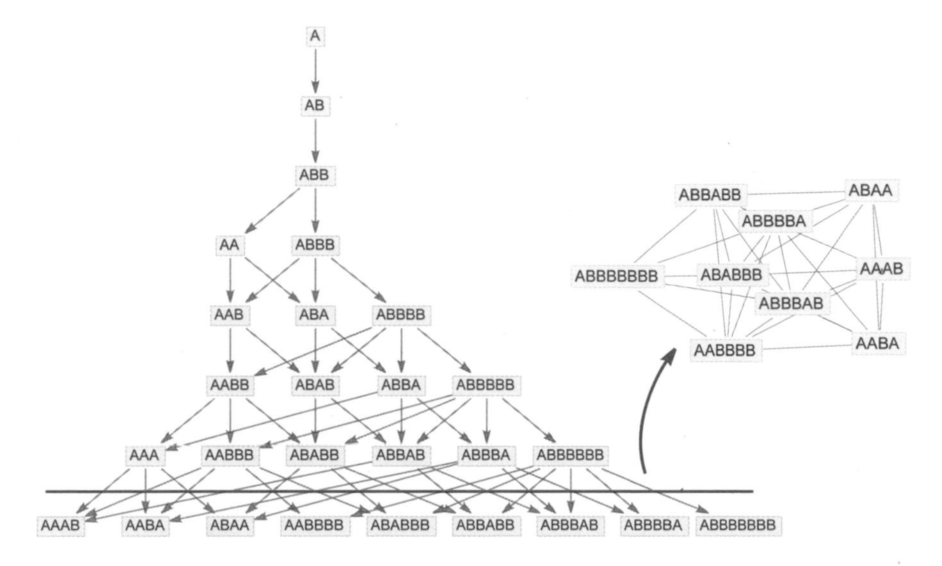

One can think of branchial space as being defined by the pattern of entanglements between different branches of history. And in our Physics Project it turns out that the fundamental laws of quantum mechanics seem to just be a direct translation of the fundamental laws of spacetime into branchial space. And just like the speed of light governs the maximum rate at which effects can propagate in physical space, so similarly in our models there's a "maximum entanglement speed" at which effects can propagate in branchial space.

So what are we like as observers in branchial space? Just like in physical space, we can presumably be thought of as having a certain size in branchial space. We don't yet know quite how to measure this size, but it's surely related to the effective number of quantum degrees of freedom we involve.

In our everyday experience of things like gases, we're sufficiently large compared to individual molecules that we normally just perceive the gas as some kind of continuum fluid—and in normal circumstances we can't even tell that it's made of molecules. Well, it's presumably the same kind of thing for physical space—where we're even much larger compared to the atoms of space, and it's a major challenge to figure out how to detect their presence. What about for branchial space? As the underlying rules for the system get applied, different branches of history will in effect "move around" in branchial space in complex and computationally irreducible ways. And just like when we observe molecules in a gas, we'll mostly just observe overall aggregate effects analogous to fluid mechanics—and only in special circumstances will we notice "quantum effects" that reveal the presence of multiple independent threads of history.

Living in Rulial Space

We've discussed how "observers like us" perceive models of physics of the type that arise in our Physics Project. But how will we perceive the whole ruliad? It begins with a generalization of the story for branchial space. Because now as well as having different branches associated with different updatings according to a particular rule, we have different branches associated with updatings according to different rules.

And just as we can slice an ordinary multiway system at a particular time to get an instantaneous version of branchial space, so now we can slice a rulial multiway system to get an instantaneous version of what we can call rulial space—a space in which different branches can correspond not just to different histories, but to different rules for history.

It's a fairly complicated setup, with "pure branchial space" inevitably being deeply interwoven with rulial space. But as a first approximation, we can think of rulial space as being somewhat separate, and laid out so that different places in it correspond to the results of applying different rules—with nearby places effectively being associated with "nearby" rules.

And just as we can think of effects propagating in branchial space, so also we can think of them propagating in rulial space. In branchial space we can talk about entanglement cones as the analog of light cones, and a maximum entanglement speed as the analog of the speed of light. In rulial space we can instead talk about "emulation cones"—and a "maximum emulation speed".

In our rough approximation of rulial space, each point is in effect associated with a particular rule. So how do we "move" from one point to another? Effectively we have to be emulating the behavior of one rule by another. But why should it even be possible to do this? The answer is the Principle of Computational Equivalence, which states that, in effect, most rules will be equivalent in their computational capabilities—and in particular they will be capable of universal computation, so that any given rule can always "run a program" that will make it emulate any other rule.

One can think of the program as an interpreter or translator that goes from one rule to another. The Principle of Computational Equivalence tells one that such a translator must essentially always exist. But how fast will the translator run? Effectively that's what distance in rulial space measures. Because to "do a certain translation", branches in the rulial multiway system have to reach from one rule to another. But they can only do that at the maximum emulation speed.

What does the maximum emulation speed measure? Effectively it corresponds to the raw computational processing speed of the universe. We can think of representing computations in some language—say the Wolfram Language. Then the processing speed will be measured in "Wolfram Language tokens processed per second" ("WLT/s"). In some sense, of course, giving a value for this speed is just a way of relating our human units of time (say, seconds) to the "intrinsic unit of time" associated with the computational processing that's going on in the universe. Or, in other words, it's a kind of ultimate definition of a second relative to purely formal constructs.

OK, but how does this relate to us as observers embedded within the ruliad? Well, just as we imagine that—along with the rest of the universe—we're continually branching and merging in branchial space, so also this will be what happens in rulial space. In other words—like the rest of the universe—our brains aren't following a particular rule; they're following branching and merging paths that represent all possible rules.

But "from inside" we can still potentially imagine that we have a single thread of experience—effectively conflating what happens on all those different branches. And once again we can ask whether doing this will be consistent. And the answer seems to be that, yes, it can be. And what guarantees this is again a kind of "rulial relativity" that's a consequence of causal invariance. There are many details here, which we'll address to some extent later. But the broad outline is that causal invariance can be thought of as being associated with paths of history that diverge, eventually converging again. But since the ruliad contains paths corresponding to all possible rules, it's basically inevitable that it will contain what's needed to "undo" whatever divergence occurs.

So what does this mean? Basically it's saying that even though the universe is in some sense intrinsically "following all possible rules"—as represented by paths in the ruliad—we as observers of the universe can still "take the point of view" that the universe follows a particular rule. Well, actually, it's not quite a particular rule. Because just as we're in some sense "quite big" in physical and presumably branchial space, so also we're potentially "quite big" in rulial space.

And being extended in rulial space is basically saying that we consider not just one, but a range of possible rules to be what describe the universe. How can it work this way? Well, as observers of the universe, we can try to deduce what the "true rule for the universe" is. But inevitably we have to do this by performing physical experiments, and then using inductive inference to try to figure out what the "rule for the universe is". But the issue is that as entities embedded within the universe, we can only ever do a finite number of experiments—and with these we'll never be able to precisely nail down the "true rule"; there'll always be some uncertainty.

When we think of ourselves as observers of the universe, there's in a sense lots of "arbitrariness" in the way we're set up. For example, we exist at a particular location in physical space—in our particular solar system and so on. Presumably we also exist at a particular location in branchial space, though it's less clear how to "name" that. And in addition we exist at a particular location in rulial space.

What determines that location? Essentially it's determined by how we operate as observers: the particular sensory system we have, and the particular means of description that we've developed in our language and in the history of knowledge in our civilization. In principle we could imagine sensing or describing our universe differently. But the way we do it defines the particular place in rulial space at which we find ourselves.

But what does all this mean in terms of the ruliad? The ruliad is the unique limiting structure formed by following all possible rules in all possible ways. But when we "observe the ruliad" we're effectively "paying attention to" only particular aspects of it. Some of that "paying attention" we can conveniently describe in terms of our particular "location in the ruliad". But some is more naturally described by thinking about equivalence classes in the ruliad.

Given two states that exist in the ruliad, we have to ask whether as observers we want to consider them distinct, or whether we want to conflate them, and consider them "the same". When we discussed the construction of the ruliad, we already had many versions of this issue. Indeed, whenever we said that two paths in the ruliad "merge", that's really just saying that we treat the outcomes as equivalent.

"Viewed from the outside", one could imagine that absolutely nothing is equivalent. Two hypergraphs produced in two different ways (and thus, perhaps, with differently labeled nodes) are "from the outside" in some sense different. But "viewed from the inside", they pretty much have to be viewed as "the same", in essence because all their effects will be the same. But at some level, even such conflation of differently labeled hypergraphs can be thought of as an "act of the observer"; something that one can only see works that way if one's "observing it from inside the system".

But all the way through our description of the observer, it's very much the same story: it's a question of what should be considered equivalent to what. In sequentializing time, we're effectively saying that "all of space" (or "all of branchial space", or rulial space) should be considered "equivalent". There are many subtle issues of equivalence that also arise in the construction of states in the ruliad from underlying tokens, in defining what rules and initial conditions should be considered the same, and in many other places.

The ruliad is in some sense the most complicated constructible object. But if we as computationally bounded observers are going to perceive things about it, we have to find some way to "cut it down to size". And we do that by defining equivalence classes, and then paying attention only to those whole classes, not all the details of what's going on inside them. But a key point is that because we are computationally bounded observers who imagine a certain coherence in their experience, there are strong constraints on what kinds of equivalence classes we can use.

If we return again to the situation of molecules in a gas, we can say that we form equivalence classes in which we look only coarsely at the positions of molecules, in "buckets" defined by simple, bounded computations—and we don't look at their finer details, with all the computational irreducibility they involve. And it's because of this way of looking at the system that we conclude that it follows the Second Law of thermodynamics, exhibits fluid behavior, etc.

And it's very much the same story with the ruliad—and with the laws of physics. If we constrain the kind of way that we observe—or "parse"—the ruliad, then it becomes inevitable that the effective laws we'll see will have certain features, which turns out apparently to be exactly what's needed to reproduce known laws of physics. The full ruliad is in a sense very wild; but as observers with certain characteristics, we see a much tamer version of it, and in fact what we see is capable of being described in terms of laws that we can largely write just in terms of existing mathematical constructs.

At the outset, we might have imagined that the ruliad would basically just serve as a kind of dictionary of possible universes—a "universe of all possible universes" in which each possible universe has different laws. But the ruliad is in a sense a much more complicated object. Rather than being a "dictionary" of possible separate universes, it is something that entangles together all possible universes. The Principle of Computational Equivalence implies a certain homogeneity to this entangled structure. But the crucial point is that we don't "look at this structure from the outside": we are instead observers embedded within the structure. And what we observe then depends on our characteristics. And it turns out that even very basic features of our consciousness and sensory apparatus in a sense inevitably lead to known laws of physics—and in a sense do so generically, independent of details of just where in rulial space we are, or exactly what slice of the ruliad we take.

So far we've primarily talked about the ruliad in terms of physics and the fundamental structure of our physical universe. But the ruliad is actually something still more general than that. Because ultimately it is just created from the abstract concept of following all possible computational rules. And, yes, we can interpret these rules as representing things going on in our universe. But we can also interpret them as representing things going on in some other, less immediately physically realizable system. Or, for that matter, representing something purely formal, and, say, mathematical.

This way of talking about the ruliad might make one think that it should be "considered a possible model" for our universe, or for other things. But the bizarre and surprising point is that it is more than that. It's not just a possible model that might be one of many. Rather, it is the unique ultimate representation of all possible models, entangled together. As we've discussed, there are many subtle choices about how we observe the ruliad. But the ultimate ruliad itself is a unique thing, with no choice about what it is.

As I've discussed at more length elsewhere, the ruliad is in a sense a representation of all possible necessary truths—a formal object whose structure is an inevitable consequence of the very notion of formalization. So how does this relate to the idea that the ruliad also at an ultimate level represents our physical universe? What I've argued elsewhere is that it means that the ultimate structure of our universe is a formal necessity. In other words, it's a matter of formal necessity that the universe must exist, and have an ultimate ruliad structure. The fact that we perceive the universe to operate in a certain way—with our standard laws of physics, for example—is then a consequence of the particular way observers like us perceive it, which in turn depends on things like where in rulial space we happen to find ourselves.

But beyond physics, what else might the ruliad represent? The ruliad is an ultimate example of multicomputation, and of what I've characterized as the fourth major paradigm for theoretical science. Often in multicomputation, what's of interest is multiway systems with specific underlying rules. And already at this level, much of the apparatus that we've described in connection with the ruliad also applies—and in a sense "trickles down" to give various universal results.

But there are also definitely cases of multicomputation (other than physics) where the full notion of applying all possible rules is relevant. The global structures of metamathematics, economics, linguistics and evolutionary biology seem likely to provide examples—and in each case we can expect that at the core is the ruliad, with its unique structure. Of course, this doesn't mean that what we observe must always be the same, because what we observe depends on our characteristics as an observer—and the characteristics of "being an observer" in metamathematics, for example, are surely different from those for economics or evolutionary biology, or, for that matter, physics.

For sure, the "sensory apparatus" that we effectively use is different in different cases. But there are certain similar human-based features that still seem to apply. Whatever the domain, we always act as computationally bounded observers. And it seems that we also always have a certain coherence, consistently maintaining our "observerhood" through time or across some form of space. And it seems likely that these "human-induced" characteristics alone are sufficient to yield some very global implications for observed behavior.

The View from Mathematics

How should we think about the ruliad mathematically? In many ways, the ruliad is more an object of metamathematics than of mathematics itself. For in talking about the effects of all possible rules, it in a sense transcends individual mathematical theories—to describe a kind of metatheory of all possible theories.

Given a particular mathematical axiom system, it's rather easy to see correspondence with a multiway system. There are a variety of ways to set it up, but one approach is to think of states in the multiway system as being expressions in the language used for the axiom system, and then to think of rules in the multiway system as applying transformations on these expressions that implement axioms in the axiom system.

For example, with the (Abelian semigroup) axioms

$$\{a \cdot (b \cdot c) = (a \cdot b) \cdot c, a \cdot b = b \cdot a\}$$

here's a multiway system generated from the expression $y \cdot (x \cdot (y \cdot x))$ by applying the (two-way) transformations defined by the axioms in all possible ways to each expression:

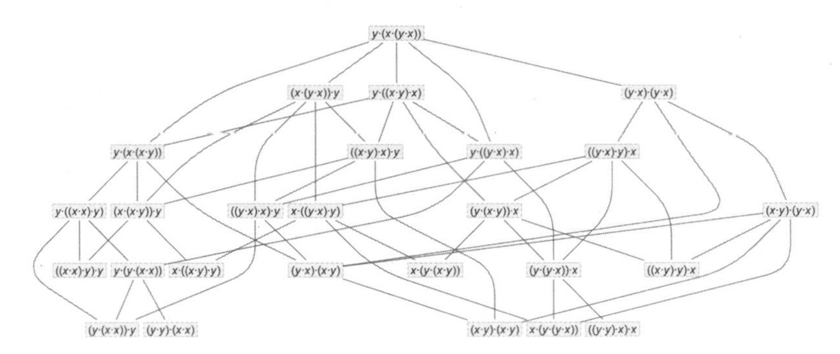

But now from this graph we can read off the "theorem":

$$y \cdot (x \cdot (y \cdot x)) = (x \cdot y) \cdot (x \cdot y)$$

A proof of this theorem

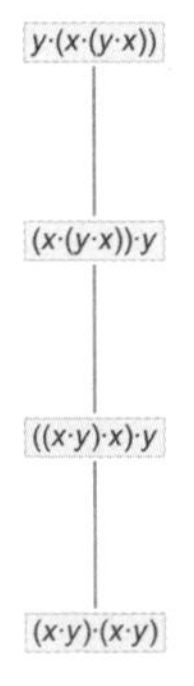

is just a path in the multiway graph:

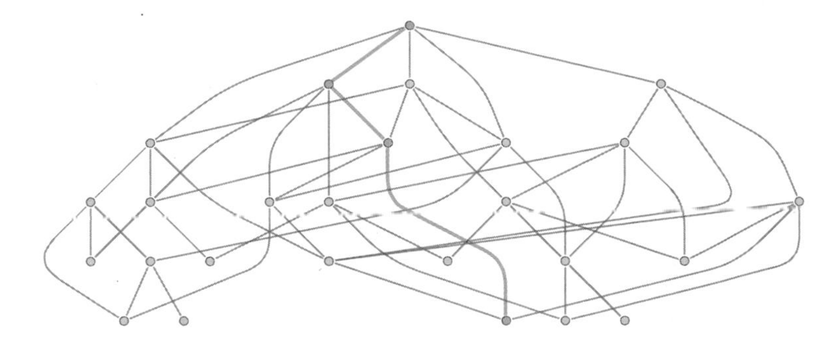

A significantly less direct but still perfectly valid proof would correspond to the 13-step path:

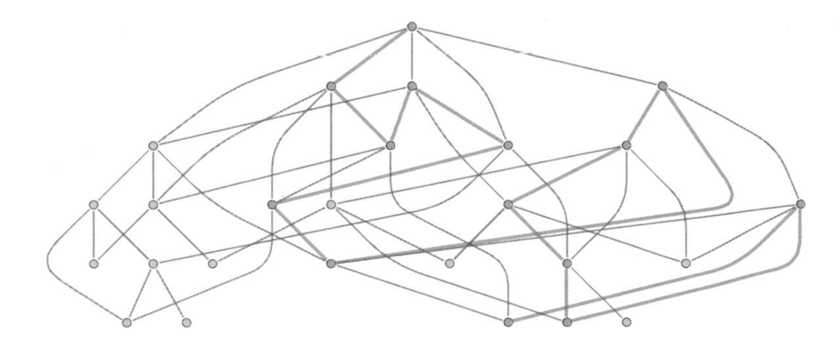

It's a slightly technical point, but perhaps worth mentioning that there are alternative ways to set up the correspondence between axiomatic mathematical systems and multiway systems. One such way is to take the states in the multiway system to be not expressions (like $a \cdot b$) but whole propositions (like $a \cdot b = b \cdot a$). Axioms then show up as states, and the rules for the multiway system are "rules of inference" or "rules of entailment" that define how, say, pairs of propositions "entail" (or "imply") other ones. (And, yes, this requires a generalized multiway system with not just 1 state → many states, but for example 2 states → 1 state.) Typical automated theorem provers (like FindEquationalProof) operate in this kind of setup, attempting to find paths that lead from initial propositions and axioms to some final state that corresponds to an "obviously true" proposition, like $a = a$.

But whatever the detailed setup, the basic picture is that an axiomatic mathematical system has an associated multiway graph, in which paths correspond to proofs. Given the rules for the multiway system, there is in general no way to guarantee that the path (if it exists) corresponding to the proof of some particular result will be of bounded length, leading to the possibility of undecidability. But even when a path exists, it may require an irreducibly large amount of computation to find it. Still, finding such paths is what automated theorem provers do. For example, we know (as I discovered in 2000) that $((b \cdot c) \cdot a) \cdot (b \cdot ((b \cdot a) \cdot b)) = a$ is the minimal axiom system for Boolean algebra, because FindEquationalProof finds a path that proves it.

But this path—and the corresponding proof—is a very "non-human" construct (and, for example, in 21 years essentially no progress has been made in finding a "human-understandable narrative" for it). And we can make an analogy here to the situation in physics. The individual rule applications in the multiway graph (or the proof) are like individual updating events applied to the atoms of space—and they show all kinds of complexity and computational irreducibility. But in physics, human observers work at a higher level. And the same, one suspects, is true in mathematics.

Rather than looking at every detail of the multiway graph, human "mathematical observers" (i.e. pure mathematicians) in effect define all sorts of equivalences that conflate together different parts of the graph. If the individual updates in the multiway graph are like molecular dynamics, human pure mathematics seems to operate much more at the "fluid dynamics level", concentrating on "broad mathematical constructs", not the "machine code" of specific low-level axiomatic representations. (Of course, there are some situations, for example related to undecidability, where the "molecular dynamics" effectively "breaks through".)

We've outlined above (and discussed at length elsewhere) how physical observers like us "parse" the low-level structure of the physical universe (and the ruliad). How might mathematical observers do it? A large part has to do with the identification of equivalences. And the key idea is that things which are considered equivalent should be assumed to be "the same", and therefore "conflated for mathematical purposes".

The most elementary example of something like this is the statement (already present in Euclid) that if $x=y$ and $x=z$, then $y=z$. The extensionality axiom of set theory is a more sophisticated example. And the univalence axiom of homotopy type theory is perhaps the most sophisticated current version.

There's a very operational version of this that appears in automated theorem proving. Imagine that you've proved that $x=y$ and $x=z$. Then (by the assumed properties of equality) it follows that $y=z$. One way we could use this result is just to merge the nodes for y and z. But a "bigger" thing we can do is to add the "completion" $y=z$ as a general rule for generating the multiway system.

Consider, for example, the string substitution multiway system A ↔ AB:

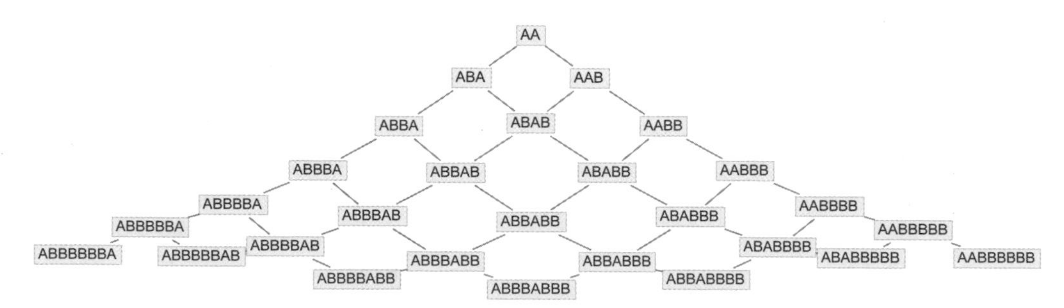

But notice here that both ABA ↔ ABBA and ABA ↔ ABAB. So now add the "completion" ABBA ↔ ABAB. Here's the resulting multiway graph:

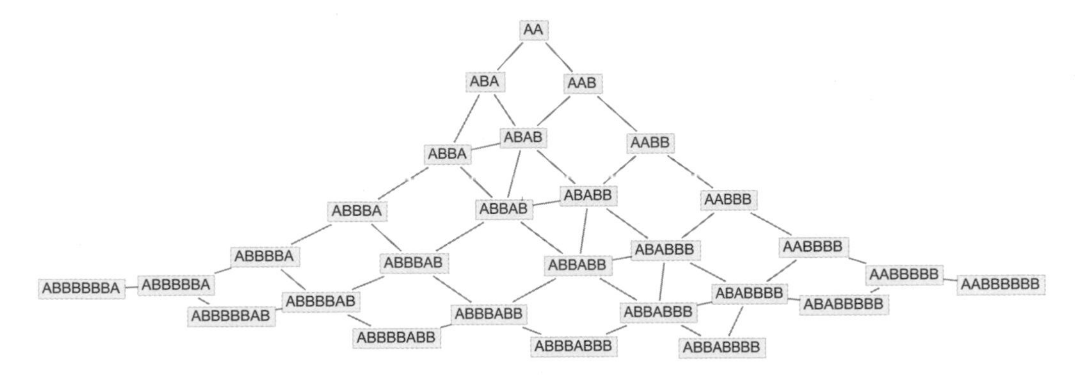

Adding the completion has generated lots of new "direct equivalences". But we can also think of it as having done something else: not only has it defined equivalences between states; it's also defined equivalences between paths—or in effect between proofs. (Or, put another way, it's implementing homotopic equivalence between proofs. By the way, it's an important feature of human mathematics that progress is typically measured in theorems proved; different proofs of the same theorem are normally implicitly considered equivalent in terms of the progress of mathematics.)

In a category theory interpretation, transformations between states in the original multiway graph are like ordinary morphisms (1-morphisms). But when we're making transformations between "proof paths", this is like 2-morphisms. And just as we can add transformations between proofs, we can also add transformations between proofs-between-proofs, and so on. The result is that we can build up a whole hierarchy of higher and higher categories, eventually ending with an ∞-category.

But now we can begin to see the connection with the ruliad. The ruliad is in a sense the result of taking the limit of adding more and more possible rules. Above we did this quite explicitly in terms of the original underlying rules, for example by enumerating possible strings or possible integer multipliers.

But we can view successive completions as doing something very similar. Yes, the rules are enumerated in a different order. But in the end there'll still be an infinite number of distinct rules being used.

Of course there are many mathematical details missing here. But in broad outline, it seems that one can think of the approach to the ruliad as some kind of limit of successively higher categories. But this limit is something that's been studied (albeit in the upper reaches of pure mathematics), and it's an object called the ∞-groupoid. (It's a groupoid because when all the rules are included things inevitably "go both ways"). So, OK, is the ruliad "just" the ∞-groupoid? Not quite. Because there are more rules and more initial conditions in the ruliad, even beyond those added by completions. And in the end the ruliad actually seems to be the ∞-category of ∞-groupoids, or what's called the (∞,1)-category.

But knowing that the ruliad can be thought of as composed of ∞-groupoids means that we can apply mathematical ideas about the ∞-groupoid to the ruliad.

Probably the most important is Grothendieck's hypothesis, which asserts that the ∞-groupoid inevitably has a topological and (with a few other conditions) ultimately geometric structure. In other words, even though one might have imagined that one constructed the ∞-groupoid from "pure logic" (or from pure formal axiomatic structures), the assertion is that the limiting object one obtains inevitably exhibits some kind of geometrical or "spatial" structure.

Viewed in terms of the ruliad—and our explicit finite examples of it—this might not seem surprising. And indeed in our Physics Project, the whole concept of the emergence of space from large-scale hypergraphs is closely related. But here from Grothendieck's hypothesis we're basically seeing a general claim that the ruliad must have "inevitable geometry"—and we can then view things like the emergence of space in our Physics Project as a kind of "trickle down" from results about the ruliad. (And in general, a big "application" of geometrical structure is the possibility of "pure motion".)

What does all this mean about the ruliad and mathematics? In a sense the ruliad represents all possible mathematics—the application of all possible rules, corresponding to all possible axiom systems. And from this "ultimate metamathematics", human "mathematical observers" are sampling pieces that correspond to the pure mathematics they consider of interest.

Perhaps these will align with particular axiom systems of the kind automated theorem provers (or proof assistants) use. But things may be "sloppier" than that, with human mathematical observers effectively being extended in rulial space—and capable of making "fluid-dynamics-level" conclusions, even if not "molecular-dynamics-level" ones.

But a key (and in some ways very surprising) point is that the ruliad can be viewed as the basis of both physics and mathematics. In some sense, physics and mathematics are at their core the same thing. They only "appear different" to us because the way we "observe" them is different.

I plan to discuss the implications for mathematics at greater length elsewhere. But suffice it to say here that the existence of a common underlying core—namely the ruliad—for both physics and mathematics immediately allows one to start importing powerful results from physics into mathematics, and vice versa. It also allows one, as I have done elsewhere, to start comparing the existence of the universe with the (Platonic-style) concept of the fundamental existence of mathematics.

The View from Computation Theory

The ruliad can be thought of as an encapsulation of doing all possible computations in all possible ways. What we might think of as a "single computation" might consist of repeatedly applying the rules for a Turing machine to "deterministically" generate a sequence of computational steps:

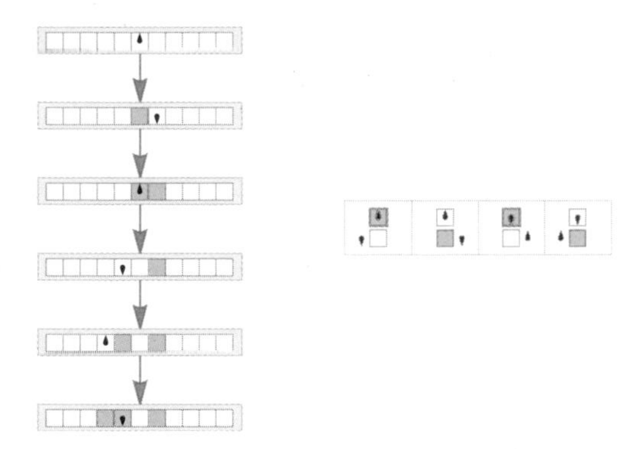

But one can also consider a "multicomputational" system, in which rules can generate multiple states, and the whole evolution of the system can be represented by a multiway graph:

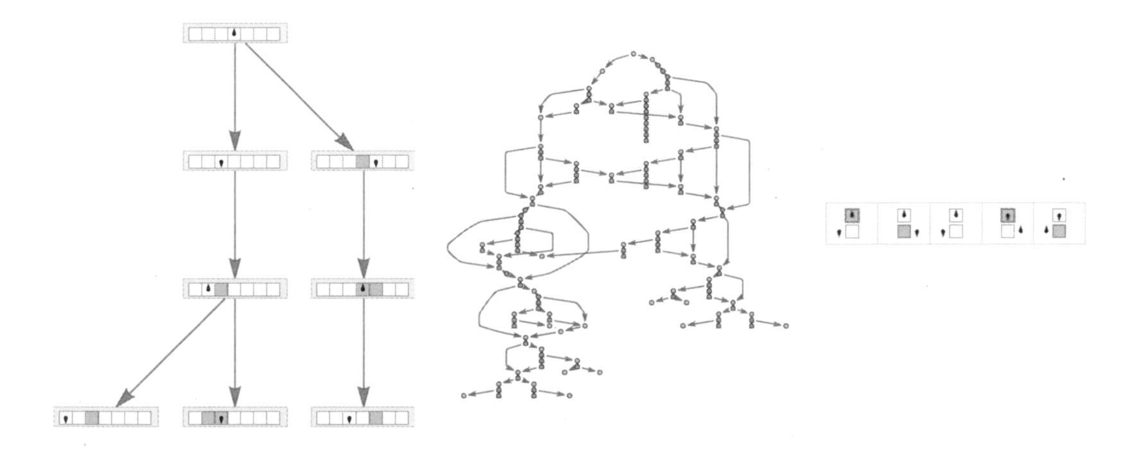

In traditional computation theory, one talks about rules like these as "nondeterministic", because they can have multiple outcomes—though one usually imagines that the final answer one wants from a computation can be found as the result of some particular path. (In what I now call the multicomputational paradigm—that I believe is important for modeling in physics and other places—one instead considers the complete multiway graph of entangled possible histories.)

In constructing the ruliad, one is in a sense going to a more extreme version of multicomputation, in which one uses not just a particular rule with multiple outcomes, but all possible rules. In effect, the concept is to use "maximal nondeterminism", and at each step to independently "pick whatever rule one wants", tracing out a rulial multiway system that includes all the different possible paths this generates.

For the kind of Turing machines illustrated above, the rulial multiway graph one gets after one step is:

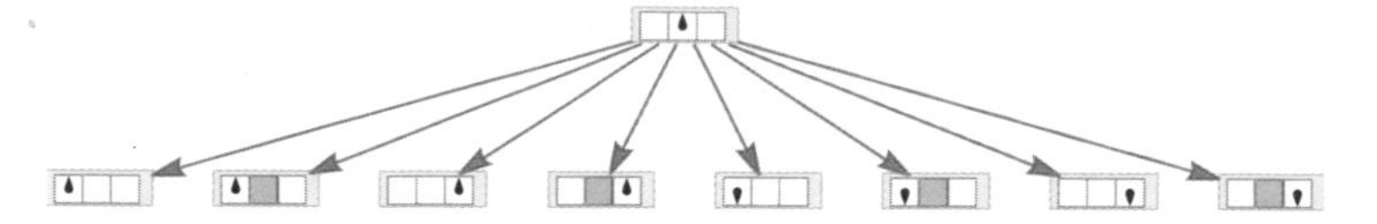

After 2 steps the result is:

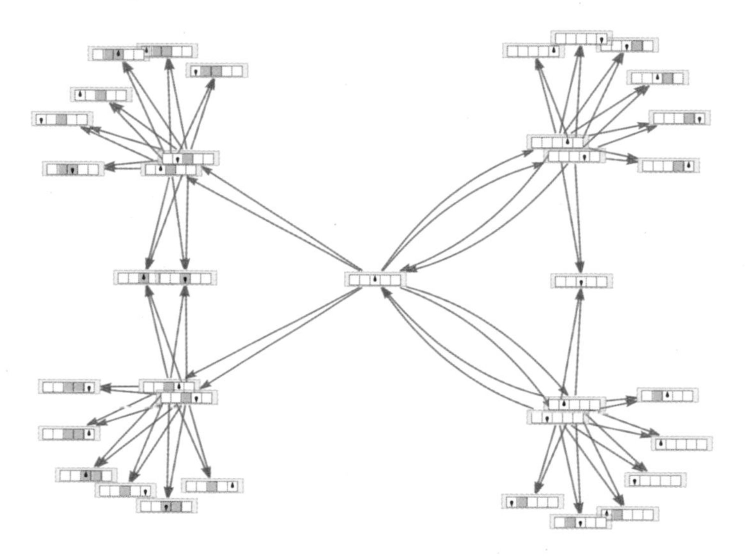

The full ruliad is then some kind of infinite limit of this process. But as before, there's lots of subtlety in how this limit is taken. But we can at least characterize some ways of approaching it using ideas from computational complexity theory. Increasing the number of steps of evolution is like increasing the time complexity one allows. Increasing the "size of states" (e.g. the width of nonzero Turing machine tape) that one includes is like increasing the space complexity one allows. And increasing the complexity of the rule (as measured in the number of bits needed to specify it) is like increasing the algorithmic complexity one allows. The ruliad is what is obtained by taking all these computational resource measures to infinity.

And a critical claim is that regardless of how this is done, the final ruliad construct one gets will always—at least in some sense—be the same. There will be many ways to coordinatize it, or to sample it, but the claim is that it's always the same object that one's dealing with. And ultimately the reason for this is the Principle of Computational Equivalence. Because it implies that whatever "computational parametrization" or "computational description language" one uses for the ruliad, one will almost always get something that can be viewed as "computationally equivalent".

We've talked about building up the ruliad using Turing machines. But what about other models of computation—like cellular automata or register machines or lambda calculus? As soon as there's computation universality we know that we'll get results that are at least in principle equivalent, because in a sense there's only a "finite translation cost" associated with setting up an interpreter from one model of computation to another. Or, put another way, we can always emulate the application of the rule for one system by just a finite number of rule applications for the other system.

But from computation universality alone we have no guarantee that there won't be "extreme deformations" introduced by this deformation. What the Principle of Computational Equivalence says, however, is that almost always the deformations won't have to be extreme. And indeed we can expect that particularly when multiple rules are involved, there'll be rapid convergence almost always to a kind of "uniform equivalence" that ensures that the final structure of the ruliad is always the same.

But the Principle of Computational Equivalence appears to say still more about the ruliad: it says that not only will the ruliad be the same independent of the "computational basis" used to construct it, but also that there'll be a certain uniformity across the ruliad. Different "regions of the ruliad" might involve different specific rules or different patterns of their application. But the Principle of Computational Equivalence implies that almost always the computations that happen will be equivalent, so that—at least at a certain scale—the structure associated with them will also be equivalent.

Knowing that the ruliad contains so many different computations, one might imagine that it would show no particular uniformity or homogeneity. But the Principle of Computational Equivalence seems to imply that it necessarily does, and moreover that there must be a certain coherence to its structure—that one can interpret (in the style of Grothendieck's hypothesis) as an inevitable emergent geometry.

An individual computation corresponds to a path in the ruliad, going from its "input state" to its "output state". In an ordinary deterministic computation, the path is restricted to always use the same rule at each step. In a nondeterministic computation, there can be different rules at different steps. But now we can formulate things like the P vs. NP problem essentially in terms of the geometry of the ruliad.

Here's a picture of the same finite Turing-machine-based approximation to the ruliad as above—but now with the paths that correspond to deterministic Turing machine computations marked in red:

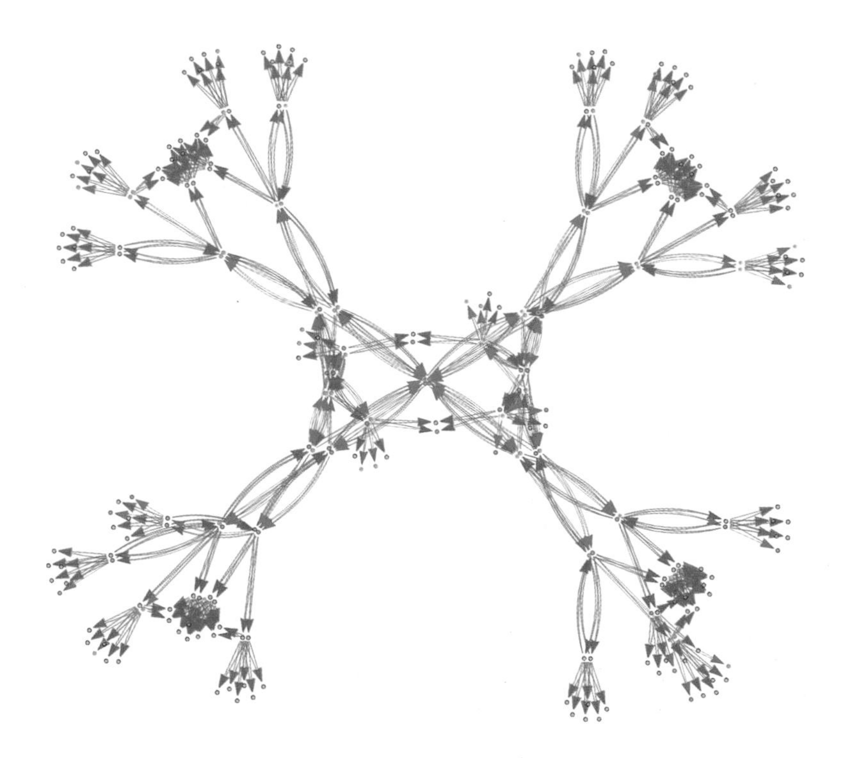

The P vs. NP problem basically asks roughly whether the deterministic computations (shown here in red) will eventually "fill the ruliad", or whether the general nondeterministic computations that are part of the ruliad will always "reach further". Once again, there are many complicated and subtle issues here. But it's interesting to see how something like the P vs. NP problem might play out in the ruliad.

In physics (and mathematics), we as human observers tend to sample the ruliad in a coarse-grained way, "noticing" only certain aspects of it. So is there an analog of this in computation theory—perhaps associated with certain characteristics of the "computation-theoretic observer"? There's a potential answer, rather similar to what we've already seen in both physics and mathematics.

The basic point is that in computation theory we tend to study classes of computations (say P or NP) rather than individual computations. And in doing this we are in a sense always conflating many different possible inputs and possible outputs—which we assume we do in a computationally bounded way (e.g. through polynomial-time transformations, etc.).

Another thing is that we tend to focus more on the "experience of the end user" than the detailed activities of the "programmer". In other words, we're concerned more with what computational results are obtained, with what computational resources, rather than on the details of the program constructed to achieve this. Or, put another way, we tend to think about computation in terms of things like the successive evaluation of functions—and we conflate the different paths by which this is achieved.

Most likely this means that there are "effective laws" that can be derived in this computational view of the ruliad, analogous to laws of physics like general relativity. So what might some other analogies be?

A computation, as we've mentioned, corresponds to a path in the ruliad. And whenever there's a possibility for an infinite path in the ruliad, this is a sign of undecidability: that there may be no finite way to determine whether a computation can reach a particular result. But what about cases when many paths converge to a point at which no further rules apply, or effectively "time stops"? This is the analog of a spacelike singularity—or a black hole—in the ruliad. And in terms of computation theory, it corresponds to something decidable: every computation one does will get to a result in finite time.

One can start asking questions like: What is the density of black holes in rulial space? If we construct the ruliad using Turing machines, this is basically analogous to asking "What's the density of halting Turing machines (+initial conditions) in rulial space?" And this is essentially given by Chaitin's Ω.

But so is there some number Ω that we can just compute for the ruliad? Well, actually, no. Because the undecidability of the halting problem makes Ω noncomputable. One can get approximations to it, but—in the language of the ruliad—those will correspond to using particular samplings or particular reference frames. Or in other words, even the perceived density of "decidability black holes" in the ruliad depends on features of the observer.

What's beyond the Ruliad?

In our Physics Project we usually talk of the universe "evolving through time" (albeit with many entangled threads of history). But if the ruliad and its structure is a matter of formal necessity, doesn't that mean that the whole ruliad effectively "already exists"—"outside of time"? Well, in some sense it does. But ultimately that would only be relevant to us if we could "look at the ruliad from the outside".

And as observers like us within the ruliad, we necessarily have a different perception. Because our consciousness—with its computational boundedness—only gets to sample a certain sequence of pieces of the ruliad. If it were not for computational irreducibility, we might get to "jump around" in time. But computational irreducibility, together with our own computational boundedness, implies that our perception must necessarily just experience the passage of time through an irreducible process of computation.

In other words, while in some sense the ruliad may all "already be there" when viewed from the outside, our own perception of it "from the inside" is necessarily a progressive one, that effectively corresponds to the passage of time.

Could we experience the ruliad differently, even while being computationally bounded? If we think of the ruliad as a graph, then our usual "sequence of configurations of space at successive times" way of experiencing the ruliad is like a breadth-first traversal. But could we for example instead do a depth-first traversal, exploring all time before investigating different parts of space? (And, yes, something like this can happen in general relativity near an event horizon, or in connection with timelike singularities.)

Later, we'll discuss different ways to perceive the ruliad and the universe. But it seems to be a feature of anything we might call a coherent observer that there needs to be some form of progression in the perception. And so while we might not call it the passage of time, there'll still be some way in which our exploration of the ruliad has a computationally irreducible process underneath.

A very important claim about the ruliad is that it's unique. Yes, it can be coordinatized and sampled in different ways. But ultimately there's only one ruliad. And we can trace the argument for this to the Principle of Computational Equivalence. In essence there's only one ruliad because the Principle of Computational Equivalence says that almost all rules lead to computations that are equivalent. In other words, the Principle of Computational Equivalence tells us that there's only one ultimate equivalence class for computations.

But what if we just imagine a "hypercomputation" not in that class? For example, imagine a hypercomputation (analogous, for example, to an oracle for a Turing machine) that in a finite number of steps will give us the result from an infinite number of steps of a computationally irreducible process. Such a hypercomputation isn't part of our usual ruliad. But we could still formally imagine a hyperruliad that includes it—and indeed we could imagine a whole infinite hierarchy of successively larger and more powerful hyperruliads.

But it's a fundamental claim that we're making—that can be thought of as a matter of natural science—that in our universe only computation can occur, not hypercomputation.

At a purely formal level, there's nothing wrong with hyperruliads. They exist as a matter of formal necessity just like the ordinary ruliad does. But the key point is that an observer embedded within the ruliad can never perceive a hyperruliad. As a matter of formal necessity there is, in a sense, a permanent event horizon that prevents anything from any hyperruliad from affecting anything in the ordinary ruliad.

So now we can be a bit more precise about our statement that "hypercomputation doesn't happen in our universe". Really we should say that we assert that we as observers operate purely computationally and not hypercomputationally. And this means that we are embedded within the ordinary ruliad, and not the hyperruliad.

Yes, we could imagine some other entity that's embedded within the hyperruliad, and perceives what it considers to be the universe to operate hypercomputationally. But in a statement that's in a sense more "about us" than "about the universe", we assert that that can't be us, and that we in a sense live purely within the ruliad—which means that for us the Principle of Computational Equivalence holds, and we perceive only computation, not hypercomputation.

Communicating across Rulial Space

What observers can there be embedded in the ruliad, and how should we characterize them? In physical spacetime we're used to characterizing observers by their locations in physical space and by things like the spacetime reference frames they construct. And it's very much the same for observers in the ruliad: we can characterize them by where they are in rulial space, and what rulial reference frames they use.

The Principle of Computational Equivalence tells us that it's almost always possible to "encode" one "model of how the ruliad works" in any other model—effectively just by setting up a program that emulates the rules for one model using the rules for the other model. But we can think of these different models as being associated with different possible observers in the ruliad.

In other words, we can say that observers "at different places in rulial space" (or "using different rulial reference frames") are using different description languages for what's happening in the ruliad. And when an observer "moves" in rulial space, they're effectively doing a translation from one description language to another. (And, yes, there's a maximum rate of motion ρ in rulial space—which is the rulial analog of the speed of light—and which is effectively determined by the fundamental processing speed of the universe.)

So far this might all seem quite abstract. But there are immediate, everyday examples that effectively correspond to being at different places in rulial space. A simple concrete one is computers with different instruction sets. Another one is different brains with different consciousnesses.

We can think of a single human consciousness as having a certain thread of experience of the universe. Part of that experience is determined by the physical location of the consciousness and by the sensory apparatus with which it samples the world. But part is determined by the "internal description language" that it uses. And inevitably this internal description language depends both on the detailed physiology of the brain in which it is implemented, and on the past history of experiences that have "defined its way of looking at the world". In the analogy of artificial neural networks, different networks will tend to have different "internal representations" because this depends not only on the network architecture, but also on the particular training data that the network has "experienced".

Why can't one human consciousness "get inside" another? It's not just a matter of separation in physical space. It's also that the different consciousnesses—in particular by virtue of their different histories—are inevitably at different locations in rulial space. In principle they could be brought together; but this would require not just motion in physical space, but also motion in rulial space.

But why then do different consciousnesses seem to have compatible views about "what happens in the universe"? Essentially this can be seen as a consequence of rulial relativity—which in turn depends on the inevitable causal invariance of the ruliad, which follows from the Principle of Computational Equivalence. There are certainly many issues to be worked out, but basically what seems to be going on is that because of causal invariance, different rulial reference frames will ultimately yield the same rulial multiway causal graphs, and therefore the same "fundamental description of reality".

We've talked about different consciousnesses. But what about just "different ways of thinking"? Well, it's definitely more than an analogy to say that different ways of thinking correspond to different positions in rulial space. If there's lots of common history then there'll be common ancestry in the rulial multiway graph and one will necessarily end up close in rulial space. But without common history, one can end up with different description languages—or different ways of thinking—that are not nearby in rulial space.

In physical space we expect to effectively use momentum to move our location. And it's potentially a bizarrely similar story in rulial space. In our models of fundamental physics, energy and momentum are essentially related to the density of activity (i.e. elementary updating events) in physical space. And we can similarly define a rulial analog of energy and momentum in terms of activity in rulial space. But it's exactly this activity that provides connections between different parts of rulial space, or in effect "enables motion" in rulial space.

In other words, if you want to move in rulial space, you can do it by putting in the appropriate computational work to change your conceptual point of view (or, essentially equivalently, your language for describing things). So what about curvature (or the analog of gravity) in rulial space—say generated through an analog of Einstein's equations from density of activity in rulial space? Presumably this relates to the difficulty—or time it takes—to get from one place in rulial space, and one way of thinking, to another. And conceivably things like "paradigm shifts" between different ways of thinking might be associated with features of rulial space like event horizons.

But let's say you're at one place in rulial space, and you want to get to another—or at least "send a signal" there. A typical microscopic change at one point in rulial space will tend to just "spread out in all directions" and "decay quickly". But if you want to "coherently communicate", you need some kind of structure that will persist as it propagates through rulial space. And by analogy with the case of physical space, what this presumably means is that you effectively need a "rulial particle".

In terms of the ruliad, a rulial particle would presumably be some kind of "topological obstruction" or "topologically stable structure" that is at any moment effectively localized in rulial space and maintains its identity as it propagates across rulial space. But what might a rulial particle be in more everyday terms?

Potentially it's like what we'd normally consider a concept—or something to which we might assign a word in human language. If we have ways of thinking—or consciousnesses—whose details are different, the issue is what will be robust enough to be able to be transported between them. And what everyday experience seems to suggest is that the answer is concepts. Even though one might have a slightly different way of thinking, what one calls "a fish" (or essentially, the concept of a fish) is something that can still robustly be communicated.

It's interesting to notice that for an observer like us, there seem to be only a finite set of types of "elementary particles" that exist in physical space. And perhaps that's not unrelated to the fact that observers like us also seem to imagine that there are in some sense only a finite number of "basic concepts" (associated, say, with distinct words in human languages). There's lots more detail that exists in rulial space—or in the ruliad—but for observers like us, with our type of way of sampling the ruliad, these might be core coherent structures that we perceive.

So Is There a Fundamental Theory of Physics?

The concept of the ruliad arose from our efforts to find a fundamental theory of physics. But now that we know about the ruliad, what does it tell us about a fundamental theory?

At the outset, we might have imagined that the end point of our project would be the identification of some particular rule of which we could say "This is the rule for the universe". But of course then we'd be faced with the question: "Why that rule, and not another?" And perhaps we would imagine just having to say "That's something that you have to go beyond science to answer".

But the ruliad implies a quite different—and in my view ultimately much more satisfying—picture. The ruliad itself is a construct of abstract necessity—that in a sense represents the entangled behavior of all possible rules for the universe. But instead of imagining that some particular rule out of all these possibilities is "picked from outside" as "the choice for our universe", what we suppose is that—as observers embedded within the ruliad—we're the ones who are implicitly picking the rule by virtue of how we sample and perceive the ruliad.

At first this might seem like it's a wimp out. We want to know how our universe works. Yet we seem to be saying "we just pick whatever rule we feel like". But that's not really the story at all. Because in fact observers that are even vaguely like us are in effect deeply constrained in what rules they can attribute to the universe. There's still some freedom, but a fundamental result is that for observers like us it seems to be basically inevitable that any rule we can pick will on a large scale reproduce the central known general laws of physics, in particular general relativity and quantum mechanics.

In other words, for observers generally like us it's a matter of abstract necessity that we must observe general laws of physics that are the ones we know. But what about more specific things, like the particular spectrum of elementary particles, or the particular distribution of matter in the universe? It's not clear how far "the general" goes—in other words, what is a matter of abstract necessity purely from the structure of the ruliad and general features of observers like us.

But inevitably at some point we will run out of "the general". And then we'll be down to specifics. So where do those specifics enter? Ultimately they must be determined by the details of how we sample the ruliad. And a prominent part of that is simply: Where in the ruliad are we? We can ask that about our location in physical space. And we can also ask it about our location in rulial space.

What does all this mean? At some level it's saying that the way we are as observers is what makes us attribute certain rules to our universe. The ruliad is in a sense the only thing that fundamentally exists—and in fact its existence is a matter of abstract necessity. And our universe as we experience it is some "slice of the ruliad", with what slice it is being determined by what we're like as observers.

Let's look at the logical structure of what we're saying. First, we're describing the ruliad, which at the outset doesn't have anything specifically to do with physics: it's just a formal construct whose structure is a matter of abstract necessity, and which relates as much to mathematics as it does to physics. But what "puts the physics in" is that we in effect "live in the ruliad", and our perception of everything is based on "experiencing the ruliad". But that experience—and the effective laws of physics it entails—inevitably depends on "where we are in the ruliad" and how we're able to sample it.

And this is where our pieces of "falsifiable natural science" come in. The first "assertion of natural science" that we make is that we are embedded only within the ordinary ruliad, and not a hyperruliad—or in other words that our experience encompasses only computation, and not hypercomputation.

This is closely related to a second assertion, which may in fact be considered to subsume this: that we are computationally bounded observers, or, in other words, that our processes of perception involve bounded computation. Relative to the whole ruliad—and all the computation it entails—we're asserting that we as observers occupy only a tiny part.

There's one more assertion as well, again related to computational boundedness: that we as observers have a certain coherence or persistence. In general the ruliad contains all sorts of wild and computationally irreducible behavior. But what we're asserting is that that part of the ruliad that is associated with us as observers has a certain simplicity or computational reducibility: and that as we evolve through time or move in space, we somehow maintain our identity.

These assertions seem very general, and in some ways almost self-evident—at least as they apply to us. But the important and surprising discovery is that they alone seem to lead us inexorably to crucial features of physics as we know it.

Where does this physics "come from"? It comes partly from the formal structure of the ruliad, and formal features of the multicomputational processes it involves. And it comes partly from the nature of us as observers. So if we ask "Why is the physics of our universe the way it is?", an important part of the answer is "Because we observe the universe the way we do".

One might imagine that in some sense physics would give us no choice about how we observe the universe. But that's not the case. Because in the end our "observation" of the universe is about the "abstract conceptual model" we build up for the universe. And, yes, that's certainly informed by the particular sensory apparatus we have, and so on. But it's something we can certainly imagine being different.

We can think of ourselves as using some particular description language for the universe. The structure of that language is constrained by the assertions we gave above. But within such a description language, the laws of physics necessarily work out the way they do. But if we chose a different description language, we'd end up with different laws of physics.

Much of our perception of the universe is based on our raw biological structure—the way our sensory organs (like our eyes) work, as well as the way our brains integrate the inputs we get. But that's not all there is to it. There's also a certain base of knowledge in our civilization that informs how we parse our "raw perception"—and in effect what description language we use. Once we have the idea of periodic behavior, say, we can use it to describe things that we'd previously have to talk about in a less "economical" way.

But what if our knowledge changed? Or we had different sensory capabilities? Or we used technology to integrate our sensory input in different ways? Then we'd be able to perceive and describe the universe in different ways.

One's first impression might be that the ruliad effectively contains many possible "parallel universes", and that we have selected ourselves into one of these, perhaps as a result of our particular characteristics. But in fact the ruliad isn't about "parallel universes", it's about universes that are entangled at the finest possible level. And an important consequence of this is that it means we're not "stuck in a particular parallel universe". Instead, we can expect that by somehow "changing our point of view", we can effectively find ourselves in a "different universe".

Put another way, a given description of the universe is roughly represented by being at a certain location in rulial space. But it's possible to move in rulial space—and end up with a different description, and different effective laws for the universe.

But how difficult is motion in rulial space? It could be that some impressive future technology would allow us to "move far enough" to end up with significantly different laws of physics. But it seems more likely that we'd be able to move only comparatively little—and never be able to "escape the box" of things like computational boundedness, and coherence of the observer.

Of course, even changing a little might lead us to different detailed laws of physics—say attributing a different mass to the electron, or a different value of the electromagnetic coupling constant α. But actually, even in traditional physics, this is already something that happens. When viewed at different energy scales—or in a sense with different technology—these quantities have different effective values (as characterized by the renormalization group).

At first it might seem a little strange to say that as our knowledge or technology change, the laws of physics change. But the whole point is that it's really our perceived laws of physics. At the level of the raw ruliad there aren't definite laws of physics. It's only when we "sample our slice" of the ruliad that we perceive definite laws.

What does all this mean operationally for the search for a fundamental theory of physics? At some level we could just point to the ruliad and declare victory. But this certainly wouldn't give us specific predictions about the particulars of our perceived universe. To get that we have to go further—and we have to be able to say something about what "slice of the ruliad" we're dealing with. But the good news is that we don't seem to have to make many assumptions about ourselves as observers to be able to identify many physical laws that observers like us should perceive.

So can we ever expect to nail down a single, specific rule for the universe, say one a particular observer would attribute to it? Given our characteristics as observers, the answer is undoubtedly no. We're simply not that small in rulial space. But we're not that big, either. And, importantly, we're small enough that we can expect to "do science" and consider the universe to "behave in definite ways". But just as in physical space we're vastly larger than the scale associated with the atoms of space, so similarly we're also undoubtedly vastly larger in rulial space than the individual components of the ruliad—so we can't expect our experience to all be "concentrated in one thread" of the ruliad, following one particular rule.

As we discussed above, by doing experiments we can use scientific inference to attempt to localize ourselves in rulial space. But we won't be able to do enough to say "from our point of view, the universe is operating according to this one specific rule, and not another". Instead, there'll be a whole collection of rules that are "good enough", in the sense that they'll be sufficient to predict the results of experiments we can realistically do.

People have often imagined that, try as we might, we'd never be able to "get to the bottom of physics" and find a specific rule for our universe. And in a sense our inability to localize ourselves in rulial space supports this intuition. But what our Physics Project seems to rather dramatically suggest is that we can "get close enough" in rulial space to have vast predictive power about how our universe must work, or at least how observers like us must perceive it to work.

Alien Views of the Ruliad

We've discussed how "observers like us" will necessarily "parse the ruliad" in ways that make us perceive the universe to follow the laws of physics as we know them. But how different could things get? We have a definite sense of what constitutes a "reasonable observer" based on our 21st-century human experience—and in particular our biology, our technology and our ways of thinking.

But what other kinds of observers can we imagine? What about, for example, animals other than humans—in particular say ones whose sensory experience emphasizes olfaction or echolocation or fluid motion? We can think of such animals as operating in a different rulial reference frame or at a different place in rulial space. But how far away in rulial space will they be? How similar or not will their "world views" (and perceived laws of physics) be to ours? It's hard to know. Presumably our basic assertions about computational boundedness and coherence still apply. But just how the specifics of something like sequentialization in time play out, say, for an ant colony, seems quite unclear.

Maybe one day we'll be able to systematically "think like other animals". But as of now we haven't been able to "travel that far" in rulial space. We've quite thoroughly explored physical space, say on the surface of our planet, but we haven't explored very far at all in rulial space. We don't have a way to translate our thinking into some kind of "thinking differently"—and we don't, for example, have a common language to get there.

There's often an assumption (a kind of "human exceptionalism") that if it wasn't for details of the human experience—like brains and words—then we'd necessarily be dealing with something fundamentally simpler, that could not, for example, show features that we might identify as intelligence. But the Principle of Computational Equivalence tells us this isn't correct. Because it says that there's a certain maximal computational sophistication that's achieved not just by us humans but also by a vast range of other systems. The restrictions of what we've chosen to study (in science and elsewhere) have often made us miss this, but in fact computational sophistication—and the direct generalization of our notion of intelligence that's associated with it—seems quite ubiquitous across many different kinds of systems.

So can those other kinds of systems act as "observers like us"? To do so, they need not just computational sophistication, but also a certain alignment with the features we have that lead to our coherent thread of "conscious experience". And even given that, to actually "connect with" such systems, we need to be able to reach far enough in rulial space to sufficiently make a translation.

Imagine the weather (sometimes said to "have a mind of its own"). It's got plenty of computational sophistication. But is there any sense in which it sequentializes time like we do? Or can one only think of all those different parts of our atmosphere "running in their own time"? To know things like this, we effectively have to have a way to "translate" from the operation of the weather to our (current) way of thinking.

And in some sense we can consider the whole enterprise of natural science as being an effort to find a method of translation—or a common language—between nature and our way of thinking.

We as observers in effect trace out particular trajectories in rulial space; the challenge of natural science is to "reach out" in rulial space and "pull in" more of the ruliad; to be able to define a way to translate more parts of the ruliad to our processes of thinking. Every time we do an experiment, we can think of this as representing a moment of "connection" or "communication" between us and some aspect of nature. The experiment in effect defines a small piece of "common history" between us and nature—which helps "knit together" the parts of rulial space associated with us and with nature.

One of the great mysteries of science has been why—in the vastness of physical space—we've never detected something we identify as "alien intelligence", or an "alien civilization". We might have thought that it was because we humans have either achieved a unique pinnacle of intelligence or computational ability—or have fundamentally not gotten far enough. But the Principle of Computational Equivalence explodes the idea of this kind of cosmic computational pecking order.

So what could actually be going on? Thinking in terms of the ruliad suggests an answer. Our radio telescopes might be able to detect signals from far away in physical space. But our putative aliens might not only live far away in physical space, but also in rulial space.

Put another way, the "alien civilization" might be sampling aspects of the ruliad—and in effect the universe—that are utterly different from those we're used to. That different sampling might be happening right down at the level of atoms of space, or it might be that the rulial distance from us to the aliens is small enough that there's enough "shared description language" that the alien civilization might rise to the level of seeming like some kind of "noise" relative to our view of "what's important in the universe".

We might wonder how far apart what we could consider "alien civilizations" would be in physical space. But what we now realize is that we also have to consider how far apart they might be in rulial space. And just like in exploring physical space we can imagine building better spacecraft or better telescopes, so also we can imagine building better ways to reach across rulial space.

We're so used to physical space that it seems to us very concrete to reach across it. Of course, in our Physics Project, things like motion in physical space end up—like everything else—being pure computational processes. And from this point of view, reaching across rulial space is ultimately no more abstract—even though today we would describe it in terms of "doing (abstract) computations" rather than "moving in space".

Relative to our own physical size, the universe already seems like a vast place. But the full ruliad is even incredibly more vast. And we are likely much tinier in rulial space relative to the whole universe than we are in physical space. From the Principle of Computational

Equivalence we can expect that there's ultimately no lack of raw computational sophistication out there—but thinking in terms of the ruliad, the issue is whether what's going on is close enough to us in rulial space that we can successfully see it as an "alien civilization".

One test of rulial distance might be to ask whether our putative aliens perceive the same laws of physics for the universe that we do. We know that at least the general forms of those laws depend only on what seem to us rather loose conditions. But to get good alignment presumably requires at the very least that we and the aliens are somehow "comparable in size" not only in physical space (and branchial space), but also in rulial space.

It's humbling how difficult it is to imagine the universe from the point of view of an alien at a different place in rulial space. But for example if the alien is big compared to us in rulial space, we can say that they'll inevitably have a version of science that seems to us much "vaguer" than ours. Because if they maintain a coherent thread of experience, they'll have to conflate more distant paths in rulial space, on which the universe will do things that are "more different" than what we're used to. (And, yes, there should be rulial analogs of quantum phenomena, associated for example with conflated paths that diverge far in rulial space.)

What would it mean operationally for there to be an alien civilization perhaps nearby in physical space but at a distance in rulial space? Basically the alien civilization will be "operating" in features of the universe that our parsing of the universe just doesn't pick up. As a simple analogy, our view of, for example, a box of gas might be that it's something with a certain temperature and pressure. But a different "parsing" of that system might identify a whole world of detailed motions of molecules that with respect to that parsing can be viewed as a vast "alien civilization". Of course, the situation is much more extreme when it comes to the whole ruliad, and all the paths of history and configurations of atoms of space that it represents.

Relative to the whole ruliad, our civilization and our experience have carved out an extremely tiny piece. And what we're thinking of as "alien civilizations" might also have carved out their own tiny pieces. And while we're all "living in the same ruliad", we might no more be able to detect each other or communicate (and likely very much less) than we can across vast distances in physical space.

What of the future? The future of our civilization might well be a story of mapping out more of rulial space. If we continue to invent new technology, explore new ideas and generally broaden our ways of thinking and perceiving, we will gradually—albeit in tiny steps—map out more of rulial space. How far can we get? The ultimate limit is determined by the maximum rulial speed. But if we expect to maintain our character as "observers like us", we'll no doubt be limited to something much less.

Among other issues, moving in rulial space involves doing computation. (The ultimate scale is set by the "processing power" of the universe—which defines the maximum rulial speed.) But "density of computation" effectively corresponds to a generalized version of mass—and is for example a source of "generalized gravity". And it could be that to "move any significant distance" in rulial space, we'd have to "experience enough generalized gravity" that we could never maintain things like the kind of coherence we need to be an "observer like us".

Put another way: yes, it might in principle be possible to "reach out in rulial space" and "contact the rulial aliens". But it might be that doing so would require us to be so different from the way we currently are that we wouldn't recognize anything like consciousness or anything that really makes us "identifiably us". And if this is so, we are in a sense limited to experiencing the ruliad "on our own" from our particular place in rulial space, forever isolated from "alien civilizations" elsewhere in rulial space.

Conceptual Implications of the Ruliad

What does the concept of the ruliad mean for the fundamental way we think about things like science? The typical conception of "what science does" is that it's about us figuring out—as "objectively" as we can—how the world happens to be. But the concept of the ruliad in a sense turns this on its head.

Because it says that at some ultimate level, everything is a matter of abstract necessity. And it's just our "parsing" of it that defines the subject matter of what we call science. We might have thought that the science of the universe was just something that's "out there". But what we're realizing is that instead in some fundamental sense, it's all "on us".

But does that mean that there's no "objective truth", and nothing that can robustly be said about the universe without "passing it through us"? Well, no. Because what we've discovered through our Physics Project is that actually there are quite global things that can ("objectively") be said about our universe and the laws it follows, as perceived by observers like us.

We don't have to know in detail about us humans and the particular ways we perceive things. All we need are some general features—particularly that we are computationally bounded, and that we have a certain persistence and coherence. And this is all it takes to deduce some quite specific statements about how our universe operates, at least as we perceive it.

So in a sense what this means is that there is a large "zone of objectivity"; a large set of choices for how we could be that will still lead us to the same "objective truth" about our universe. But if we go far enough away in our mechanism for "parsing the ruliad", this will no longer be the case. From our current vantage point, we'd no doubt then be hard-pressed to recognize how we're "doing the parsing", but the results we'd get would no longer give us the same laws of physics or general perception of the universe that we're used to.

This view of things has all sorts of implications for various long-discussed philosophical issues. But it's also a view that has precise scientific consequences. And these don't just relate to physics. Because the ruliad is really a general object that represents the entangled behavior of all possible abstract rules. When we think of ourselves as observers embedded within this object, it means that for us things are actualized, and we have what we call physics. But we can also imagine sampling the ruliad in different ways.

Some of those ways correspond to mathematics (or metamathematics). Some correspond to theoretical computer science. The ruliad is the single object that underlies all of them. And which of them we're talking about just depends on how we imagine we're sampling or parsing the ruliad, and how we're describing what we're observing.

With this degree of generality and universality, it's inevitable that the ruliad must be a complicated object; in fact, in a sense it must encapsulate all possible achievable complexity. But what's important is that we now have a definite concept of the ruliad, as something we can study and analyze.

It's not simple to do this. The ruliad is at some level an object of great and perhaps supremely elegant abstract regularity. But for us to get any concrete handle on it and its structure, we need to break it down into some kind of "digestible slices" which inevitably lose much of its abstract regularity.

And we're just at the beginning of seeing how best to "unpack" and "pick through" the ruliad. With explicit computations, we can only chip away at the very simplest approximations to the ruliad. In a sense it's a tribute to the naturalness and inevitability of the ruliad that it's so closely related to some of the most advanced abstract mathematical methods we know so far. But again, even with these methods we're barely scratching the surface of the ruliad and what it contains.

The theoretical exploration of the ruliad will be a long and difficult journey. But the incredible generality and universality of the ruliad means that every piece of progress is likely to have exceptionally powerful consequences. In some sense the exploration of the ruliad can be seen as the encapsulated expression of everything it means to do theoretical investigation: a kind of ultimately abstract limit of theoretical science and more.

For me, the ruliad in a sense builds on a tower of ideas, that include the computational paradigm in general, the exploration of the computational universe of simple programs, the Principle of Computational Equivalence, our Physics Project and the notion of multicomputation. But even with all of these it's still a significant further jump in abstraction. And one whose consequences will take considerable time to unfold.

But for now it's exciting to have at least been able to define this thing I call the ruliad, and to start seeing some of its unprecedentedly broad and deep implications.

Appendix: The Case of the "Multiplicad"

As a very simple example of something like the ruliad, we can consider what we might call the "multiplicad": a rulial multiway system based on integers, in which the rules simply multiply by successive integers:

$n \longmapsto \{2\,n, 3\,n, 4\,n, 5\,n, \ldots\}$

(Note that this kind of pure multiplication is presumably not computation universal, so the limiting object here will not be a coordinatization of the actual full ruliad.)

Just like with the full ruliad, there are many different "directions" in which to build up the multiplicad. We could allow as many multipliers and steps as we want, but limit the total size of numbers generated, here say to 30:

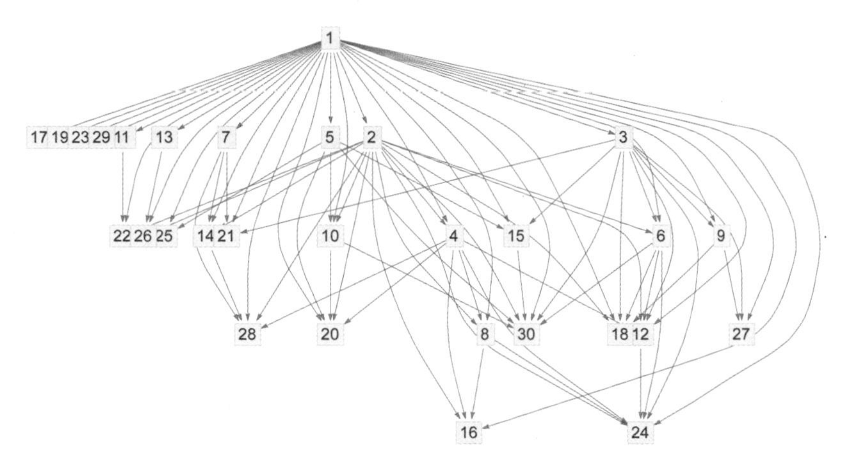

As an alternative, we can limit the number of multipliers s, say to $s = 4$. Then the multiplicad would build up like this:

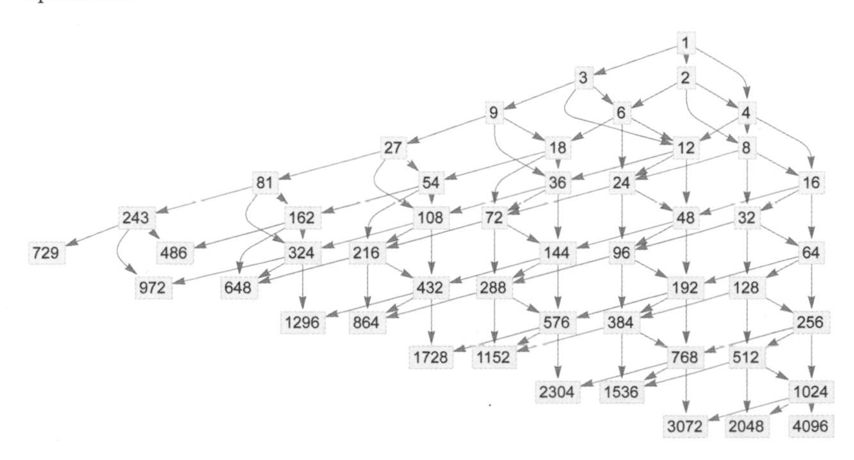

In the pictures we've drawn so far, we're effectively always deduplicating different occurrences of the same integer. So, for example, the integer 12 can be generated as $1 \times 3 \times 4$ or $1 \times 6 \times 2$ or $1 \times 3 \times 2 \times 2$, etc. And in principle we could show each of these "different 12s" separately. But in our deduplicated graph, only a single 12 appears—with the different possible decompositions of 12 being reflected in the presence of multiple paths that lead to the 12.

Sometimes the structure we get is richer—if much bigger—when we don't immediately do deduplication. For example, if we allow any number of multipliers (i.e. take $s=\infty$) then after just 1 step we will get all integers—and if we do deduplication, then this will be the end of our graph, because we "already have all the integers". But if we don't do deduplication, we'll get a slightly more complicated picture, that begins like this:

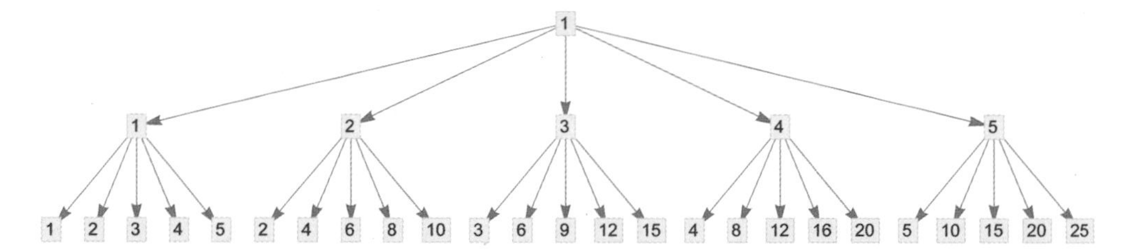

The "topological" structure of this graph is now straightforward, but its "labeling" with numbers is less so—and if we ask, for example, where a particular number appears after t steps, this can be more complicated.

Imagine that we are looking only at the subtrees associated with up to s multipliers at the first step—or, equivalently, that we are looking at the rulial multiway system "truncated" with only s rules. Which numbers will appear after $t=2$ steps? The answer is that it will be precisely those numbers that show up in an $s\times s$ multiplication table where we start from $n=2$:

4	6	8	10	12	14	16	18	20	22	24
6	9	12	15	18	21	24	27	30	33	36
8	12	16	20	24	28	32	36	40	44	48
10	15	20	25	30	35	40	45	50	55	60
12	18	24	30	36	42	48	54	60	66	72
14	21	28	35	42	49	56	63	70	77	84
16	24	32	40	48	56	64	72	80	88	96
18	27	36	45	54	63	72	81	90	99	108
20	30	40	50	60	70	80	90	100	110	120
22	33	44	55	66	77	88	99	110	121	132
24	36	48	60	72	84	96	108	120	132	144

Clearly no primes appear here, but some numbers can appear multiple times (e.g. 12 appears 4 times). In general, the number of times that the number $n\leq s$ will show up is the number of proper divisors it has, or DivisorSigma[0, n]−2:

2	3	4	5	6	7	8	9	10	11	12	13	14	15	16	17	18	19	20	21	22	23	24	25	26	27	28	29	30	31	32
0	0	1	0	2	0	2	1	2	0	4	0	2	2	3	0	4	0	4	2	2	0	6	1	2	2	4	0	6	0	4

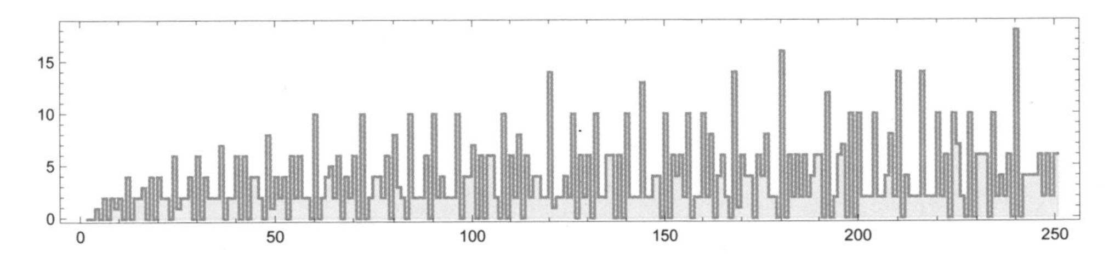

We can continue this, to ask how many times a given number n will occur at a particular step t:

2	3	4	5	6	7	8	9	10	11	12	13	14	15	16	17	18	19	20	21	22	23	24	25	26	27	28	29	30	31	32
1	1	1	1	1	1	1	1	1	1	1	1	1	1	1	1	1	1	1	1	1	1	1	1	1	1	1	1	1	1	1
		1		2		2	1	2		4		2	2	3		4		4	2	2		6	1	2	2	4		6		4
						1				3				3		3		3				9			1	3		6		6
														1								4								4
																														1

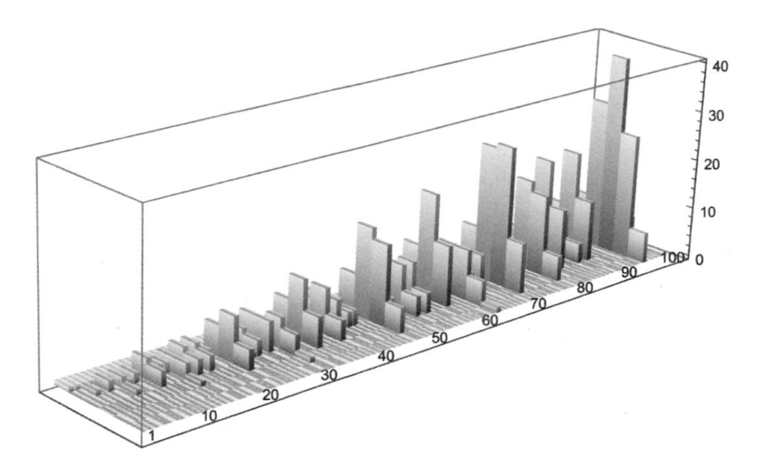

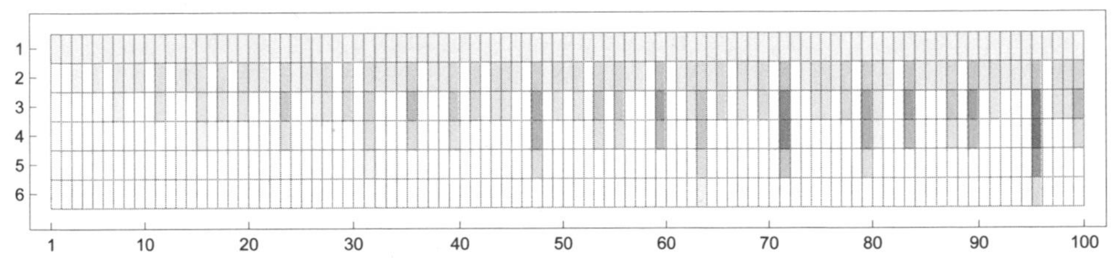

We can think of these results as being determined by the number of times that n appears in an $s \times s \times s \ldots$ (t times) multiplication array. Alternatively, to know the results for a given number n, we can look at all the ways n can be decomposed into factors. For $n = 12$, for example, we would have:

{{12}, {2, 6}, {3, 4}, {4, 3}, {6, 2}, {2, 2, 3}, {2, 3, 2}, {3, 2, 2}}

And from this we can deduce that 12 appears once at $t=1$ (i.e. with 1 factor), 4 times at $t=2$ (i.e. with 2 factors) and 3 times at $t=3$ (i.e. with 3 factors).

The full multiplicad is formed by taking the limits $s \to \infty$ and $t \to \infty$ (as well as what is essentially the limit $n \to \infty$ for an infinite set of possible initial conditions). As we can see, our "finite perception" of the multiplicad will be different depending on how we sample it in s and t.

As an example, let's consider what happens for given s as a function of t. For $s=2$, we simply have powers of 2:

For $s=3$, where can multiply by both 2 and 3, we get:

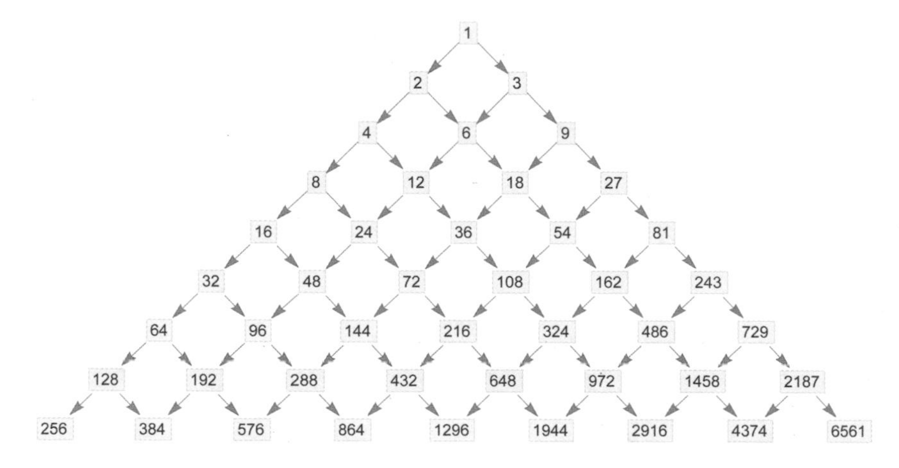

In studying multiway systems, it is often of interest to ask about the growth rates of the number of states reached over the course of t steps (i.e. the growth rates of volumes of geodesic balls). In the case $s=2$, the number of states reached by step t is just t. For $s=3$, it's the triangular numbers $t\,(t-1)/2$:

{1, 3, 6, 10, 15, 21, 28, 36, 45, 55, 66, 78, 91, 105, 120, 136, 153, 171, 190}

Here are some results for larger s:

2	1	2	3	4	5	6	7	8	9	10	11	12	13	14	15
3	1	3	6	10	15	21	28	36	45	55	66	78	91	105	120
4	1	4	9	16	25	36	49	64	81	100	121	144	169	196	225
5	1	5	14	30	55	91	140	204	285	385	506	650	819	1015	1240
6	1	6	18	40	75	126	196	288	405	550	726	936	1183	1470	1800
7	1	7	25	65	140	266	462	750	1155	1705	2431	3367	4550	6020	7820
8	1	8	30	80	175	336	588	960	1485	2200	3146	4368	5915	7840	10 200
9	1	9	36	100	225	441	784	1296	2025	3025	4356	6084	8281	11 025	14 400
10	1	10	42	120	275	546	980	1632	2565	3850	5566	7800	10 647	14 210	18 600

Each of these sequences is generated by a linear recurrence relation with a kernel given by a sequence of signed binomial coefficients. The values for successive t can be represented by polynomials:

2	t
3	$\frac{1}{2} t (t + 1)$
4	t^2
5	$\frac{1}{6} t (t+1) (2 t + 1)$
6	$\frac{1}{2} t^2 (t+1)$
7	$\frac{1}{24} t (t+1) (t+2) (3 t+1)$
8	$\frac{1}{6} t^2 (t+1) (t+2)$
9	$\frac{1}{4} t^2 (t+1)^2$
10	$\frac{1}{6} t^2 (t+1) (2 t+1)$

The leading term in the growth of number of states is then determined by the orders of these polynomials, which turn out to be just PrimePi[s]:

2	3	4	5	6	7	8	9	10	11	12	13	14	15	16	17	18	19	20	21	22	23	24	25	26	27	28	29	30	31	32
1	2	2	3	3	4	4	4	4	5	5	6	6	6	6	7	7	8	8	8	8	9	9	9	9	9	9	10	10	11	11

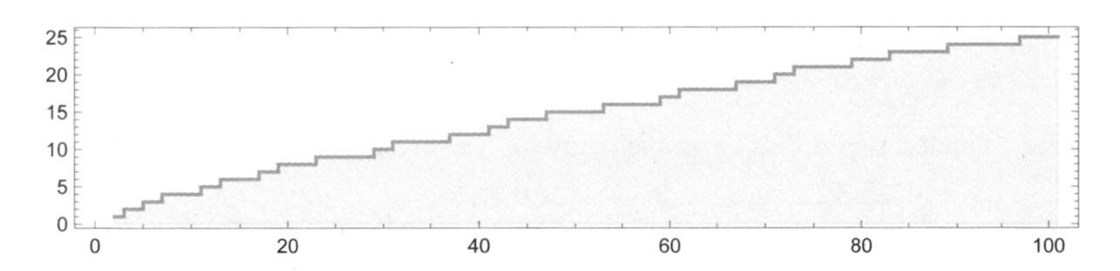

For the case $s=3$, we saw above that the multiway graph essentially forms a simple 2D grid. For larger s, the graph becomes more complicated, though it still approximates a grid—but in dimension PrimePi[s]. (The reason PrimePi[s] appears is that in a sense the combining of primes less than s are the largest "drivers" of structure in the multiway graph.)

In our general analysis of multiway graphs, it is common to consider branchial graphs—or for a rulial multiway system what we can call rulial graphs—obtained by looking at a slice of the multiway graph, effectively for a given t, and asking what states are connected by having a common ancestor. The results for $s=2$ are rather trivial (here shown for $t=1, 2, 3$):

For $s=3$ we get:

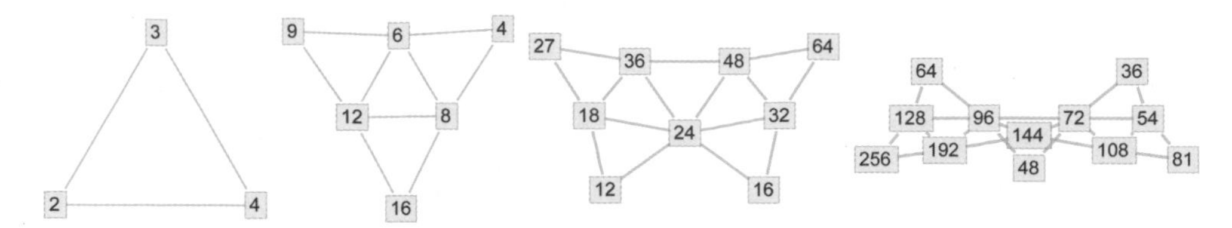

And for $s=4$ we have:

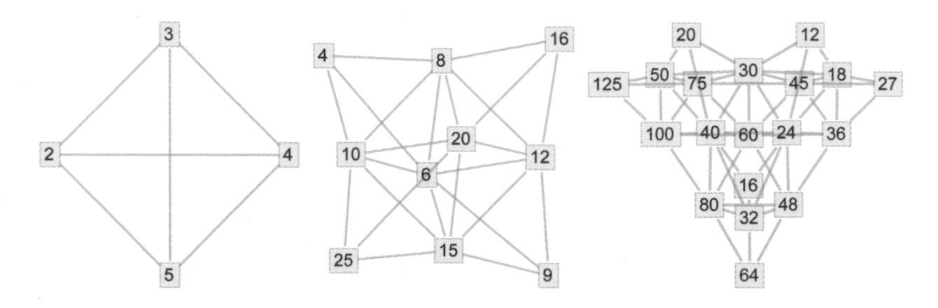

In a sense these pictures show how numbers in the multiplicad can be "laid out in rulial space". For $s=3$, the "large-t graph" has a very linear form

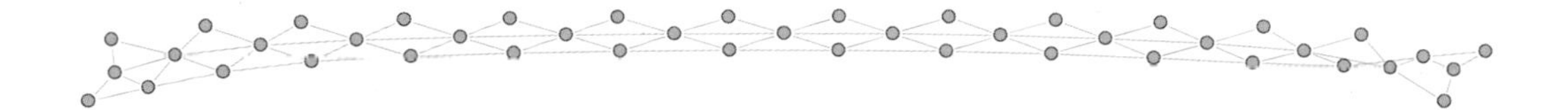

and the numbers that appear "from left to right" are arranged more or less in numerical order:

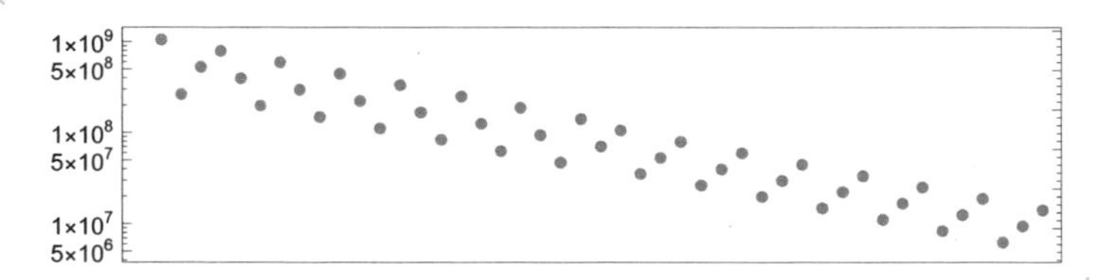

For $s=4$, the result is a 2D-like structure:

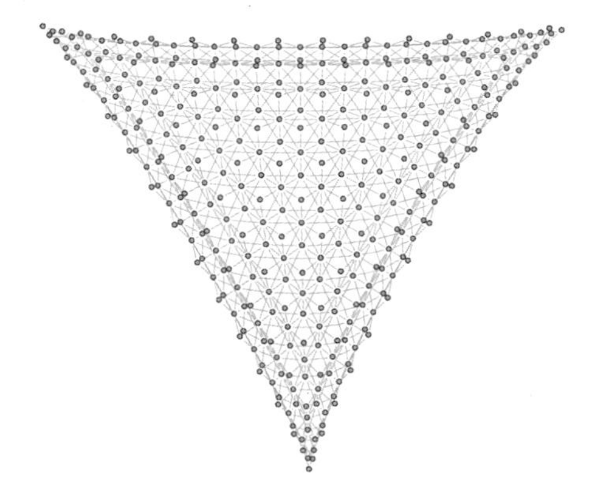

And again the numbers that appear are roughly arranged in a kind of "numerical sequence":

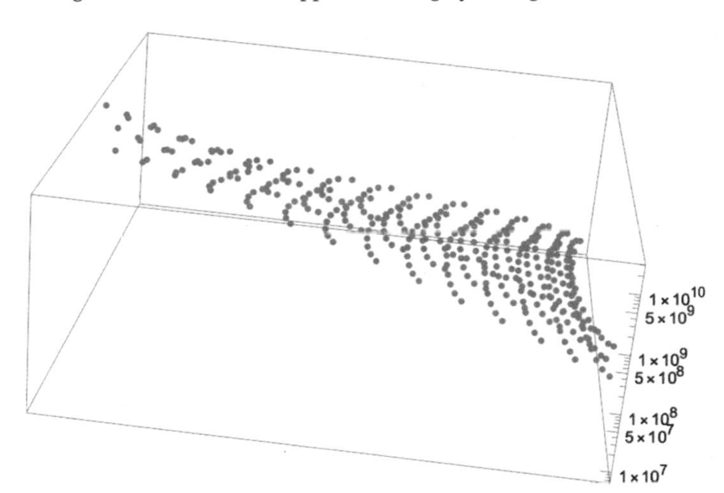

We can then think of this as suggesting that the multiplicad "generates a rulial space" that can be approximately coordinatized purely by the values of the numbers that appear in it. How all this works in the limit $s \to \infty$ is not clear, though somehow the "1D" sequence of numerical values presumably "snakes through" PrimePi[*s*]-dimensional space as some kind of approximation to a space-filling curve.

It should be noted that we've only considered one particular way of sampling the rulial multiway graph as a function of *t*. In general there are many different possible foliations that could be used, all of them giving us in effect a different view of the multiplicad, from a different "reference frame".

As mentioned at the beginning, the multiplicad is presumably not on its own capable of giving us the full ruliad. But if we change the underlying rules—probably even just inserting addition as well as multiplication—we'll potentially get a system that is capable of universal computation, and which can therefore generate the full ruliad. Needless to say, the particular representation of the ruliad obtained by the kind of "numerical processes" that we've used here may be utterly different from any representation that we would recognize from our perception of the physical universe.

Thanks & Note

Thanks for discussions of various aspects of the ruliad to Xerxes Arsiwalla, James Boyd, Elise Cawley, Hatem Elshatlawy, Jonathan Gorard and Nik Murzin. Thanks also to Ed Pegg and Joseph Stocke for input about the multiplicad. A new paper by Xerxes Arsiwalla and Jonathan Gorard discusses in a more technical way some ideas and results related to the ruliad (arXiv: 2111.03460).

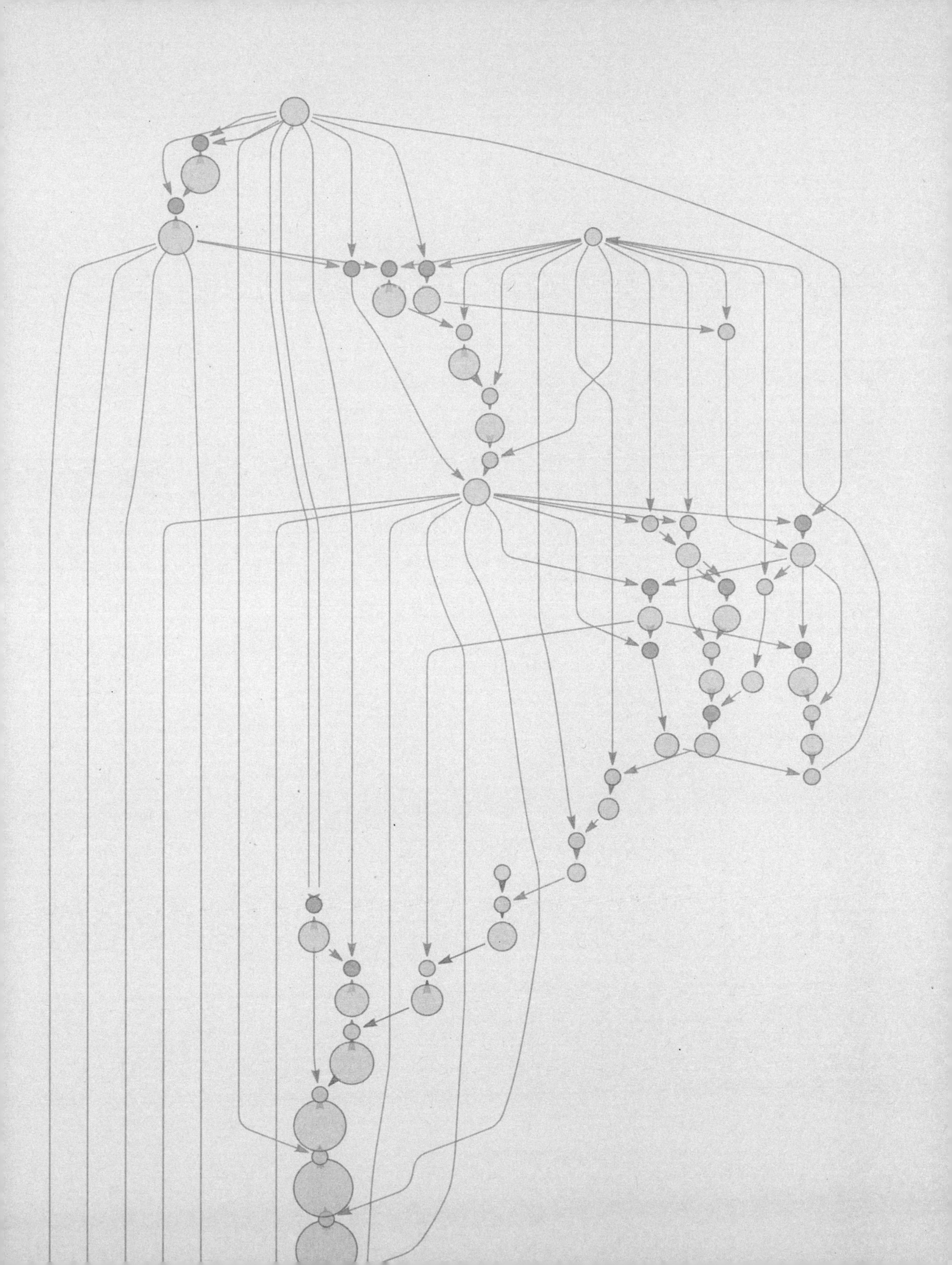

The Empirical Metamathematics of Euclid and Beyond

(September 28, 2020)

Towards a Science of Metamathematics

One of the many surprising things about our Wolfram Physics Project is that it seems to have implications even beyond physics. In our effort to develop a fundamental theory of physics it seems as if the tower of ideas and formalism that we've ended up inventing are actually quite general, and potentially applicable to all sorts of areas.

One area about which I've been particularly excited of late is metamathematics—where it's looking as if it may be possible to use our formalism to make what might be thought of as a "bulk theory of metamathematics".

Mathematics itself is about what we establish about mathematical systems. Metamathematics is about the infrastructure of how we get there—the structure of proofs, the network of theorems, and so on. And what I'm hoping is that we're going to be able to make an overall theory of how that has to work: a formal theory of the large-scale structure of metamathematics—that, among other things, can make statements about the general properties of "metamathematical space".

Like with physical space, however, there's not just pure underlying "geometry" to study. There's also actual "geography": in our human efforts to do mathematics over the last few millennia, where in metamathematical space have we gone, and "colonized"? There've been a few million mathematical theorems explicitly published in the history of human mathematics. What does the "empirical metamathematics" of them reveal? Some of it presumably reflects historical accidents, but some may instead reflect general features of metamathematics and metamathematical space.

I've wondered about empirical metamathematics for a long time, and tucked away on page 1176 at the end of the Notes for the section about "Implications for Mathematics and Its Foundations" in *A New Kind of Science* is something I wrote more than 20 years ago about it:

▪ **Page 820 · Empirical metamathematics.** One can imagine a network representing some field of mathematics, with nodes corresponding to theorems and connections corresponding to proofs, gradually getting filled in as the field develops. Typical rough characterizations of mathematical results—independent of their detailed history—might then end up being something like the following:

- lemma: short theorem appearing at an intermediate stage in a proof

The picture below shows the network of theorems associated with Euclid's *Elements*. Each stated theorem is represented by a node connected to the theorems used in its stated proof. (Only the shortest connection from each theorem is shown explicitly.) The axioms (postulates and common notions) are given in the first column on the left, and successive columns then show theorems with progressively longer proofs. (Explicit annotations giving theorems used in proofs were apparently added to editions of Euclid only in the past few centuries; the picture below extends the usual annotations in a few cases.) The theorem with the longest proof is the one that states that there are only five Platonic solids.

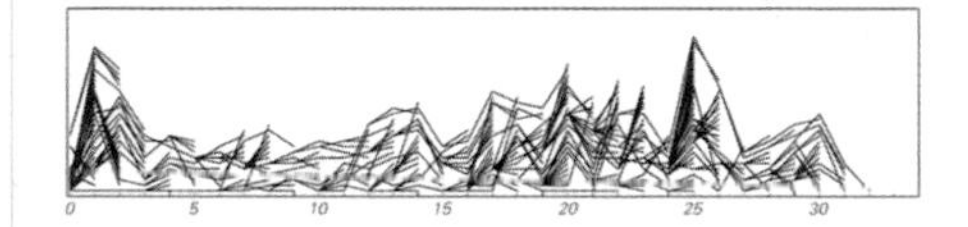

This note is mostly about what a descriptive theory of empirical metamathematics might be like—for example characterizing what one might mean by a powerful theorem, a deep theorem, a surprising theorem and so on. But at the end of the note is a graph: an actual piece of quantitative empirical metamathematics, based on the best-known structured piece of mathematics in history—Euclid's *Elements*.

The graph shows relationships between theorems in the *Elements*: a kind of causal graph of how different theorems make use of each other. As presented in *A New Kind of Science*, it's a small "footnote item" that doesn't look like much. But for more than 20 years, I've kept wondering what more there might be to learn from it. And now that I'm trying to make a general theory of metamathematics, it seemed like it was a good time to try to find out...

The Most Famous Math Book in History

Euclid's *Elements* is an impressive achievement. Written in Greek around 300 BC (though presumably including many earlier results), the *Elements* in effect defined the way formal mathematics is done for more than two thousand years. The basic idea is to start from certain axioms that are assumed to be true, then—without any further "input from outside"—use purely deductive methods to establish a collection of theorems.

Euclid effectively had 10 axioms (5 "postulates" and 5 "common notions"), like "one can draw a straight line from any point to any other point", or "things which equal the same thing are also equal to one another". (One of his axioms was his fifth postulate—that parallel lines never meet—which might seem obvious, but which actually turns out not to be true for physical curved space in our universe.)

On the basis of his axioms, Euclid then gave 465 theorems. Many were about 2D and 3D geometry; some were about arithmetic and numbers. Among them were many famous results, like the Pythagorean theorem, the triangle inequality, the fact that there are five Platonic solids, the irrationality of $\sqrt{2}$ and the fact that there are an infinite number of primes. But certainly not all of them are famous—and some seem to us now pretty obscure. And in what has remained a (sometimes frustrating) tradition of pure mathematics for more than two thousand years, Euclid never gives any narrative about why he's choosing the theorems he does, out of all the infinitely many possibilities.

We don't have any original Euclids, but versions from a few centuries later exist. They're written in Greek, with each theorem explained in words, usually by referring to a diagram. Mathematical notation didn't really start getting invented until the 1400s or so (i.e. a millennium and a half later)—and even the notation for numbers in Euclid's time was pretty unwieldy. But Euclid had basically modern-looking diagrams, and he even labeled points and angles with (Greek) letters—despite the fact that the idea of variables standing for numbers wouldn't be invented until the end of the 1500s.

There's a stylized—almost "legalistic"—way that Euclid states his theorems. And so far as we can tell, in the original version, all that was done was to state theorems; there was no explanation for why a theorem might be true—no proof offered. But it didn't take long before people started filling in proofs, and there was soon a standard set of proofs, in which each particular theorem was built up from others—and ultimately from the axioms.

There've been more than a thousand editions of Euclid printed (probably more than any other book except the Bible), and reading Euclid was until quite recently part of any serious education. (At Eton—where I went to high school—it was only in the 1960s that learning "mathematics" began to mean much other than reading Euclid, in the original Greek of course.) Here's an edition of Euclid from the 1800s that I happen to own, with the proof of every theorem giving little references to other theorems that are used:

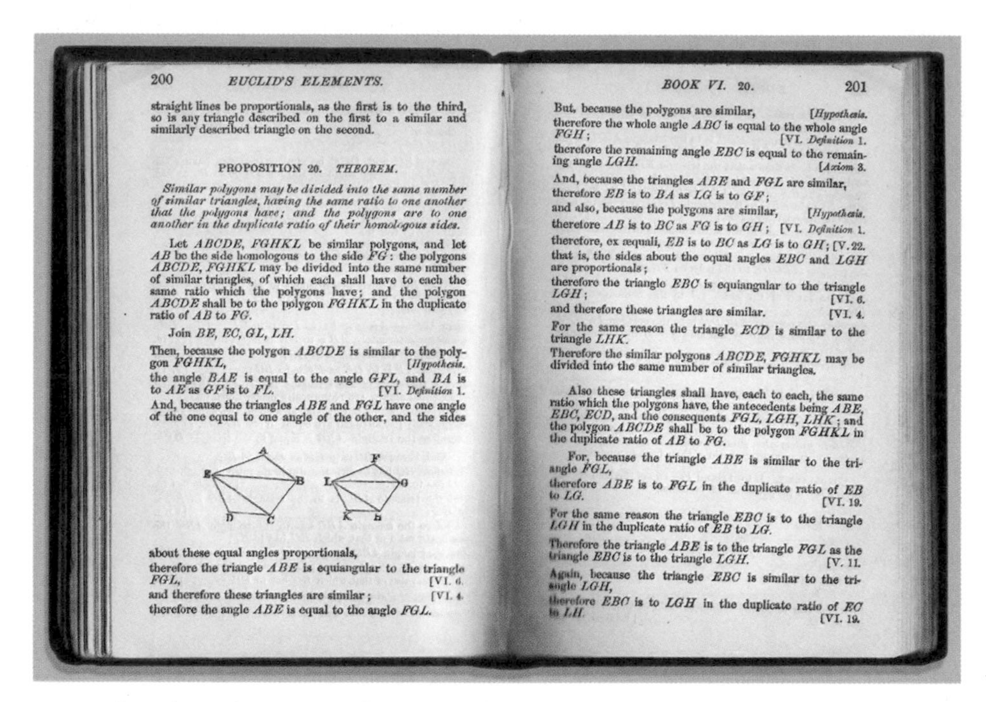

200 *EUCLID'S ELEMENTS.*

straight lines be proportionals, as the first is to the third, so is any triangle described on the first to a similar and similarly described triangle on the second.

PROPOSITION 20. *THEOREM.*

Similar polygons may be divided into the same number of similar triangles, having the same ratio to one another that the polygons have; and the polygons are to one another in the duplicate ratio of their homologous sides.

Let *ABCDE*, *FGHKL* be similar polygons, and let *AB* be the side homologous to the side *FG*: the polygons *ABCDE*, *FGHKL* may be divided into the same number of similar triangles, of which each shall have to each the same ratio which the polygons have; and the polygon *ABCDE* shall be to the polygon *FGHKL* in the duplicate ratio of *AB* to *FG*.

Join *BE*, *EC*, *GL*, *LH*.

Then, because the polygon *ABCDE* is similar to the polygon *FGHKL*, [*Hypothesis.*
the angle *BAE* is equal to the angle *GFL*, and *BA* is to *AE* as *GF* is to *FL*. [VI. *Definition* 1.
And, because the triangles *ABE* and *FGL* have one angle of the one equal to one angle of the other, and the sides

about these equal angles proportionals,
therefore the triangle *ABE* is equiangular to the triangle *FGL*, [VI. 6.
and therefore these triangles are similar; [VI. 4.
therefore the angle *ABE* is equal to the angle *FGL*.

BOOK VI. 20. 201

But, because the polygons are similar, [*Hypothesis.*
therefore the whole angle *ABC* is equal to the whole angle *FGH*; [VI. *Definition* 1.
therefore the remaining angle *EBC* is equal to the remaining angle *LGH*. [*Axiom* 3.
And, because the triangles *ABE* and *FGL* are similar,
therefore *EB* is to *BA* as *LG* is to *GF*;
and also, because the polygons are similar, [*Hypothesis.*
therefore *AB* is to *BC* as *FG* is to *GH*; [VI. *Definition* 1.
therefore, ex æquali, *EB* is to *BC* as *LG* is to *GH*; [V. 22.
that is, the sides about the equal angles *EBC* and *LGH* are proportionals;
therefore the triangle *EBC* is equiangular to the triangle *LGH*; [VI. 6.
and therefore these triangles are similar. [VI. 4.
For the same reason the triangle *ECD* is similar to the triangle *LHK*.
Therefore the similar polygons *ABCDE*, *FGHKL* may be divided into the same number of similar triangles.

Also these triangles shall have, each to each, the same ratio which the polygons have, the antecedents being *ABE*, *EBC*, *ECD*, and the consequents *FGL*, *LGH*, *LHK*; and the polygon *ABCDE* shall be to the polygon *FGHKL* in the duplicate ratio of *AB* to *FG*.

For, because the triangle *ABE* is similar to the triangle *FGL*,
therefore *ABE* is to *FGL* in the duplicate ratio of *EB* to *LG*. [VI. 19.
For the same reason the triangle *EBC* is to the triangle *LGH* in the duplicate ratio of *EB* to *LG*.
Therefore the triangle *ABE* is to the triangle *FGL* as the triangle *EBC* is to the triangle *LGH*. [V. 11.
Again, because the triangle *EBC* is similar to the triangle *LGH*,
therefore *EBC* is to *LGH* in the duplicate ratio of *EC* to *LH*. [VI. 19.

But so what about the metamathematics of Euclid? Given all those theorems—and proofs—can we map out the structure of what Euclid did? That's what the graph in *A New Kind of Science* was about. A few years ago, we put the data for that graph into our Wolfram Data Repository—and I looked at it again, but nothing immediately seemed to jump out about it; it still just seemed like a complicated mess:

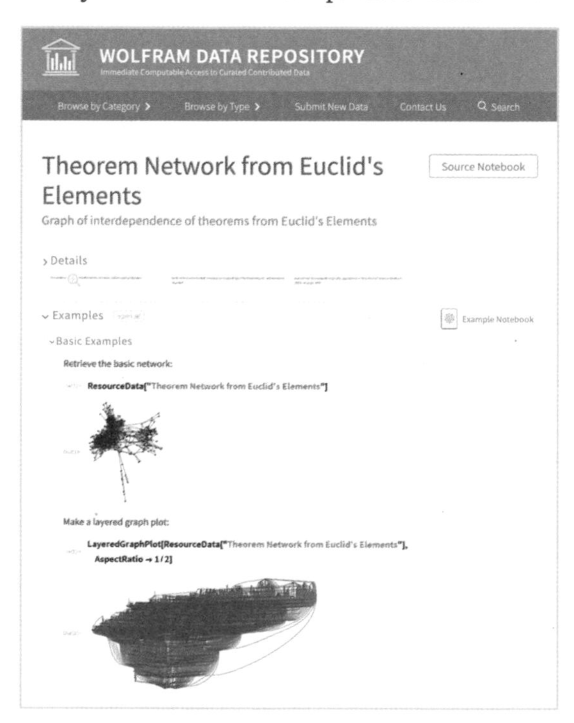

What else happened? One thing is that we added automated theorem proving to Mathematica and the Wolfram Language. Enter a potential theorem, and axioms from which to derive it, and FindEquationalProof will try to generate a proof. This works well for "structurally simple" mathematical systems (like basic logic), and indeed one can generate proofs with complex networks of lemmas that go significantly beyond what humans can do (or readily understand):

```
In[ ]:= FindEquationalProof[p·q == q·p, ∀{a,b,c} ((a·b)·c)·(a·((a·c)·a)) == c]["ProofGraph"]
```

Out[ ]=

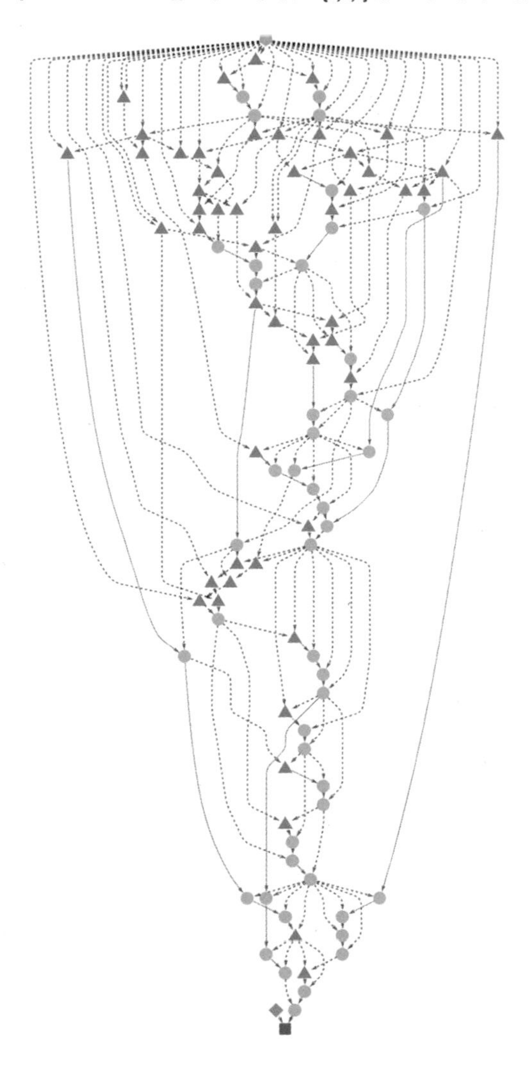

It's in principle possible to use these methods to prove theorems in Euclidean geometry too. But it's a different problem to make the proofs readily understandable to humans (like the step-by-step solutions of Wolfram|Alpha). So at least for now—even after 2000 years—the most effective source of information about the empirical metamathematics of proofs of Euclid's theorems is still basically going to be Euclid's *Elements*.

But when it comes to representing Euclid's theorems there's something new. The whole third-of-a-century story of the Wolfram Language has been about finding ways to represent more and more things in the world computationally. I had long wondered what it would take to represent Euclid-style geometry computationally. And in April I was excited to announce that we'd managed to do it:

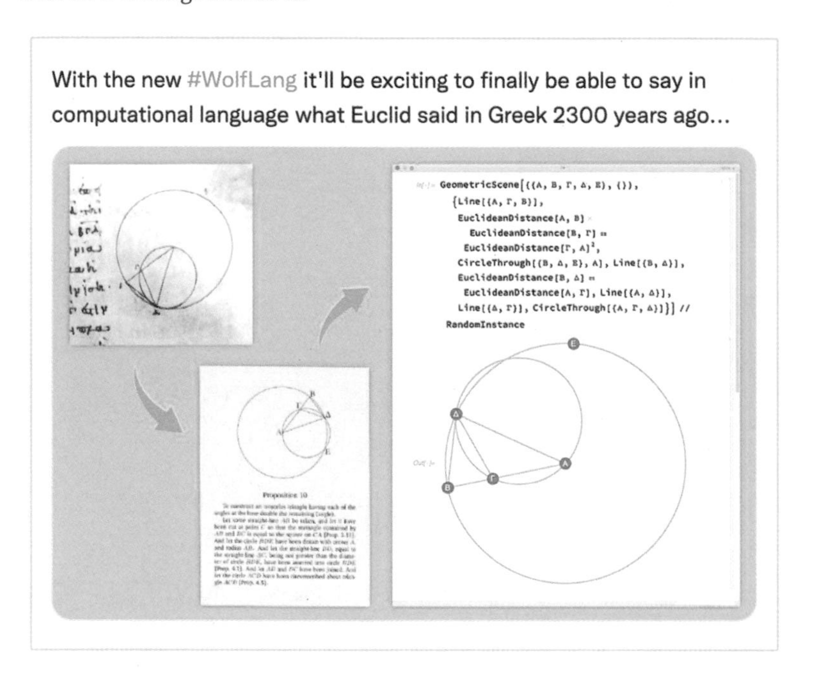

Basic Statistics of Euclid

Euclid's *Elements* is divided into 13 "books", containing a total of 465 theorems (and 131 definitions):

subjects	*2D geometry*						*numbers*				*3D geometry*		
books	*1*	*2*	*3*	*4*	*5*	*6*	*7*	*8*	*9*	*10*	*11*	*12*	*13*
theorems	48	14	37	16	25	33	39	27	36	115	39	18	18
totals	*(173)*						*(217)*				*(75)*		
definitions	23	2	11	7	18	4	22	0	0	16	28	0	0
totals	*(65)*						*(38)*				*(28)*		

Stating the theorems takes 9589 words (about 60k characters) of Greek (about 13,000 words in a standard English translation). (The 10 axioms take another 115 words in Greek or about 140 in English, and the definitions another 2369 words in Greek or about 3300 in English.)

A typical theorem (or "proposition")—in this case Book 1, Theorem 20—is stated as:

παντὸς τριγώνου αἱ δύο πλευραὶ τῆς λοιπῆς μείζονές εἰσι πάντῃ μεταλαμβανόμεναι.

In any triangle two sides taken together in any manner are greater than the remaining one.

(This is what we now call the triangle inequality. And of course, to make this statement we have to have defined what a triangle is, and Euclid does that earlier in Book 1.)

If we look at the statements of Euclid's theorems in Greek (or in English), there's a distribution of lengths (colored here by subjects, and reasonably fit by a Pascal distribution):

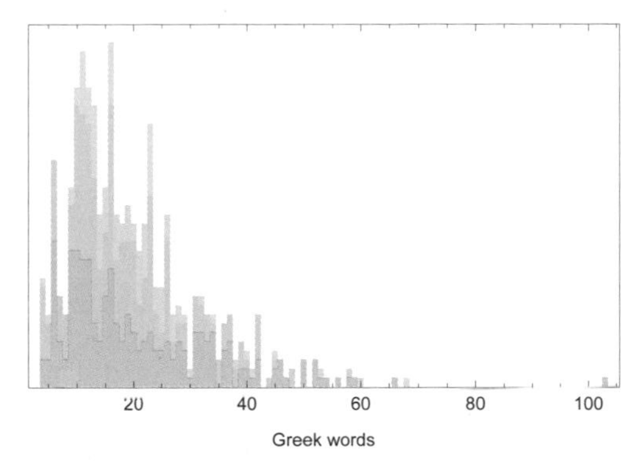

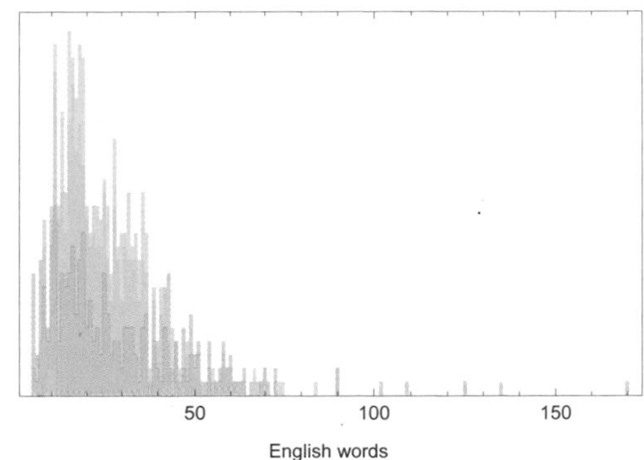

The "outlier" longest-to-state theorem (in both Greek and English) is the rather unremarkable 103-Greek-word 3.8

ἐὰν κύκλου ληφθῇ τι σημεῖον ἐκτός, ἀπὸ δὲ τοῦ σημείου πρὸς τὸν κύκλον διαχθῶσιν εὐθεῖαί τινες, ὧν μία μὲν διὰ τοῦ κέντρου, αἱ δὲ λοιπαί, ὡς ἔτυχεν, τῶν μὲν πρὸς τὴν κοίλην περιφέρειαν προσπιπτουσῶν εὐθειῶν μεγίστη μέν ἐστιν ἡ διὰ τοῦ κέντρου, τῶν δὲ ἄλλων ἀεὶ ἡ ἔγγιον τῆς διὰ τοῦ κέντρου τῆς ἀπώτερον μείζων ἐστίν, τῶν δὲ πρὸς τὴν κυρτὴν περιφέρειαν προσπιπτουσῶν εὐθειῶν ἐλαχίστη μέν ἐστιν ἡ μεταξὺ τοῦ τε σημείου καὶ τῆς διαμέτρου, τῶν δὲ ἄλλων ἀεὶ ἡ ἔγγιον τῆς ἐλαχίστης τῆς ἀπώτερόν ἐστιν ἐλάττων, δύο δὲ μόνον ἴσαι ἀπὸ τοῦ σημείου προσπεσοῦνται πρὸς τὸν κύκλον ἐφ' ἑκάτερα τῆς ἐλαχίστης.

If a point be taken outside a circle and from the point straight lines be drawn through to the circle, one of which is through the centre and the others are drawn at random, then, of the straight lines which fall on the concave circumference, that through the centre is greatest, while of the rest the nearer to that through the centre is always greater than the more remote, but, of the straight lines falling on the convex circumference, that between the point and the diameter is least, while of the rest the nearer to the least is always less than the more remote, and only two equal straight lines will fall on the circle from the point, one on each side of the least.

which can be illustrated as:

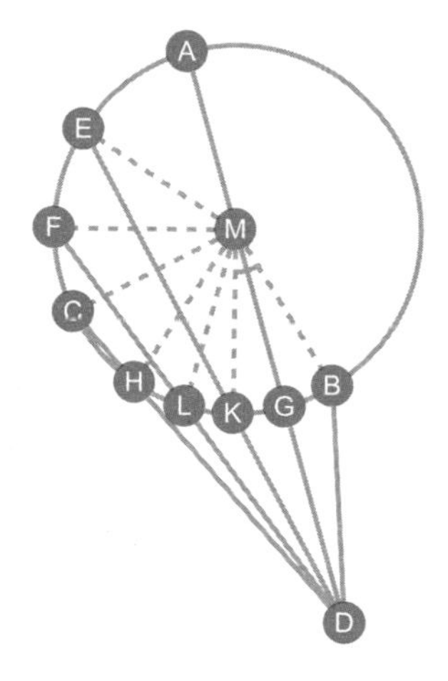

(The runner-up, at about two-thirds the length, is the also rather unremarkable 11.35.)

The nominally shortest-to-state theorems are in Book 10, Theorems 85 through 90, and all have just 4 Greek words:

εὑρεῖν τὴν πρώτην ἀποτομήν.

To find the first apotome.

⋮

εὑρεῖν τὴν ἕκτην ἀποτομήν.

To find the sixth apotome.

The shortness of these theorems is a bit of a cheat, since the successive "apotomes" (pronounced /əˈpɒtəmi/ like "hippopotamus") actually have quite long definitions that are given elsewhere. And, yes, some emphasis in math has changed in the past 2000+ years; you don't hear about apotomes these days. (An apotome is a number $x - y$ where $\frac{x}{y}$ isn't rational, but $\frac{x^2}{y^2}$ is—as for $x = \sqrt{2}$, $y = 1$. It's difficult enough to describe even this without math notation. But then for a "first apotome" Euclid added the conditions that both $\frac{\sqrt{x^2-y^2}}{x}$ and x must be rational—all described in words.)

At five words, we've got one more familiar theorem (3.30) and another somewhat obscure one (10.26):

τὴν δοθεῖσαν περιφέρειαν δίχα τεμεῖν.
To bisect a given circumference.

μέσον μέσου οὐχ ὑπερέχει ῥητῷ.
A medial area does not exceed a medial area by a rational area.

In our modern Wolfram Language representation, we've got a precise, symbolic way to state Euclid's theorems. But Euclid had to rely on natural language (in his case, Greek). Some words he just assumed people would know the meanings of. But others he defined. Famously, he started at the beginning of Book 1 with his Definition 1—and in a sense changing how we think about this is what launched our whole Physics Project:

σημεῖόν ἐστιν, οὗ μέρος οὐθέν.
A point is that which has no part.

There is at least an implicit network of dependencies among Euclid's definitions. Having started by defining points and lines, he moves on to defining things like triangles, and equilaterality, until eventually, for example, by Book 11 Definition 27 he's saying things like "An icosahedron is a solid figure contained by twenty equal and equilateral triangles".

Of course, Euclid didn't ultimately have to set up definitions; he could just have repeated the content of each definition every time he wanted to refer to that concept. But like words in natural language—or functions in our computational language—definitions are an important form of compression for making statements. And, yes, you have to pick the right definitions to make the things you want to say easy to say. And, yes, your definitions will likely play at least some role in determining what kinds of things you choose to talk about. (Apotomes, anyone?)

The Interdependence of Theorems

All the theorems Euclid states represent less than 10,000 words of Greek. But the standard proofs of them are perhaps 150,000 words of Greek. (They're undoubtedly not minimal proofs—but the fact that the same ones are being quoted after more than 2000 years presumably tells us at least something.)

Euclid is very systematic. Every theorem throughout the course of his *Elements* is proved in terms of earlier theorems (and ultimately in terms of his 10 axioms). Thus, for example, the proof of 1.14 (i.e. Book 1, Theorem 14) uses 1.13 as well as the axioms P2 (i.e. Postulate 2), P4, CN1 (i.e. Common Notion 1) and CN3. By the time one's got to 12.18 the proof is written only in terms of other theorems (in this case 12.17, 12.2, 5.14 and 5.16) and not directly in terms of axioms.

The total number of theorems (or axioms) directly referenced in a given proof varies from 0 (for axioms) to 21 (for 12.17, which is about inscribing polyhedra in spheres); the average is 4.3:

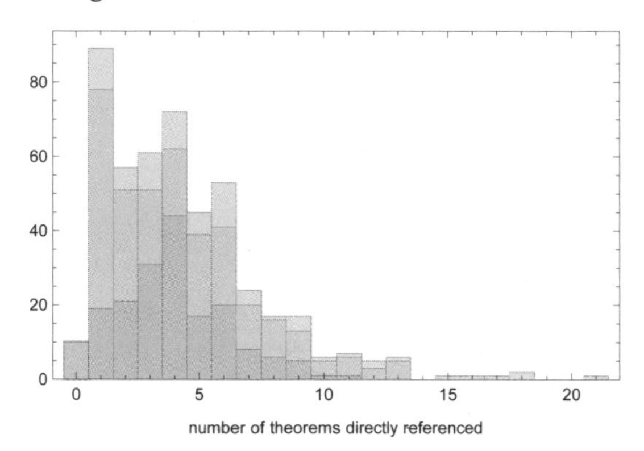

If we put Euclid's axioms and theorems in order, we can represent which axioms or theorems occur in a given proof by an arrangement of dots across the page. For example, for 1.12 through 1.17 we have:

	CN1	CN2	CN3	CN4	CN5	P1	P2	P3	P4	P5	1.1	1.2	1.3	1.4	1.5	1.6	1.7	1.8	1.9	1.10	1.11	1.12	1.13	1.14	1.15	1.16
1.12						•		•										•		•						
1.13	•	•																			•					
1.14	•		•				•		•														•			
1.15	•		•						•														•			
1.16					•	•	•						•	•						•					•	
1.17							•																•			

Doing this for all the theorems we get:

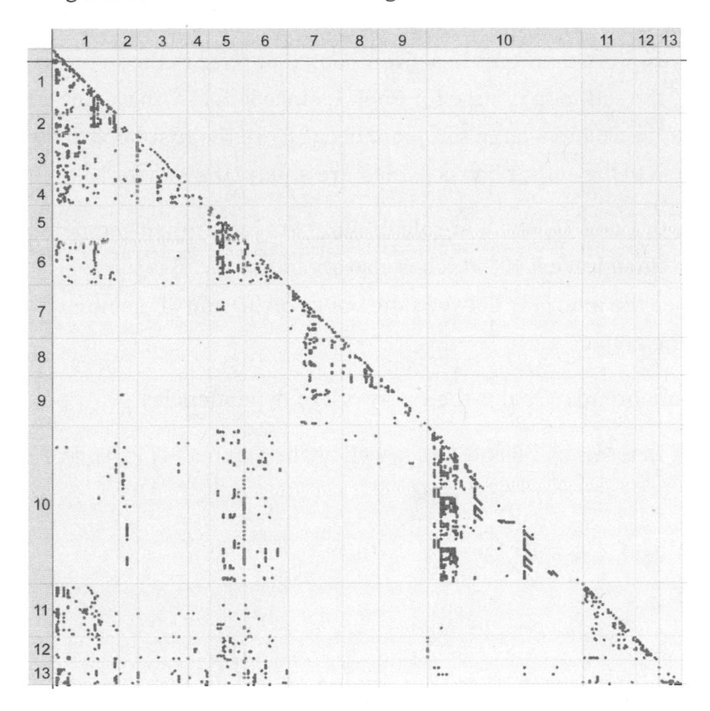

We can see there's lots of structure here. For example, there are clearly "popular" theorems near the beginning of Book 6 and Book 10, to which lots of at least "nearby" theorems refer. There are also "gaps": ranges of theorems that no theorems in a given book refer to.

At a coarse level, something we can do is to look at cross-referencing within and between books:

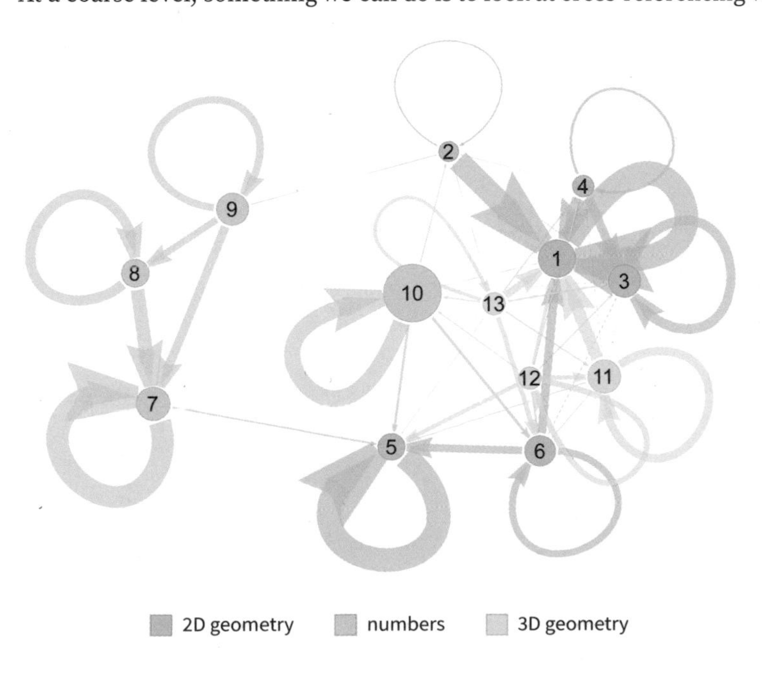

The size of each node represents the number of theorems in each book. The thickness of each arrow represents the fraction of references in the proofs of those theorems going to different books. The self-loops are from theorems in a given book that refer to theorems in the same book. Needless to say, the self-loop is large for Book 1, since it doesn't have any previous book to refer to. Book 7 again has a large self-loop, because it's the first book about numbers, and doesn't refer much to the earlier books (which are about 2D geometry).

It's interesting to see that Books 7, 8 and 9—which are about numbers rather than geometry—"keep to themselves", even though Book 10, which is also about numbers, is more central. It's also interesting to see the interplay between the books on 2D and 3D geometry over on the right-hand side of the graph.

But, OK, what about individual theorems? What is their network of dependencies?

Here's 1.5, whose proof is given in terms of 1.3 and 1.4, as well as the axioms P1, P2 and CN3:

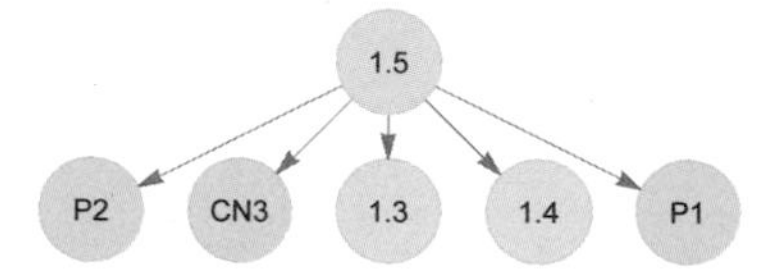

But now we can continue this, and show what 1.3 and 1.4 depend on—all the way down to the axioms:

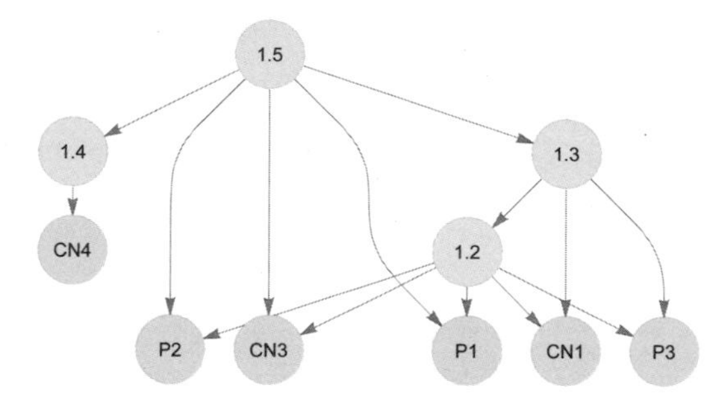

Later theorems depend on much more. Here are the direct dependencies for 12.18:

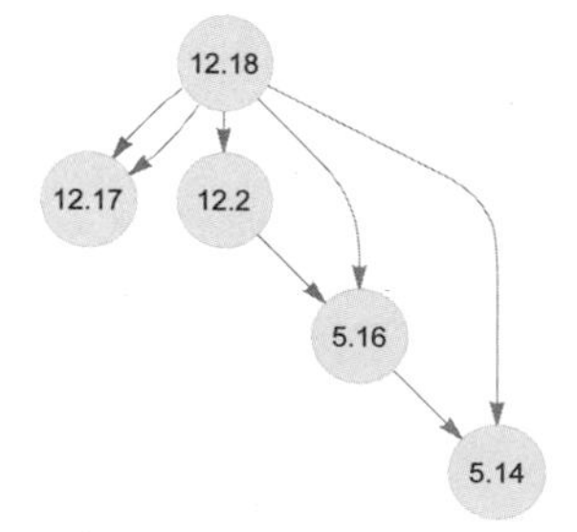

Here's what happens if one goes another step:

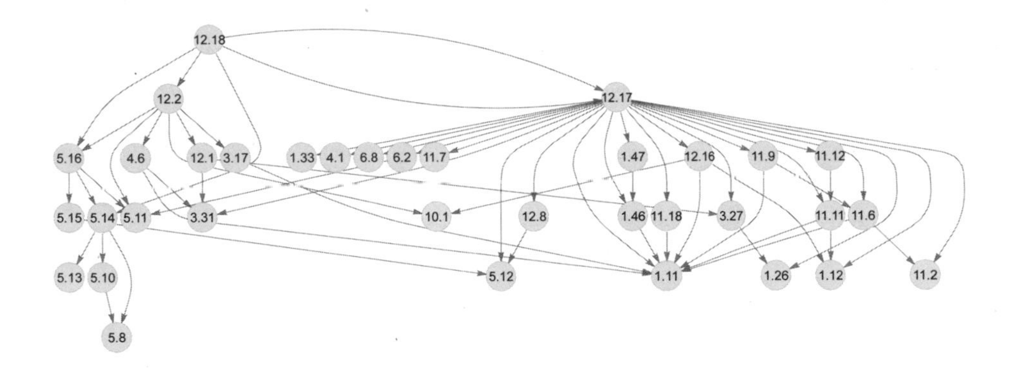

Here's 3 steps:

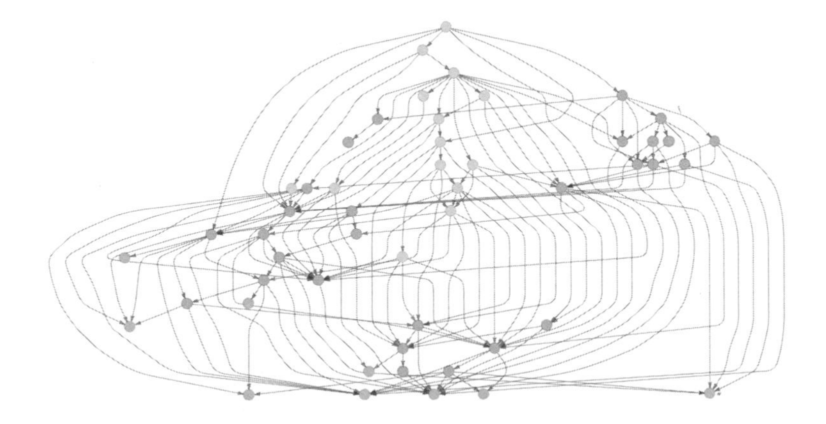

And here's what happens if one goes all the way down to the axioms (which in this case takes 5 steps):

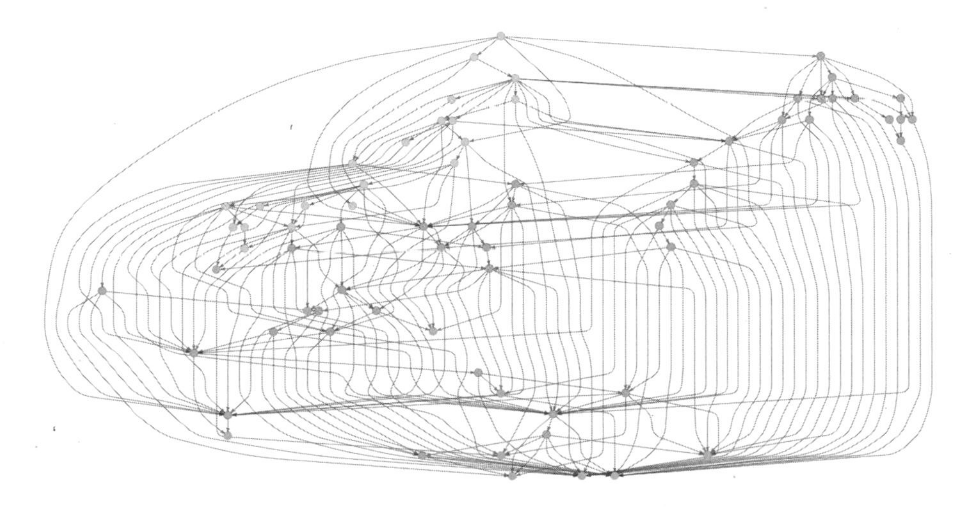

Things look a little simpler if we consider the transitive reduction of this graph. We're no longer faithfully representing what's in the text of Euclid, but we're still capturing the core dependency information. If theorem A in Euclid refers to B, and B refers to C, then even if in Euclid A refers to C we won't mention that. And, yes, graph theoretically A → C is just the transitive closure of A → B and B → C. But it could still be that the pedagogical structure of the proof of theorem A makes it desirable to refer to theorem B, even if in principle one could rederive theorem B from theorem C.

Here's the original 1-step graph for 12.18, along with its transitive reduction:

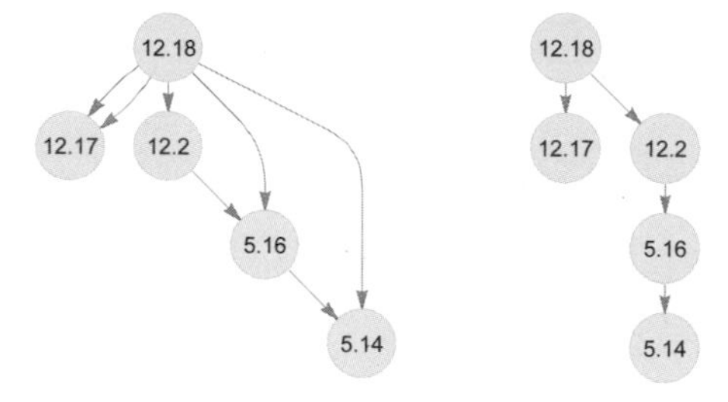

And here, by the way, is also the "fully pedantic" transitive closure, including all indirect connections, whether they're mentioned by Euclid or not:

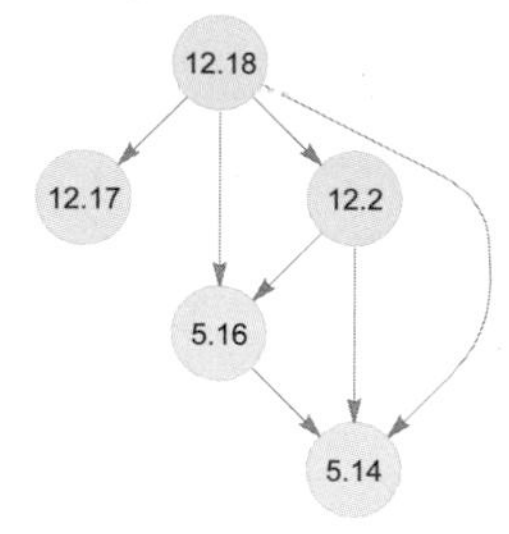

And now here's the transitive reduction of the full 12.8 dependency graph, all the way down to the axioms:

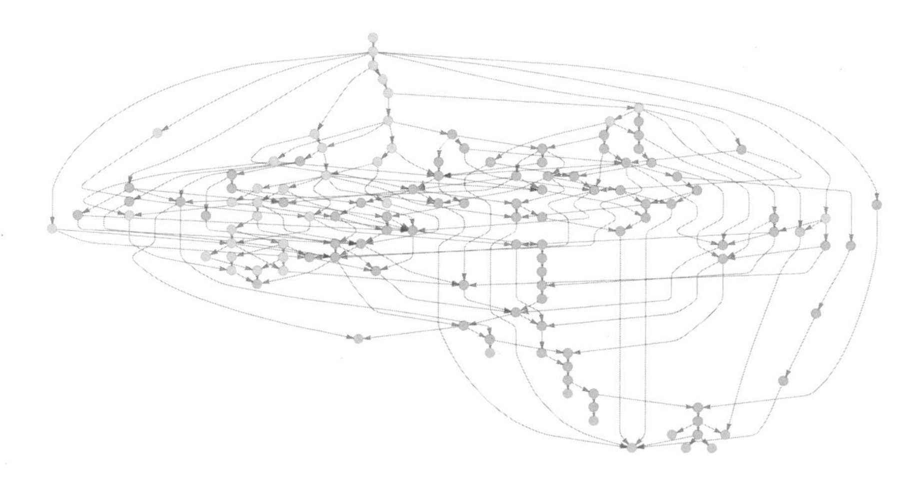

And what all these graphs show is that even to prove one theorem, one's making use of lots of other theorems. To make this quantitative, we can plot the total number of theorems that appear anywhere in the "full proof" of a given theorem, ultimately working all the way down to the axioms:

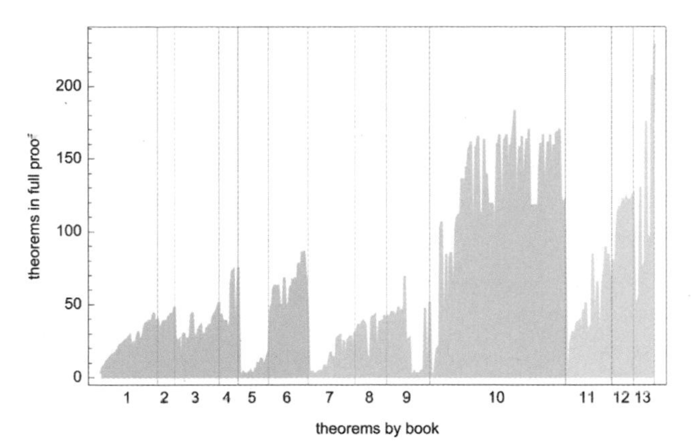

At the beginnings of many of the books, there tend to be theorems that are proved more directly from the axioms, so they don't depend on as much. But as one progresses through the books, one's relying on more and more theorems—sometimes, as we saw above, in the same book, and sometimes in earlier books.

From the picture above, we can see that Euclid in a sense builds up to a "climax" at the end—with his very last theorem (13.18) depending on more theorems than anything else. We'll be discussing "Euclid's last theorem" some more below...

The Graph of All Theorems

OK, so what is the full interdependence graph for all the theorems in Euclid? It's convenient to go the opposite way than in our previous graphs—and put the axioms at the top, and show how theorems below are derived from them. Here's the graph one gets by doing that:

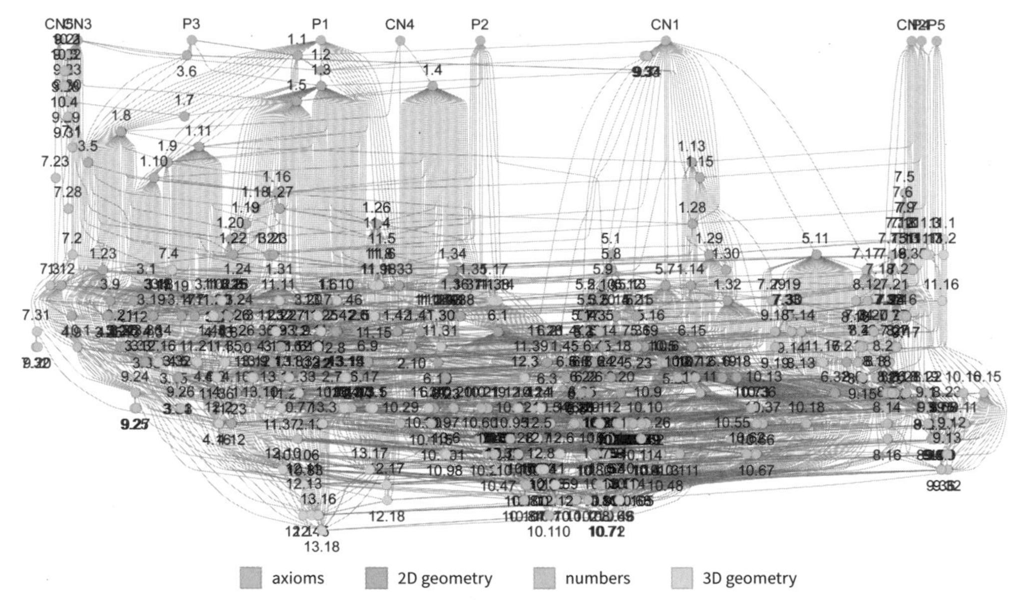

One can considerably simplify this by looking just at the transitive reduction graph (the full graph has 2054 connections; this reduction has 974, while if one went "fully pedantic" with transitive closure, one would have 25,377 connections):

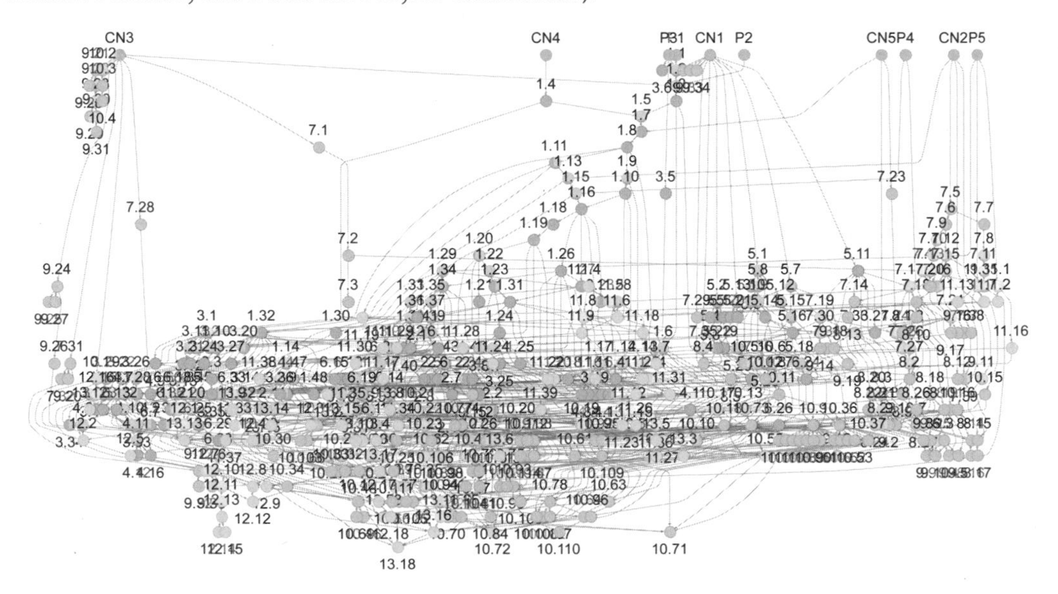

What can we see from this? Probably the most obvious thing is that the graphs start fairly sparse, then become much denser. And what this effectively means is that one starts off by proving certain "preliminaries", and then after one's done that, it unlocks a mass of other theorems. Or, put another way, if we were exploring this metamathematical space starting from the axioms, progress might seem slow at first. But after proving a bunch of preliminary theorems, we'd be able to dramatically speed up.

Here's another view of this, plotting how many subsequent theorems depend on each different theorem:

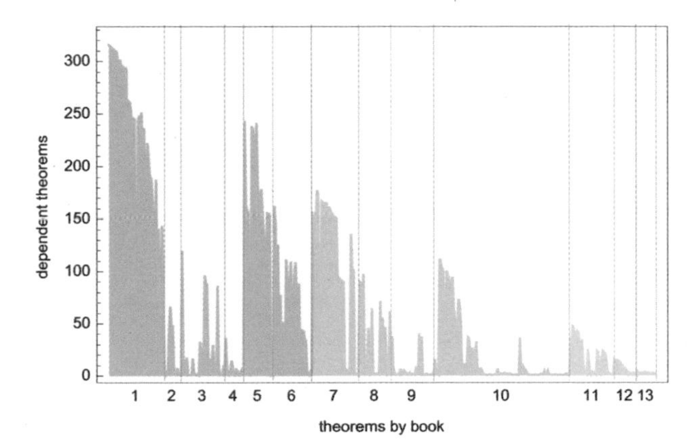

In a sense, this is complementary to the plot we made above, that showed how many theorems a given theorem depends on. (From a graph-theoretical point of view they're very directly complementary: this plot involves VertexInComponent; the previous one involved VertexOutComponent.)

And what the plot shows is that there are a bunch of early theorems (particularly in Book 1) that have lots of subsequent theorems depending on them—so that they're effectively foundational to much of what follows. The plot also shows that in most of the books the early theorems are the most "foundational", in the sense that the most subsequent theorems depend on them.

By the way, we can also look at the overall form of the basic dependency graph, not layering it starting from the axioms:

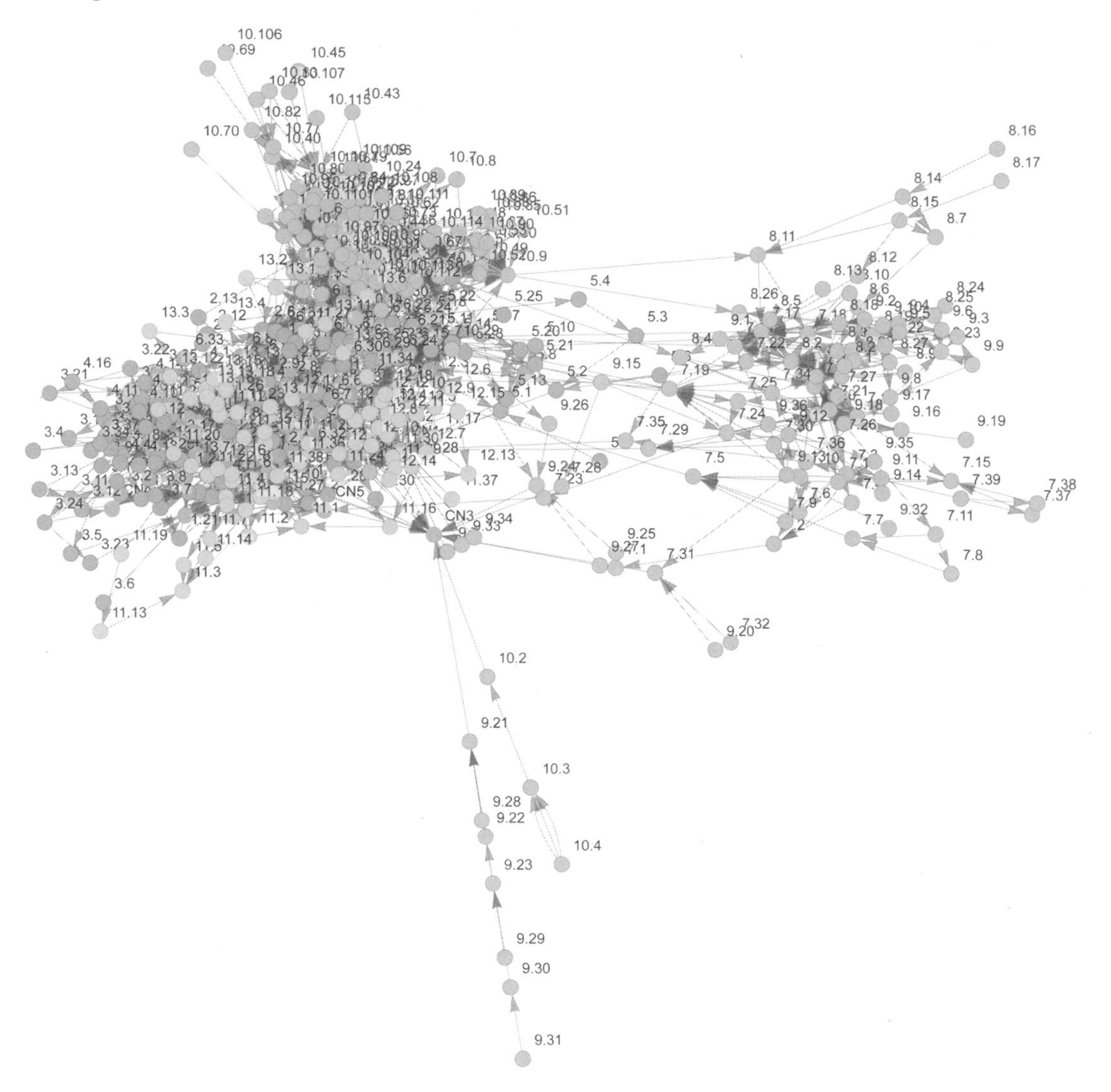

The transitive reduction is slightly easier to interpret:

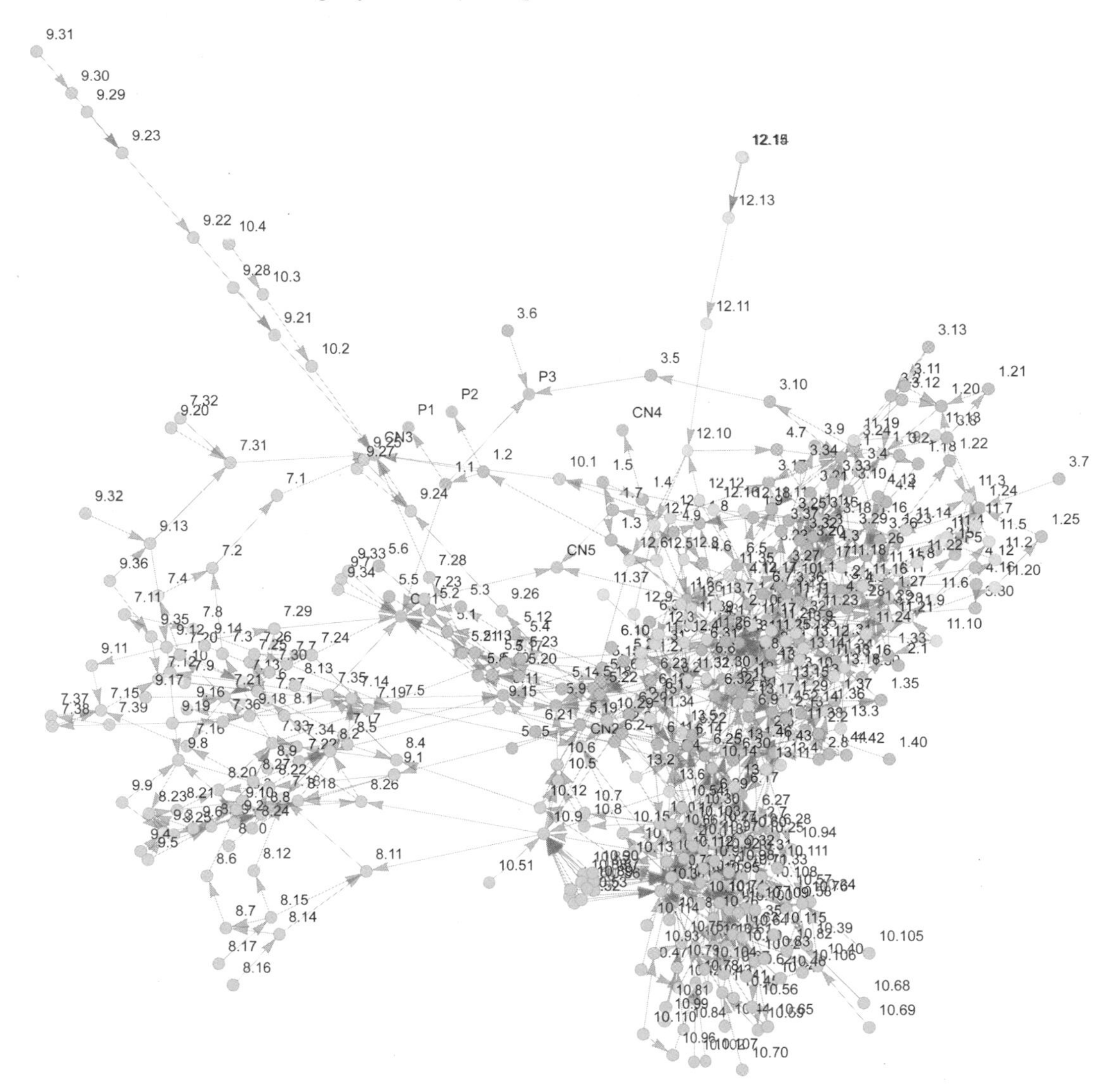

And the main notable feature is the presence of "prongs" associated, for example, with Book 9 theorems about the properties of even and odd numbers.

The Causal Graph Analogy

Knowing about the Wolfram Physics Project, there's an obvious analog of theorem dependency graphs: they're like causal graphs. You start from a certain set of "initial events" (the "big bang"), corresponding to the axioms. Then each subsequent theorem is like an event, and the theorem dependency graph is tracing out the causal connections between these events.

Just like the causal graph, the theorem dependency graph defines a partial ordering: you can't write down the proof of a given theorem until the theorems that will appear in it have been proved. Like in the causal graph, one can define light cones: there's a certain set of "future" theorems that can be affected by any given theorem. Here is the "future light cone" of Book 1, Theorem 5:

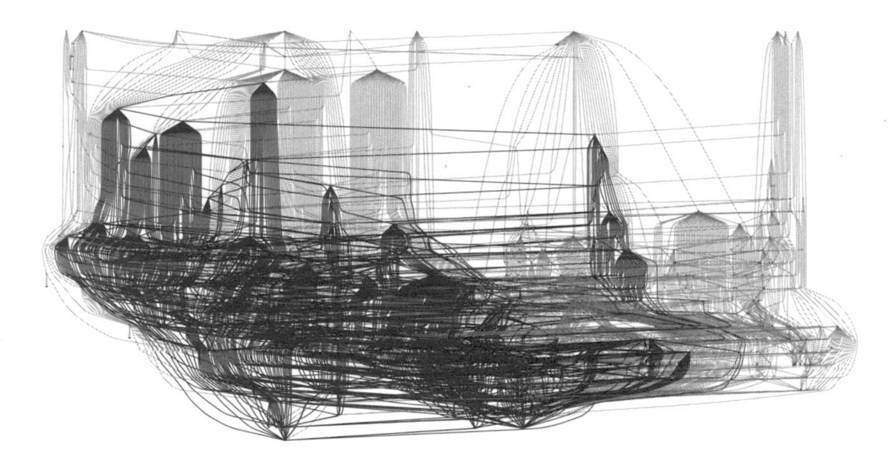

And here is the corresponding transitive reduction graph:

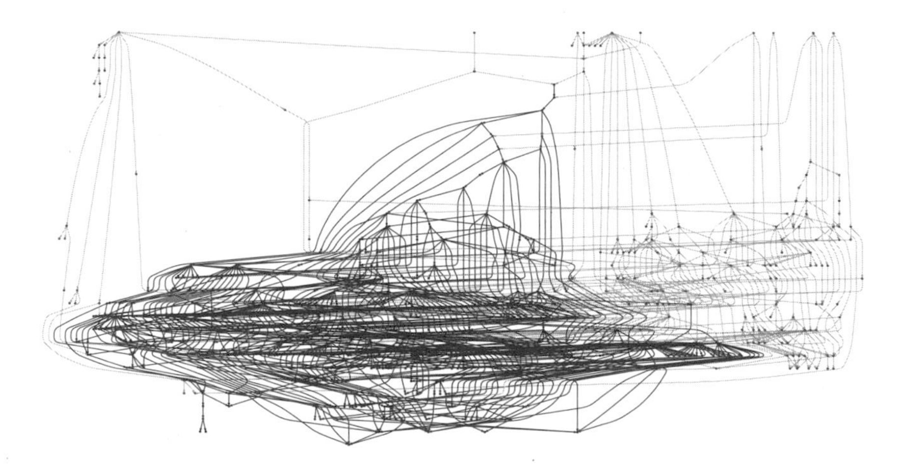

But now let's think about the notion of time in the theorem dependency graph. Imagine you were rederiving the theorems in Euclid in a series of "time steps". What would you have to do at each time step? The theorem dependency graph tells you what you will have to have done in order to derive a particular theorem. But just like for spacetime causal graphs, there are many different foliations one can use to define consistent time steps.

Here's an obvious one, effectively corresponding to a "cosmological rest frame" in which at each step one "does as much as one consistently can at that step":

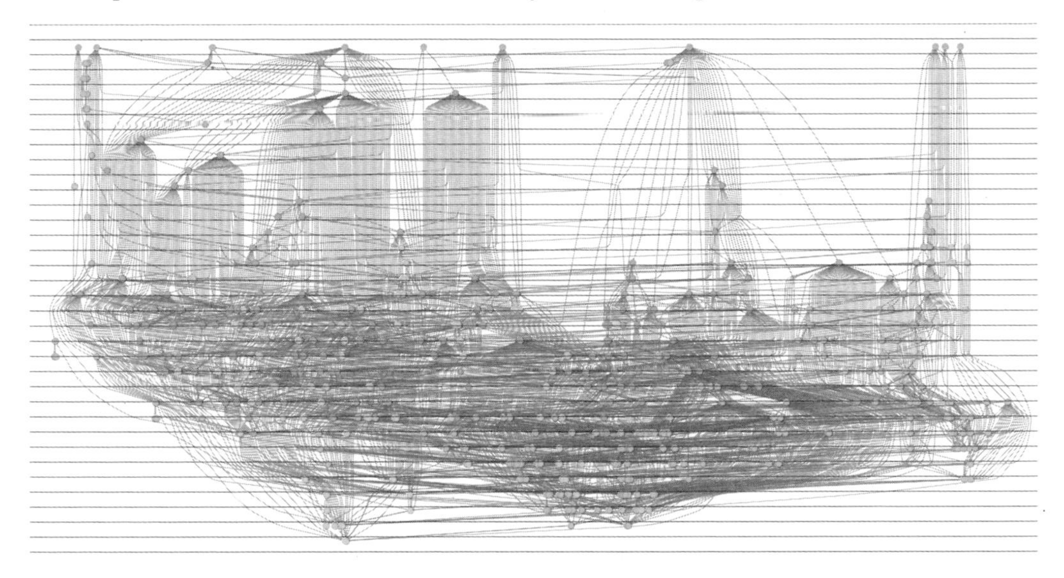

And here are the number of theorems that appear on each slice (in effect each theorem appears on the slice determined by its longest path to any axiom):

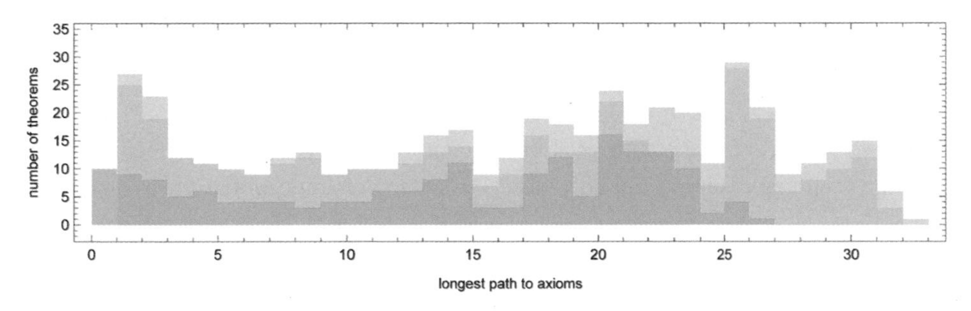

But there are many other foliations that are possible, in which one for example concentrates first on a particular group of theorems, only doing others when one "needs to".

Each choice of foliation can be thought of as corresponding to a different reference frame—and a different choice of how one explores the analog of spacetime in Euclid. But, OK, if the foliations define successive moments in time—or successive "simultaneity surfaces"—what is the analog of space? In effect, the "structure of space" is defined by the way that theorems are laid out on the slices defined by the foliations. And a convenient way to probe this is to look at branchial graphs, in which pairs of theorems on a given slice are connected by an edge if they have an immediate common ancestor on the slice before.

So here are the branchial graphs for all successive slices of Euclid in the "cosmological rest frame":

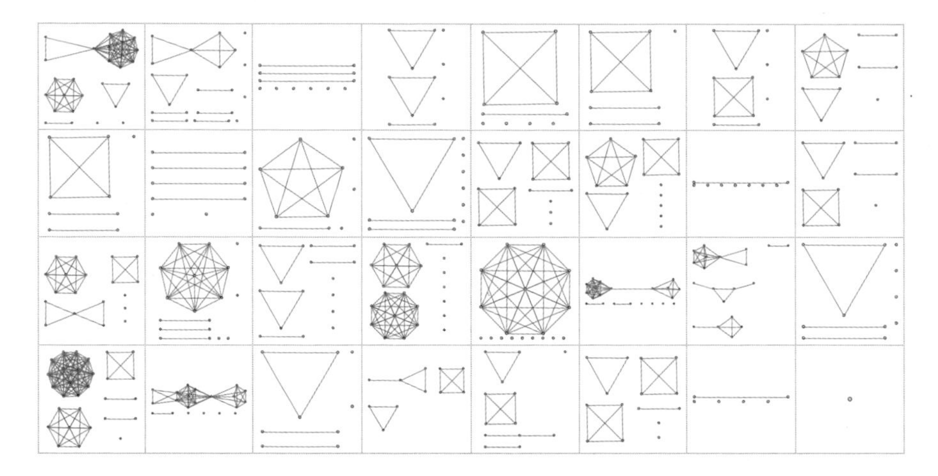

And here are the branchial graphs specifically from slices 23 and 26:

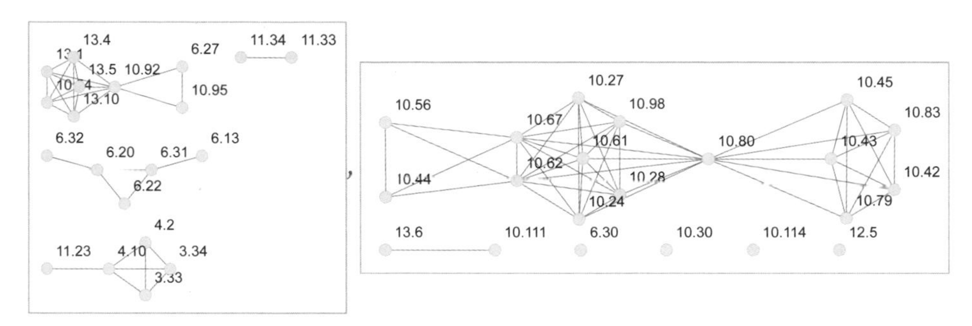

How should we interpret these graphs? Just like in quantum mechanics, they effectively define a map of "entanglements", but now these are "entanglements" not between quantum states but between theorems. But potentially we can also interpret these graphs as showing how theorems are laid out in a kind of "instantaneous metamathematical space"—or, in effect, we can use the graphs to define "distances between theorems".

We can generalize our ordinary branchial graphs by connecting theorems that have common ancestors not just one slice back, but also up to δt slices back. Here are the results for slice 26 (in the cosmological rest frame):

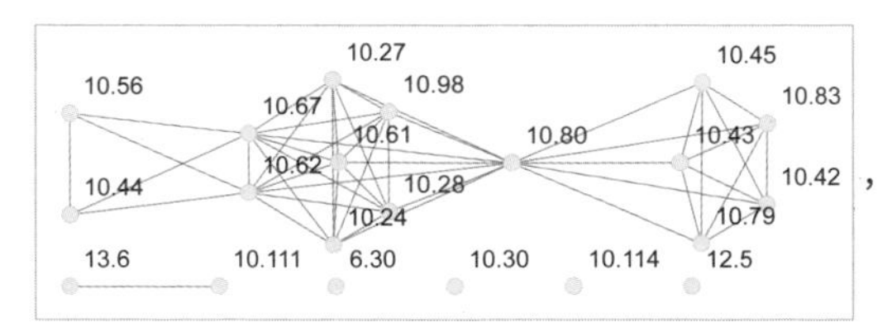

$\delta t = 1$

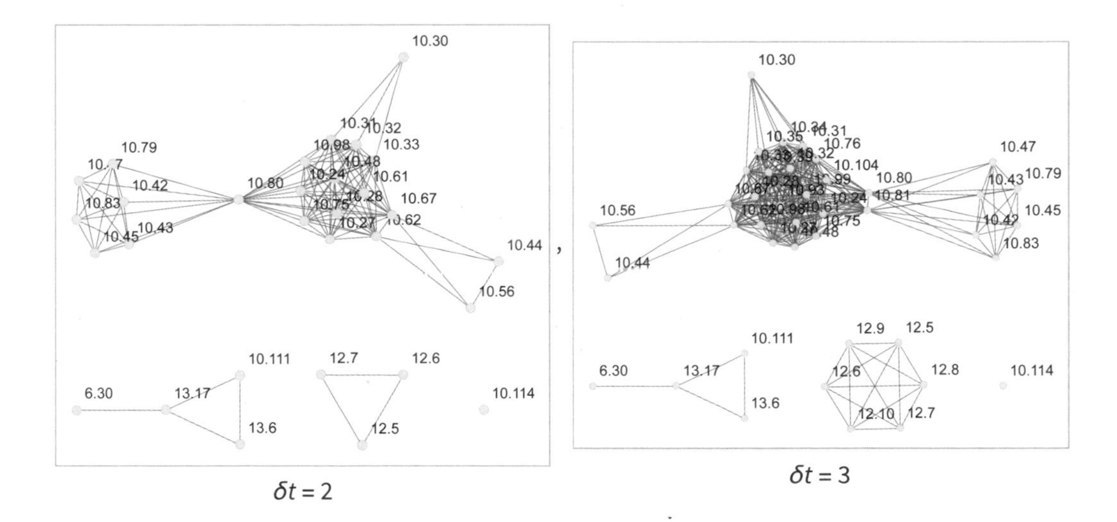

$\delta t = 2$, $\delta t = 3$

If we went all the way back to the axioms (the analog of the "big bang") then we'd just get a complete graph, connecting all the theorems on slice 26. But here we're seeing in effect "fuzzier and fuzzier" versions of how the theorems that exist at slice 26 can be thought of as being "metamathematically laid out". The disconnected components in these branchial graphs represent theorems that have no recent shared history—so that in some sense they're "causally disconnected".

In thinking about "theorem search", it's interesting to try to imagine measures of "distance between theorems"—and in effect branchial distance captures some of this. And even for Euclid there are presumably things to learn about the "layout" of theorems, and what should count as "close to" what.

There are only 465 theorems in Euclid's *Elements.* But what if there were many more? What might the "metamathematical space" they define be like? Just as for the hypergraphs—or, for that matter, the multiway graphs—in our models of physics we can ask questions about the limiting emergent geometry of this space. And—ironically enough—one thing we can immediately say is that it seems to be far from Euclidean!

But does it for example have some definite effective dimension? There isn't enough data to say much about the branchial slices we just saw. But we can say a bit more about the complete theorem dependency graph—which is the analog of the multiway graph in our physics models. For example, starting with the axioms (the analog of the "big bang") we can ask how many theorems are reached in successive steps. The result (counting the axioms) is:

{10, 81, 325, 444, 470, 475, 475, 475, 475, 475, 475}

If we were dealing with something that approximated a d-dimensional manifold, we'd expect these numbers to be of order r^d. Computing their logarithmic differences to fit for d gives

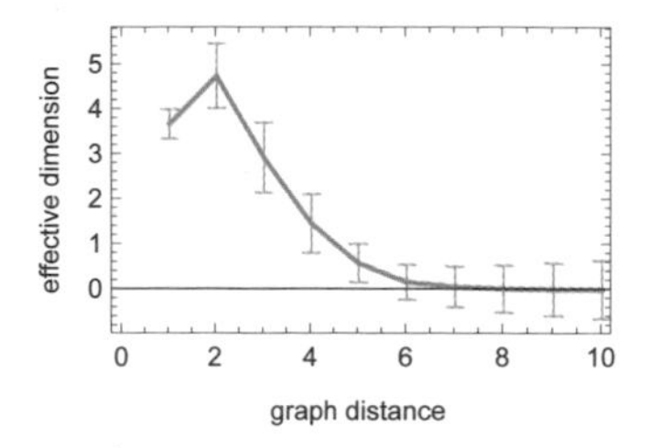

if one starts from the axioms, and

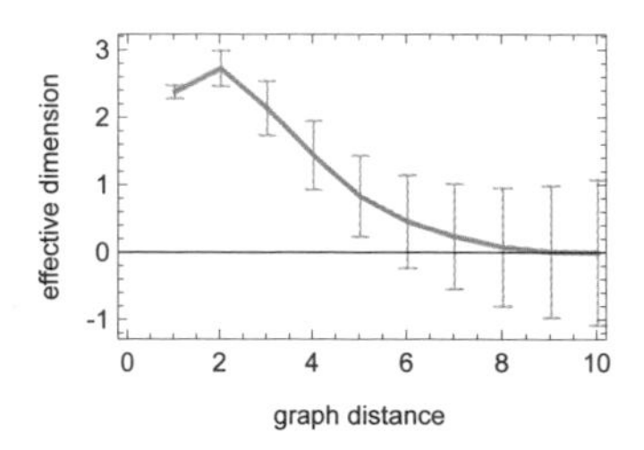

if one starts from all possible theorems in the network.

One gets somewhat different results if one deals not with the actual theorem dependency graph in Euclid, but instead with its transitive reduction—removing all "unnecessary" direct connections. Now the number of theorems reached on successive steps is:

{10, 44, 122, 267, 390, 445, 465, 470, 471, 474, 475}

The "dimension estimate" based on theorems reached starting from the axioms is

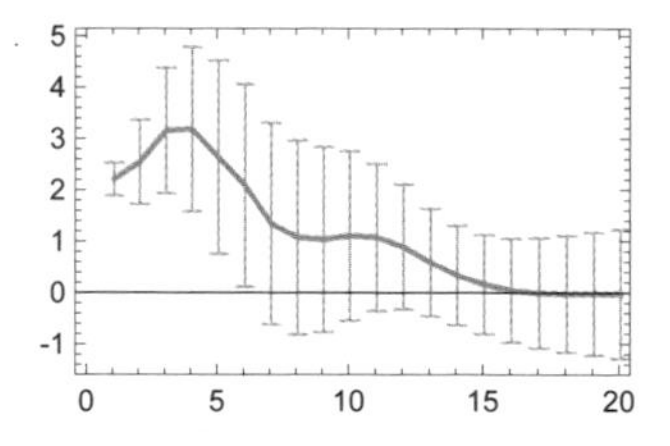

while the corresponding result starting from all theorems is:

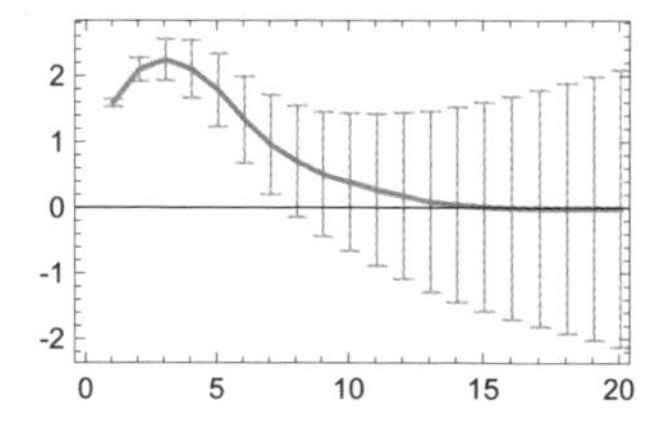

Euclid's *Elements* represents far too little data to make a definite statement, but perhaps there's a hint of 2-dimensional structure, with positive curvature.

The Most Difficult Theorem in Euclid

One way to assess the "difficulty" of a theorem is to look at what results have to have already been built up in order to prove the theorem. And by this measure, the most difficult theorem in Euclid's *Elements* is the very last theorem in the last book—what one might call "Euclid's last theorem", the climax of the *Elements*—Book 13, Theorem 18, which amounts to the statement that there are five Platonic solids, or more specifically:

τὰς πλευρὰς τῶν πέντε σχημάτων ἐκθέσθαι καὶ συγκρῖναι πρὸς ἀλλήλας.

This theorem uses all 10 axioms, and 219 of the 464 previous theorems. Here's its graph of dependencies:

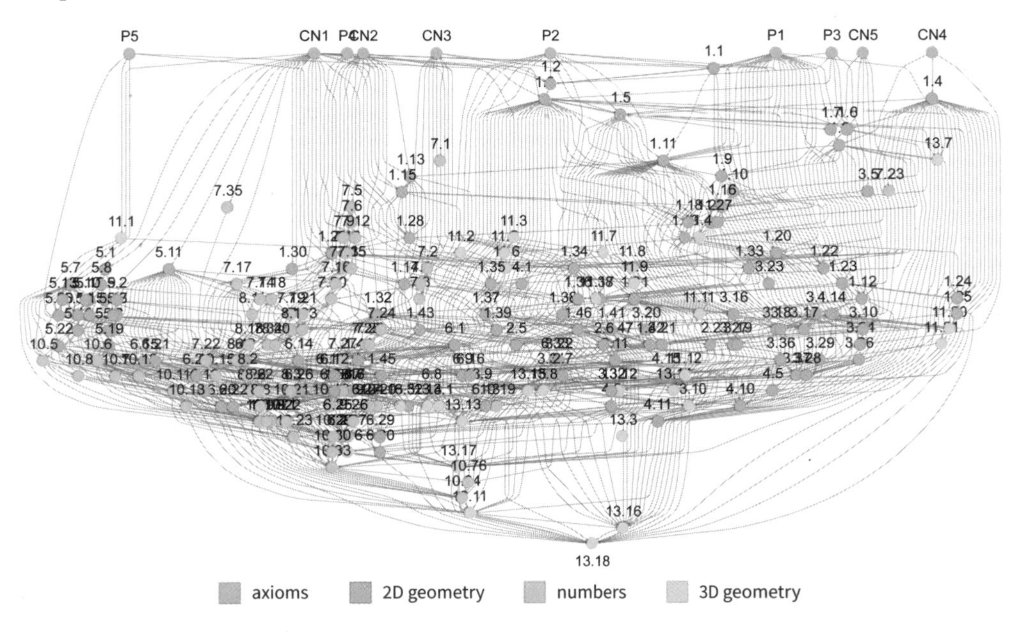

And here is the transitive reduction of this—notably with different subject areas being more obviously separated:

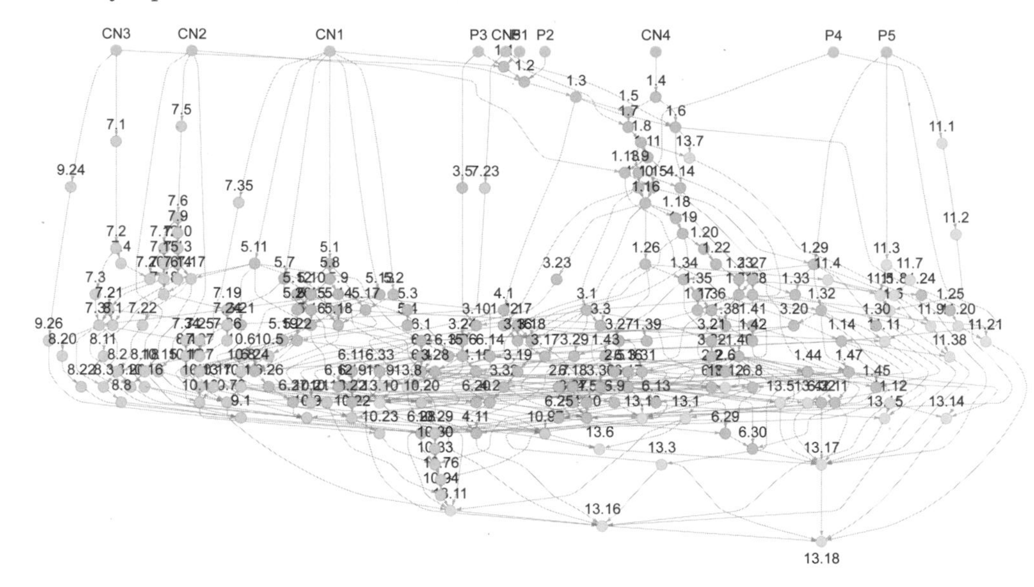

This shows how 13.18 and its prerequisites (its "past light cone") sit inside the whole theorem dependency graph:

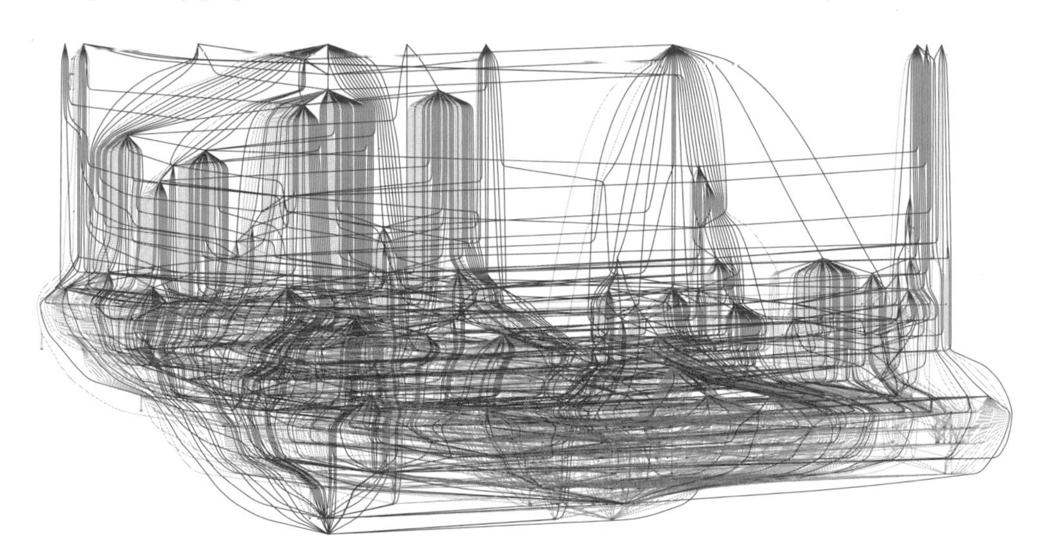

If we started from the axioms, the longest chains of theorems we'd have to prove to get to 13.18 are:

CN1 → 1.1 → 1.2 → 1.3 → 1.5 → 1.7 → 1.8 → 1.11 → 1.13 → 1.15 → 1.16 → 1.18 →
1.19 → 1.20 → 1.22 → 1.23 → 1.31 → 1.37 → 1.41 → 6.1 → 6.2 → 6.11 → 6.19 →
6.20 → 10.9 → 10.29 → 10.30 → 10.33 → 10.76 → 10.94 → 13.11 → 13.16 → 13.18

CN2 → 1.13 → 1.15 → 1.16 → 1.18 → 1.19 → 1.20 → 1.22 →
1.23 → 1.31 → 1.37 → 1.41 → 6.1 → 6.2 → 6.11 → 6.19 → 6.20 →
10.9 → 10.29 → 10.30 → 10.33 → 10.76 → 10.94 → 13.11 → 13.16 → 13.18

CN3 → 1.2 → 1.3 → 1.5 → 1.7 → 1.8 → 1.11 → 1.13 → 1.15 → 1.16 → 1.18 →
1.19 → 1.20 → 1.22 → 1.23 → 1.31 → 1.37 → 1.41 → 6.1 → 6.2 → 6.11 → 6.19 →
6.20 → 10.9 → 10.29 → 10.30 → 10.33 → 10.76 → 10.94 → 13.11 → 13.16 → 13.18

CN4 → 1.4 → 1.5 → 1.7 → 1.8 → 1.11 → 1.13 → 1.15 → 1.16 → 1.18 → 1.19 →
1.20 → 1.22 → 1.23 → 1.31 → 1.37 → 1.41 → 6.1 → 6.2 → 6.11 → 6.19 →
6.20 → 10.9 → 10.29 → 10.30 → 10.33 → 10.76 → 10.94 → 13.11 → 13.16 → 13.18

CN5 → 1.7 → 1.8 → 1.11 → 1.13 → 1.15 → 1.16 → 1.18 → 1.19 → 1.20 →
1.22 → 1.23 → 1.31 → 1.37 → 1.41 → 6.1 → 6.2 → 6.11 → 6.19 → 6.20 →
10.9 → 10.29 → 10.30 → 10.33 → 10.76 → 10.94 → 13.11 → 13.16 → 13.18

P1 → 1.1 → 1.2 → 1.3 → 1.5 → 1.7 → 1.8 → 1.11 → 1.13 → 1.15 → 1.16 → 1.18 →
1.19 → 1.20 → 1.22 → 1.23 → 1.31 → 1.37 → 1.41 → 6.1 → 6.2 → 6.11 → 6.19 →
6.20 → 10.9 → 10.29 → 10.30 → 10.33 → 10.76 → 10.94 → 13.11 → 13.16 → 13.18

P2 → 1.2 → 1.3 → 1.5 → 1.7 → 1.8 → 1.11 → 1.13 → 1.15 → 1.16 → 1.18 →
1.19 → 1.20 → 1.22 → 1.23 → 1.31 → 1.37 → 1.41 → 6.1 → 6.2 → 6.11 → 6.19 →
6.20 → 10.9 → 10:29 → 10.30 → 10.33 → 10.76 → 10.94 → 13.11 → 13.16 → 13.18

P3 → 1.1 → 1.2 → 1.3 → 1.5 → 1.7 → 1.8 → 1.11 → 1.13 → 1.15 → 1.16 → 1.18 →
1.19 → 1.20 → 1.22 → 1.23 → 1.31 → 1.37 → 1.41 → 6.1 → 6.2 → 6.11 → 6.19 →
6.20 → 10.9 → 10.29 → 10.30 → 10.33 → 10.76 → 10.94 → 13.11 → 13.16 → 13.18

P4 → 1.15 → 1.16 → 1.18 → 1.19 → 1.20 → 1.22 → 1.23 →
1.31 → 1.37 → 1.41 → 6.1 → 6.2 → 6.11 → 6.19 → 6.20 → 10.9 →
10.29 → 10.30 → 10.33 → 10.76 → 10.94 → 13.11 → 13.16 → 13.18

P5 → 1.29 → 1.34 → 1.35 → 1.36 → 1.38 → 6.1 → 6.2 → 6.11 → 6.19 →
6.20 → 10.9 → 10.29 → 10.30 → 10.33 → 10.76 → 10.94 → 13.11 → 13.16 → 13.18

Or in other words, from CN1 and from P1 and P3 we'd have to go 33 steps to reach 13.18. If we actually look at the paths, however, we see that after different segments at the beginning, they all merge at Book 6, Theorem 1, and then are the same for the last 14 steps:

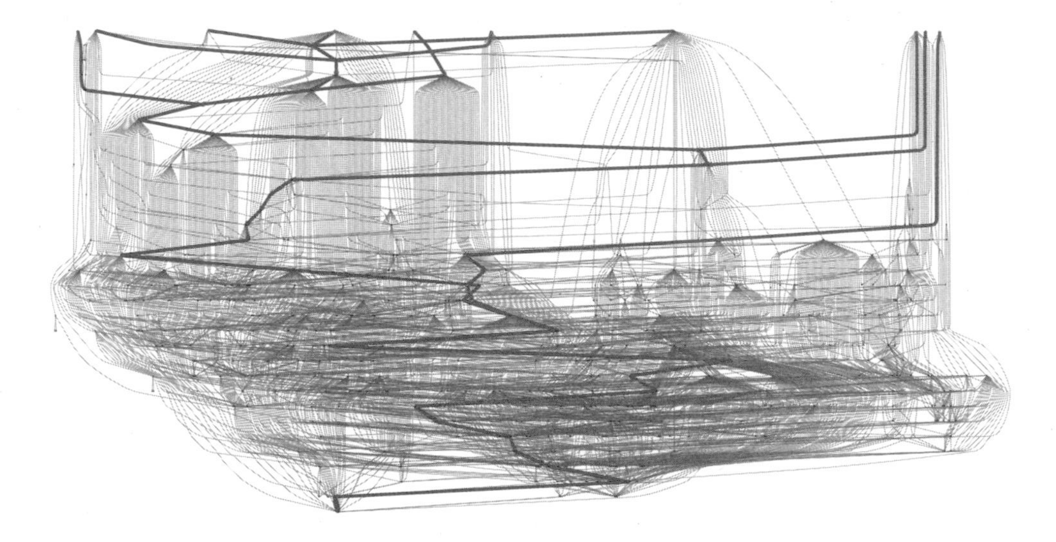

(Theorem 6.1 is the statement that both triangles and parallelograms that have the same base and same height have the same area, i.e. one can skew a triangle or parallelogram without changing its area.)

How much more difficult than other theorems is 13.18? Here's a histogram of maximum path lengths for all theorems (ignoring cases to be discussed later where a particular theorem does not use a given axiom at all):

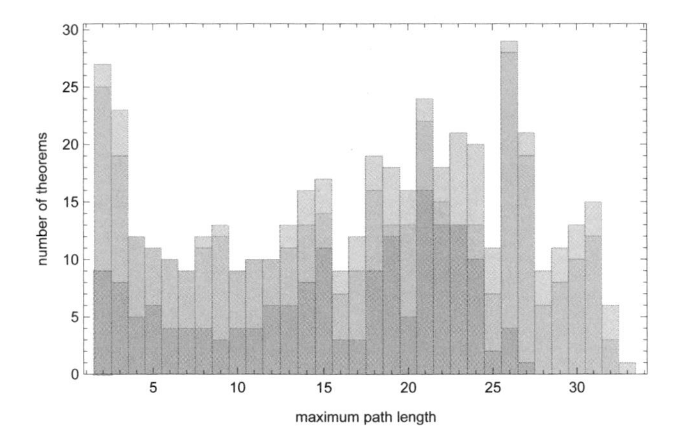

And here's how the maximum path length varies through the sequence of all 465 theorems:

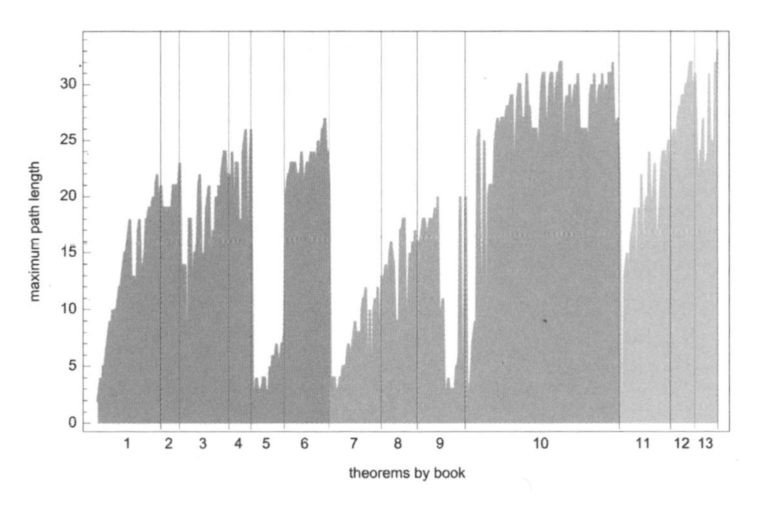

In the causal graph interpretation, and using the "flat foliation" (i.e. the "cosmological rest frame") what this basically shows is at what "time slice" a given theorem first emerges from Euclid's proofs. Or, in other words, if one imagines exploring the "metamathematical space of Euclid" by going "one level of theorems at a time", the order in which one will encounter theorems is:

CN1 CN2 CN3 CN4 CN5 P1 P2 P3 P4 P5
1.1 1.4 3.5 3.6 5.1 5.2 5.7 5.11 5.13 7.1 7.5 7.23 7.28 7.29 7.31 7.35 9.7 9.21 9.24 9.33 9.34 10.1 10.2 10.15 10.16 11.1 11.3
1.2 5.3 5.5 5.6 5.8 5.12 5.17 6.21 7.2 7.6 7.7 7.32 9.20 9.22 9.25 9.26 9.27 9.28 10.3 11.2 11.7 11.13 11.16
1.3 5.4 5.9 5.10 5.15 7.3 7.4 7.8 7.9 7.12 9.23 10.4
1.5 1.6 4.1 5.14 5.20 5.21 7.10 7.11 7.15 9.29 9.30
1.7 5.16 5.18 5.22 7.13 7.16 7.37 7.38 9.11 9.31
1.8 5.19 5.23 5.24 7.14 7.17 7.20 9.35 10.5
1.9 1.11 4.9 5.25 7.18 7.21 7.22 8.13 8.18 10.6 10.8 13.7
1.10 1.13 4.14 7.19 8.1 8.10 8.11 8.12 8.19 9.16 9.17 10.7 11.19
1.12 1.14 1.15 3.1 7.24 7.30 7.33 9.18 10.11
1.16 1.29 3.9 3.10 7.25 7.26 7.34 8.20 9.14 9.19
1.17 1.18 1.26 1.27 1.30 3.23 7.27 7.36 8.4 8.22
1.19 1.28 1.33 1.34 3.3 3.24 7.39 8.2 8.5 8.24 10.12 11.4 11.14
1.20 1.35 1.43 3.2 3.4 3.16 3.18 3.26 8.3 8.9 8.26 8.27 10.13 11.5 11.8 11.28
1.21 1.22 1.36 3.11 3.12 3.17 3.19 3.28 4.4 4.7 4.13 8.6 8.8 8.21 11.6 11.18 12.16
1.23 3.13 3.30 8.7 8.23 9.1 9.2 11.9 11.29
1.24 1.31 3.25 8.14 8.15 8.25 9.3 9.6 9.8 11.10 11.30 11.38
1.25 1.32 1.37 1.38 1.46 2.1 3.7 3.8 4.8 8.10 8.17 9.4 9.5 9.9 9.10 9.12 11.11 11.22 11.24
1.39 1.40 1.41 2.2 2.3 2.4 2.5 2.6 2.7 2.8 3.20 4.3 9.13 11.12 11.15 11.20 11.25 13.3
1.42 1.47 3.21 3.27 6.1 9.15 9.32 9.36 10.17 10.36 10.73 10.74 10.77 11.21 11.26 11.31
1.44 1.48 2.9 2.10 2.11 2.12 2.13 3.14 3.22 3.29 3.35 3.36 6.2 6.14 6.15 6.33 10.18 10.19 10.20 10.21 10.37 10.106 11.39 13.2
1.45 3.15 3.31 3.37 4.15 6.3 6.4 6.9 6.10 6.11 6.12 6.16 6.24 10.55 10.115 11.17 11.35 12.3
2.14 3.32 4.5 4.6 6.5 6.6 6.7 6.8 6.17 6.18 6.19 6.23 6.26 11.27 11.32 11.36 13.8 13.9 13.12 13.14 13.15
3.33 3.34 4.2 4.10 6.13 6.20 6.22 6.27 6.31 6.32 10.54 10.92 10.95 11.23 11.33 11.34 13.1 13.4 13.5 13.10
4.11 6.25 10.9 10.14 10.22 10.91 10.109 11.37 12.1 12.4 13.13
4.12 4.16 6.28 6.29 10.10 10.23 10.25 10.26 10.29 10.38 10.49 10.50 10.51 10.52 10.53 10.60 10.66 10.85 10.86 10.87 10.88 10.89 10.90 10.97 10.101 10.103 10.112 10.113 12.2
6.30 10.24 10.27 10.28 10.30 10.42 10.43 10.44 10.45 10.56 10.61 10.62 10.67 10.79 10.80 10.83 10.98 10.111 10.114 12.5 13.6
10.31 10.32 10.33 10.47 10.48 10.75 12.6 12.7 13.17
10.34 10.35 10.39 10.76 10.81 10.93 10.99 10.104 12.8 12.9 12.10
10.40 10.41 10.57 10.63 10.68 10.78 10.82 10.94 10.100 10.105 12.11 12.12 12.17
10.46 10.58 10.59 10.64 10.65 10.69 10.70 10.84 10.96 10.102 10.107 10.108 12.13 12.18 13.11
10.71 10.72 10.110 12.14 12.15 13.16
13.18

A question one might ask is whether "short-to-state" theorems are somehow "easier to prove" than longer-to-state ones. This shows the maximum path length to prove theorems as a function of the length of their statements in Euclid's Greek. Remarkably little correlation is seen.

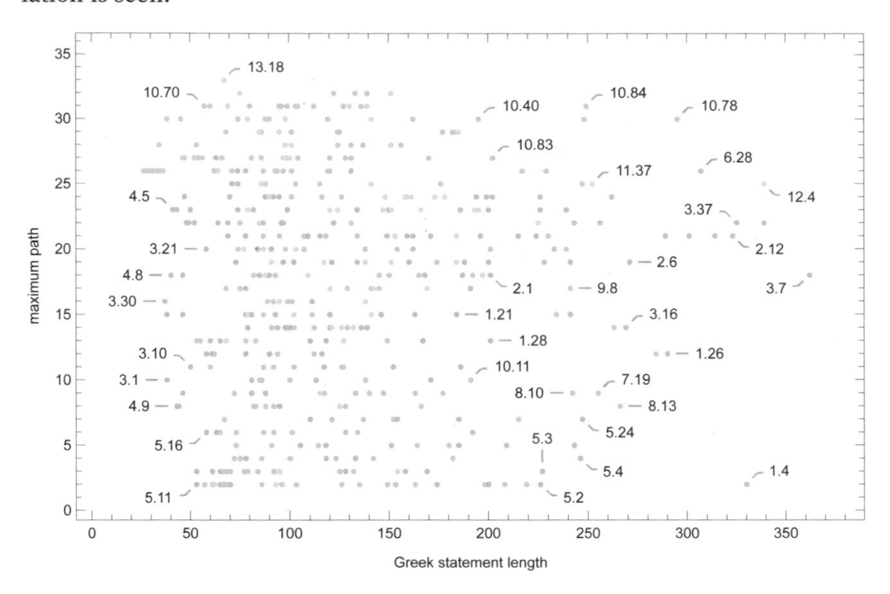

This plot shows instead the number of "prerequisite theorems" as a function of statement length:

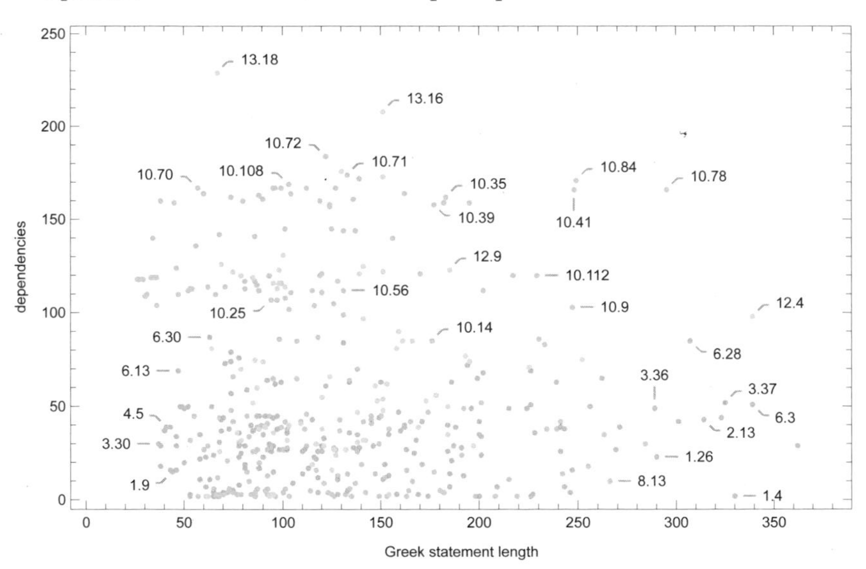

Once again there is poor correlation.

The Most Popular Theorems in Euclid

How often do particular theorems get used in the proofs of other theorems? The "most popular" theorems in terms of being directly quoted in the proofs of other theorems are:

10.11	*59*	CN1	*29*	5.19	*22*	5.22	*16*	5.9	*13*
6.1	*52*	10.73	*27*	10.15	*21*	3.31	*16*	CN3	*13*
5.11	*46*	10.22	*26*	10.36	*21*	10.9	*16*	10.18	*13*
1.3	*46*	5.7	*26*	1.32	*21*	2.7	*16*	7.20	*12*
10.6	*42*	P1	*25*	1.46	*20*	1.12	*16*	P2	*11*
1.4	*41*	1.34	*23*	1.5	*20*	10.20	*16*	2.4	*11*
1.31	*39*	1.47	*23*	1.29	*18*	7.17	*15*	CN2	*11*
10.13	*32*	5.16	*23*	6.17	*18*	6.4	*14*	3.16	*11*
1.11	*32*	10.21	*22*	10.23	*18*	1.10	*14*	10.17	*10*
3.1	*29*	1.8	*22*	1.23	*18*	10.12	*13*	6.2	*10*

Notably, all but one of 10.11's direct mentions are in other theorems in Book 10. Theorem 6.1 (which we already encountered above) is used in 4 books.

By the way, there is some subtlety here, because 26 theorems reference a particular theorem more than once in their proofs: for example, 10.4 references 10.3 three times, while 13.18 references both 13.17 and 13.16 twice:

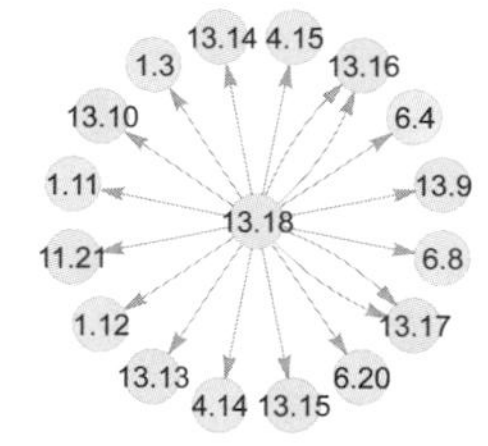

But looking simply at the distribution of the number of direct uses (here on a log scale), we see that the vast majority of theorems are very rarely used—with just a few being quite widely used:

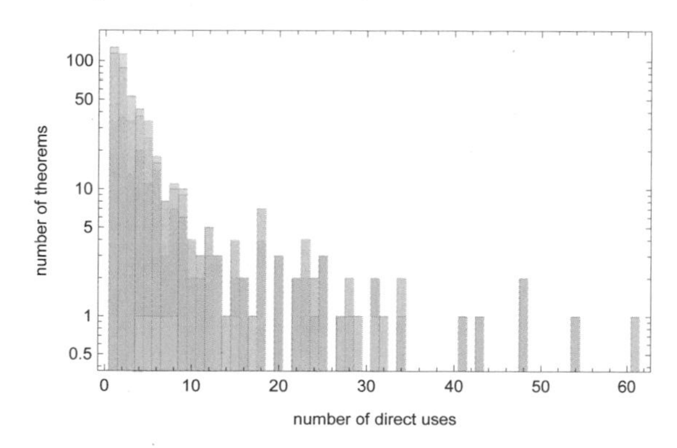

Indicating the number of direct uses by size, here are the "directly popular" theorems:

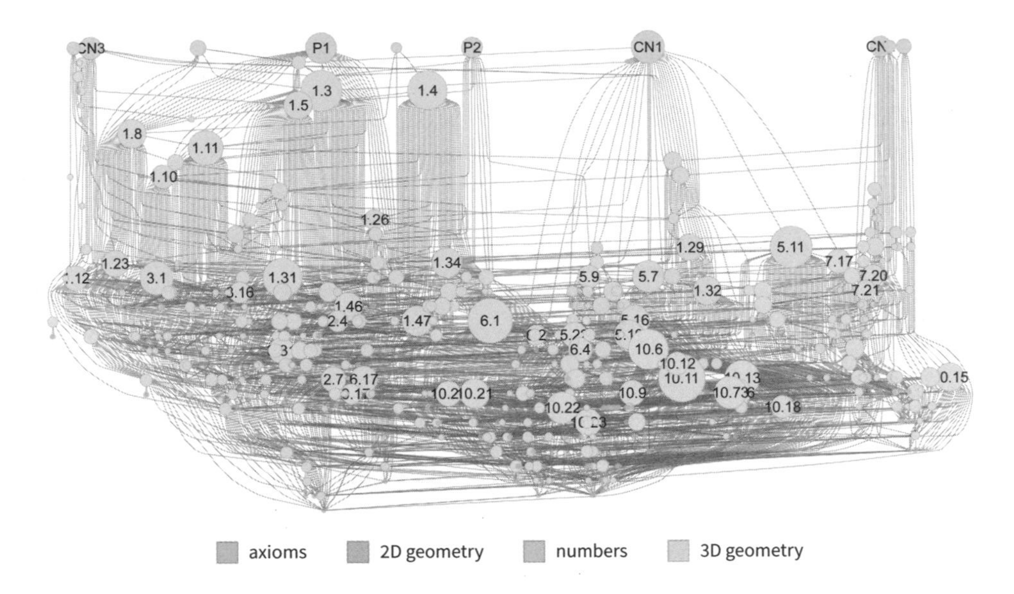

If we ask also about indirect uses, the results are as follows:

CN1	411	1.3	313	1.15	294	1.23	244	1.37	189
CN3	390	1.4	312	1.16	291	5.1	243	1.41	187
CN2	377	1.5	310	P5	262	5.11	241	1.36	187
CN5	344	1.7	309	1.18	261	5.7	238	5.12	179
P3	318	1.8	308	1.19	260	1.31	236	5.15	178
P1	316	1.11	301	1.29	251	5.8	235	7.5	177
P2	315	1.9	298	1.20	250	1.34	222	7.6	173
1.1	315	P4	296	1.27	248	5.9	221	1.38	169
1.2	314	1.13	296	1.22	245	1.35	202	5.13	166
CN4	313	1.10	296	1.26	244	1.33	192	5.10	166

Not too surprisingly, the axioms and early theorems are the most popular. But overall, the distribution of total number of uses is somewhat broader than the distribution of direct uses:

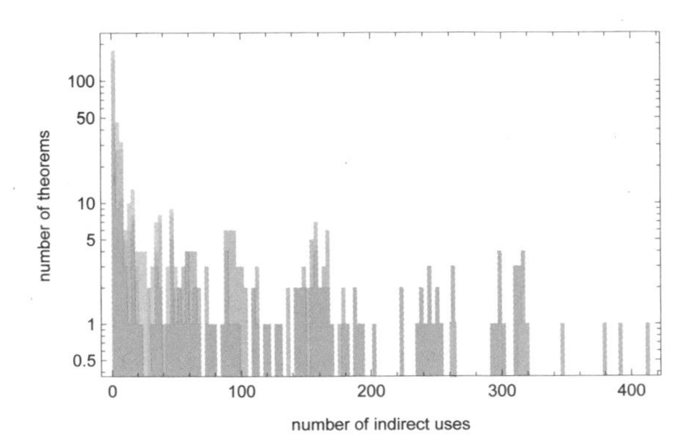

This shows all theorems, with their sizes in the graph essentially determined by the sizes of their "future light cone" in the theorem dependency graph:

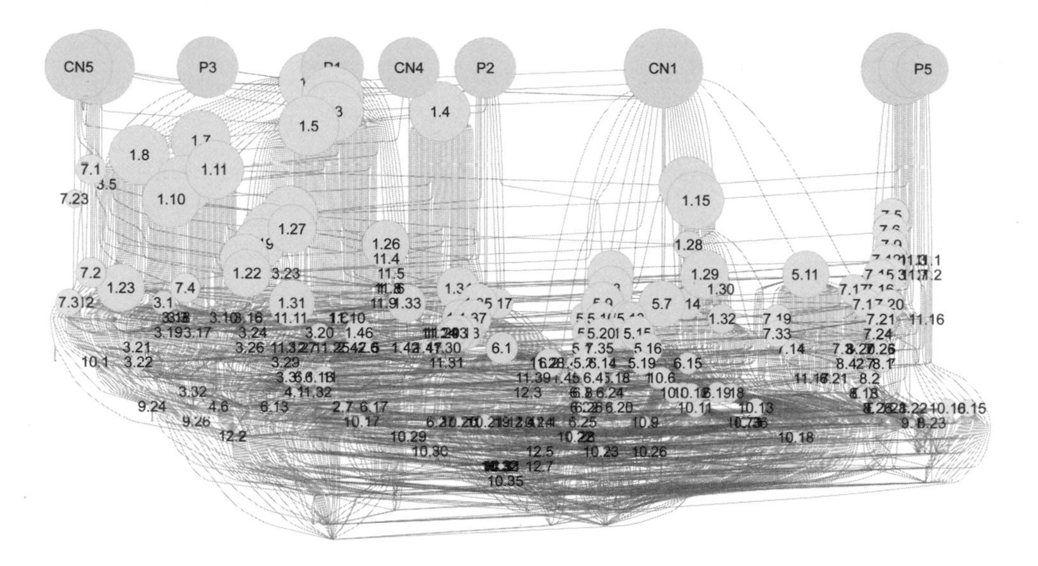

In addition to asking about direct and indirect uses, one can also assess the "centrality" of a given theorem by various graph-theoretical measures. One example is betweenness centrality (the fraction of shortest paths that pass through a given node):

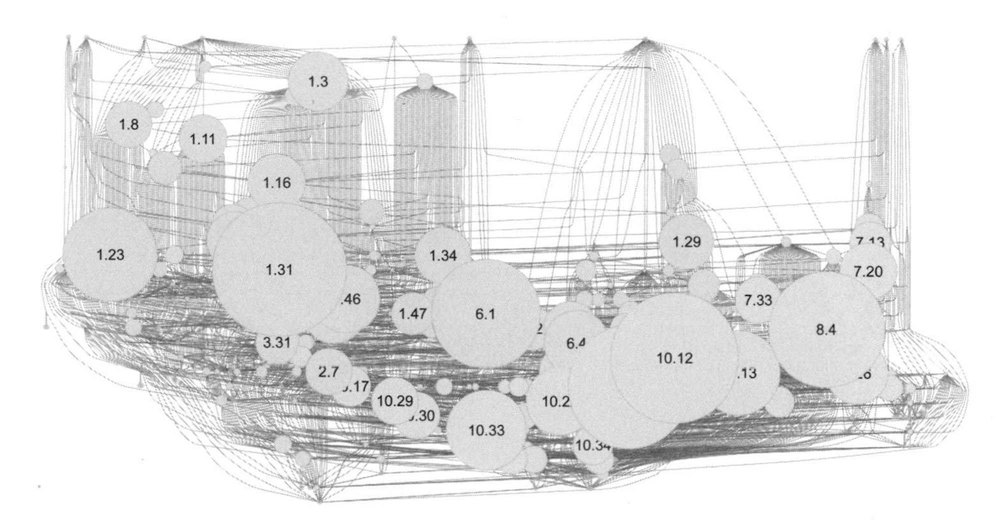

The theorems with top betweenness centralities are 1.31 (construction of parallel lines), 10.12 (transitivity of commensurability), 10.9 (commensurabilty in squares), 8.4 (continued ratios in lowest terms), etc.

For closeness centrality (average inverse distance to all other nodes) one gets:

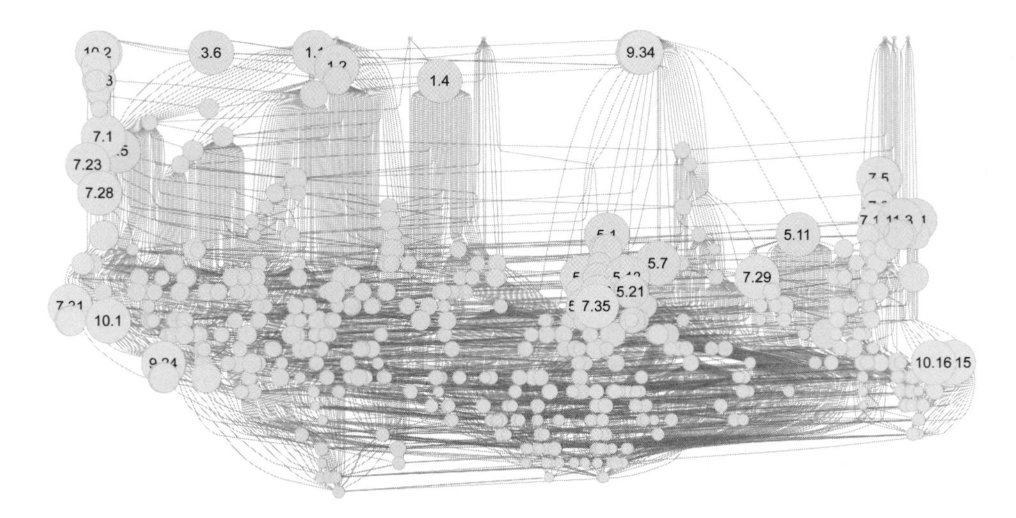

What Really Depends on What?

Euclid's *Elements* starts with 10 axioms, from which all the theorems it contains are derived. But what theorems really depend on what axioms? This shows how many of the 465 theorems depend on each of the Common Notions and Postulates according to the proofs given in Euclid:

CN1	CN2	CN3	CN4	CN5	P1	P2	P3	P4	P5
411	377	390	313	344	316	315	318	296	262

The famous fifth postulate (that parallel lines do not cross) has the fewest theorems depending on it. (And actually, for many centuries there was a suspicion that no theorems really depended on it—so people tried to find proofs that didn't use it, although ultimately it became clear it actually was needed.)

Interestingly, at least according to Euclid, more than half (255 out of 465) of the theorems actually depend on all 10 axioms, though one sees definite variation through the course of the *Elements*:

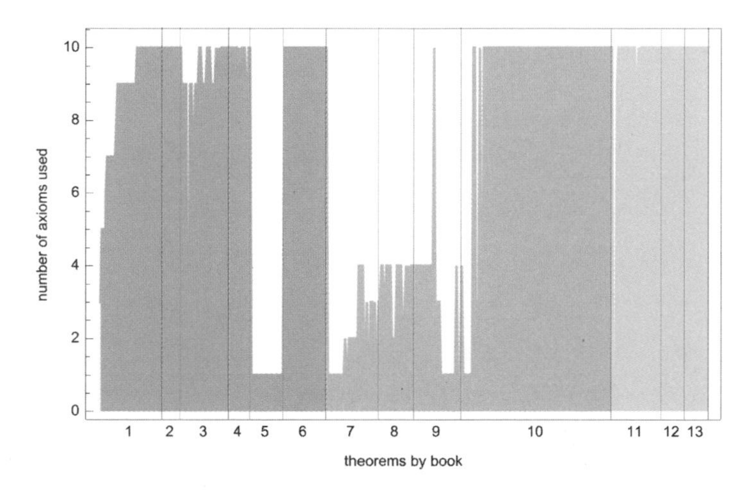

The number of theorems depending on different numbers of axioms is:

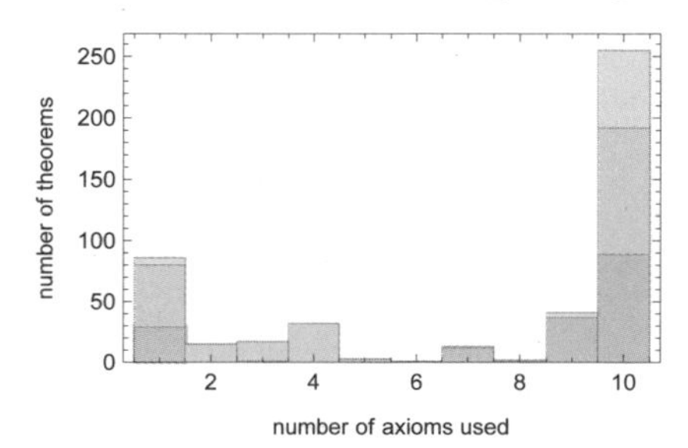

Scattered through the *Elements* there are 86 theorems that depend only on one axiom, most often CN1 (which is transitivity of equality):

CN1	CN3	CN2	P5	P3	CN5	CN4
36	23	17	6	2	1	1

In most cases, the dependence is quite direct, but there are cases in which it is actually quite elaborate, such as:

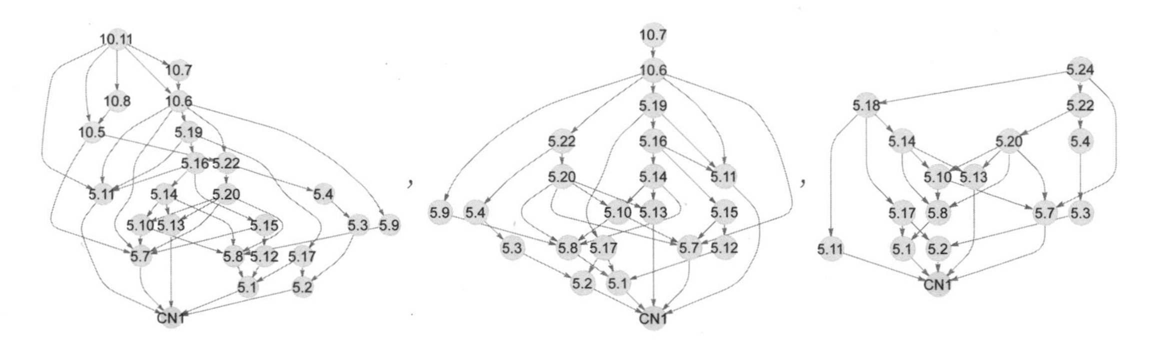

These get slightly simpler after transitive reduction:

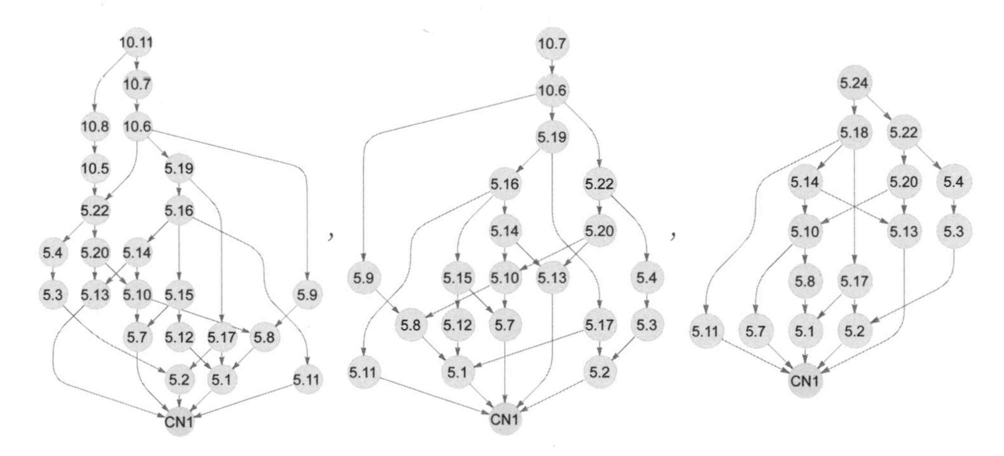

We can now also ask the opposite question of how many theorems don't depend on any given axiom (and, yes, this immediately follows from what we listed above):

CN1	CN2	CN3	CN4	CN5	P1	P2	P3	P4	P5
63	97	84	161	130	158	159	156	178	212

And in general we can ask what subsets of the axioms different theorems depend on. Interestingly, of the 1024 possible such subsets, only 19 actually occur, suggesting some considerable correlation between the axioms. Here is a representation of the partial ordering of the subsets that occur, indicating in each case for how many theorems that subset of dependencies occurs:

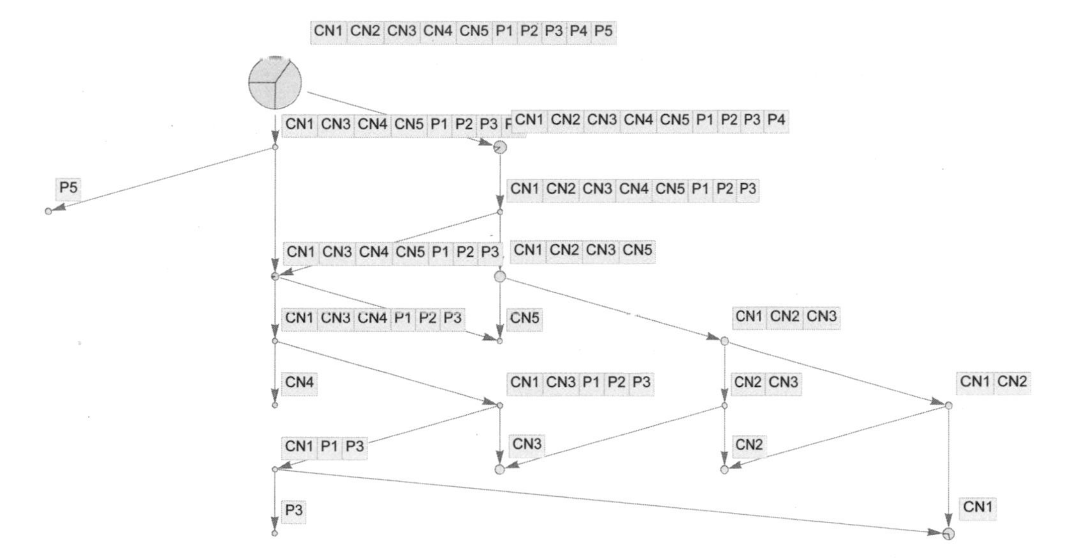

The Machine Code of Euclid: All the Way Down to Axioms

Any theorem in Euclid can ultimately be proved just by using Euclid's axioms enough times. In other words, the proofs Euclid gave were stated in terms of "intermediate theorems"—but we can always in principle just "compile things down" so we just get a sequence of axioms. And here for example is how that works for Book 1, Theorem 5:

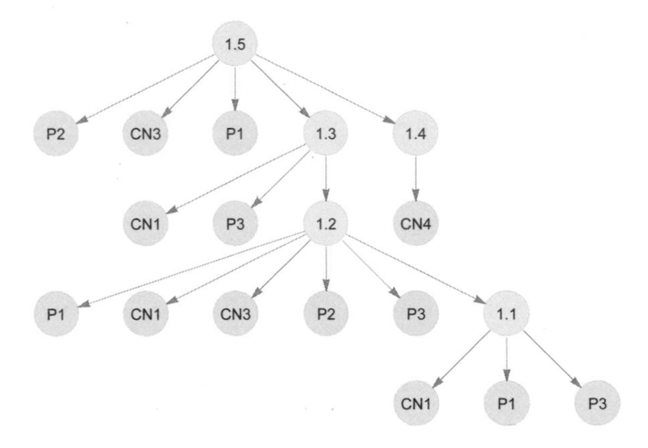

Of course it's much more efficient to "share the work" by using intermediate theorems:

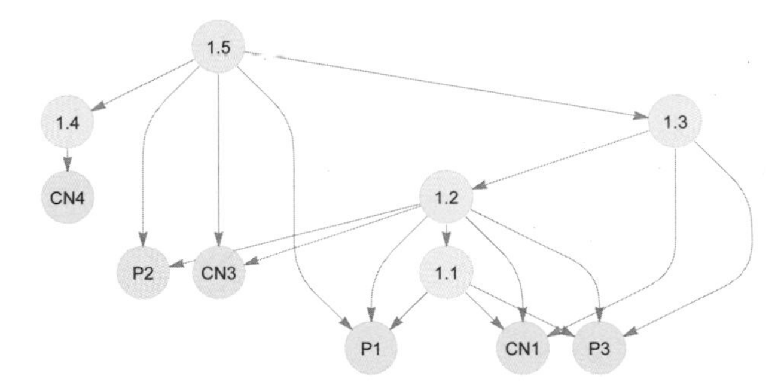

This doesn't change the "depth"—i.e. the length of any given path to get to the axioms. But it reduces the number of independent paths that have to be followed, because every time one reaches the same theorem (or axiom) one just "uses what one already knows about it".

But to get a sense of the "axiomatic machine code" of Euclid we can just "compile" the proof of every theorem down to its underlying sequence of axioms. And for example if we do this for 3.18 the final sequence of axioms we get has length 835,416. These are broken down among the various axioms according to:

CN1	CN2	CN3	CN4	CN5	P1	P2	P3	P4	P5
203 000	21 240	80 232	54 446	24 060	190 249	80 254	170 675	8091	3169

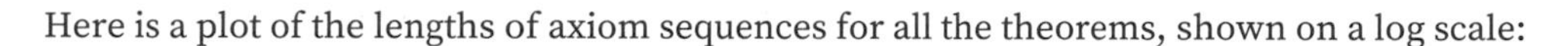

Here is a plot of the lengths of axiom sequences for all the theorems, shown on a log scale:

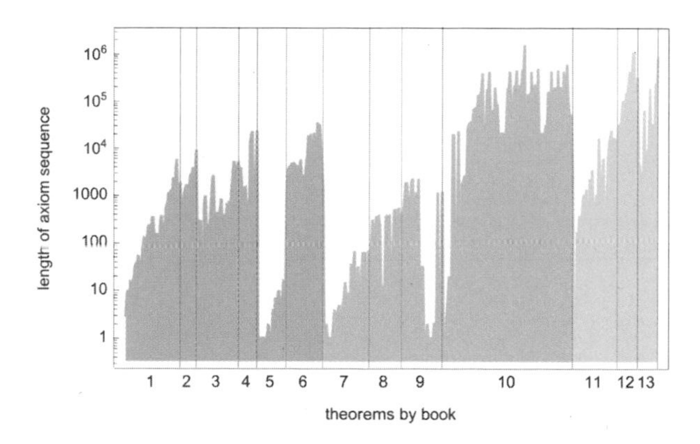

Interestingly, 3.18 isn't the theorem with the longest axiom sequence; it's in 4th place, and the top 10 are (in gray are the results with intermediate theorems allowed):

10.72	12.14	12.15	13.18	10.110	10.84	10.102	10.65	10.96	10.59
1 584 084	1 048 323	1 048 316	835 416	616 779	493 513	461 107	458 109	438 610	432 526
184	*122*	*122*	*229*	*172*	*171*	*167*	*167*	*167*	*167*

(10.72 is about addition of incommensurable medial areas, and is never referenced anywhere; 12.14 says the volumes of cones and cylinders with equal bases are proportional to their heights; 12.15 says the heights and bases of cones and cylinders with equal volumes are inversely proportional; etc.)

Here's the distribution of the lengths of axiom sequences across all theorems:

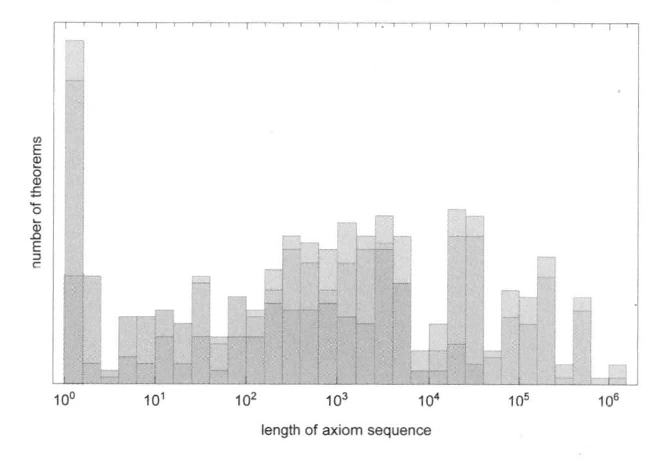

We can get some sense of the dramatic value of "remembering intermediate theorems" by comparing the total number of "intermediate steps" obtained with and without merging different instances of the same theorem:

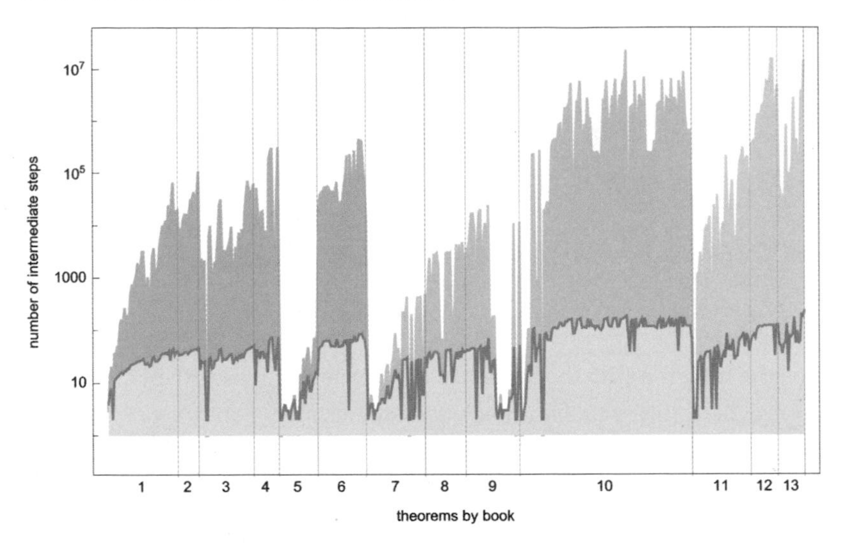

For example, for 8.13, 229 steps are needed when intermediate theorems are remembered, while 14,412,576 steps are needed otherwise. (For 10.72, it's 184 vs. 23,921,481 steps.)

Superaxioms, or What Are the Most Powerful Theorems?

Euclid's 10 axioms are ultimately all we need in order to prove all the 465 theorems in the *Elements*. But what if we supplement these axioms with some of the theorems? Are there small sets of theorems we can add that will make the proofs of many theorems much shorter? To get a full understanding of this, we'd have to redo all the proofs. But we can get some sense of it just from the theorem dependency graph.

Consider the graph representing the proof of 1.12, with 1.7 highlighted:

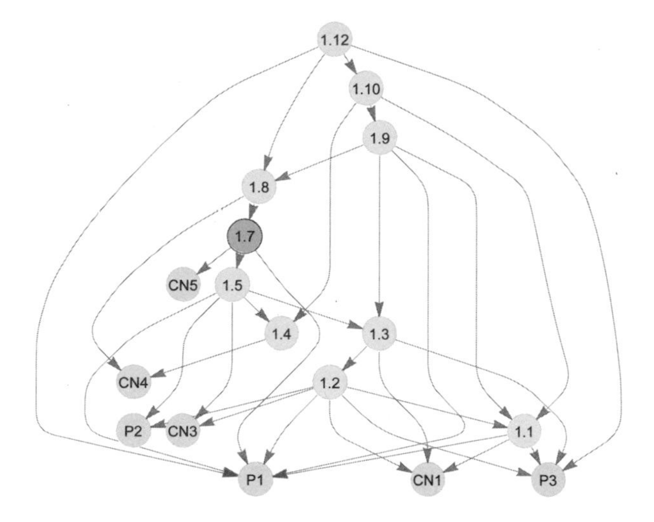

Now imagine adding 1.7 as a "superaxiom". Doing this, we can get a smaller proof graph for 1.12—with 4 nodes (and 14 connections) fewer:

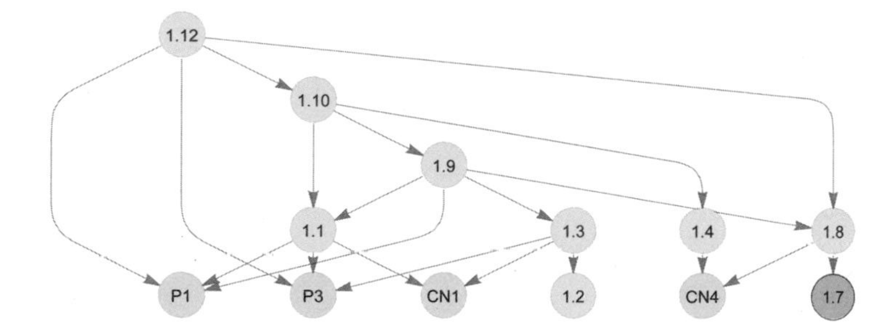

What does adding 1.7 as a superaxiom do for the proofs of other theorems? Here's how much it shortens each of them:

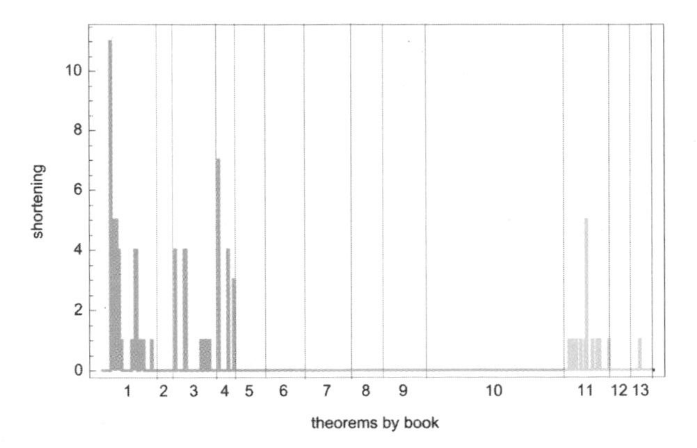

(The largest shortening is for 1.8, followed by 4.1.)

So what are the "best" superaxioms to add? Here's a plot of the average amount of shortening achieved by adding each possible individual theorem as a superaxiom:

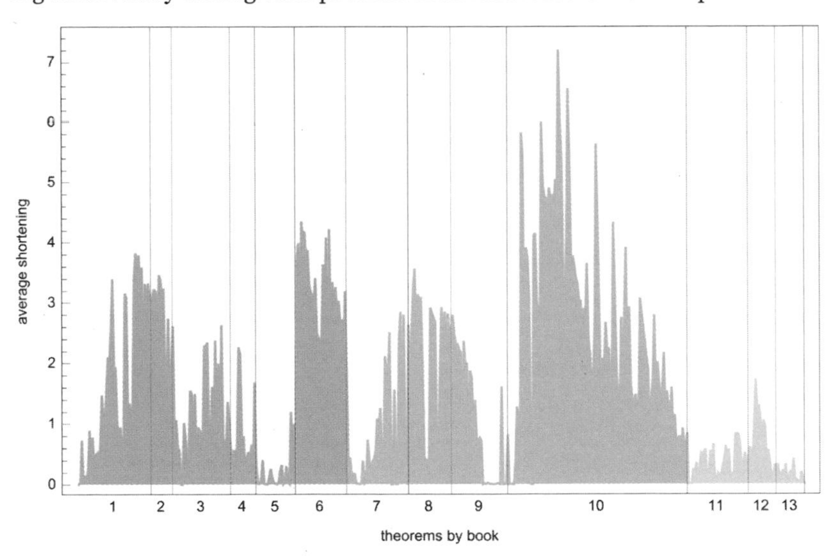

The rather unimpressive best result—an average shortening of 7.2—is achieved with 10.33 (which says that it's possible to come up with numbers x and y such that $\frac{x}{y}$ and $\sqrt{x\,y}$ are irrational, while $x\,y$ and $x+y$ are rational).

The maximum shortenings are more impressive—with 10.41 and 10.78 achieving the maximum shortening of 165

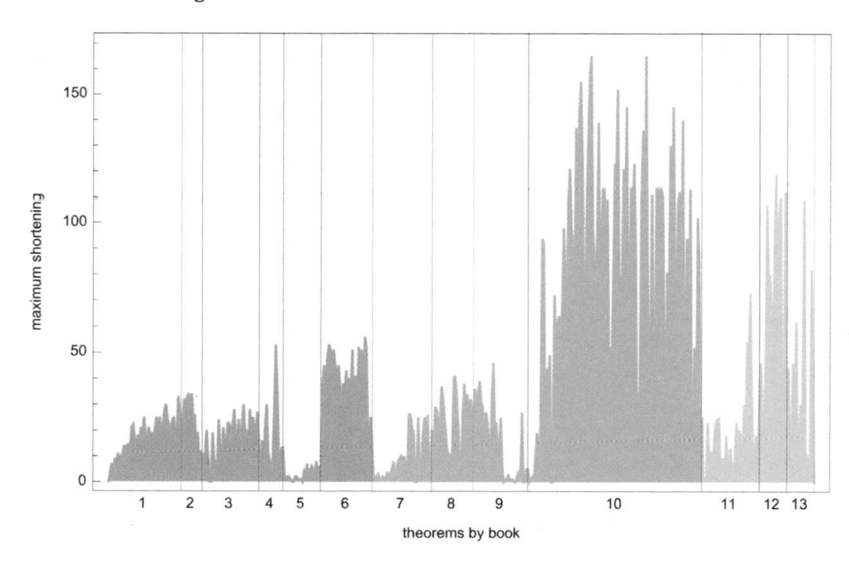

although this shortening is very concentrated around "nearby theorems":

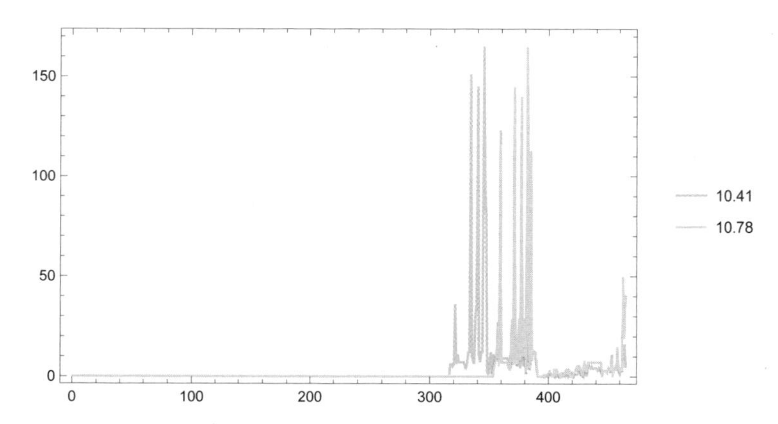

By the way, adding a superaxiom can not only decrease the number of intermediate theorems used in a proof, it can also decrease the "depth" of the proof, i.e. the longest path needed to reach an axiom (or superaxiom). Here is the average depth reduction achieved by adding each possible theorem as a superaxiom:

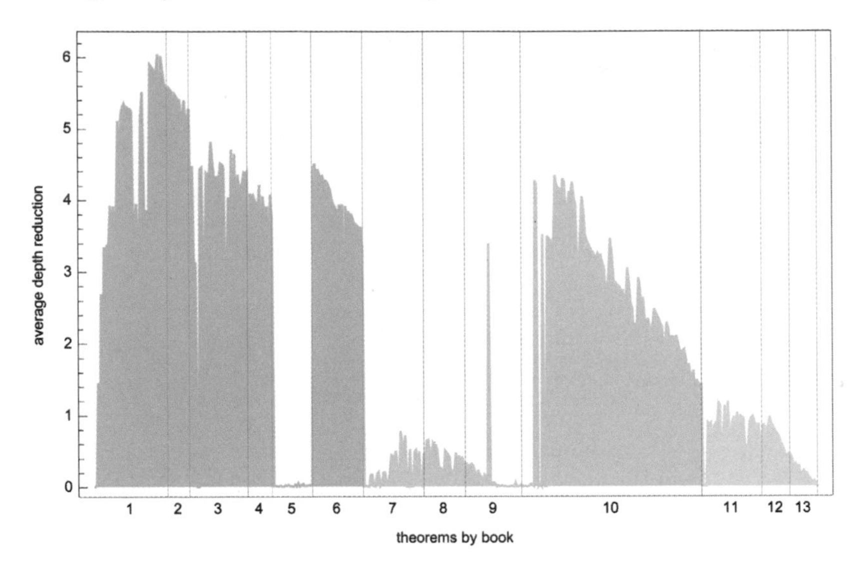

(The peak in Book 9 is 9.15, which reduces the depth of many subsequent theorems by 10 steps, though—in a possible goof—is not actually used by Euclid in the proofs of any of them.)

Here is the maximum depth reduction achieved by adding each possible theorem:

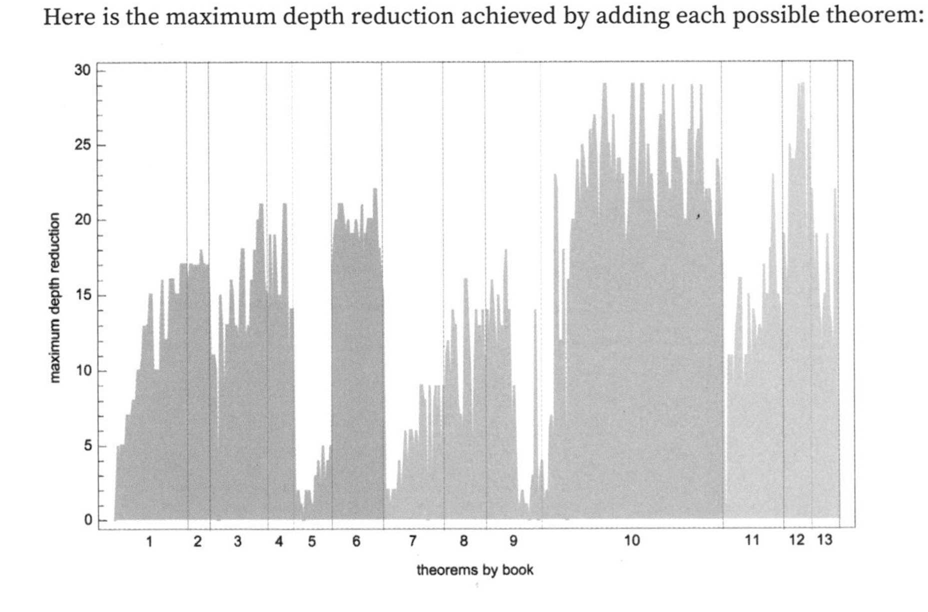

Formalizing Euclid

Everything we've discussed so far is basically derived from the original text of Euclid's *Elements.* But what if we look instead at the pure "mathematical content" of Euclid? We've now got a way to represent this in the Wolfram Language. Consider Euclid's 3.16. It asserts that:

The perpendicular ($\overline{BI}$) to the diameter ($\overline{AB}$) of a circle at one of its endpoints (B) is tangent to the circle at that point.

Well, we can now give a "computational translation" of this:

```
In[ ]:= Euclid book 3 proposition 16 GEOMETRIC SCENE ["Scene"]

Out[ ]= GeometricScene[{{A, B, C, I}, {}}, {CircleThrough[{A, B}, C],
      Line[{A, C, B}], GeometricAssertion[{InfiniteLine[{B, I}], Line[{A, B}]}, Perpendicular]},
   {GeometricAssertion[{CircleThrough[{A, B}, C], InfiniteLine[{B, I}]}, Tangent]}]
```

And this is all we need to say to define that theorem in Euclid. Given the definition of the Wolfram Language, this is completely self-contained, and ready to be understood by both computers and humans. And from this form, we can now for example compute a random instance of the theorem:

```
In[ ]:= RandomInstance[%]
```

Out[ ]=

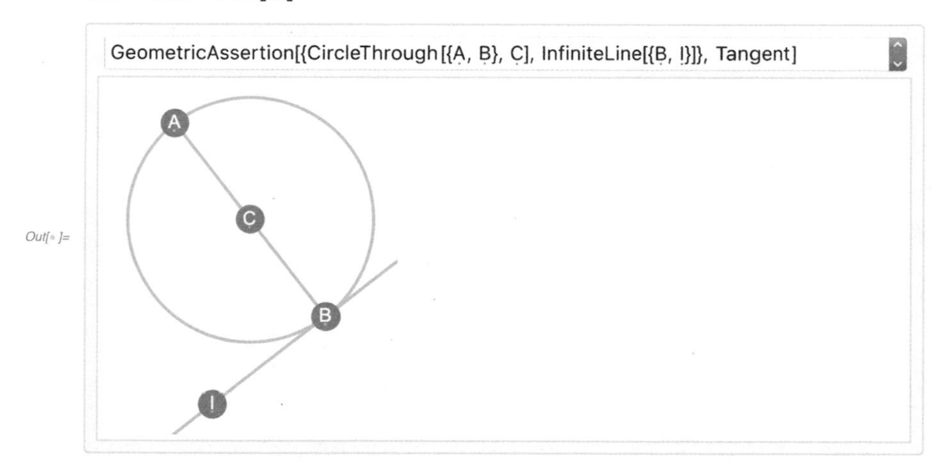

As another example, here's Euclid's 4.2:

Out[ ]= To inscribe inside a given circle a triangle similar to a given triangle.

This is now asking for a construction—or, effectively, stating the theorem that it's possible to do such a construction with ruler and compass. And again we can give a computable version of this in the Wolfram Language, including the construction:

```
In[•]:= Euclid book 4 proposition 2 GEOMETRIC SCENE ["Scene"]

Out[•]= GeometricScene[{{A, O, B, C, D, E, F, G, H}, {}}, {{CircleThrough[{A}, O], Triangle[{D, E, F}]},
    {GeometricAssertion[{Line[{G, A, H}], CircleThrough[{A}, O]}, Tangent]},
    {C ∈ CircleThrough[{A}, O], Line[{A, C}], PlanarAngle[{C, A, H}] == PlanarAngle[{D, E, F}]},
    {B ∈ CircleThrough[{A}, O], Line[{A, B}], PlanarAngle[{B, A, G}] == PlanarAngle[{D, F, E}]}, {Line[{B, C}]}}
  {GeometricAssertion[{Triangle[{A, B, C}], Triangle[{D, E, F}]}, Similar]}]

In[•]:= RandomInstance[%]
```

Out[•]=

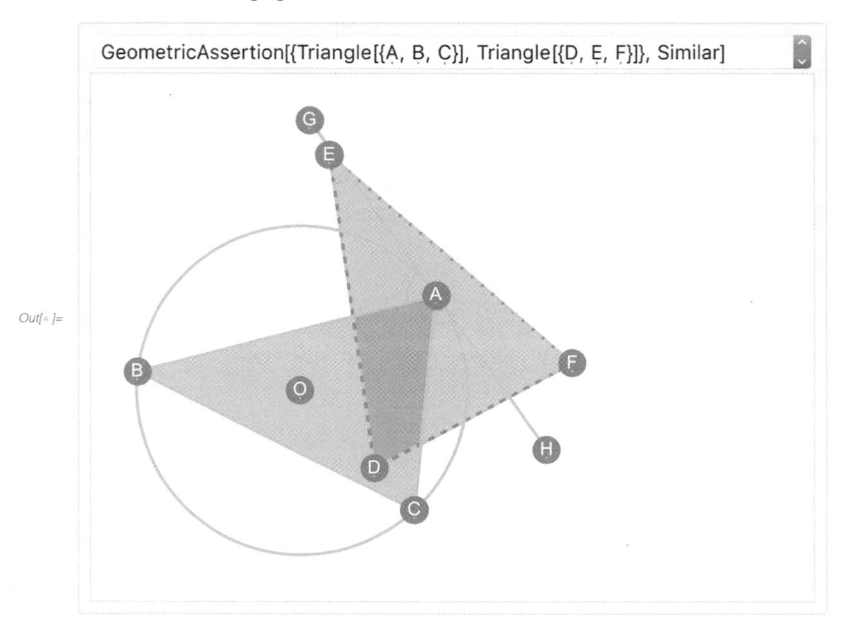

It's interesting to see, though, how the computable versions of theorems compare to their textual ones. Here are length comparisons for 2D geometry theorems:

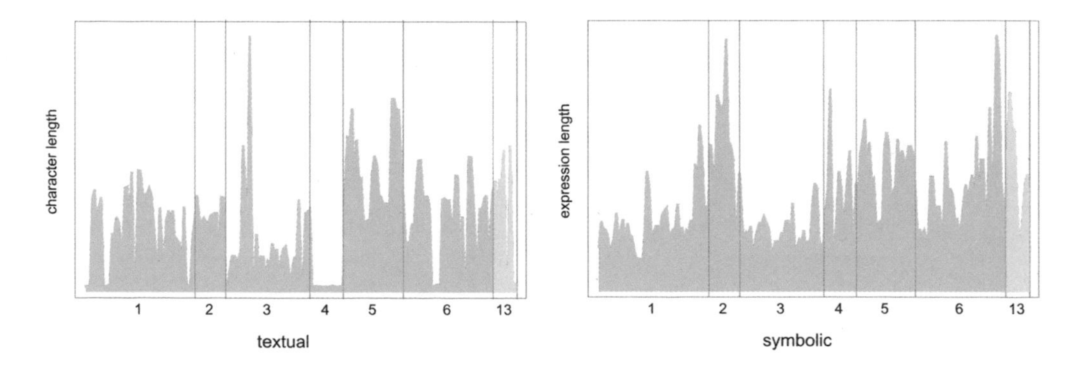

And we see that there is indeed at least some correlation between the lengths of textual and symbolic representations of theorems (the accumulation of points on the left is associated with constructions, where the text just says what's wanted, and the symbolic form also says how to do it):

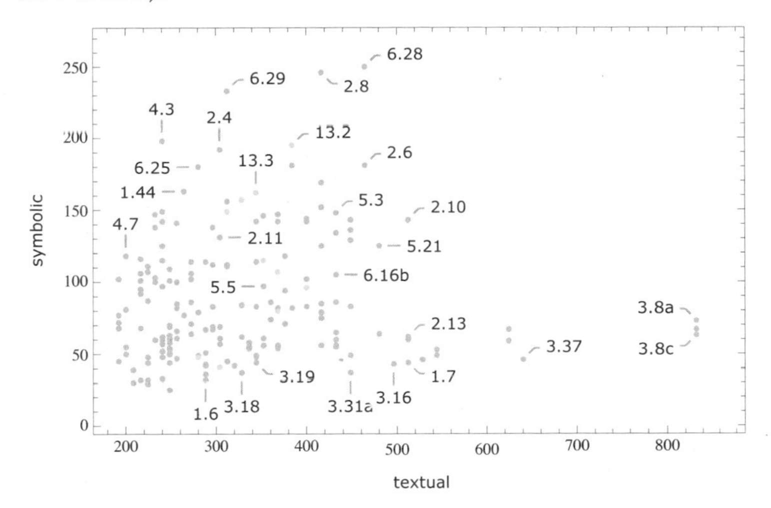

In the Wolfram Language representation we've just been discussing, there's a built-in Wolfram Language meaning to things like CircleThrough and PlanarAngle—and we can in a sense do general computations with these.

But at some level we can view what Euclid did as something purely formal. Yes, he talks about lines and planes. But we can think of these things just as formal constructs, without any externally known properties. Many centuries after Euclid, this became a much more familiar way to think about mathematics. And in the Wolfram Language we capture it with AxiomaticTheory and related functions.

For example, we can ask for an axiom system for Boolean algebra, or group theory:

In[∘]:= **AxiomaticTheory["BooleanAxioms"]**

Out[∘]= $\{\forall_{\{a,b\}}\, a\otimes b == b\otimes a,\ \forall_{\{a,b\}}\, a\oplus b == b\oplus a,\ \forall_{\{a,b\}}\, a\otimes(b\oplus\overline{b}) == a,\ \forall_{\{a,b\}}\, a\oplus b\otimes\overline{b} == a,\ \forall_{\{a,b,c\}}\, a\otimes(b\oplus c) == a\otimes b\oplus a\otimes c,\ \forall_{\{a,b,c\}}\, a\oplus b\otimes c == (a\oplus b)\otimes(a\oplus c)\}$

In[∘]:= **AxiomaticTheory["GroupAxioms"]**

Out[∘]= $\{\forall_{\{a,b,c\}}\, a\otimes(b\otimes c) == (a\otimes b)\otimes c,\ \forall_{a}\, a\otimes\tilde{1} == a,\ \forall_{a}\, a\otimes\overline{a} == \tilde{1}\}$

What does the ⊗ mean? We're not saying. We're just formally defining certain properties it's supposed to have. In the case of Boolean algebra, we can interpret it as AND. In the case of group theory, it's group multiplication—though we're not saying what particular group it's for. And, yes, we could as well write the group theory axioms for example as:

$\{\forall_{\{a,b,c\}}\, f[a, f[b, c]] == f[f[a, b], c],\ \forall_{a}\, f[a, e] == a,\ \forall_{a}\, f[a, c[a]] == e\}$

OK, so can we do something similar for Euclid's geometry? It's more complicated, but thanks particularly to work by David Hilbert and Alfred Tarski in the first half of the 1900s, we can—and here's a version of the result:

$\forall_{\{x,y,z\}}$ implies[congruent[line[x, y], line[z, z]], congruent[x, y]]
$\forall_{\{x,y,z,u,v,w\}}$ implies[and[congruent[line[x, y], line[z, u]], congruent[line[x, y], line[v, w]]], congruent[line[z, u], line[v, w]]]
$\forall_{\{x,y,z\}}$ implies[between[x, y, z], equal[x, y]]
$\forall_{\{x,y,z,u,v\}}$ implies[and[between[x, u, z], between[y, v, z]], $\exists_a$ and[between[u, a, y], between[v, a, x]]]
$\forall_{\{x,y,z,u,v\}}$ implies[and[and[and[congruent[line[x, u], line[x, v]], congruent[line[y, u], line[y, v]]], congruent[line[z, u], line[z, v]]], not[equal[u, v]]], or[or[between[x, y, z], between[y, z, x]], between[z, x, y]]]
$\forall_{\{x,y,z,u,v,w\}}$ implies[and[and[and[between[x, y, w], congruent[line[x, y], line[y, w]]], and[between[x, u, v], congruent[line[x, u], line[u, v]]]], and[between[y, u, z], congruent[line[y, u], line[z, u]]]], congruent[line[y, z], line[v, w]]]
$\forall_{\{x,y,z,a,b,c,u,v\}}$ implies[and[and[and[and[and[and[not[equal[x, y]], between[x, y, z]], between[a, b, c]], congruent[line[x, y], line[a, b]]], congruent[line[y, z], line[b, c]]], congruent[line[x, u], line[a, v]]], congruent[line[y, u], line[b, v]]], congruent[line[z, u], line[c, v]]]
$\forall_{\{x,y\}}$ implies[equal[x, y], equal[y, x]]
$\forall_{\{x,y,z\}}$ implies[and[equal[x, y], equal[y, z]], equal[x, z]]
$\forall_x$ equal[x, x]
$\forall_{\{a,b\}}$ and[a, b] == and[b, a]
$\forall_{\{a,b\}}$ or[a, b] == or[b, a]
$\forall_{\{a,b\}}$ and[a, or[b, not[b]]] == a
$\forall_{\{a,b\}}$ or[a, and[b, not[b]]] == a
$\forall_{\{a,b,c\}}$ and[a, or[b, c]] == or[and[a, b], and[a, c]]
$\forall_{\{a,b,c\}}$ or[a, and[b, c]] == and[or[a, b], or[a, c]]
$\forall_{\{a,b\}}$ implies[a, b] == or[not[a], b]
$\forall_{\{\alpha,\beta,y,z\}}$ implies[$\exists_x$ implies[and[α[y], β[z]], between[x, y, z]], $\exists_u$ implies[and[α[y], β[z]], between[y, u, z]]]

Once again, this is all just a collection of formal statements. The fact that we're calling an operator between is just for our convenience and understanding. All we can really say for sure is that this is some ternary operator; any properties it has have to be defined by the axioms.

To get to this formalization of Euclid, quite a bit of tightening up had to be done. Euclid's theorems often had implicit assumptions, and it sometimes wasn't even clear exactly what their logical structure was supposed to be. But the mathematical content is presumably the same, and indeed some of Euclid's axioms (like CN1) say basically exactly the same as these. (An important addition to what Euclid explicitly said is the last axiom above, which states Euclid's implicit assumption—that I now believe to be incorrect for the physical universe—that space is continuous. Unlike other axioms, which just make statements "true for all values of...", this axiom makes a statement "true for all functions...".)

So what can we do with these axioms? Well, in principle we can prove any theorem in Euclidean geometry. Appending to the axioms (that we refer to—ignoring the last axiom—as geometry) an assertion that we can interpret as saying that if a point y is between x and z and between x and w, then either z is between y and w or w is between y and z:

```
In[ ]:= FindEquationalProof[or[between[y, z, w], between[y, w, z]],
          Append[geometry, and[between[x, y, z], between[x, y, w]]]]
```

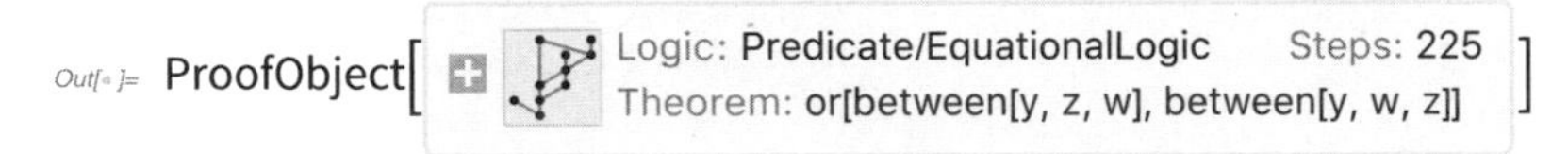

Here's a graph representing this proof:

```
In[ ]:= %["ProofGraph"]
```

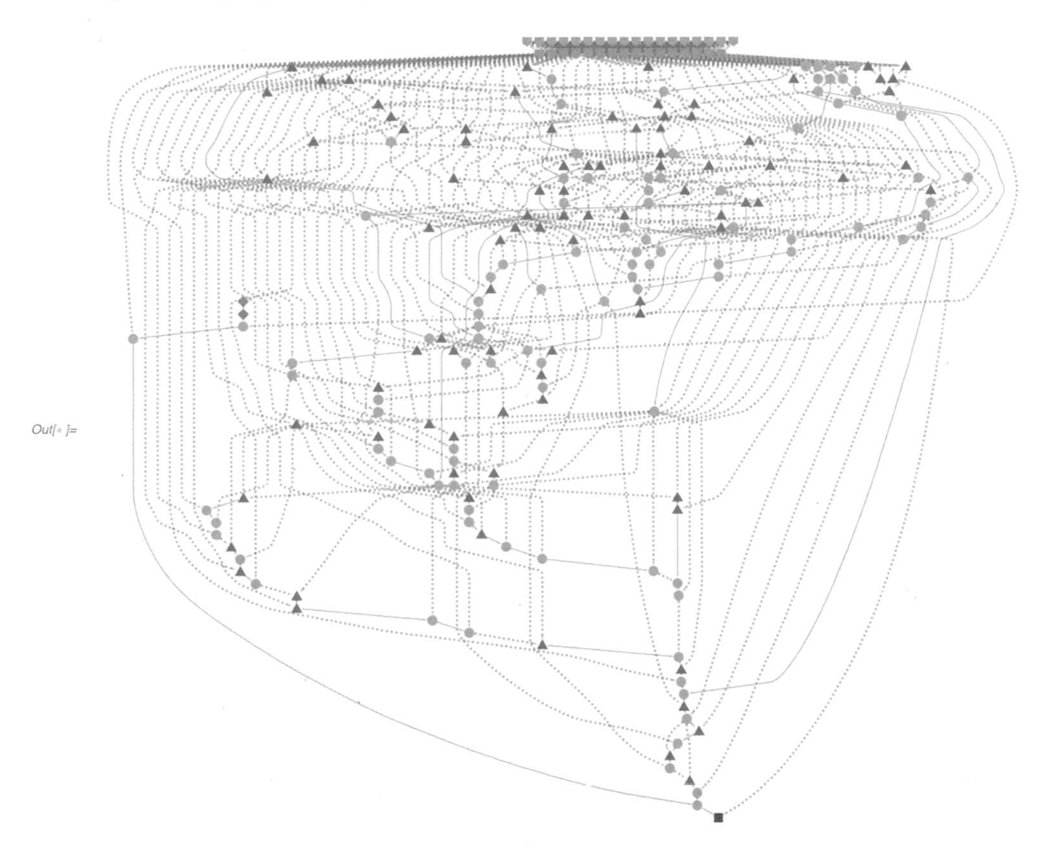

The axioms (including the "setup assertion") are at the top—and the proof, with all its various intermediate lemmas, establishes that our "hypothesis" (represented by a little purple diamond on the left) eventually leads to "true" at the bottom.

As a more complicated example, we can look at Euclid's very first theorem, 1.1, which asserts that there's a ruler-and-compass way to construct an equilateral triangle on any line segment. In the Wolfram Language, the construction is:

In[ ]:= Euclid book 1 proposition 1 GEOMETRIC SCENE ["Scene"]

Out[ ]= GeometricScene[{{A, B, C}, {}},
 {{Line[{A, B}]}, {GeometricAssertion[{CircleThrough[{B}, A], CircleThrough[{A}, B]}, {Concurrent, C}]},
 {Line[{C, A}], Line[{C, B}]}}, {GeometricAssertion[Triangle[{A, B, C}], Equilateral]}]

In[ ]:= **RandomInstance[%]**

Out[ ]=

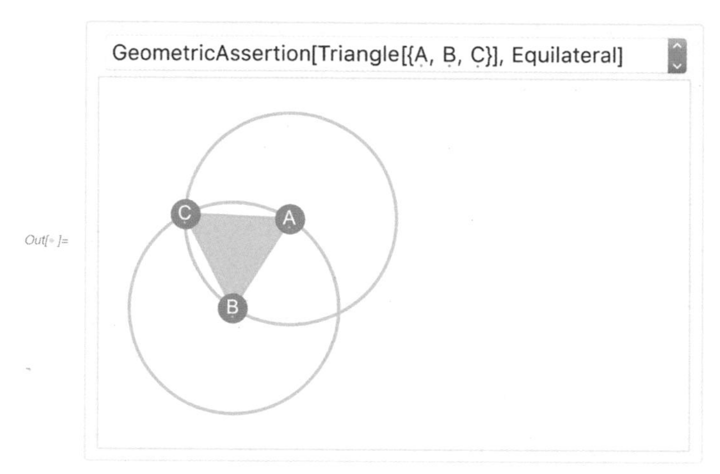

And now we can write this directly in terms of our low-level constructs. First we need a definition of what circles are (Euclid has this as Definition 1.15)—basically saying that two circles centered at *a* that go through *b* and *c* are equal if the lines from *a* to *b* and *a* to *c* are congruent:

$\forall_{\{a,b,c\}}$ implies[equal[circle[a, b], circle[a, c]], congruent[line[a, b], line[a, c]]]

We'll call this definition circles. We're going to do a construction that involves having circles that overlap, as specified by the assertions:

{equal[circle[a, b], circle[a, c]], equal[circle[b, a], circle[b, c]]}

And then our goal is to show that we get an equilateral triangle, for which the following is true:

and[congruent[line[a, b], line[a, c]], congruent[line[b, a], line[b, c]]]

Putting this all together we can prove Euclid's 1.1:

In[ ]:= **FindEquationalProof[**
and[congruent[line[a, b], line[a, c]], congruent[line[b, a], line[b, c]]], Join[geometry,
{$\forall_{\{a,b,c\}}$ implies[equal[circle[a, b], circle[a, c]], congruent[line[a, b], line[a, c]]]},
{equal[circle[a, b], circle[a, c]], equal[circle[b, a], circle[b, c]]}]]

Out[ ]= ProofObject[Logic: Predicate/EquationalLogic Steps: 272
Theorem: and[congruent[line[a, b], line[a, c]], congruent[line[b, a], line[b, c]]]]

And, yes, it took 272 steps—and here's a graphical representation of the proof that got generated, with all its intermediate lemmas:

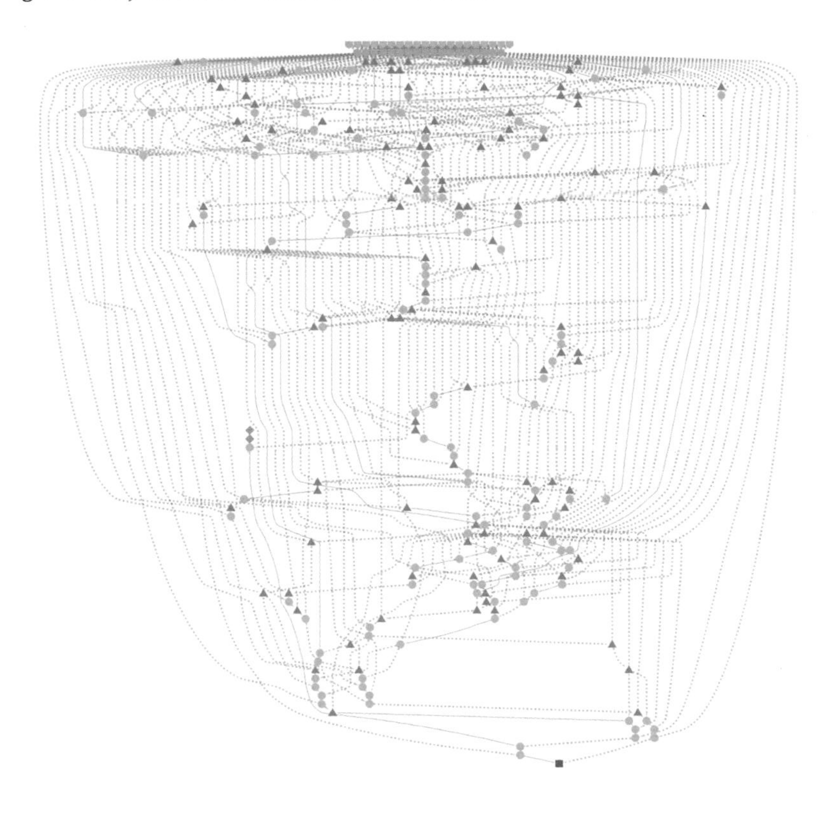

We can go on and prove Euclid's 1.2 as well, all the way from the lowest-level axioms. This time it takes us 330 steps, with proof graph:

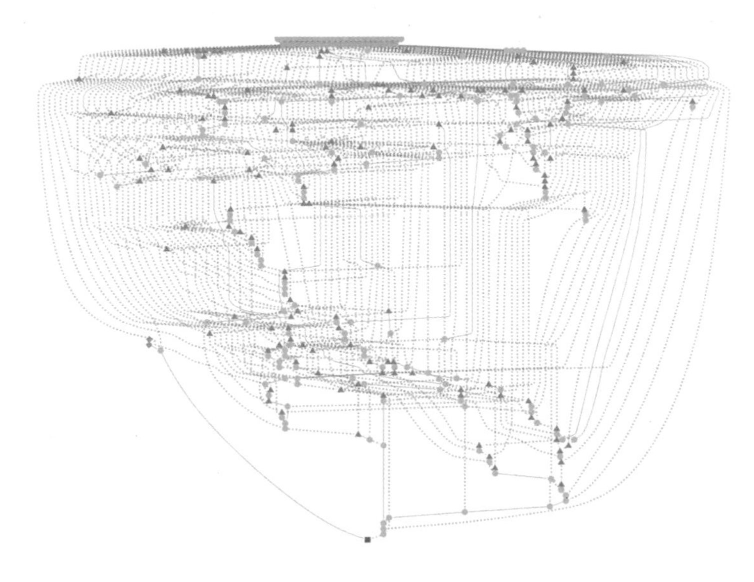

These graphs are conceptually similar to, but concretely rather different from, our "empirical metamathematics" graphs above. There are differences at the level of how interdependence of theorems is defined. But, more important, this graph is generated by automated theorem proving methods; the intermediate theorems (or lemmas) it involves are produced "on the fly" for the convenience of the computer, not because they help in any way to explain the proof to a human. In our empirical metamathematics on Euclid's *Elements*, however, we're dealing with the theorems that Euclid chose to define, and that have served as a basis for explaining his proofs to humans for more than two thousand years.

By the way, if our goal is simply to find out what's true in geometry—rather than to write out step-by-step proofs—then we now know how to do that. Essentially it involves turning geometric assertions into algebraic ones—and then systematically solving the polynomial equations and inequalities that result. It can be computationally expensive, but in the Wolfram Language we now have one master function, CylindricalDecomposition, that ultimately does the job. And, yes, given Gödel's theorem, one might wonder whether this kind of finite procedure for solving any Euclid-style geometry problem was even possible. But it turns out that—unlike arithmetic, for which Gödel's theorem was originally proved—Euclid-style geometry, like basic logic, is decidable, in the sense that there is ultimately a finite procedure for deciding whether any given statement is true or not. In principle, this procedure could be based on theorem proving from the axioms, but CylindricalDecomposition effectively leverages a tower of more sophisticated mathematics to provide a much more efficient approach.

All Possible Theorems

From the axioms of geometry one can in principle derive an infinite number of true theorems—of which Euclid picked just 465 to include in his *Elements*. But why these theorems, and not others? Given a precise symbolic representation of geometry—as in the axioms above—one can just start enumerating true theorems.

One way to do this is to use a multiway system, with the axioms defining transformation rules that one can apply in all possible ways. In effect this is like constructing every possible proof, and seeing what gets proved. Needless to say, the network that gets produced quickly becomes extremely large—even if its structure is interesting for our attempt to find a "bulk theory of metamathematics".

Here's an example of doing it, not for the full geometry axioms above, but for basic logic (which is actually part of the axiom system we've used for geometry). We can either start with expressions, or with statements. Here we start with the expression $x \wedge y$, and then progressively find all expressions equal to it. Here's the first, rather pedantic step:

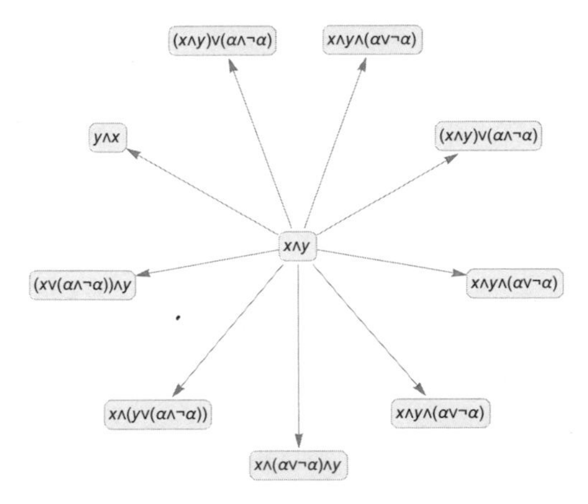

And here's the second step:

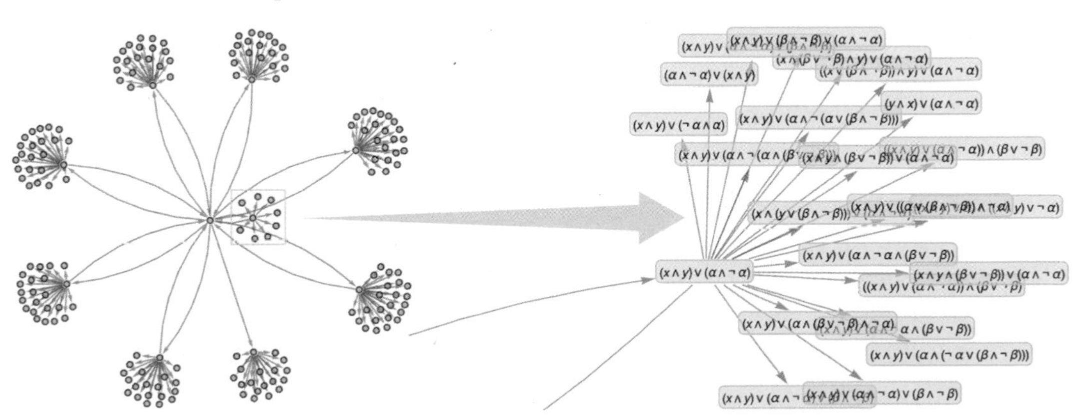

Every path in this graph is a proof that its endpoint expressions are equal. And while eventually this approach will give us every possible theorem (in this case about equalities involving $x \wedge y$), it'll obviously take a while, generating huge numbers of long and uninteresting results on its way to anything interesting.

As a different approach, we can consider just enumerating short possible statements, then picking out ones that we determine are true. In principle we could determine truth by explicitly proving theorems using the axioms (and, yes, if there was undecidability we wouldn't always be able to do this). But in practice for the case of basic logic that we're using as an example here, we can basically just explicitly construct truth tables to find out what's true and what's not.

Here are some statements in logic, sorted in increasing order of complexity (as measured by depth and number of symbols):

$$
\begin{aligned}
&\{a=b, a=(\neg a), a=(\neg b), (\neg a)=(\neg b), a=(a \wedge a), (\neg a)=(a \wedge a), a=(a \vee a), (\neg a)=(a \vee a), (a \wedge a)=(a \vee a),\\
&a=(a \wedge b), (\neg a)=(a \wedge b), (a \wedge a)=(a \wedge b), (a \vee a)=(a \wedge b), a=(a \vee b), (\neg a)=(a \vee b), (a \wedge a)=(a \vee b),\\
&(a \vee a)=(a \vee b), (a \wedge b)=(a \vee b), (a \wedge b)=(a \wedge c), (a \vee b)=(a \wedge c), (a \wedge b)=(a \vee c), (a \vee b)=(a \vee c),\\
&a=(b \wedge a), (\neg a)=(b \wedge a), (a \wedge a)=(b \wedge a), (a \vee a)=(b \wedge a), (a \wedge b)=(b \wedge a), (a \vee b)=(b \wedge a),\\
&a=(b \vee a), (\neg a)=(b \vee a), (a \wedge a)=(b \vee a), (a \vee a)=(b \vee a), (a \wedge b)=(b \vee a), (a \vee b)=(b \vee a),\\
&a=(b \wedge b), (\neg a)=(b \wedge b), (a \wedge a)=(b \wedge b), (a \vee a)=(b \wedge b), (a \wedge b)=(b \wedge b), (a \vee b)=(b \wedge b),\\
&a=(b \vee b), (\neg a)=(b \vee b), (a \wedge a)=(b \vee b), (a \vee a)=(b \vee b), (a \wedge b)=(b \vee b), (a \vee b)=(b \vee b),\\
&a=(b \wedge c), (\neg a)=(b \wedge c), (a \wedge a)=(b \wedge c), (a \vee a)=(b \wedge c), (a \wedge b)=(b \wedge c), (a \vee b)=(b \wedge c), a=(b \vee c),\\
&(\neg a)=(b \vee c), (a \wedge a)=(b \vee c), (a \vee a)=(b \vee c), (a \wedge b)=(b \vee c), (a \vee b)=(b \vee c), (a \wedge b)=(c \wedge a),\\
&(a \vee b)=(c \wedge a), (a \wedge b)=(c \vee a), (a \vee b)=(c \vee a), (a \wedge b)=(c \wedge b), (a \vee b)=(c \wedge b), (a \wedge b)=(c \vee b),\\
&(a \vee b)-(c \vee b), (a \wedge b)=(c \wedge c), (a \vee b)=(c \wedge c), (a \wedge b)=(c \vee c), (a \vee b)=(c \vee c), a=(\neg(\neg a)),\\
&(\neg a)=(\neg(\neg a)), (a \wedge a)=(\neg(\neg a)), (a \vee a)=(\neg(\neg a)), (a \wedge b)=(\neg(\neg a)), (a \vee b)=(\neg(\neg a)), a=(\neg(\neg b)),\\
&(\neg a)=(\neg(\neg b)), (a \wedge a)=(\neg(\neg b)), (a \vee a)=(\neg(\neg b)), (a \wedge b)=(\neg(\neg b)), (a \vee b)=(\neg(\neg b)), (\neg(\neg a))=(\neg(\neg b)),\\
&(a \wedge b)=(\neg(\neg c)), (a \vee b)=(\neg(\neg c)), a=(\neg a \wedge a), (\neg a)=(\neg a \wedge a), (a \wedge a)=(\neg a \wedge a), (a \vee a)=(\neg a \wedge a),\\
&(a \wedge b)=(\neg a \wedge a), (a \vee b)=(\neg a \wedge a), (\neg(\neg a))=(\neg a \wedge a), a=(a \wedge \neg a), (\neg a)=(a \wedge \neg a), (a \wedge a)=(a \wedge \neg a),\\
&(a \vee a)=(a \wedge \neg a), (a \wedge b)=(a \wedge \neg a), (a \vee b)=(a \wedge \neg a), (\neg(\neg a))=(a \wedge \neg a), (\neg a \wedge a)=(a \wedge \neg a)\}
\end{aligned}
$$

Many (like $a = b$) are very obviously not true, at least not for all possible values of each variable. But—essentially by using truth tables—we can readily pick out ones that are always true:

$$
\begin{aligned}
&\{a=b, a=(\neg a), a=(\neg b), (\neg a)=(\neg b), \boxed{a=(a \wedge a)}, (\neg a)=(a \wedge a), \boxed{a=(a \vee a)}, (\neg a)=(a \vee a), \boxed{(a \wedge a)=(a \vee a)}, a=(a \wedge b),\\
&(\neg a)=(a \wedge b), (a \wedge a)=(a \wedge b), (a \vee a)=(a \wedge b), a=(a \vee b), (\neg a)=(a \vee b), (a \wedge a)=(a \vee b), (a \vee a)=(a \vee b),\\
&(a \wedge b)=(a \vee b), (a \wedge b)=(a \wedge c), (a \vee b)=(a \wedge c), (a \wedge b)=(a \vee c), (a \vee b)=(a \vee c), a=(b \wedge a), (\neg a)=(b \wedge a),\\
&(a \wedge a)=(b \wedge a), (a \vee a)=(b \wedge a), \boxed{(a \wedge b)=(b \wedge a)}, (a \vee b)=(b \wedge a), a=(b \vee a), (\neg a)=(b \vee a), (a \wedge a)=(b \vee a),\\
&(a \vee a)=(b \vee a), (a \wedge b)=(b \vee a), \boxed{(a \vee b)=(b \vee a)}, a=(b \wedge b), (\neg a)=(b \wedge b), (a \wedge a)=(b \wedge b), (a \vee a)=(b \wedge b),\\
&(a \wedge b)=(b \wedge b), (a \vee b)=(b \wedge b), a=(b \vee b), (\neg a)=(b \vee b), (a \wedge a)=(b \vee b), (a \vee a)=(b \vee b), (a \wedge b)=(b \vee b),\\
&(a \vee b)=(b \vee b), a=(b \wedge c), (\neg a)=(b \wedge c), (a \wedge a)=(b \wedge c), (a \vee a)=(b \wedge c), (a \wedge b)=(b \wedge c), (a \vee b)=(b \wedge c), a=(b \vee c),\\
&(\neg a)=(b \vee c), (a \wedge a)=(b \vee c), (a \vee a)=(b \vee c), (a \wedge b)=(b \vee c), (a \vee b)=(b \vee c), (a \wedge b)=(c \wedge a), (a \vee b)=(c \wedge a),\\
&(a \wedge b)=(c \vee a), (a \vee b)=(c \vee a), (a \wedge b)=(c \wedge b), (a \vee b)=(c \wedge b), (a \wedge b)=(c \vee b), (a \vee b)=(c \vee b), (a \wedge b)=(c \wedge c),\\
&(a \vee b)=(c \wedge c), (a \wedge b)=(c \vee c), (a \vee b)=(c \vee c), \boxed{a=(\neg(\neg a))}, (\neg a)=(\neg(\neg a)), \boxed{(a \wedge a)=(\neg(\neg a))}, \boxed{(a \vee a)=(\neg(\neg a))},\\
&(a \wedge b)=(\neg(\neg a)), (a \vee b)=(\neg(\neg a)), a=(\neg(\neg b)), (\neg a)=(\neg(\neg b)), (a \wedge a)=(\neg(\neg b)), (a \vee a)=(\neg(\neg b)), (a \wedge b)=(\neg(\neg b)),\\
&(a \vee b)=(\neg(\neg b)), (\neg(\neg a))=(\neg(\neg b)), (a \wedge b)=(\neg(\neg c)), (a \vee b)=(\neg(\neg c)), a=(\neg a \wedge a), (\neg a)=(\neg a \wedge a), (a \wedge a)=(\neg a \wedge a),\\
&(a \vee a)=(\neg a \wedge a), (a \wedge b)=(\neg a \wedge a), (a \vee b)=(\neg a \wedge a), (\neg(\neg a))=(\neg a \wedge a), a=(a \wedge \neg a), (\neg a)=(a \wedge \neg a),\\
&(a \wedge a)=(a \wedge \neg a), (a \vee a)=(a \wedge \neg a), (a \wedge b)=(a \wedge \neg a), (a \vee b)=(a \wedge \neg a), (\neg(\neg a))=(a \wedge \neg a), \boxed{(\neg a \wedge a)=(a \wedge \neg a)}\}
\end{aligned}
$$

OK, so now we can get a list of true theorems:

$$\{a=(a\wedge a),\ a=(a\vee a),\ (a\wedge a)=(a\vee a),\ (a\wedge b)=(b\wedge a),\ (a\vee b)=(b\vee a),\ a=(\neg(\neg a)),\ (a\wedge a)=(\neg(\neg a)),\ (a\vee a)=(\neg(\neg a)),$$
$$(\neg a\wedge a)=(a\wedge\neg a),\ (\neg a\vee a)=(a\vee\neg a),\ (\neg a)=(\neg(a\wedge a)),\ (\neg a)=(\neg(a\vee a)),\ (\neg(a\wedge a))=(\neg(a\vee a)),\ (a\wedge\neg b)=(\neg b\wedge a),$$
$$(\neg a\wedge b)=(b\wedge\neg a),\ (a\vee\neg b)=(\neg b\vee a),\ (\neg a\vee b)=(b\vee\neg a),\ (\neg(a\wedge b))=(\neg(b\wedge a)),\ (\neg(a\vee b))=(\neg(b\vee a)),$$
$$(\neg a\wedge a)=(\neg b\wedge b),\ (a\wedge\neg a)=(\neg b\wedge b),\ (\neg a\wedge a)=(b\wedge\neg b),\ (a\wedge\neg a)=(b\wedge\neg b),\ (\neg a\vee a)=(\neg b\vee b),\ (a\vee\neg a)=(\neg b\vee b),$$
$$(\neg a\vee a)=(b\vee\neg b),\ (a\vee\neg a)=(b\vee\neg b),\ (\neg a)=(\neg a\wedge\neg a),\ (\neg(a\wedge a))=(\neg a\wedge\neg a),\ (\neg(a\vee a))=(\neg a\wedge\neg a),$$
$$(\neg a)=(\neg a\vee\neg a),\ (\neg(a\wedge a))=(\neg a\vee\neg a),\ (\neg(a\vee a))=(\neg a\vee\neg a),\ (\neg a\wedge\neg a)=(\neg a\vee\neg a),\ (\neg(a\vee b))=(\neg a\wedge\neg b),$$
$$(\neg(a\wedge b))=(\neg a\vee\neg b),\ (\neg(a\vee b))=(\neg b\wedge\neg a),\ (\neg a\wedge\neg b)=(\neg b\wedge\neg a),\ (\neg(a\wedge b))=(\neg b\vee\neg a),\ (\neg a\vee\neg b)=(\neg b\vee\neg a),$$
$$a=((a\wedge a)\wedge a),\ (a\wedge a)=((a\wedge a)\wedge a),\ (a\vee a)=((a\wedge a)\wedge a),\ (\neg(\neg a))=((a\wedge a)\wedge a),\ a=((a\vee a)\wedge a),$$
$$(a\wedge a)=((a\vee a)\wedge a),\ (a\vee a)=((a\vee a)\wedge a),\ (\neg(\neg a))=((a\vee a)\wedge a),\ ((a\wedge a)\wedge a)=((a\vee a)\wedge a),\ a=(a\wedge(a\wedge a)),$$
$$(a\wedge a)=(a\wedge(a\wedge a)),\ (a\vee a)=(a\wedge(a\wedge a)),\ (\neg(\neg a))=(a\wedge(a\wedge a)),\ ((a\wedge a)\wedge a)=(a\wedge(a\wedge a)),\ ((a\vee a)\wedge a)=(a\wedge(a\wedge a)),$$
$$a=(a\wedge(a\vee a)),\ (a\wedge a)=(a\wedge(a\vee a)),\ (a\vee a)=(a\wedge(a\vee a)),\ (\neg(\neg a))=(a\wedge(a\vee a)),\ ((a\wedge a)\wedge a)=(a\wedge(a\vee a))\}$$

Some are "interesting". Others seem repetitive, overly complicated, or otherwise not terribly interesting. But if we want to "channel Euclid" we somehow have to decide which are the interesting theorems that we're going to write down. And although Euclid himself didn't explicitly discuss logic, we can look at textbooks of logic from the last couple of centuries—and we find that there's a very consistent set of theorems that they end up picking out from the list, and giving names to:

a = a ∧ a [idempotence of AND]	a = a ∨ a [idempotence of OR]	a ∧ a = a ∨ a	a ∧ b = b ∧ a [commutativity of AND]	a ∨ b = b ∨ a [commutativity of OR]	a = ¬¬a [law of double negation]
a ∧ a = ¬¬a	a ∨ a = ¬¬a	¬a = ¬(a ∧ a)	¬a = ¬(a ∨ a)	a = (a ∧ a) ∧ a	a = (a ∨ a) ∧ a
a = a ∧ (a ∧ a)	a = a ∧ (a ∨ a)	a = a ∧ a ∨ a	a = (a ∨ a) ∨ a	a = a ∨ a ∧ a	a = a ∨ (a ∨ a)
a = a ∧ (a ∨ b) [absorption laws of AND]	a = a ∨ a ∧ b [absorption laws of OR]	a = (a ∨ b) ∧ a	a = a ∧ (b ∨ a)	a = a ∧ b ∨ a	a = a ∨ b ∧ a
[illegible]	[illegible]	¬a = ¬a ∧ ¬a	[illegible]	¬a ∧ a = a ∧ ¬a	¬a ∨ a = a ∨ ¬a
[illegible]	[illegible]	¬¬a = (a ∨ a) ∧ a	[illegible]	¬¬a = a ∧ (a ∨ a)	¬¬a = a ∧ a ∨ a
¬¬a = (a ∨ a) ∨ a	¬¬a = a ∨ a ∧ a	¬¬a = a ∨ (a ∨ a)	[illegible]	¬¬a = a ∨ a ∧ b	¬¬a = (a ∨ b) ∧ a
¬¬a = a ∧ (b ∨ a)	¬¬a = a ∧ b ∨ a	¬¬a = a ∨ b ∧ a	¬a ∧ a = ¬b ∧ b [law of noncontradiction (definition of False)]	a ∧ ¬a = ¬b ∧ b	¬a ∧ a = b ∧ ¬b
a ∧ ¬a = b ∧ ¬b	¬a ∨ a = ¬b ∨ b [law of noncontradiction (definition of True)]	a ∨ ¬a = ¬b ∨ b	¬a ∨ a = b ∨ ¬b	a ∨ ¬a = b ∨ ¬b	¬¬a = (b ∨ a) ∧ a
¬¬a = b ∧ a ∨ a	[illegible]	¬a ∧ b = b ∧ ¬a	a ∨ ¬b = ¬b ∨ a	¬a ∨ b = b ∨ ¬a	¬(a ∧ b) = ¬(b ∧ a)
¬(a ∨ b) = ¬(b ∨ a)	[illegible]	a ∨ a = (a ∧ a) ∧ a	a ∧ a = (a ∨ a) ∧ a	a ∨ a = (a ∨ a) ∧ a	a ∧ a = a ∧ (a ∧ a)
a ∨ a = a ∧ (a ∧ a)	[illegible]	a ∨ a = a ∧ (a ∨ a)	a ∧ a = a ∧ a ∨ a	a ∨ a = a ∧ a ∨ a	a ∧ a = (a ∨ a) ∨ a
a ∨ a = (a ∨ a) ∨ a	a ∧ a = a ∨ a ∧ a	a ∨ a = a ∨ a ∧ a	a ∧ a = a ∨ (a ∨ a)	a ∨ a = a ∨ (a ∨ a)	a ∧ a = a ∧ (a ∨ b)
a ∨ a = a ∧ (a ∨ b)	a ∧ a = a ∨ a ∧ b	a ∨ a = a ∨ a ∧ b	a ∧ a = (a ∨ b) ∧ a	a ∨ a = (a ∨ b) ∧ a	a ∧ a = a ∧ (b ∨ a)
a ∨ a = a ∧ (b ∨ a)	a ∧ a = a ∧ b ∨ a	a ∨ a = a ∧ b ∨ a	a ∧ a = a ∨ b ∧ a	a ∨ a = a ∨ b ∧ a	a ∧ a = (b ∨ a) ∧ a
a ∨ a = (b ∨ a) ∧ a	a ∧ a = b ∧ a ∨ a	a ∨ a = b ∧ a ∨ a	a ∧ b = (a ∧ a) ∧ b	a ∧ b = (a ∨ a) ∧ b	a ∧ b = a ∧ (a ∧ b)
a ∨ b = a ∧ a ∨ b	a ∨ b = (a ∨ a) ∨ b	a ∨ b = a ∨ (a ∨ b)	a ∧ b = (a ∧ b) ∧ a	a ∧ b = a ∧ (b ∧ a)	a ∨ b = (a ∨ b) ∨ a
a ∨ b = a ∨ (b ∨ a)	a ∧ b = (a ∧ b) ∧ b	a ∧ b = a ∧ (b ∧ b)	a ∧ b = a ∧ (b ∨ b)	a ∨ b = (a ∨ b) ∨ b	a ∨ b = a ∨ b ∧ b
a ∨ b = a ∨ (b ∨ b)	a ∧ b = (b ∧ a) ∧ a	a ∧ b = b ∧ (a ∧ a)	a ∧ b = b ∧ (a ∨ a)	a ∨ b = (b ∨ a) ∨ a	a ∨ b = b ∨ a ∧ a
a ∨ b = b ∨ (a ∨ a)	a ∧ b = (b ∧ a) ∧ b	a ∧ b = b ∧ (a ∧ b)	[illegible]	a ∨ b = b ∨ (a ∨ b)	a ∧ b = (b ∧ b) ∧ a
a ∧ b = (b ∨ b) ∧ a	a ∧ b = b ∧ (b ∧ a)	a ∨ b = b ∧ b ∨ a	[illegible]	a ∨ b = b ∨ (b ∨ a)	¬(a ∧ a) = ¬a ∧ ¬a
¬(a ∨ a) = ¬a ∧ ¬a	¬(a ∧ a) = ¬a ∨ ¬a	¬(a ∨ a) = ¬a ∨ ¬a	¬(a ∨ b) = ¬a ∧ ¬b [de Morgan's law]	¬(a ∧ b) = ¬a ∨ ¬b	¬(a ∨ b) = ¬b ∧ ¬a
¬(a ∧ b) = ¬b ∨ ¬a	¬a ∧ ¬a = ¬a ∨ ¬a	¬a ∧ ¬b = ¬b ∧ ¬a	¬a ∨ ¬b = ¬b ∨ ¬a	(a ∧ a) ∧ a = (a ∨ a) ∧ a	(a ∧ a) ∧ a = a ∧ (a ∧ a)
(a ∨ a) ∧ a = a ∧ (a ∧ a)	(a ∧ a) ∧ a = a ∧ (a ∨ a)	(a ∨ a) ∧ a = a ∧ (a ∨ a)	a ∧ (a ∧ a) = a ∧ (a ∨ a)	(a ∧ a) ∧ a = a ∧ a ∨ a	(a ∨ a) ∧ a = a ∧ a ∨ a
a ∧ (a ∧ a) = a ∧ a ∨ a	a ∧ (a ∨ a) = a ∧ a ∨ a	(a ∧ a) ∧ a = (a ∨ a) ∨ a	(a ∨ a) ∧ a = (a ∨ a) ∨ a	a ∧ (a ∧ a) = (a ∨ a) ∨ a	a ∧ (a ∨ a) = (a ∨ a) ∨ a
a ∧ a ∨ a = (a ∨ a) ∨ a	(a ∧ a) ∧ a = a ∨ a ∧ a	(a ∨ a) ∧ a = a ∨ a ∧ a	a ∧ (a ∧ a) = a ∨ a ∧ a	a ∧ (a ∨ a) = a ∨ a ∧ a	a ∧ a ∨ a = a ∨ a ∧ a
(a ∨ a) ∨ a = a ∨ a ∧ a	(a ∧ a) ∧ a = a ∨ (a ∨ a)	(a ∨ a) ∧ a = a ∨ (a ∨ a)	a ∧ (a ∧ a) = a ∨ (a ∨ a)	a ∧ (a ∨ a) = a ∨ (a ∨ a)	a ∧ a ∨ a = a ∨ (a ∨ a)
(a ∨ a) ∨ a = a ∨ (a ∨ a)	a ∨ a ∧ a = a ∨ (a ∨ a)	(a ∧ a) ∧ a = a ∧ (a ∨ b)	(a ∨ a) ∧ a = a ∧ (a ∨ b)	a ∧ (a ∧ a) = a ∧ (a ∨ b)	a ∧ (a ∨ a) = a ∧ (a ∨ b)
a ∧ a ∨ a = a ∧ (a ∨ b)	(a ∨ a) ∨ a = a ∧ (a ∨ b)	a ∨ a ∧ a = a ∧ (a ∨ b)	a ∨ (a ∨ a) = a ∧ (a ∨ b)	(a ∧ a) ∧ a = a ∨ a ∧ b	(a ∨ a) ∧ a = a ∨ a ∧ b
a ∧ (a ∧ a) = a ∨ a ∧ b	a ∧ (a ∨ a) = a ∨ a ∧ b	a ∧ a ∨ a = a ∨ a ∧ b	(a ∨ a) ∨ a = a ∨ a ∧ b	a ∨ a ∧ a = a ∨ a ∧ b	a ∨ (a ∨ a) = a ∨ a ∧ b

⋮ *50 lines*

a ∧ b ∨ b = b ∧ (b ∨ a)	(a ∨ b) ∨ b = b ∧ b ∨ a	a ∨ b ∧ b = b ∧ b ∨ a	a ∨ (b ∨ b) = b ∧ b ∨ a	(a ∨ b) ∨ b = (b ∨ b) ∨ a	a ∨ b ∧ b = (b ∨ b) ∨ a
a ∨ (b ∨ b) = (b ∨ b) ∨ a	(a ∨ b) ∧ b = b ∨ b ∧ a	a ∧ b ∨ b = b ∨ b ∧ a	(a ∨ b) ∨ b = b ∨ (b ∨ a)	a ∨ b ∧ b = b ∨ (b ∨ a)	a ∨ (b ∨ b) = b ∨ (b ∨ a)
(a ∨ b) ∧ b = (b ∧ b) ∧ b	a ∧ b ∨ b = (b ∧ b) ∧ b	(a ∨ b) ∧ b = (b ∨ b) ∧ b	a ∧ b ∨ b = (b ∨ b) ∧ b	(a ∨ b) ∧ b = b ∧ (b ∧ b)	a ∧ b ∨ b = b ∧ (b ∧ b)
(a ∨ b) ∧ b = b ∧ (b ∨ b)	a ∧ b ∨ b = b ∧ (b ∨ b)	(a ∨ b) ∧ b = b ∧ b ∨ b	a ∧ b ∨ b = b ∧ b ∨ b	(a ∨ b) ∧ b = (b ∨ b) ∨ b	a ∧ b ∨ b = (b ∨ b) ∨ b
(a ∨ b) ∧ b = b ∨ b ∧ b	a ∧ b ∨ b = b ∨ b ∧ b	(a ∨ b) ∧ b = b ∨ (b ∨ b)	a ∧ b ∨ b = b ∨ (b ∨ b)	(a ∨ b) ∧ b = b ∧ (b ∨ c)	a ∧ b ∨ b = b ∧ (b ∨ c)
(a ∨ b) ∧ b = b ∨ b ∧ c	a ∧ b ∨ b = b ∨ b ∧ c	[illegible]	[illegible]	(a ∨ b) ∧ b = b ∧ (c ∨ b)	a ∧ b ∨ b = b ∧ (c ∨ b)
(a ∨ b) ∧ b = b ∧ c ∨ b	a ∧ b ∨ b = b ∧ c ∨ b	[illegible]	[illegible]	(a ∨ b) ∧ b = (c ∨ b) ∧ b	a ∧ b ∨ b = (c ∨ b) ∧ b
(a ∨ b) ∧ b = c ∧ b ∨ b	a ∧ b ∨ b = c ∧ b ∨ b	(a ∧ b) ∧ c = a ∧ (b ∧ c) [associativity of AND]	(a ∨ b) ∨ c = a ∨ (b ∨ c) [associativity of OR]	(a ∧ b) ∧ c = (a ∧ c) ∧ b	a ∧ (b ∧ c) = (a ∧ c) ∧ b
(a ∧ b) ∧ c = a ∧ (c ∧ b)	a ∧ (b ∧ c) = a ∧ (c ∧ b)	a ∧ (b ∨ c) = a ∧ (c ∨ b)	(a ∨ b) ∨ c = (a ∨ c) ∨ b	a ∨ (b ∨ c) = (a ∨ c) ∨ b	a ∨ b ∧ c = a ∨ c ∧ b
(a ∨ b) ∨ c = a ∨ (c ∨ b)	a ∨ (b ∨ c) = a ∨ (c ∨ b)	(a ∧ b) ∧ c = (b ∧ a) ∧ c	a ∧ (b ∧ c) = (b ∧ a) ∧ c	(a ∨ b) ∧ c = (b ∨ a) ∧ c	(a ∧ b) ∧ c = b ∧ (a ∧ c)

⋮ *392 lines*

[illegible]	[illegible]	a ∨ b ∧ c = a ∧ a ∨ c ∧ b	a ∨ b ∧ c = (a ∨ a) ∨ c ∧ b	(a ∨ b) ∨ c = a ∧ a ∨ (c ∨ b)
[illegible]	[illegible]	a ∨ (b ∨ c) = (a ∨ a) ∨ (c ∨ b)	(a ∧ b) ∧ c = (a ∧ b) ∧ (a ∧ c)	a ∧ (b ∧ c) = (a ∧ b) ∧ (a ∧ c)
a ∧ (b ∨ c) = a ∧ b ∨ a ∧ c [distributive laws of AND]	a ∨ b ∧ c = (a ∨ b) ∧ (a ∨ c) [distributive laws of OR]	(a ∨ b) ∨ c = (a ∨ b) ∨ (a ∨ c)	a ∨ (b ∨ c) = (a ∨ b) ∨ (a ∨ c)	(a ∧ b) ∧ c = (a ∧ b) ∧ (b ∧ c)
a ∧ (b ∧ c) = (a ∧ b) ∧ (b ∧ c)	(a ∨ b) ∨ c = (a ∨ b) ∨ (b ∨ c)	a ∨ (b ∨ c) = (a ∨ b) ∨ (b ∨ c)	(a ∧ b) ∧ c = (a ∧ b) ∧ (c ∧ a)	a ∧ (b ∧ c) = (a ∧ b) ∧ (c ∧ a)

One might assume that these named theorems were just the result of historical convention. But when I was writing *A New Kind of Science* I discovered something quite surprising. With all the theorems written out in "order of complexity", I tried seeing which theorems I could prove just from theorems earlier in the list. Many were easy to prove. But some simply couldn't be proved. And it turned out that these were essentially precisely the "named theorems":

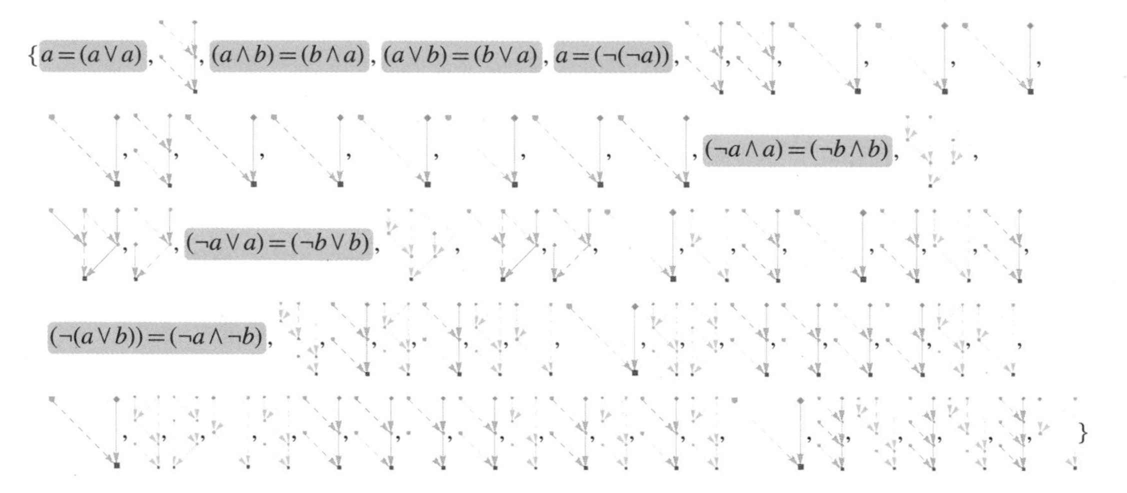

In other words, the "named theorems" are basically the simplest statements of new facts about logic, that can't be established from "simpler facts". Eventually as one's going through the list of theorems, one will have accumulated enough to fill out what can serve as full axioms for logic—so that then all subsequent theorems can be proved from "existing facts".

Now of course the setup we've just used relies on the idea that one's separately got a list of true theorems. To do something more like Euclid, we'd have to pick certain theorems to serve as axioms, then derive all others from these.

Back in 2000 I figured out the very simplest possible axiom system for logic, written in terms of NAND, just the single axiom:

$$\{\forall_{\{a,b,c\}}((b\cdot c)\cdot a)\cdot(b\cdot((b\cdot a)\cdot b))=a\}$$

So now writing AND, OR and NOT in terms of NAND according to

$$\{(\neg a)=a\cdot a,(a\land b)=(a\cdot b)\cdot(a\cdot b),(a\lor b)=(a\cdot a)\cdot(b\cdot b)\}$$

we can, for example, derive the notable theorems of logic from my axiom. FindEquationalProof gives automated proofs of these theorems, though most of them involve quite a few steps (the — indicates a theorem that is trivially true after substituting the forms for AND, OR and NOT):

$a=(a\wedge a)$	54	$(\neg a\wedge a)=(\neg b\wedge b)$	95	$a=(a\wedge(a\vee b))$	328	$(a\vee b)=(a\vee(\neg a\wedge b))$	56
$a=(a\vee a)$	54	$(\neg a\vee a)=(\neg b\vee b)$	92	$a=(a\vee(a\wedge b))$	274	$a=(a\wedge(b\vee\neg b))$	131
$(a\wedge b)=(b\wedge a)$	103	$(\neg(a\vee b))=(\neg a\wedge\neg b)$	132	$(\neg a\vee b)=(\neg(\neg b)\vee\neg a)$	958	$a=(a\vee(b\wedge\neg b))$	130
$(a\vee b)=(b\vee a)$	102	$(\neg(a\wedge b))=(\neg a\vee\neg b)$	143	$((a\wedge b)\wedge c)=(a\wedge(b\wedge c))$	1502	$(a\vee(b\wedge c))=((a\vee b)\wedge(a\vee c))$	120
$(\neg(\neg a))=a$	54	$(a\wedge b)=(a\wedge(a\wedge b))$	91	$((a\vee b)\vee c)=(a\vee(b\vee c))$	—	$(a\wedge(b\vee c))=((a\wedge b)\vee(a\wedge c))$	103

The longer cases here involve first proving the lemma $a\cdot b=b\cdot a$ which takes 102 steps. Including this lemma as an axiom, the minimal axiom system (as I also found in 2000) is:

$$\{\forall_{\{a,b,c\}}(a\cdot b)\cdot(a\cdot(b\cdot c))=a,\forall_{\{a,b\}}a\cdot b=b\cdot a\}$$

And with this axiom system FindEquationalProof succeeds in finding shorter proofs for the notable theorems of logic, even though now the definitions for And, Or and Not are just treated as theorems:

$a=(a\wedge a)$	21	$(\neg a\wedge a)=(\neg b\wedge b)$	130	$a=(a\wedge(a\vee b))$	32	$(a\vee b)=(a\vee(\neg a\wedge b))$	89
$a=(a\vee a)$	15	$(\neg a\vee a)=(\neg b\vee b)$	119	$a=(a\vee(a\wedge b))$	26	$a=(a\wedge(b\vee\neg b))$	129
$(a\wedge b)=(b\wedge a)$	8	$(\neg(a\vee b))=(\neg a\wedge\neg b)$	9	$(\neg a\vee b)=(\neg(\neg b)\vee\neg a)$	20	$a=(a\vee(b\wedge\neg b))$	129
$(a\vee b)=(b\vee a)$	9	$(\neg(a\wedge b))=(\neg a\vee\neg b)$	28	$((a\wedge b)\wedge c)=(a\wedge(b\wedge c))$	249	$(a\vee(b\wedge c))=((a\vee b)\wedge(a\vee c))$	328
$(\neg(\neg a))=a$	17	$(a\wedge b)=(a\wedge(a\wedge b))$	43	$((a\vee b)\vee c)=(a\vee(b\vee c))$	239	$(a\wedge(b\vee c))=((a\wedge b)\vee(a\wedge c))$	338

Actually looking at these proofs is not terribly illuminating; they certainly don't have the same kind of "explanatory feel" as Euclid. But combining the graphs for all these proofs is more interesting, because it shows us the common lemmas that were used in these proofs, and effectively defines a network of interdependencies between theorems:

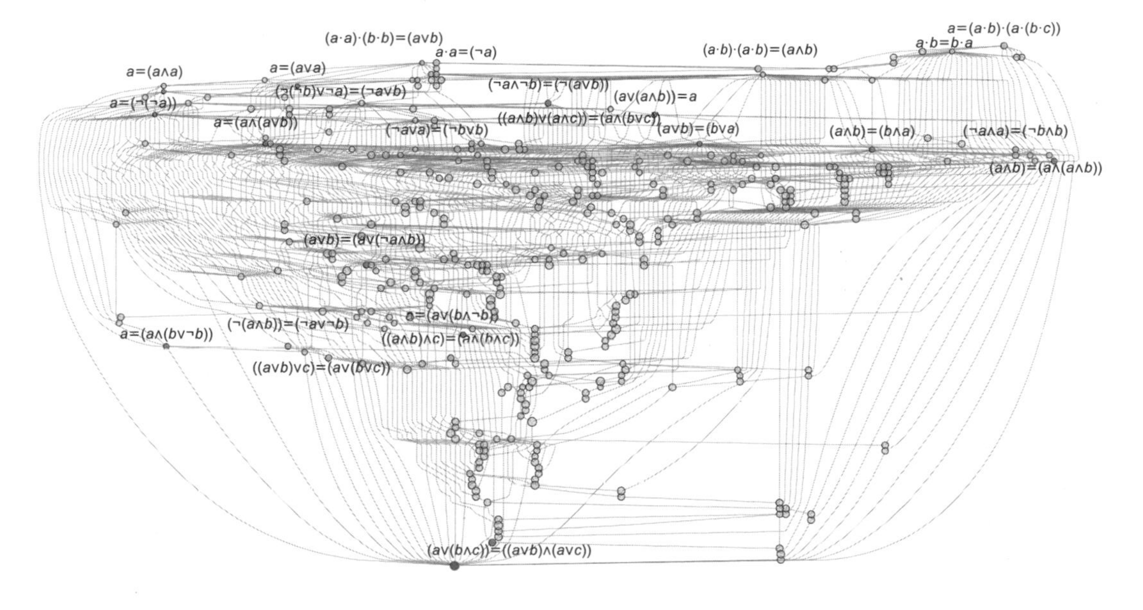

There are 361 lemmas (i.e. automatically generated intermediate theorems) here. It's a fair number, given that we're only proving 20 theorems—but it's definitely much less than the total of 1978 that would be involved in proving each of the theorems separately.

In our graph here—like in our Euclid theorem-dependency graphs above—the axioms are shown (in yellow) at the top. The "notable theorems" that we're proving are shown in pink. But the structure of the graph is a little different from our earlier Euclid theorem-dependency graphs, and this alternative layout makes it clearer:

In Euclid, a given theorem is proved on the basis of other theorems, and ultimately on the basis of axioms. But here the automated theorem-proving process creates lemmas that ultimately allow one to show that the theorems one's trying to prove are equivalent to "true" (i.e. to a tautology)—shown as a red node.

We can ask other questions, such as how long the lemmas are. Here are the distributions of lengths of the final notable theorems, and of the intermediate lemmas used to prove them:

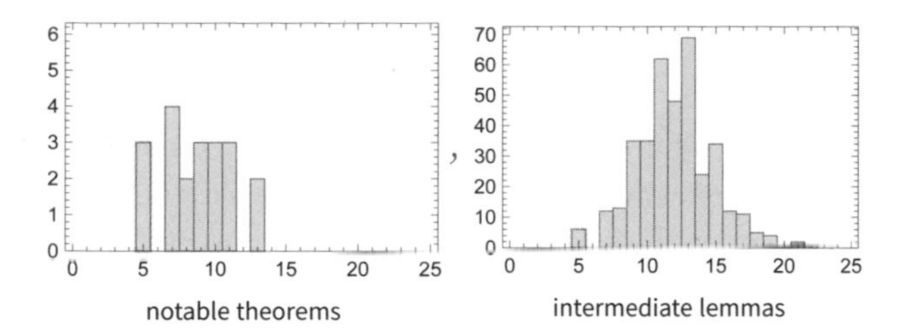

We get something slightly more in the spirit of Euclid if we elide the lemmas, and just find the implied effective dependency graph between notable theorems:

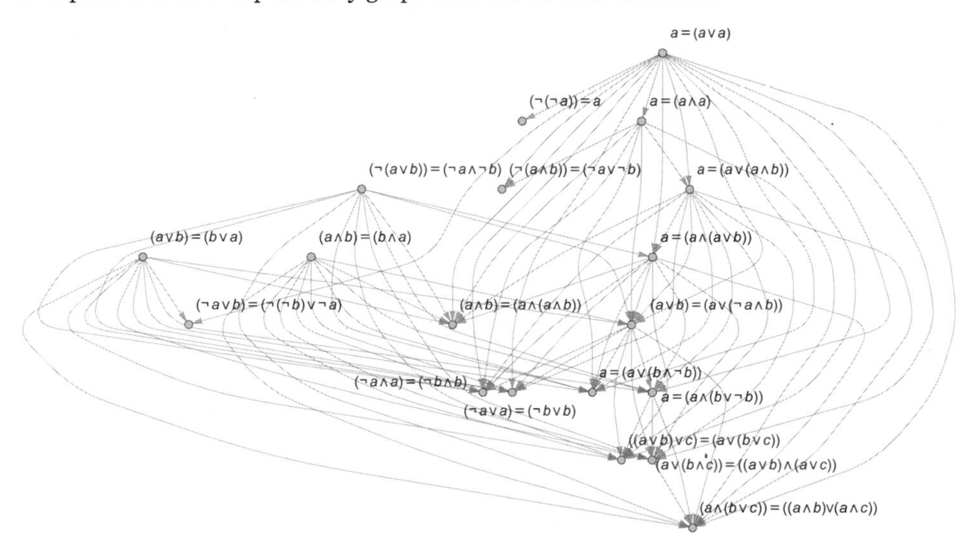

Transitive reduction then gives:

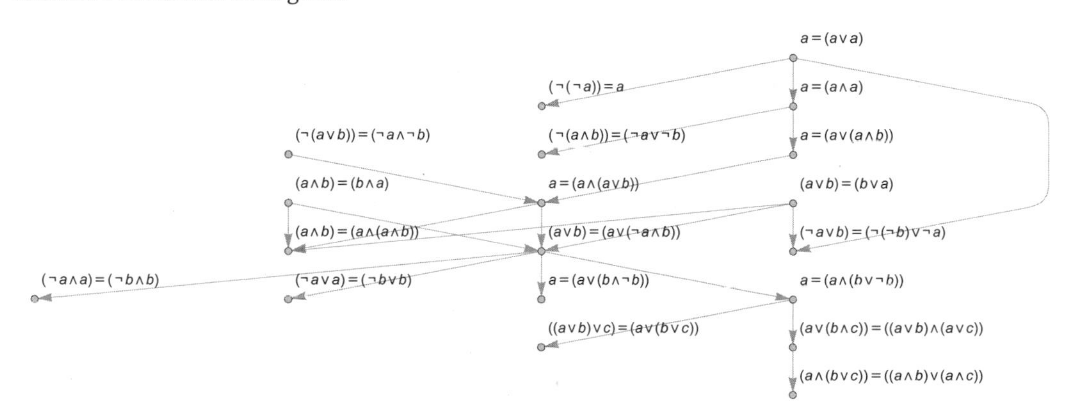

By omitting intermediate lemmas, we're in a sense just getting a shadow of the dependencies of the notable theorems, in the "environment" defined by our particular choice of axioms. But with this setup, it's interesting to see the distributive law be the "hardest theorem"—kind of the metamathematical analog of Euclid's 13.18 about the Platonic solids.

OK, but what we're doing so far with logic is still fundamentally a bit different from how most of Euclid works. Because what Euclid typically does is to say something like "imagine such-and-such a geometrical setup; then the following theorem will be true about it". And the analog of that for logic would be to take axioms of logic, then append some logical assertion, and ask if with the axioms and this assertion some particular statement is true. In other words, there are some statements—like the axioms—that will be true in "pure logic", but there are more statements that will be true with particular setups (or, in the case of logic, particular possible values for variables).

For example, in "pure logic" $a \lor b == b \lor b$ is not necessarily true (i.e. it is not a tautology). But if we assert that $a == (a \land b)$ is true, then this implies the following possible choices for a and b

```
In[ ]:= SatisfiabilityInstances[a == (a∧b), {a, b}, All]
Out[ ]= {{True, True}, {False, True}, {False, False}}
```

and in all these cases $a \lor b == b \lor b$ is true. So, in a Euclid tradition, we could say "imagine a setup where $a == (a \land b)$; then we can prove from the axioms of logic the theorem that $a == (a \land b)$".

Above we looked at which statements in logic are true for all values of variables:

$$\{a=b,\ a=(\neg a),\ a=(\neg b),\ (\neg a)=(\neg b),\ a=(a\land a),\ (\neg a)=(a\land a),\ a=(a\lor a),\ (\neg a)=(a\lor a),\ (a\land a)=(a\lor a),\ a=(a\land b),\ (\neg a)=(a\land b),\ (a\land a)=(a\land b),\ (a\lor a)=(a\land b),\ a=(a\lor b),\ (\neg a)=(a\lor b),\ (a\land a)=(a\lor b),\ (a\lor a)=(a\lor b),\ (a\land b)=(a\lor b),\ (a\land b)=(a\land c),\ (a\lor b)=(a\land c),\ (a\land b)=(a\lor c),\ (a\lor b)=(a\lor c),\ a=(b\land a),\ (\neg a)=(b\land a),\ (a\land a)=(b\land a),\ (a\lor a)=(b\land a),\ (a\land b)=(b\land a),\ (a\lor b)=(b\land a),\ a=(b\lor a),\ (\neg a)=(b\lor a),\ (a\land a)=(b\lor a),\ (a\lor a)=(b\lor a),\ (a\land b)=(b\lor a),\ (a\lor b)=(b\lor a),\ a=(b\land b),\ (\neg a)=(b\land b),\ (a\land a)=(b\land b),\ (a\lor a)=(b\land b),\ (a\land b)=(b\land b),\ (a\lor b)=(b\land b),\ a=(b\lor b),\ (\neg a)=(b\lor b),\ (a\land a)=(b\lor b),\ (a\lor a)=(b\lor b),\ (a\land b)=(b\lor b),\ (a\lor b)=(b\lor b),\ a=(b\land c),\ (\neg a)=(b\land c),\ (a\land a)=(b\land c),\ (a\lor a)=(b\land c)\}$$

Now let's look at the ones that aren't always true. If we assume that some particular one of these statements is true, we can see which other statements it implies are true:

$a=b$	$\{a=b, a=(a\vee b), a=(b\vee a), a=(a\wedge b), a=(b\wedge a), (\neg a)=(\neg b), (a\vee a)=(a\vee b), (a\vee a)=(a\wedge b), (a\vee a)=(b\wedge a), (a\vee b)=(b\wedge a), (a\wedge a)=(a\vee b), (a\wedge a)=(a\wedge b), (a\wedge a)=(b\wedge a), (a\wedge b)=(a\vee b)\}$
$a=(\neg a)$	$\{\}$
$a=(\neg b)$	$\{a=(\neg b)\}$
$a=(a\vee b)$	$\{a=(a\vee b), a=(b\vee a), (a\vee a)=(a\vee b), (a\wedge a)=(a\vee b)\}$
$a=(b\vee a)$	$\{a=(a\vee b), a=(b\vee a), (a\vee a)=(a\vee b), (a\wedge a)=(a\vee b)\}$
$a=(a\wedge b)$	$\{a=(a\wedge b), a=(b\wedge a), (a\vee a)=(a\wedge b), (a\vee a)=(b\wedge a), (a\wedge a)=(a\wedge b), (a\wedge a)=(b\wedge a)\}$
$a=(b\wedge a)$	$\{a=(a\wedge b), a=(b\wedge a), (a\vee a)=(a\wedge b), (a\vee a)=(b\wedge a), (a\wedge a)=(a\wedge b), (a\wedge a)=(b\wedge a)\}$
$(\neg a)=(\neg b)$	$\{a=b, a=(a\vee b), a=(b\vee a), a=(a\wedge b), a=(b\wedge a), (\neg a)=(\neg b), (a\vee a)=(a\vee b), (a\vee a)=(a\wedge b), (a\vee a)=(b\wedge a), (a\vee b)=(b\wedge a), (a\wedge a)=(a\vee b), (a\wedge a)=(a\wedge b), (a\wedge a)=(b\wedge a), (a\wedge b)=(a\vee b)\}$
$(\neg a)=(a\vee a)$	$\{\}$
$(\neg a)=(a\vee b)$	$\{a=(\neg b), a=(a\wedge b), a=(b\wedge a), (\neg a)=(a\vee b), (\neg a)=(b\vee a), (a\vee a)=(a\wedge b), (a\vee a)=(b\wedge a), (a\wedge a)=(a\wedge b), (a\wedge a)=(b\wedge a), (a\wedge b)=(a\wedge c)\}$
$(\neg a)=(b\vee a)$	$\{a=(\neg b), a=(a\wedge b), a=(b\wedge a), (\neg a)=(a\vee b), (\neg a)=(b\vee a), (a\vee a)=(a\wedge b), (a\vee a)=(b\wedge a), (a\wedge a)=(a\wedge b), (a\wedge a)=(b\wedge a), (a\wedge b)=(a\wedge c)\}$
$(\neg a)=(a\wedge a)$	$\{\}$
$(\neg a)=(a\wedge b)$	$\{a=(\neg b), a=(a\vee b), a=(b\vee a), (\neg a)=(a\wedge b), (\neg a)=(b\wedge a), (a\vee a)=(a\vee b), (a\vee b)=(a\vee c), (a\wedge a)=(a\vee b)\}$
$(\neg a)=(b\wedge a)$	$\{a=(\neg b), a=(a\vee b), a=(b\vee a), (\neg a)=(a\wedge b), (\neg a)=(b\wedge a), (a\vee a)=(a\vee b), (a\vee b)=(a\vee c), (a\wedge a)=(a\vee b)\}$
$(a\vee a)=(a\vee b)$	$\{a=(a\vee b), a=(b\vee a), (a\vee a)=(a\vee b), (a\wedge a)=(a\vee b)\}$
$(a\vee a)=(a\wedge b)$	$\{a=(a\wedge b), a=(b\wedge a), (a\vee a)=(a\wedge b), (a\vee a)=(b\wedge a), (a\wedge a)=(a\wedge b), (a\wedge a)=(b\wedge a)\}$
$(a\vee a)=(b\wedge a)$	$\{a=(a\wedge b), a=(b\wedge a), (a\vee a)=(a\wedge b), (a\vee a)=(b\wedge a), (a\wedge a)=(a\wedge b), (a\wedge a)=(b\wedge a)\}$
$(a\vee b)=(a\vee c)$	$\{(a\vee b)=(a\vee c)\}$

Or on a larger scale, with a black dot when one statement implies another:

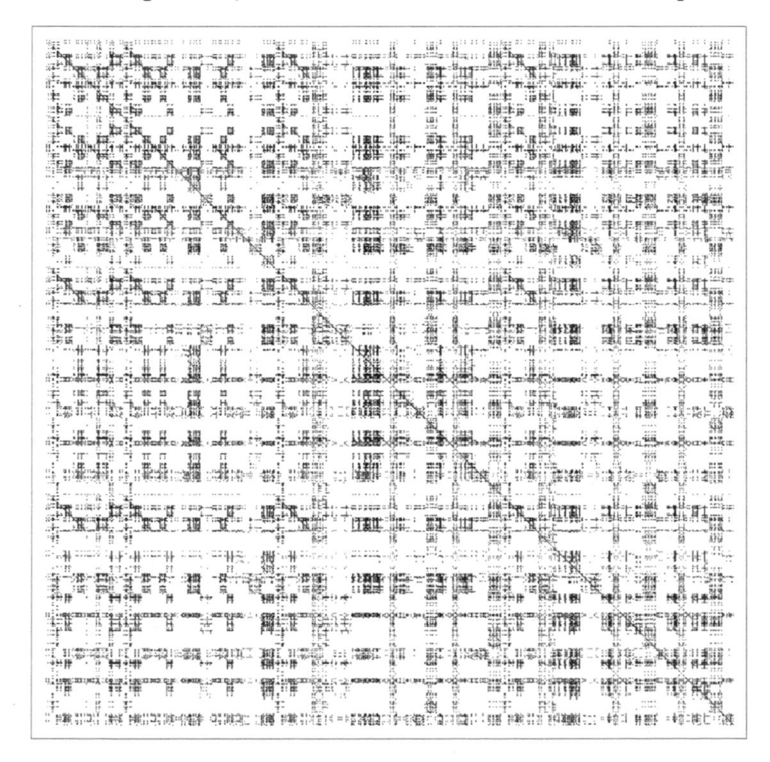

For each of these theorems we can in principle construct a proof, using the axioms:

```
In[•]:= FindEquationalProof[(a⊕b) == (b⊕b),
          Append[AxiomaticTheory["BooleanAxioms"], a == (a⊗b)], "ProofGraph"]
```

Out[•]=

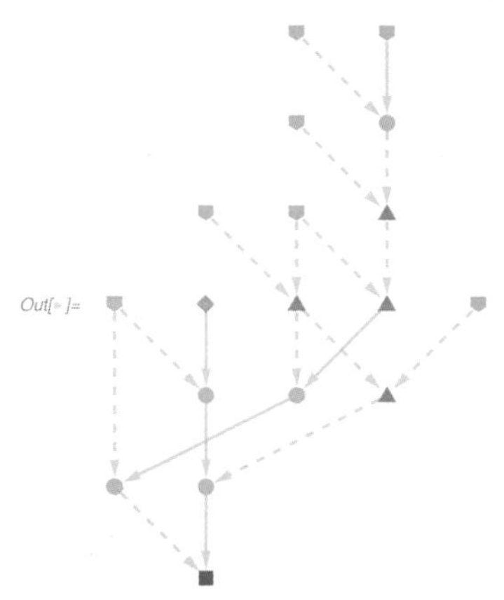

And now we could go through and find out which theorems are useful in proving other theorems—and in principle this would allow us to build up a theorem dependency network. But there are undoubtedly many ways to do this, and so we'd need additional criteria to find ones that have whatever attributes would make us say "that might have been how someone like Euclid would have done it".

OK, so could one look at geometry the same way? Basically, yes. Using the formalization we had above in terms of line, between, congruent, etc. we can again start by just enumerating possible statements. Unlike for logic, many of them won't even make "structural sense"; for example they might contain line[congruent[...],...], but it makes no sense to have a line whose endpoint is a truth value. But we can certainly get a list of "structurally meaningful" statements.

And then we can ask which are "tautologically true"—though it's in practice considerably harder to do this than for logic (the best known methods involve all sorts of elaborate algebraic computations, which Mathematica can certainly do, but which quickly become quite unwieldy). And after that, we can proceed like Euclid, and start saying "assert this, then you can prove this". And, yes, it's nice that after 2000+ years, we can finally imagine automating the process of producing generalizations of Euclid's *Elements*. Though this just makes it more obvious that part of what Euclid did was in a sense a matter of art—picking in some kind of aesthetic way which possible sequence of theorems would best "tell his story" of geometry.

Math beyond Euclid

We've looked here at some of the empirical metamathematics of what Euclid did on geometry more than 2000 years ago. But what about more recent mathematics, and all those other areas of mathematics that have now been studied? In the history of mathematics, there have been perhaps 5 million research papers published, as well as probably hundreds of thousands of textbooks (though few quite as systematic as Euclid).

And, yes, in modern times almost all mathematics that's published is on the web in some form. A few years ago we scraped arXiv and identified about 2 million things described as theorems there (the most popular being the central limit theorem, the implicit function theorem and Fubini's theorem); we also scraped as much as we could of the visible web and found about 30 million theorems there. No doubt many were duplicates (though it's hard—and in principle undecidable!—which they are). But it's a reasonable estimate that there are a few million distinct theorems for which proofs have been published in the history of human mathematics.

It's a remarkable piece of encapsulated intellectual achievement—perhaps the largest coherent such one produced by our species. And I've long been interested in seeing just what it would take to make it computable, and to bring it into the whole computational knowledge framework we have in the Wolfram Language. A few years ago I hoped that we could mobilize the mathematics community to help make this happen. But formalization is hard work, and it's not at the center of what most mathematicians aspire to. Still, we've at least been slowly working—much as we have for Euclid-style geometry—to define the elements of computational language needed to represent theorems in various areas of mathematics.

For example, in the area of point-set topology, we have under development things like

```
In[•]:= [is Hausdorff  TOPOLOGY CONCEPT]["Output"] // InputForm

Out[•]= ForAll[{x, y}, Element[x, X["Elements"]] && Element[y, X["Elements"]] && x != y,
  Exists[{U, V}, Element[U, X["Topology"]] && Element[V, X["Topology"]] &&
 SetIntersection[U, V] == EmptySet &&
   Element[x, U] && Element[y, V]]]
```

which in traditional mathematical notation becomes:

$$\forall_{\{x,y\},x\in\mathcal{X}\wedge y\in\mathcal{X}\wedge x\neq y}\exists_{\{U,V\}}\left(U\in\tau_{\mathcal{X}}\wedge V\in\tau_{\mathcal{X}}\wedge U\bigcap V=\emptyset\wedge x\in U\wedge y\in V\right)$$

So far we have encoded in computable form 742 "topology concepts", and 1687 theorems about them. Here are the connections recorded between concepts (dropping the concept of "topological spaces" that a third of all concepts are connected to, and labeling concepts with high betweenness centrality):

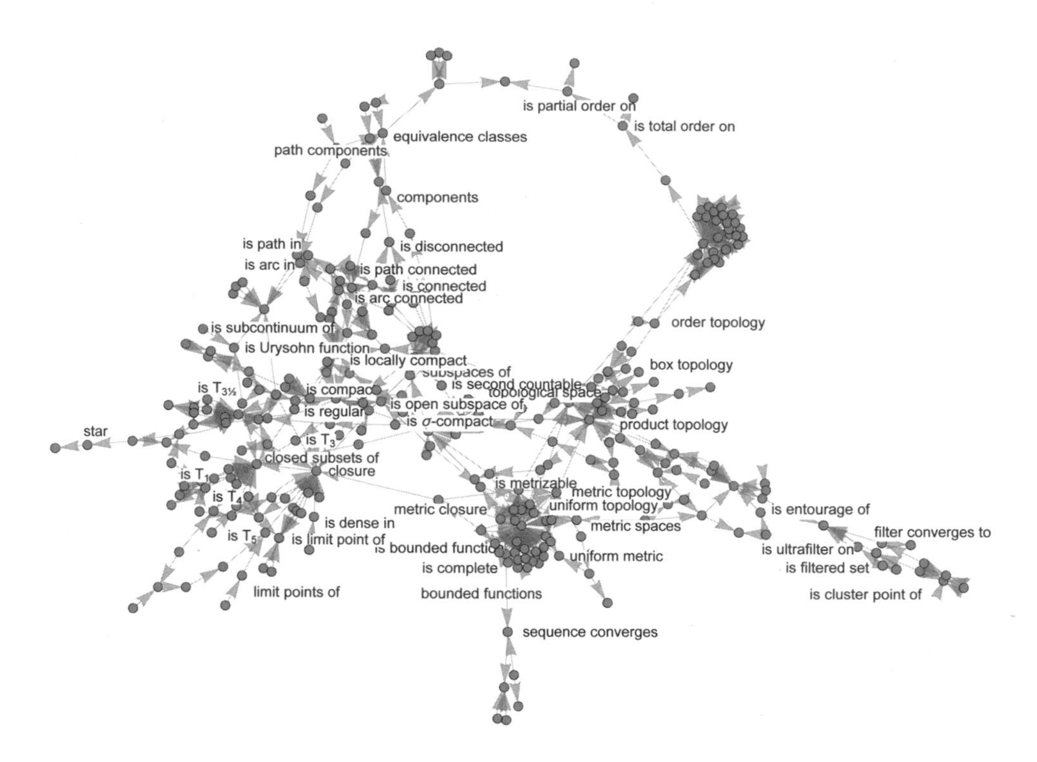

And here is the graph of what theorem references what in its description:

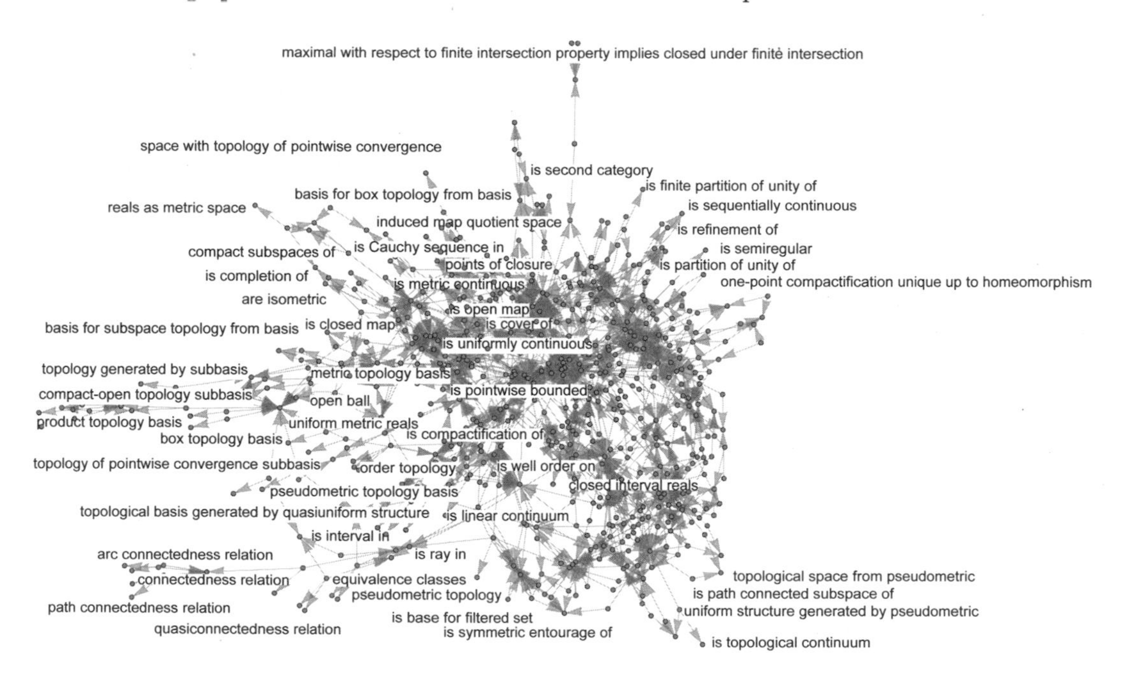

We haven't encoded proofs for these theorems, so we can't yet make the kind of theorem dependency graph that we did for Euclid. But we do have the dependency graph for 76 properties of topological spaces:

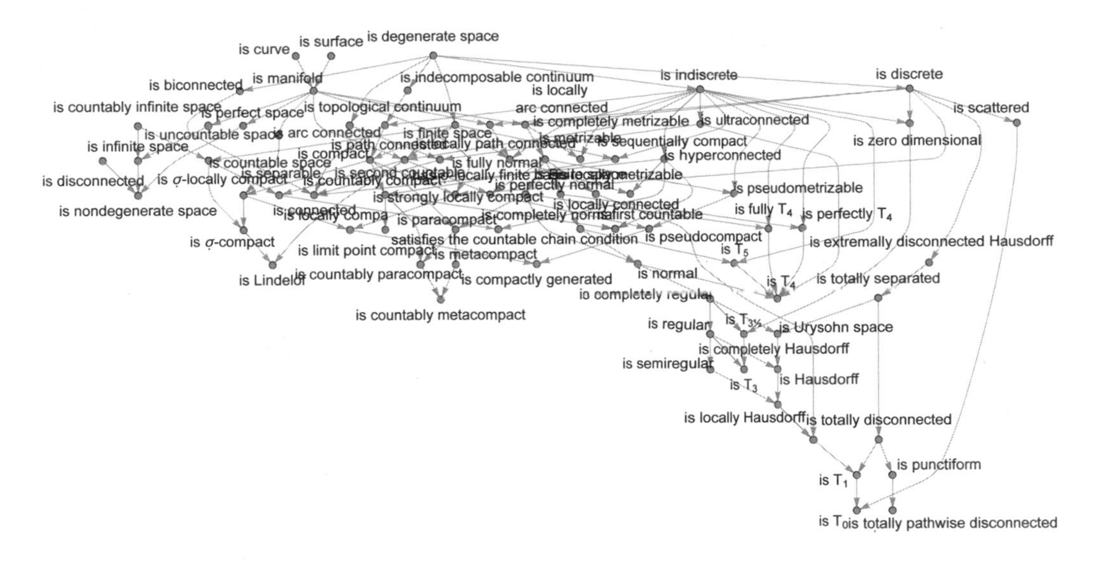

The longest path here (along with a similar one starting with is curve) is 14 steps:

is surface → is manifold → is completely metrizable → is metrizable → is perfectly normal → is completely normal → is normal → is completely regular → is regular → is semiregular → is Hausdorff → is locally Hausdorff → is T_1 → is T_0

(And, yes, this isn't particularly profound; it's just an indication of what it looks like to make specific definitions in topology computable.)

So far, what we've discussed is being able to represent pure mathematical ideas and results in a high-level computable way, understandable to both humans and computers. But what if we want to just formalize everything, from the ground up, explicitly deriving and validating every theorem from the lowest-level foundations? Over the past few decades there have been a number of large-scale projects—like Mizar, Coq, Isabelle, HOL, Metamath, Lean—that have tried to do this (nowadays often in connection with creating "proof assistants").

Ultimately each project defines a certain "machine code" for mathematics. And yes, even though people might think that "mathematics is a universal language", if one's really going to give full, precise, formal specifications there are all sorts of choices to be made. Should things be based on set theory, type theory, higher-order logic, calculus of constructions, etc.? Should the law of excluded middle be assumed? The axiom of choice? What if one's axiomatic structure seems great, but implies a few silly results, like 1 / 0 = 0? There's no perfect solution, but each of these projects has made a certain set of choices.

And the good news here is that for our purposes in doing large-scale empirical metamathematics—as in doing mathematics in the way mathematicians usually do it—it doesn't seem like the choices will matter much. But what's important for us is that these projects have accumulated tens of thousands of theorems (well, OK, some are "throwaway lemmas" or simple rearrangements), and that starting from axioms (or what amount to axioms), they've reached decently far into quite a few areas of mathematics.

Looking at them is a bit of a different experience from looking at Euclid. While the *Elements* has the feel of a "narrative textbook" (albeit from a different age), formalized mathematics projects tend to seem more like software codebases, with their theorem dependency graphs being more like function call graphs. But they still provide fascinating metamathematical corpuses, and there's undoubtedly lots about empirical metamathematics that one can learn from them.

Here I'm going to look at two examples: the Lean mathlib collection, which includes about 36,000 theorems (and 16,000 definitions) and the Metamath set.mm ("set theory") collection, which has about 44,000 theorems (and 1500 definitions).

To get a sense of what's in these collections, we can start by drawing interdependence graphs for the theorems they contain in different areas of mathematics. Just like for Euclid, we make the size of each node represent the number of theorems in a particular area, and the thickness of each edge represent the fraction of theorems from one area that directly reference another in their proof.

Leaving out theorems that effectively just do structural manipulation, rather than representing mathematical content (as well as "self-loop" connections within a single domain) here's the interdependence graph for Lean:

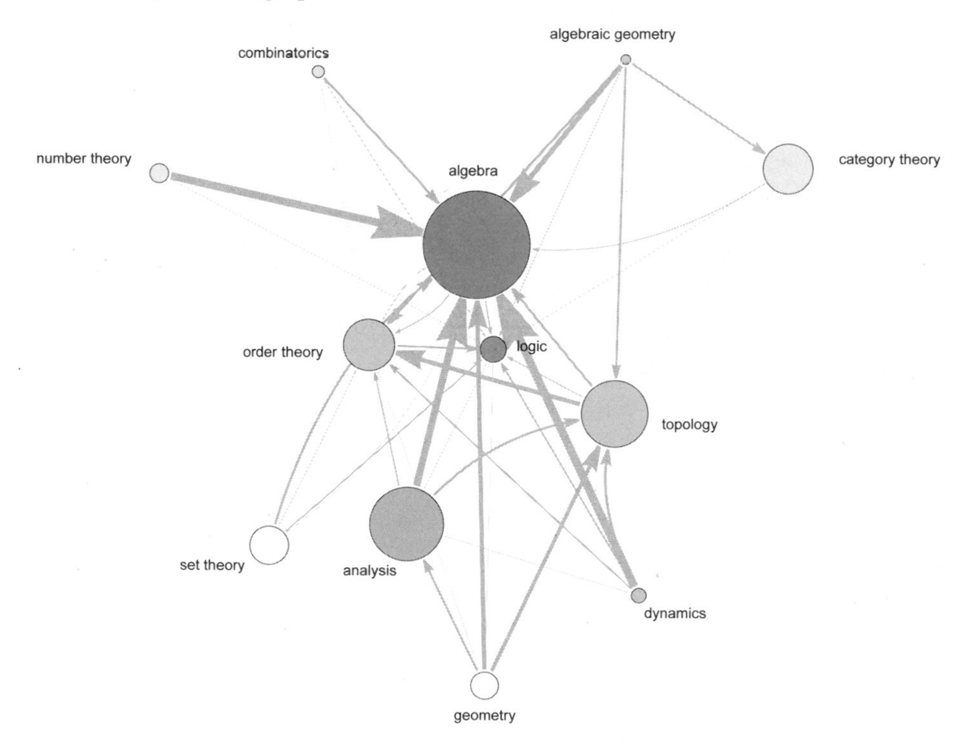

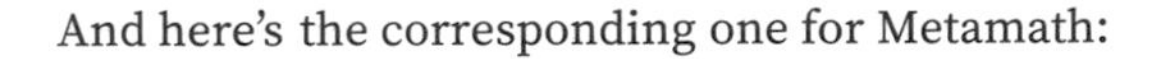

And here's the corresponding one for Metamath:

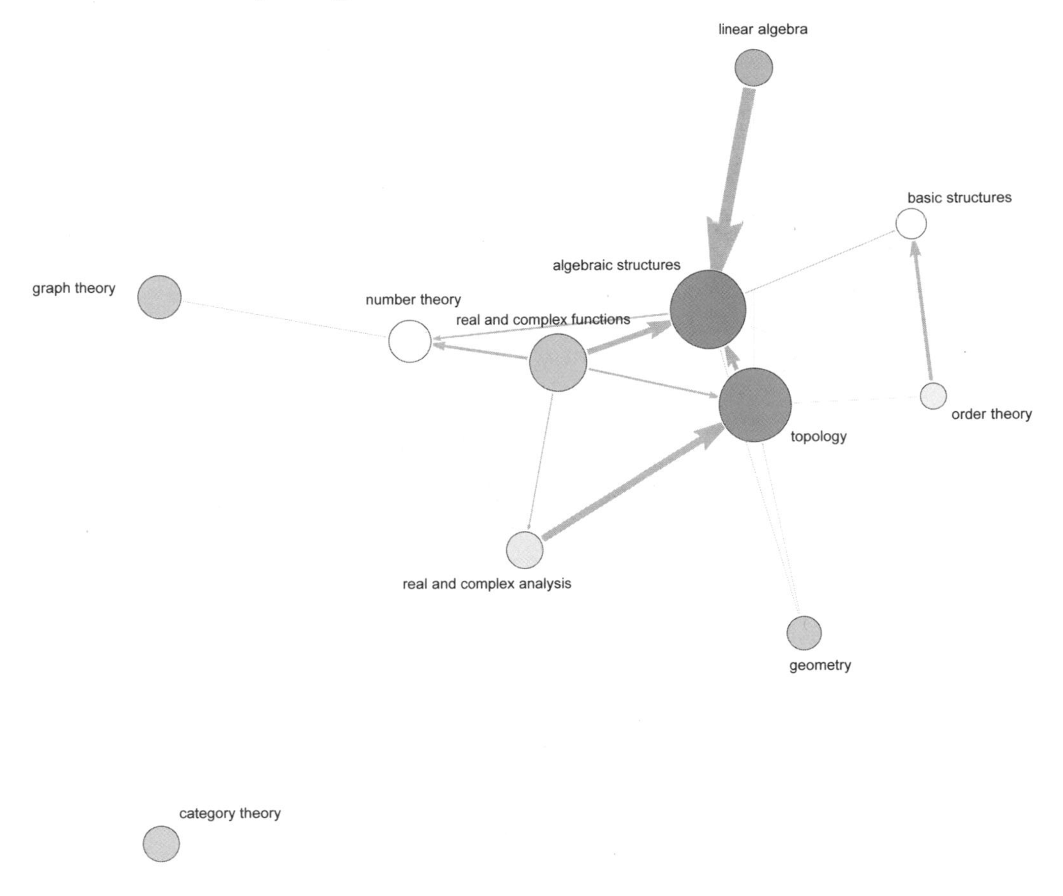

It's somewhat interesting to see how central algebra ends up being in both cases, and how comparatively "off on the side" category theory is. But it's clear that much of what one's seeing in these graphs is a reflection of the particular user communities of these systems, with some important pieces of modern mathematics (like the applications of algebraic geometry to number theory) notably missing.

But, OK, how do individual theorems work in these systems? As an example, let's consider the Pythagorean theorem. In Euclid, this is 1.47, and here's the first level of its dependency graph:

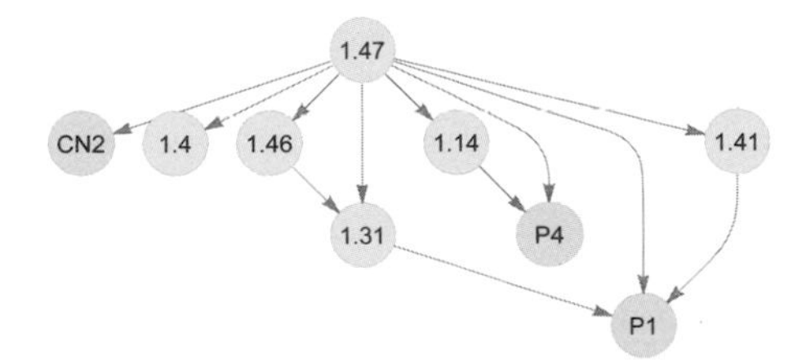

Here's the full graph involving a total of 39 elements (including, by the way, all 10 of the axioms), and having "depth" 20:

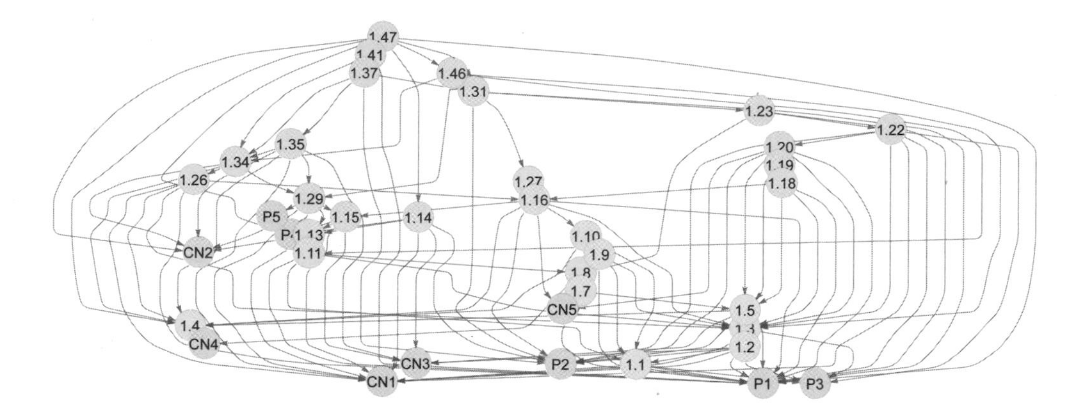

In Lean's mathlib, the theorem is called euclidean_geometry.dist_square_eq_dist_square_add_dist_square_iff_angle_eq_pi_div_two—and its stated proof directly involves 7 other theorems:

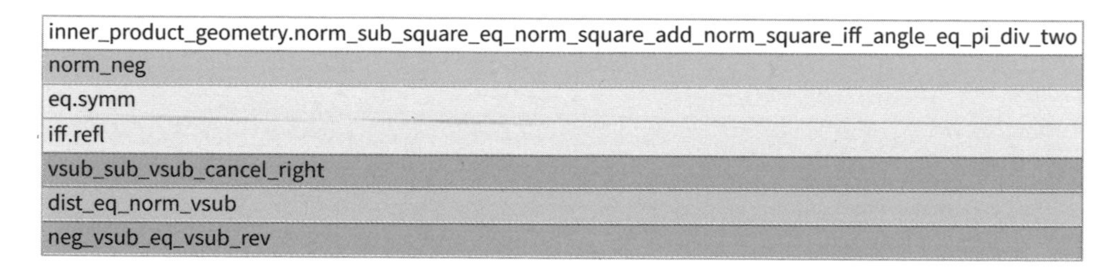

inner_product_geometry.norm_sub_square_eq_norm_square_add_norm_square_iff_angle_eq_pi_div_two
norm_neg
eq.symm
iff.refl
vsub_sub_vsub_cancel_right
dist_eq_norm_vsub
neg_vsub_eq_vsub_rev

Going 3 steps, the theorem dependency graph looks like (where "init" and "tactic" basically refer to structure rather than mathematical content):

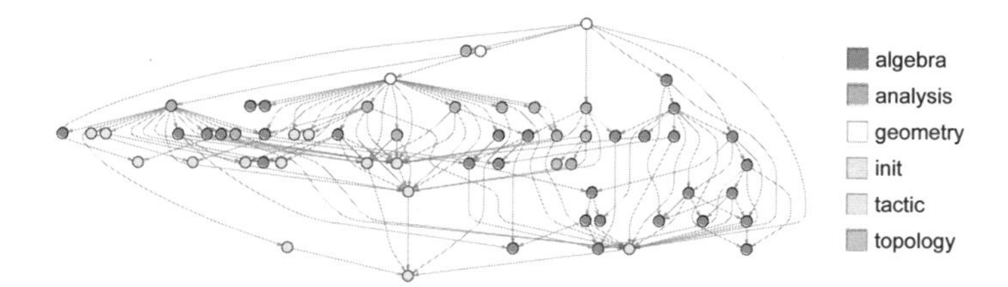

The full graph involves a total of 2850 elements (and has "depth" 84), and after transitive reduction has the form:

And, yes, this is considerably more complicated than Euclid's version—but presumably that's what happens if you insist on full formalization. Of the 2850 theorems used, 1503 are basically structural. The remainder bring mathematical content from different areas, and it's notable in the picture above that different parts of the proof seem to "concentrate" on different areas. Curiously, theorems from geometry (which is basically all Euclid used) occupy only a tiny sliver of the pie chart of all theorems used:

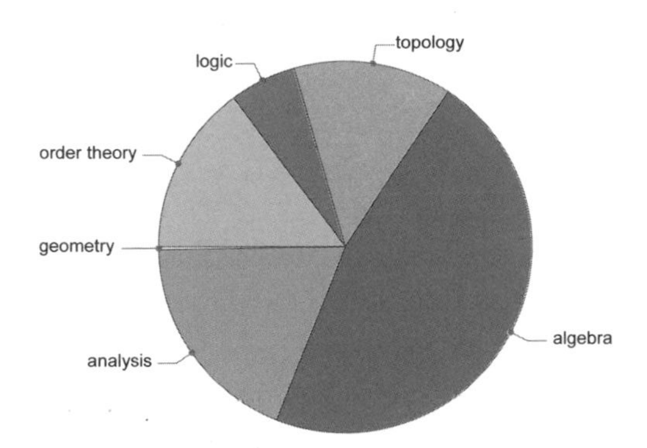

The Metamath set.mm version of the Pythagorean theorem is called pythag, and its proof directly depends on 26 other theorems:

lawcos	3adant3	elpri	fveq2	coshalfpi	syl6eq
cosneghalfpi	jaoi	syl	3ad2ant3	oveq2d	subcl
3adant1	3ad2ant1	abscld	recnd	syl5eqel	3adant2
mulcld	mul01d	eqtrd	2t0e0	sqcld	addcld
subid1d	3eqtrd				

After 1 step, the theorem dependency graph is:

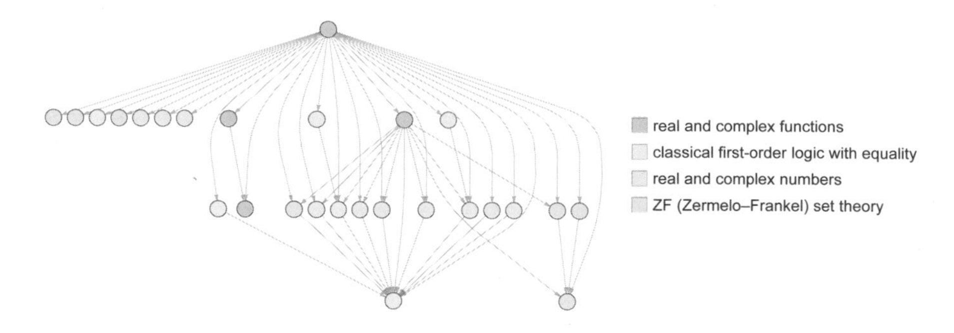

The full graph involves 7099 elements—and has depth 270. In other words, to get from the Pythagorean theorem all the way to the axioms can take as many as 270 steps.

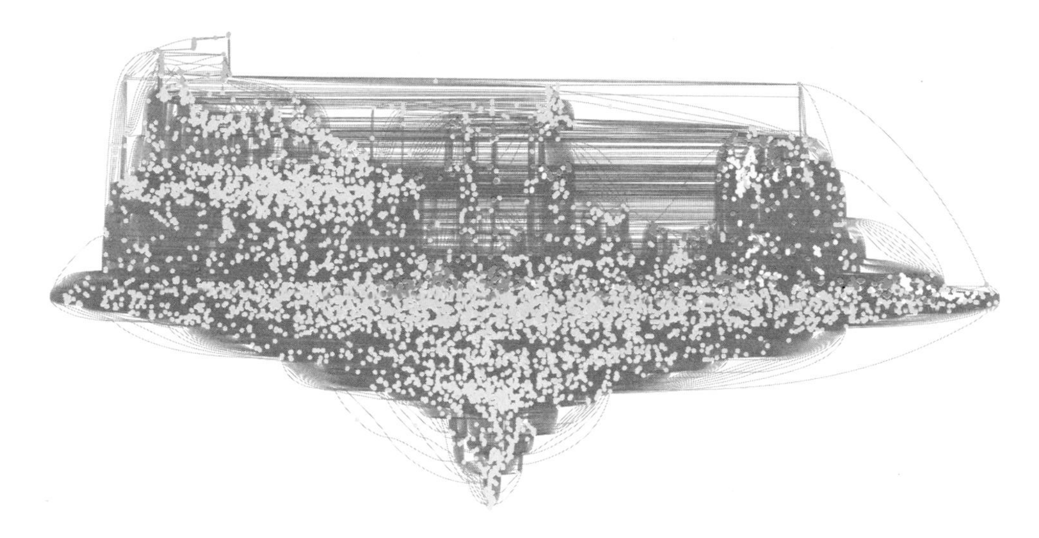

Given the complete Lean or Metamath corpuses, we can start doing the same kind of empirical metamathematics we did for Euclid's *Elements*—except now the higher level of formalization that's being used potentially allows us to go much further.

As a very simple example, here's the distribution of numbers of theorems directly referenced in the proof of each theorem in Lean, Metamath and Euclid:

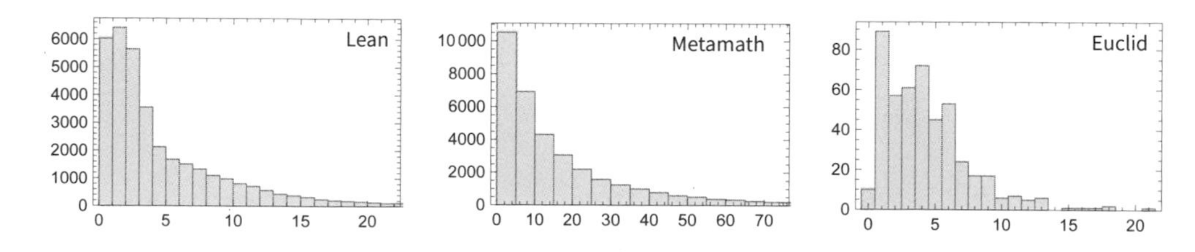

The differences presumably reflect different “hierarchical modularity conventions” in Lean and Metamath (and Euclid). But it’s interesting to note, for example, that in all three cases, the Pythagorean theorem is “above average” in terms of number of theorems referenced in its proof:

	Lean	Metamath	Euclid
	7	26	8
mean	4.9	18.7	4.3

What are the most popular theorems used in proofs? In terms of direct references, here are the top-5 lists:

congr_arg	9896
congr	9241
eq.trans	8632
eq.symm	6032
eq_self_iff_true	3711

Lean

syl	11 524
eqid	8360
adantr	7434
syl2anc	6469
a1i	5727

Metamath

10.11	60
6.1	53
5.11	47
1.3	47
10.6	43

Euclid

Not surprisingly, for Lean and Metamath these are quite “structural”. For Lean, congr_arg is the “congruency” statement that if $a = b$ then $f(a) = f(b)$; congr is a variant that says if $a = b$ and $f = g$ then $f(a) = g(b)$; eq.trans is the transitivity statement if $a = b$ and $b = c$ then $a = c$ (Euclid’s CN1); eq.symm is the statement if $a = b$ then $b = a$; etc. For Metamath, syl is “transitive syllogism”: if $x \Rightarrow y$ and $y \Rightarrow z$ then $x \Rightarrow z$; eqid is about reflexity of equality; etc. In Euclid, these kinds of low-level results—if they are even stated at all—tend to be “many levels down” in the hierarchy of theorems, leaving the single most popular theorem, 10.11, to be one about proportion and rationality.

If one looks at all theorems directly and indirectly referenced by a given theorem, the distribution of total numbers of theorems is as follows (with Lean showing the most obviously exponential decay):

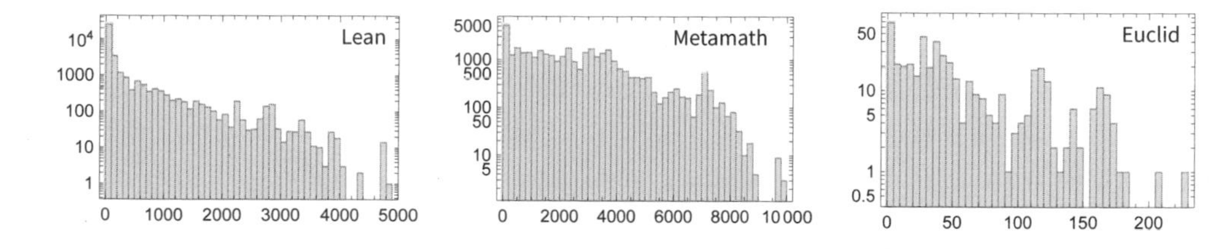

What about the overall structure of the Lean and Metamath dependency graphs? We can ask about effective dimension, about causal invariance, about “event horizons”, and much more. But right now I’ll leave that for another time...

The Future of Empirical Metamathematics

I don't think empirical metamathematics has been much of a thing in the past. In fact, looking on the web as I write this, I'm surprised to see that essentially all references to the actual term "empirical metamathematics" seem to point directly or indirectly to that one note of mine on the subject in *A New Kind of Science*.

But as I hope this piece has made clear, there's a lot that can be done in empirical metamathematics. In everything I've written here, I haven't started analyzing questions like how one can recognize a powerful or a surprising theorem. And I've barely scratched the surface even of the empirical metamathematics that can be done on Euclid's *Elements* from 2000 years ago.

But what kind of a thing is empirical metamathematics? Assuming one's looking at theorems and proofs constructed by humans rather than by automated systems, it's about analyzing large-scale human output—a bit like doing data science on literary texts, or on things like websites or legal corpuses. But it's different. Because ultimately the theorems and proofs that are the subject of empirical metamathematics are derived not from features of the world, but from a formal system that defines some area of mathematics.

With computational language the goal is to be able to describe anything in formalized, computational terms. But in empirical metamathematics, things are in a sense "born formalized". Whatever the actual presentation of theorems and proofs may be there, their "true form" is ultimately something grounded in the formal structure of the mathematics being used.

Of course there is also a strong human element to the raw material of empirical metamathematics. It is (at least for now) humans who have chosen which of the infinite number of possible theorems should be considered interesting, and worthy of presentation. And at least traditionally, when humans write proofs, they usually do it less as a way to certify correctness, and more as a form of exposition: to explain to other humans why a particular theorem is true, and what structure it fits into.

In a sense, empirical metamathematics is a quite desiccated way to look at mathematics, in which all the elegant conceptual structure of its content has been removed. But if we're to make a "science of metamathematics", it's almost inevitable that we have to think this way. Part of what we need to do is to understand some of the human aesthetics of mathematics, and in effect to see to deduce laws by which it may operate.

In this piece I've mostly concentrated on doing fairly straightforward graph-oriented data science, primarily on Euclid's *Elements*. But in moving forward with empirical metamathematics a key question is what kind of model one should be trying to fit one's observations into.

And this comes back to my current motivation for studying empirical metamathematics: as a window onto a general "bulk" theory of metamathematics—and as the foundation for a science not just of how we humans have explored metamathematical space, but of what fundamentally is out there in metamathematical space, and what its overall structure may be.

No doubt there are already clues in what I've done here, but probably only after we have the general theory will we have the paradigm that's needed to identify them. But even without this, there's much to do in studying empirical metamathematics for its own sake—and of better characterizing the remarkable human achievement that is mathematics.

And for now, it's interesting to be able to look at something as old as Euclid's *Elements* and to realize what new perspectives modern computational thinking can give us about it. Euclid was a pioneer in the notion of building everything up from formal rules—and the seeds he sowed played an important role in leading us to the modern computational paradigm. So it's something of a thrill to be able to come back two thousand years later and see that paradigm—now all grown up—applied not only to something like the fundamental theory of physics, but also to what Euclid did all those years ago.

Thanks

For help with various aspects of the content of this piece I'd like to thank Peter Barendse, Ian Ford, Jonathan Gorard, Rob Lewis, Jose Martin-Garcia, Norm Megill, James Mulnix, Nik Murzin, Mano Namuduri, Ed Pegg, Michael Trott, and Xiaofan Zhang, as well as Sushma Kini and Jessica Wong, and for past discussions about related topics, also Bruno Buchberger, Dana Scott and the various participants of our 2016 workshop on the Semantic Representation of Mathematical Knowledge.

Note Added

As I was working on this piece, I couldn't help wondering whether—in 2300 years—anyone else had worked on the empirical metamathematics of Euclid before. Turns out (as Don Knuth pointed out to me) at least one other person did—more than 400 years ago.

The person in question was Thomas Harriot (1560–1621).

The only thing Thomas Harriot published in his lifetime was the book *A Briefe and True Report of the New Found Land of Virginia,* based on a trip that he made to America in 1585. But his papers show that he did all sorts of math and science (including inventing the · notation for multiplication, < and >, as well as drawing pictures of the Moon through a telescope before Galileo, etc.). He seems to have had a well-ahead-of-his-time interest in discrete mathematics, apparently making Venn diagrams a couple of centuries before Venn

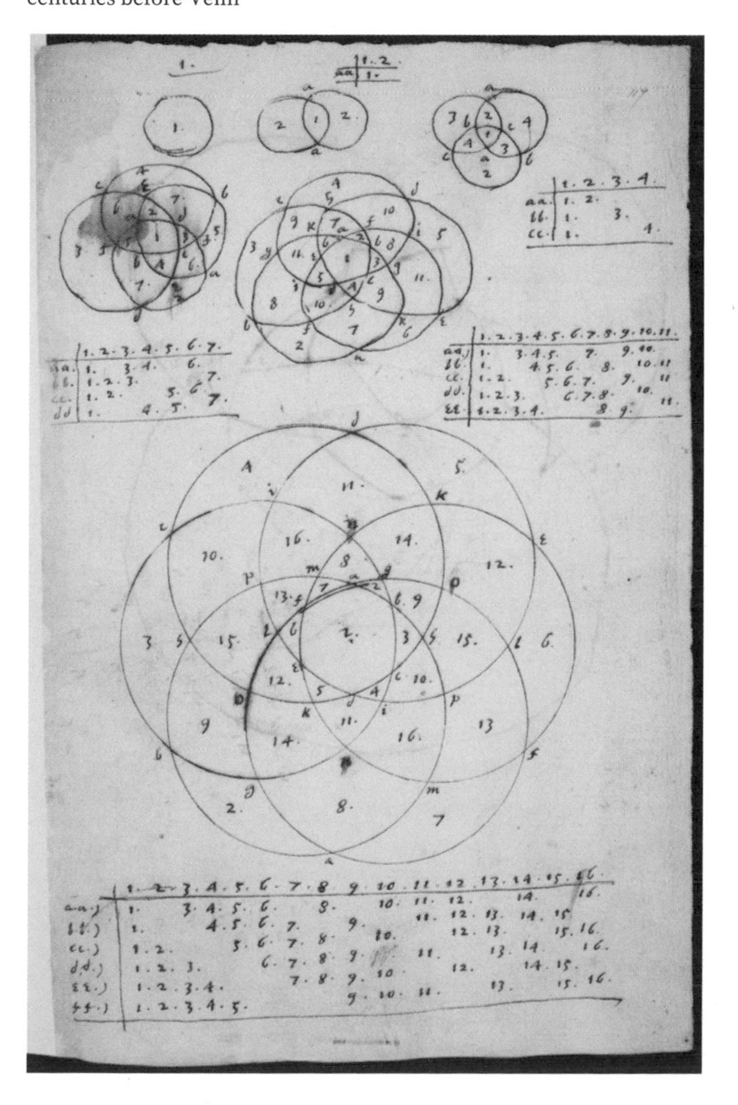

doing various enumerations of structures

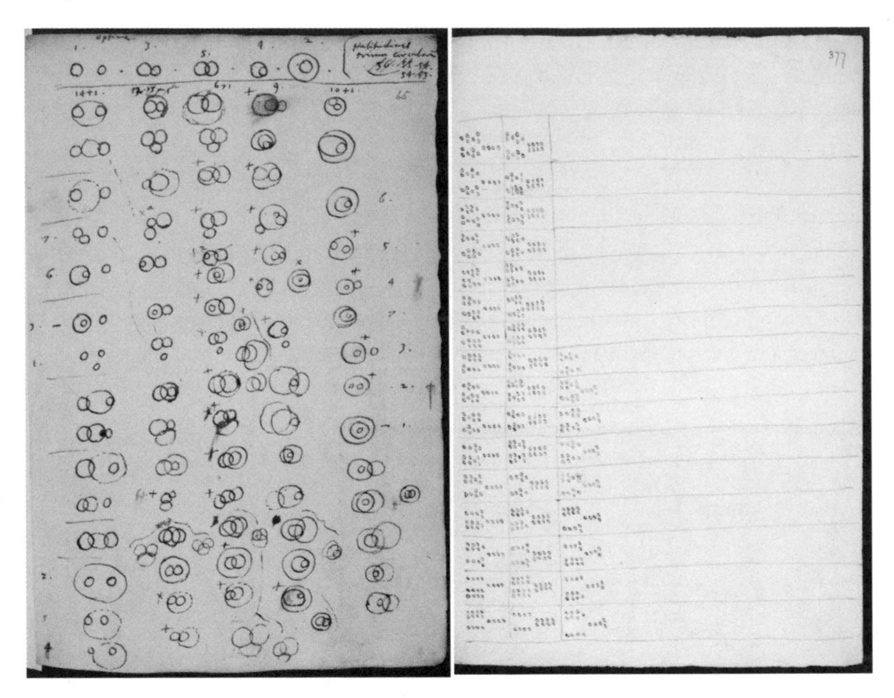

as well as various repeated computations (but no cellular automata, so far as I can tell!):

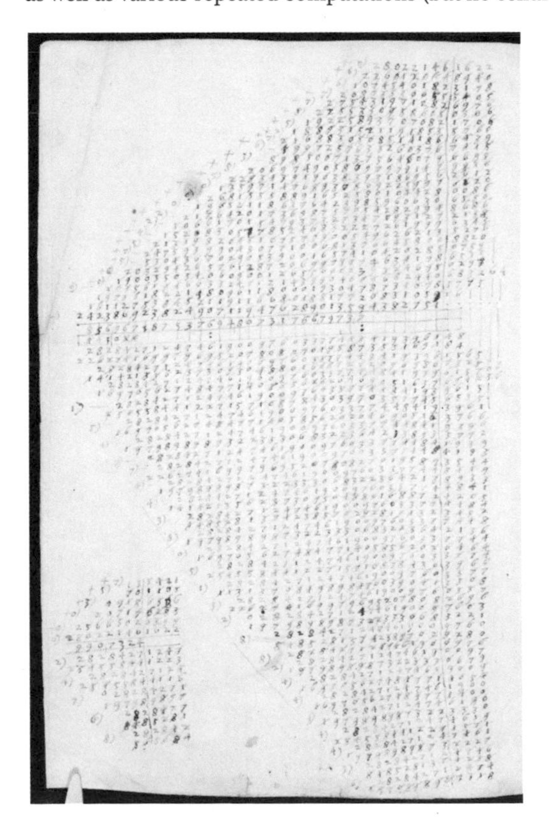

And he seems to have made a detailed study of Euclid's *Elements,* listing in detail (as I did) what theorems are used in each proof (this is for Book 1):

In Libro primo Elementorum Euclidis

Definitiones { Graeco codice et apud Comandinum. 35.
Secundum Flussatem et Clavium. 36.

Postulata { In graecis quibusdam codicibus et [illegible] } 3.
In alijs graecis codicibus et apud Comandinum } 5.
Apud Clavium. — 4.

Axiomata { In Graecis quibusdam. — 12.
Apud Comandinum. — 10.
Apud Clavium. — 20.

Problemata. 14. } propositiones. 48.
Theoremata. 34.

probl.	Theor.	prop.				
1.	—	1.	3,p: 1,p: 20,d: 15,d: 1,a:	15,d: 20,d: 1,p: 3,p: 1,a:		0.
2.	—	2.	3,p: 1,p: 1,1: 2,p: 3,p:	15,d: 1,p: 2,p: 3,p: 1,a:	1,1: —	1.
			15,d: 3,a: 15,d: 1,a: 3,p:	3,a:		
			1,p: 15,d:			
					2,1: —	2.
3.	—	3.	2,1: 3,p: 15,d: 1,a:	15,d: 3,p: 1,a:	0.	0.
	1.	4.	8,a: 10,a: 14,a: 8,a: 14,a:	8,a: 10,a: 14,a:		
			8,a:			
					3,1: 4,1:	4.
	2.	5.	3,1: 1,p: 4,1: 3,a: 4,1:	1,p: 3,a:		
			3,a:			
					3,1: 4,1:	4.
	3.	6.	3,1: 4,1:			
	4.	7.	5,1: 9,a: 5,1: 5,1: 9,a:	9,a:	5,1:	5.
			5,1:		7,1:	6.
	5.	8.	8,a: 8,a: 7,1:	8,a:		
4.	—	9.	3,1: 1,1: 8,1:		1,1: 3,1: 8,1:	7.
5.	—	10.	1,1: 9,1: 4,1:		1,1: 4,1: 9,1:	8.
6.	—	11.	3,1: 1,1: 8,1: 10,d:	10,d:	1,1: 3,1: 8,1:	7.
7.	—	12.	3,p: 10,1: 8,1: 10,d:	3,p: 10,d:	8,1: 10,1:	9.
	6.	13.	10,d: 11,1: 19,a: 2,a: 19,a:	10,d: 1,a: 2,a: 19,a:	11,1:	8.
			2,a: 1,a:			
					13,1:	9.
	7.	14.	13,1: 12,a: 3,a:	3,a: 12,a:		
	8.	15.	13,1: 12,a: 3,a: 13,1: 3,a:	3,a: 12,a:	13,1:	9.
	9.	16.	10,1: 3,1: 15,1: 4,1: 15,1:	—	3,1: 4,1: 10,1: 15,1:	12.

But then, in his "moment of empirical metamathematics" he lists out the full dependency table for theorems in Book 1, having computed what we'd now call the transitive closure:

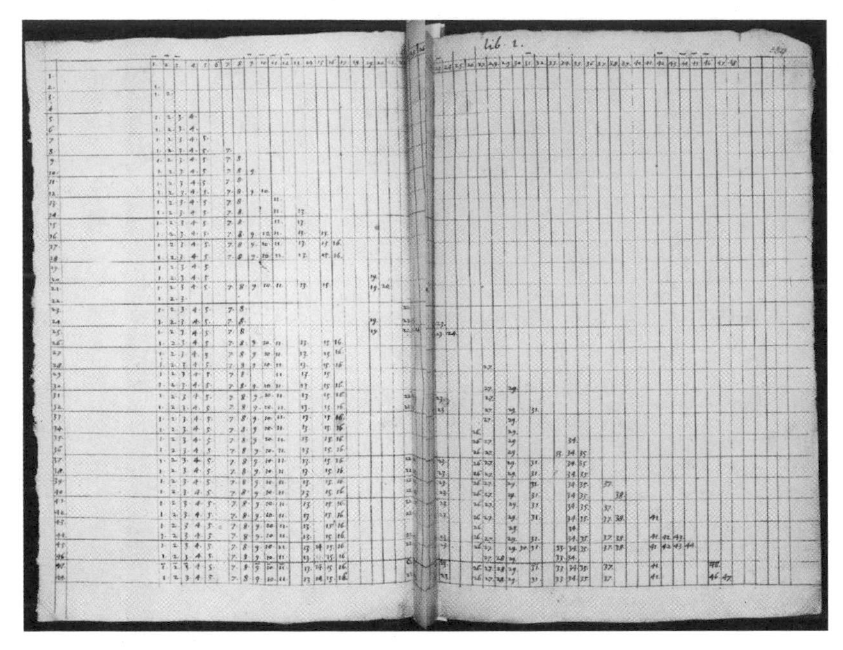

It's easy for us to reproduce this now, and, yes, he did make a few mistakes:

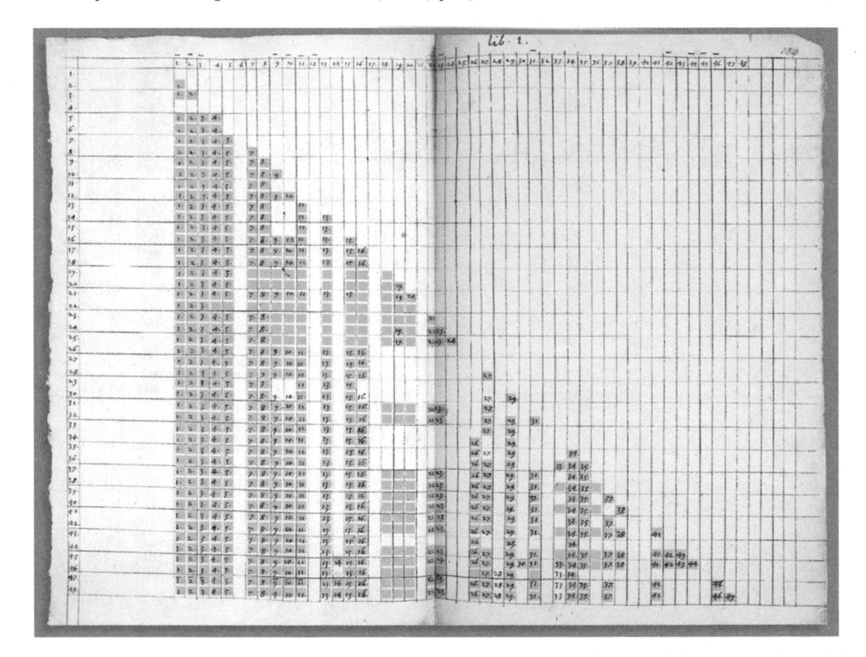

Studying the empirical metamathematics of Euclid seems (to me) like an obvious thing to do, and it's good to know I'm not the first one doing it. And actually I'm now wondering if someone actually already did it not "just" 400 years ago, but perhaps 2000 (or more) years ago...

Excerpt from

A NEW KIND OF SCIENCE (2002)

Chapter 12: The Principle of Computational Equivalence

)
a
∨ a)
∨ a
∨ a)
¬ b
∨ a
b ∨ a)
∧ a)
a) ∨ a
a ∨ b)
b ∨ a)
a) ∧ a
a ∨ b
(b ∨ a)
(b ∨ b)
(a ∨ a)
∨ b) ∧ a
¬ a ∧ ¬ a
¬ b ∨ ¬ a
= a ∧ (a ∧ a)
= a ∧ a ∨ a
(a ∨ a) ∨ a
= a ∨ a ∧ a
= a ∨ (a ∨ a)
= a ∧ (a ∨ b)
= a ∨ a ∧ b

2 a = a ∨ a
a ∨ a = ¬ ¬ a
a = a ∧ (a ∨ a)
7 a = a ∨ a ∧ b
a = b ∧ a ∨ a
¬ ¬ a = (a ∧ a) ∧ a
¬ ¬ a = a ∨ a ∧ a
¬ ¬ a = a ∧ b ∨ a
9 ¬ a ∨ a = ¬ b ∨ b
a ∧ ¬ b = ¬ b ∧ a
a ∧ a = (a ∧ a) ∧ a
a ∧ a = a ∧ (a ∨ a)
a ∧ a = a ∨ a ∧ a
a ∧ a = a ∨ a ∧ b
a ∧ a = a ∧ b ∨ a
a ∧ a = b ∧ a ∨ a
a ∨ b = (a ∨ a) ∨ b
a ∧ b = (a ∧ b) ∧ b
a ∧ b = (b ∧ a) ∧ a
a ∧ b = (b ∧ a) ∧ b
a ∧ b = b ∧ (b ∧ a)
¬ (a ∧ a) = ¬ a ∨ ¬ a
¬ a ∧ ¬ a = ¬ a ∨ ¬ a
(a ∧ a) ∧ a = a ∧ (a ∨ a)
a ∧ (a ∨ a) = a ∧ a ∨ a
(a ∧ a) ∧ a = a ∨ a ∧ a
(a ∧ a) ∧ a = a ∨ (a ∨ a)
a ∨ a ∧ a = a ∨ (a ∨ a)
(a ∨ a) ∨ a = a ∧ (a ∨ b)
a ∧ (a ∨ a) = a ∨ a ∧ b

a ∧ a = a ∨ a
¬ a = ¬ (a ∧ a)
a = a ∧ a ∨ a
a = (a ∨ b) ∧ a
¬ a = ¬ a ∧ ¬ a
¬ ¬ a = (a ∨ a) ∧ a
¬ ¬ a = a ∨ (a ∨ a)
¬ ¬ a = a ∨ b ∧ a
a ∨ ¬ a = ¬ b ∨ b
¬ a ∧ b = b ∧ ¬ a
a ∨ a = (a ∧ a) ∧ a
a ∨ a = a ∧ (a ∨ a)
a ∨ a = a ∨ a ∧ a
a ∨ a = a ∨ a ∧ b
a ∨ a = a ∧ b ∨ a
a ∨ a = b ∧ a ∨ a
a ∨ b = a ∨ (a ∨ b)
a ∧ b = a ∧ (b ∧ b)
a ∧ b = b ∧ (a ∧ a)
a ∧ b = b ∧ (a ∧ b)
a ∨ b = b ∧ b ∨ a
¬ (a ∨ a) = ¬ a ∨ ¬ a
¬ a ∧ ¬ b = ¬ b ∧ ¬ a
(a ∨ a) ∧ a = a ∧ (a ∨ a)
(a ∧ a) ∧ a = (a ∨ a) ∨ a
(a ∨ a) ∧ a = a ∨ a ∧ a
(a ∨ a) ∧ a = a ∨ (a ∨ a)
(a ∧ a) ∧ a = a ∧ (a ∨ b)
a ∨ a ∧ a = a ∧ (a ∨ b)
a ∧ a ∨ a = a ∨ a ∧ b

3 a ∧ b = b ∧ a
¬ a = ¬ (a ∨ a)
a = (a ∨ a) ∨ a
a = a ∧ (b ∨ a)
¬ a = ¬ a ∨ ¬ a
¬ ¬ a = a ∧ (a ∧ a)
¬ ¬ a = a ∧ (a ∨ b)
8 ¬ a ∧ a = ¬ b ∧ b
¬ a ∨ a = b ∨ ¬ b
a ∨ ¬ b = ¬ b ∨ a
a ∧ a = (a ∨ a) ∧ a
a ∧ a = a ∧ a ∨ a
a ∧ a = a ∨ (a ∨ a)
a ∧ a = (a ∨ b) ∧ a
a ∧ a = a ∨ b ∧ a
a ∧ b = (a ∧ a) ∧ b
a ∧ b = (a ∧ b) ∧ a
a ∧ b = a ∧ (b ∨ b)
a ∧ b = b ∧ (a ∨ a)
a ∨ b = (b ∨ a) ∨ b
a ∨ b = (b ∨ b) ∨ a
10 ¬ (a ∨ b) = ¬ a ∧ ¬ b
¬ a ∨ ¬ b = ¬ b ∨ ¬ a
a ∧ (a ∧ a) = a ∧ (a ∨ a)
(a ∨ a) ∧ a = (a ∨ a) ∨ a
a ∧ (a ∧ a) = a ∨ a ∧ a
a ∧ (a ∧ a) = a ∨ (a ∨ a)
(a ∨ a) ∧ a = a ∧ (a ∨ b)
a ∨ (a ∨ a) = a ∧ (a ∨ b)
(a ∨ a) ∨ a = a ∨ a ∧ b

4 a ∨ b = b ∨ a
a = (a ∧ a) ∧ a
a = a ∨ a ∧ a
a = a ∧ b ∨ a
¬ a ∧ a = a ∧ ¬ a
¬ ¬ a = a ∧ (a ∨ a)
¬ ¬ a = a ∨ a ∧ b
a ∧ ¬ a = ¬ b ∧ b
a ∨ ¬ a = b ∨ ¬ b
¬ a ∨ b = b ∨ ¬ a
a ∨ a = (a ∨ a) ∧ a
a ∨ a = a ∧ a ∨ a
a ∨ a = a ∨ (a ∨ a)
a ∨ a = (a ∨ b) ∧ a
a ∨ a = a ∨ b ∧ a
a ∧ b = (a ∨ a) ∧ b
a ∧ b = a ∧ (b ∧ a)
a ∨ b = (a ∨ b) ∨ b
a ∨ b = (b ∨ a) ∨ a
a ∨ b = b ∨ (a ∨ b)
a ∨ b = b ∨ (b ∨ a)
¬ (a ∧ b) = ¬ a ∨ ¬ b
(a ∧ a) ∧ a = (a ∨ a) ∧ a
(a ∧ a) ∧ a = a ∧ a ∨ a
a ∧ (a ∧ a) = (a ∨ a) ∨ a
a ∧ (a ∨ a) = a ∨ a ∧ a
a ∧ (a ∨ a) = a ∨ (a ∨ a)
a ∧ (a ∧ a) = a ∧ (a ∨ b)
(a ∧ a) ∧ a = a ∨ a ∧ b
a ∨ a ∧ a = a ∨ a ∧ b

6 a = ¬ ¬ a
a = (a ∨ a) ∧ a
a = a ∨ (a ∨ a)
a = a ∨ b ∧ a
¬ a ∨ a = a ∨ ¬ a
¬ ¬ a = a ∧ a ∨ a
¬ ¬ a = (a ∨ b) ∧ a
¬ a ∧ a = b ∧ ¬ b
¬ ¬ a = (b ∨ a) ∧ a
¬ (a ∧ b) = ¬ (b ∧ a)
a ∧ a = a ∧ (a ∧ a)
a ∧ a = (a ∨ a) ∨ a
a ∧ a = a ∧ (a ∨ b)
a ∧ a = a ∧ (b ∨ a)
a ∧ a = (b ∨ a) ∧ a
a ∧ b = a ∧ (a ∧ b)
a ∨ b = (a ∨ b) ∨ a
a ∨ b = a ∨ b ∧ b
a ∨ b = b ∨ a ∧ a
a ∧ b = (b ∧ b) ∧ a
¬ (a ∧ a) = ¬ a ∧ ¬ a
¬ (a ∨ b) = ¬ b ∧ ¬ a
(a ∧ a) ∧ a = a ∧ (a ∧ a)
(a ∨ a) ∧ a = a ∧ a ∨ a
a ∧ (a ∨ a) = (a ∨ a) ∨ a
a ∧ a ∨ a = a ∨ a ∧ a
a ∧ a ∨ a = a ∨ (a ∨ a)
a ∧ (a ∨ a) = a ∧ (a ∨ b)
(a ∨ a) ∧ a = a ∨ a ∧ b
a ∨ (a ∨ a) = a ∨ a ∧ b

⋮ 50 lines

= b ∧ (b ∨ a)
) = (b ∨ b) ∨ a
b = (b ∧ b) ∧ b
b = b ∧ (b ∨ b)
∧ b = b ∨ b ∧ b
∧ b = b ∨ b ∧ c
∧ b = b ∧ c ∨ b
∧ b = c ∧ b ∨ b
∧ c = a ∧ (c ∧ b)
∨ c = a ∨ (c ∨ b)

(a ∨ b) ∨ b = b ∧ b ∨ a
(a ∨ b) ∧ b = b ∨ b ∧ a
a ∧ b ∨ b = (b ∧ b) ∧ b
a ∧ b ∨ b = b ∧ (b ∨ b)
a ∧ b ∨ b = b ∨ b ∧ b
a ∧ b ∨ b = b ∨ b ∧ c
a ∧ b ∨ b = b ∧ c ∨ b
a ∧ b ∨ b = c ∧ b ∨ b
a ∧ (b ∧ c) = a ∧ (c ∧ b)
a ∨ (b ∨ c) = a ∨ (c ∨ b)

a ∨ b ∧ b = b ∧ b ∨ a
a ∧ b ∨ b = b ∨ b ∧ a
(a ∨ b) ∧ b = (b ∨ b) ∧ b
(a ∨ b) ∧ b = b ∧ b ∨ b
(a ∨ b) ∧ b = b ∨ (b ∨ b)
(a ∨ b) ∧ b = (b ∨ c) ∧ b
(a ∨ b) ∧ b = b ∨ c ∧ b
11 (a ∧ b) ∧ c = a ∧ (b ∧ c)
a ∧ (b ∨ c) = a ∧ (c ∨ b)
(a ∧ b) ∧ c = (b ∧ a) ∧ c

a ∨ (b ∨ b) = b ∧ b ∨ a
(a ∨ b) ∨ b = b ∨ (b ∨ a)
a ∧ b ∨ b = (b ∨ b) ∧ b
a ∧ b ∨ b = b ∧ b ∨ b
a ∧ b ∨ b = b ∨ (b ∨ b)
a ∧ b ∨ b = (b ∨ c) ∧ b
a ∧ b ∨ b = b ∨ c ∧ b
12 (a ∨ b) ∨ c = a ∨ (b ∨ c)
(a ∨ b) ∨ c = (a ∨ c) ∨ b
a ∧ (b ∧ c) = (b ∧ a) ∧ c

(a ∨ b) ∨ b = (b ∨ b) ∨ a
a ∨ b ∧ b = b ∨ (b ∨ a)
(a ∨ b) ∧ b = b ∧ (b ∧ b)
(a ∨ b) ∧ b = (b ∨ b) ∨ b
(a ∨ b) ∧ b = b ∧ (b ∨ c)
(a ∨ b) ∧ b = b ∧ (c ∨ b)
(a ∨ b) ∧ b = (c ∨ b) ∧ b
(a ∧ b) ∧ c = (a ∧ c) ∧ b
a ∨ (b ∨ c) = (a ∨ c) ∨ b
(a ∨ b) ∧ c = (b ∨ a) ∧ c

a ∨ b ∧ b = (b ∨ b) ∨ a
a ∨ (b ∨ b) = b ∨ (b ∨ a)
a ∧ b ∨ b = b ∧ (b ∧ b)
a ∧ b ∨ b = (b ∨ b) ∨ b
a ∧ b ∨ b = b ∧ (b ∨ c)
a ∧ b ∨ b = b ∧ (c ∨ b)
a ∧ b ∨ b = (c ∨ b) ∧ b
a ∧ (b ∧ c) = (a ∧ c) ∧ b
a ∨ b ∧ c = a ∨ c ∧ b
(a ∧ b) ∧ c = b ∧ (a ∧ c)

12

12.9 Implications for Mathematics and Its Foundations

Much of what I have done in this book has been motivated by trying to understand phenomena in nature. But the ideas that I have developed are general enough that they do not apply just to nature. And indeed in this section what I will do is to show that they can also be used to provide important new insights on fundamental issues in mathematics.

At some rather abstract level one can immediately recognize a basic similarity between nature and mathematics: for in nature one knows that fairly simple underlying laws somehow lead to the rich and complex behavior we see, while in mathematics the whole field is in a sense based on the notion that fairly simple axioms like those on the facing page can lead to all sorts of rich and complex results.

So where does this similarity come from? At first one might think that it must be a consequence of nature somehow intrinsically following mathematics. For certainly early in its history mathematics was specifically set up to capture certain simple aspects of nature.

But one of the starting points for the science in this book is that when it comes to more complex behavior mathematics has never in fact done well at explaining most of what we see every day in nature.

Yet at some level there is still all sorts of complexity in mathematics. And indeed if one looks at a presentation of almost any piece of modern mathematics it will tend to seem quite complex. But the point is that this complexity typically has no obvious relationship to anything we see in nature. And in fact over the past century what has been done in mathematics has mostly taken increasing pains to distance itself from any particular correspondence with nature.

So this suggests that the overall similarity between mathematics and nature must have a deeper origin. And what I believe is that in the end it is just another consequence of the very general Principle of Computational Equivalence that I discuss in this chapter.

For both mathematics and nature involve processes that can be thought of as computations. And then the point is that all these computations follow the Principle of Computational Equivalence, so

$a \wedge b = b \wedge a$
$a \vee b = b \vee a$
$a \wedge (b \vee \neg b) = a$
$a \vee (b \wedge \neg b) = a$
$a \wedge (b \vee c) = (a \wedge b) \vee (a \wedge c)$
$a \vee (b \wedge c) = (a \vee b) \wedge (a \vee c)$

basic logic (standard axioms)

$a \vee b = b \vee a$
$a \vee (b \vee c) = (a \vee b) \vee c$
$\neg(\neg a \vee b) \vee \neg(\neg a \vee \neg b) = a$

basic logic (Huntington axioms)

$a \vee b = b \vee a$
$a \vee (b \vee c) = (a \vee b) \vee c$
$\neg(\neg(a \vee b) \vee \neg(a \vee \neg b)) = a$

basic logic (Robbins axioms)

$(a \barwedge a) \barwedge (a \barwedge a) = a$
$a \barwedge (b \barwedge (b \barwedge b)) = a \barwedge a$
$(a \barwedge (b \barwedge c)) \barwedge (a \barwedge (b \barwedge c)) =$
$((b \barwedge b) \barwedge a) \barwedge ((c \barwedge c) \barwedge a)$

basic logic (Sheffer axioms)

$(a \barwedge a) \barwedge (a \barwedge b) = a$
$a \barwedge (a \barwedge b) = a \barwedge (b \barwedge b)$
$a \barwedge (a \barwedge (b \barwedge c)) = b \barwedge (b \barwedge (a \barwedge c))$

basic logic (shorter axioms)

$((a \barwedge b) \barwedge c) \barwedge (a \barwedge ((a \barwedge c) \barwedge a)) = c$

basic logic (shortest axioms)

basic logic, $x\_ \wedge y\_ \to x\_$, $x\_ \to \forall_{y\_}\, x\_$, $x\_ \to x\_ \wedge \#\, \&$, and ...

$\forall_{a\_}\, (b\_ \Rightarrow c\_) \Rightarrow (\forall_{a\_}\, b\_ \Rightarrow \forall_{a\_}\, c\_)$
$a\_ \Rightarrow \forall_{b\_}\, a\_$ /; FreeQ[a, b]
$\exists_{a\_}\, a\_ == b\_$ /; FreeQ[b, a]
$a\_ == b\_ \Rightarrow (c\_ \Rightarrow d\_)$ /; FreeQ[c, $\forall\_$ _] &&
MatchQ[d, c /. a → a | b]

predicate logic

predicate logic and ...

$0 \neq \Delta a$
$\Delta a == \Delta b \Rightarrow a == b$
$a + 0 == a$
$a + \Delta b == \Delta (a + b)$
$a \times 0 == 0$
$a \times \Delta b == (a \times b) + a$
$a \neq 0 \Rightarrow \exists_b\, a == \Delta b$

reduced arithmetic (Robinson axioms)

$0 \neq \Delta a$
$\Delta a == \Delta b \Rightarrow a == b$
$a + 0 == a$
$a + \Delta b == \Delta (a + b)$
$a \times 0 == 0$
$a \times \Delta b == (a \times b) + a$
$(\alpha\_ /_{a \to 0} \wedge \forall_b\, (\alpha\_ /_{a \to b} \Rightarrow \alpha\_ /_{a \to \Delta b})) \Rightarrow$
$\forall_b\, \alpha\_ /_{a \to b}$ /; FreeQ[α, b]

arithmetic (Peano axioms)

$a \cdot (b \cdot c) == (a \cdot b) \cdot c$

semigroup theory

$a \cdot (b \cdot c) == (a \cdot b) \cdot c$
$a \cdot 1 == a$
$1 \cdot a == a$

monoid theory

$a \cdot (b \cdot c) == (a \cdot b) \cdot c$
$a \cdot 1 == a$
$a \cdot \bar{a} == 1$

group theory (standard axioms)

$a \,\bar{\circ}\, ((((a \,\bar{\circ}\, a) \,\bar{\circ}\, b) \,\bar{\circ}\, c) \,\bar{\circ}\, (((a \,\bar{\circ}\, a) \,\bar{\circ}\, a) \,\bar{\circ}\, c)) == b$

group theory (shorter axioms)

$a \cdot \overline{b \cdot (((c \cdot \bar{c}) \cdot \overline{d \cdot b})} \cdot a) == d$

group theory (shorter axioms)

$a \cdot (b \cdot c) == (a \cdot b) \cdot c$
$a \cdot 1 == a$
$a \cdot \bar{a} == 1$
$a \cdot b == b \cdot a$

commutative group theory (standard axioms)

$a \,\bar{\circ}\, (b \,\bar{\circ}\, (c \,\bar{\circ}\, (a \,\bar{\circ}\, b))) == c$

commutative group theory (shorter axioms)

$((a \cdot b) \cdot c) \cdot \overline{a \cdot c} == b$

commutative group theory (shorter axioms)

$a \oplus (b \oplus c) == (a \oplus b) \oplus c$
$a \oplus 0 == a$
$a \oplus \bar{a} == 0$
$a \oplus b == b \oplus a$
$a \otimes (b \otimes c) == (a \otimes b) \otimes c$
$a \otimes (b \oplus c) == (a \otimes b) \oplus (a \otimes c)$
$a \otimes b == b \otimes a$

ring theory

$a \oplus (b \oplus c) == (a \oplus b) \oplus c$
$a \oplus 0 == a$
$a \oplus \bar{a} == 0$
$a \oplus b == b \oplus a$
$a \otimes (b \otimes c) == (a \otimes b) \otimes c$
$a \otimes (b \oplus c) == (a \otimes b) \oplus (a \otimes c)$
$a \otimes b == b \otimes a$
$a \otimes 1 == a$
$a \neq 0 \Rightarrow a \otimes a^{-1} == 1$
$0 \neq 1$

field theory

$a + (b + c) == (a + b) + c$	$a \neq 0 \Rightarrow a \times a^{-1} == 1$
$a + 0 == a$	$(a > b \wedge b > c) \Rightarrow a > c$
$a + (-a) == 0$	$a > b \Rightarrow a \neq b$
$a + b == b + a$	$a > b \vee a == b \vee b > a$
$a \times (b \times c) == (a \times b) \times c$	$a > b \Rightarrow a + c > b + c$
$a \times (b + c) == (a \times b) + (a \times c)$	$(a > b \wedge c > 0) \Rightarrow$
$a \times b == b \times a$	$a \times c > b \times c$
$a \times 1 == a$	$1 > 0$

$(\exists_a\, \alpha\_ \wedge \exists_b\, \forall_a\, (\alpha\_ \Rightarrow a > b)) \Rightarrow$
$\exists_b\, \forall_c\, (c > b \Leftrightarrow \exists_a\, (\alpha\_ \wedge c > a))$ /; FreeQ[α, c | b]

real algebra (Tarski axioms)

Axiom systems for traditional mathematics. It is from the axiom systems on this page and the next that most of the millions of theorems in the literature of mathematics have ultimately been derived. Note that in several cases axiom systems are given here in much shorter forms than in standard mathematics textbooks. (See also the definitions on the next page.)

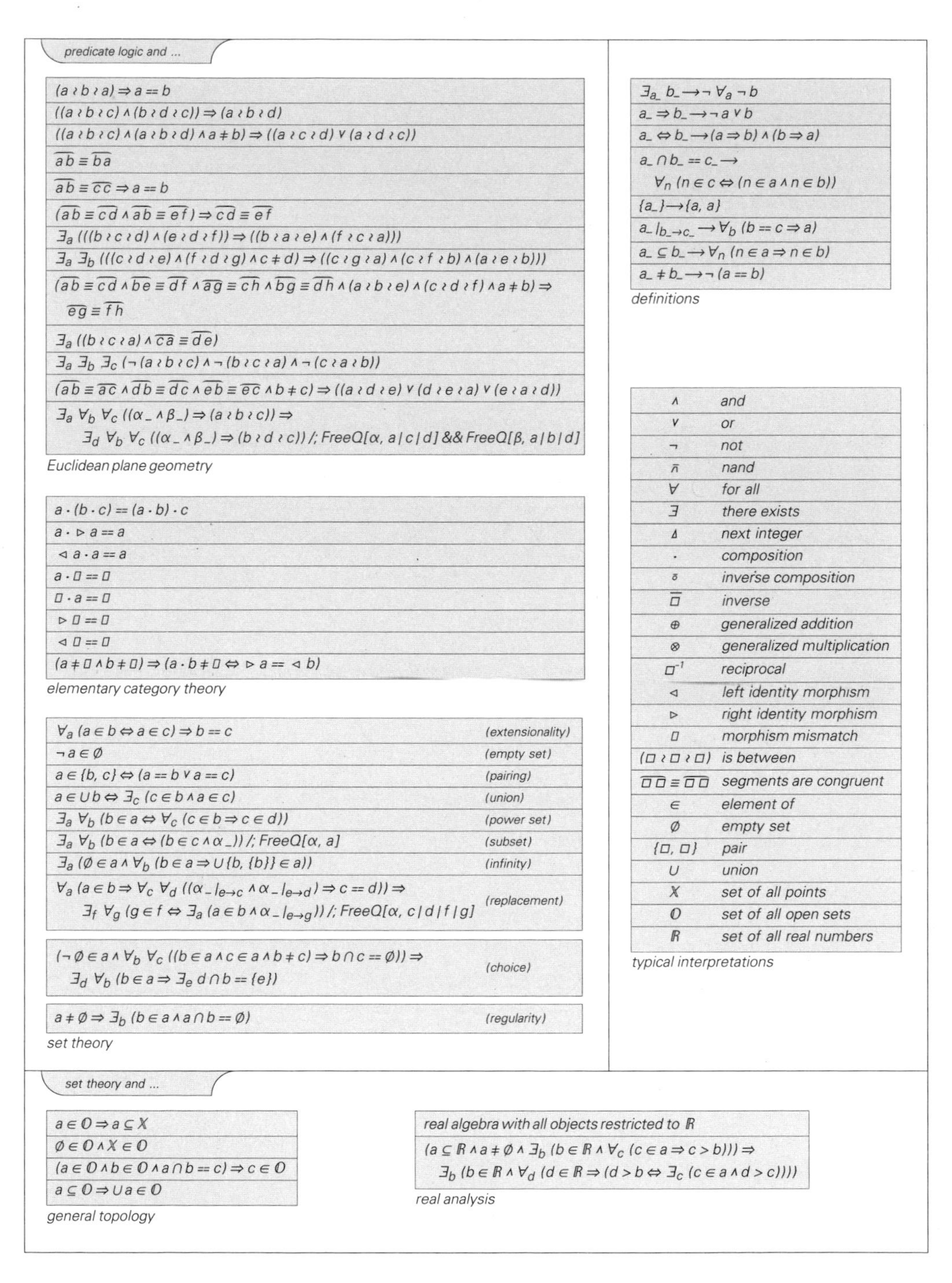

Further axiom systems for traditional mathematics. The typical interpretations are relevant for applications, though not for formal derivation of theorems. The last two axioms listed for set theory are usually considered optional.

that they ultimately tend to be equivalent in their computational sophistication—and thus show all sorts of similar phenomena.

And what we will see in this section is while some of these phenomena correspond to known features of mathematics—such as Gödel's Theorem—many have never successfully been recognized.

But just what basic processes are involved in mathematics?

Ever since antiquity mathematics has almost defined itself as being concerned with finding theorems and giving their proofs. And in any particular branch of mathematics a proof consists of a sequence of steps ultimately based on axioms like those of the previous two pages.

The picture below gives a simple example of how this works in basic logic. At the top right are axioms specifying certain fundamental equivalences between logic expressions. A proof of the equivalence $p \barwedge q = q \barwedge p$ between logic expressions is then formed by applying these axioms in the particular sequence shown.

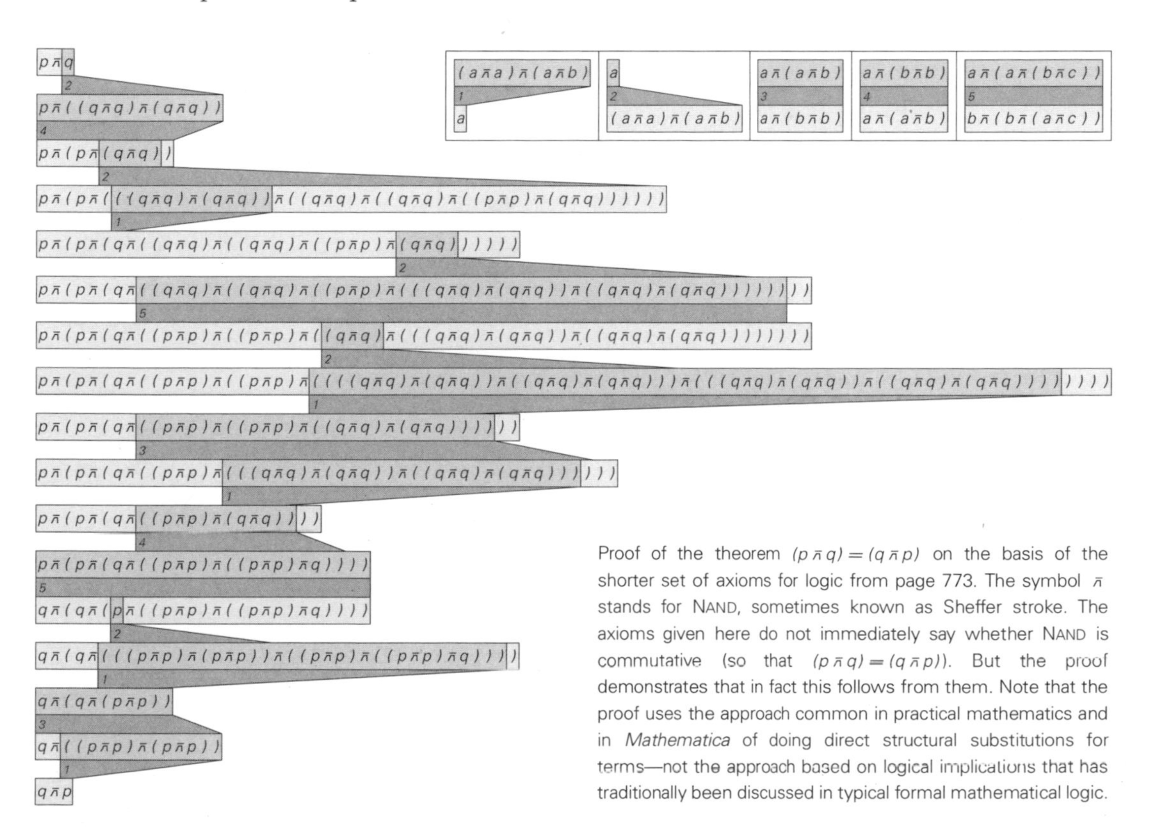

Proof of the theorem $(p \barwedge q) = (q \barwedge p)$ on the basis of the shorter set of axioms for logic from page 773. The symbol $\barwedge$ stands for NAND, sometimes known as Sheffer stroke. The axioms given here do not immediately say whether NAND is commutative (so that $(p \barwedge q) = (q \barwedge p)$). But the proof demonstrates that in fact this follows from them. Note that the proof uses the approach common in practical mathematics and in *Mathematica* of doing direct structural substitutions for terms—not the approach based on logical implications that has traditionally been discussed in typical formal mathematical logic.

In most kinds of mathematics there are all sorts of additional details, particularly about how to determine which parts of one or more previous expressions actually get used at each step in a proof. But much as in our study of systems in nature, one can try to capture the essential features of what can happen by using a simple idealized model.

And so for example one can imagine representing a step in a proof just by a string of simple elements such as black and white squares. And one can then consider the axioms of a system as defining possible transformations from one sequence of these elements to another—just like the rules in the multiway systems we discussed in Chapter 5.

The pictures below show how proofs of theorems work with this setup. Each theorem defines a connection between strings, and proving the theorem consists in finding a series of transformations—each associated with an axiom—that lead from one string to another.

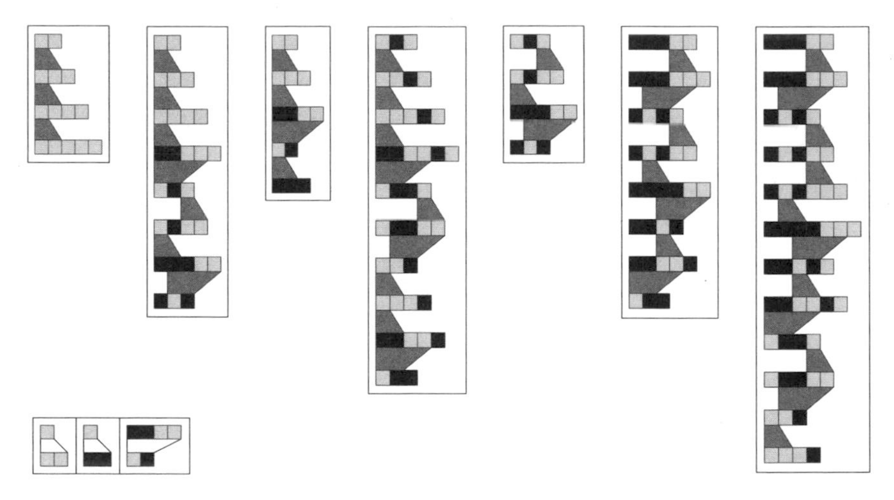

Simple idealizations of proofs in mathematics. The rules on the left in effect correspond to axioms that specify valid transformations between strings of black and white elements. The proofs above then show how one string—say —can be transformed into another—say —by using the axioms. Typically there are many different proofs that can be given of a particular theorem; here in each case the ones shown are examples of the shortest possible proofs. The system shown is an example of a general substitution system of the kind discussed on page 497. Note that the fifth theorem → occurs in effect as a lemma in the second theorem → .

But just as in the multiway systems in Chapter 5 one can also consider an explicit process of evolution, in which one starts from a particular string, then at each successive step one applies all possible transformations, so that in the end one builds up a whole network of connections between strings, as in the pictures below.

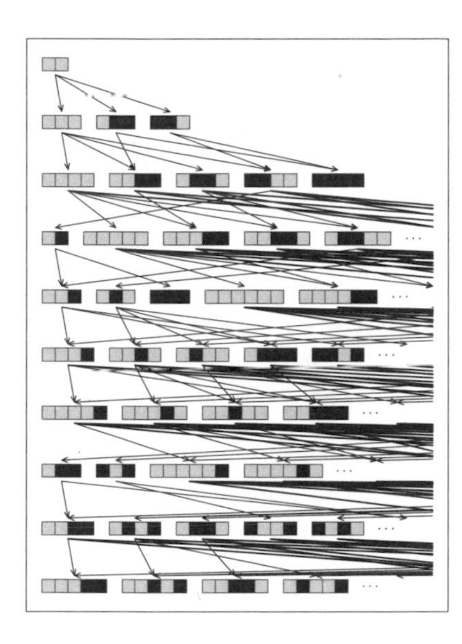
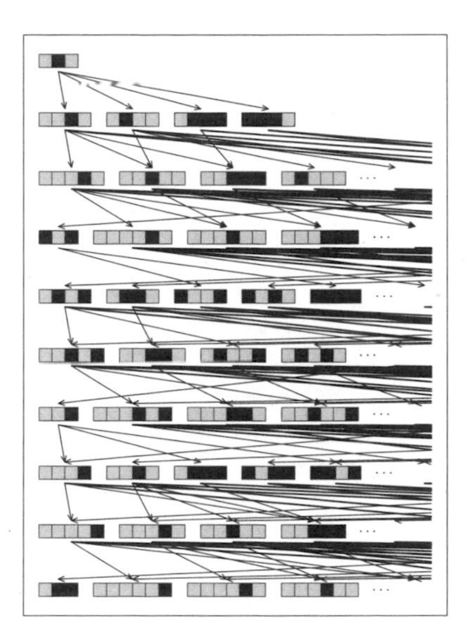

The result of applying the same transformations as on the facing page—but in all possible ways, corresponding to the evolution of a multiway system that represents all possible theorems that can be derived from the axioms. With the axioms used here, the total number of strings grows by a factor of roughly 1.7 at each step; on the last steps shown there are altogether 237 and 973 strings respectively.

In a sense such a network can then be thought of as representing the whole field of mathematics that can be derived from whatever set of axioms one is using—with every connection between strings corresponding to a theorem, and every possible path to a proof.

But can networks like the ones above really reflect mathematics as it is actually practiced? For certainly the usual axioms in every traditional area of mathematics are significantly more complicated than any of the multiway system rules used above.

But just like in so many other cases in this book, it seems that even systems whose underlying rules are remarkably simple are already able to capture many of the essential features of mathematics.

An obvious observation in mathematics is that proofs can be difficult to do. One might at first assume that any theorem that is easy

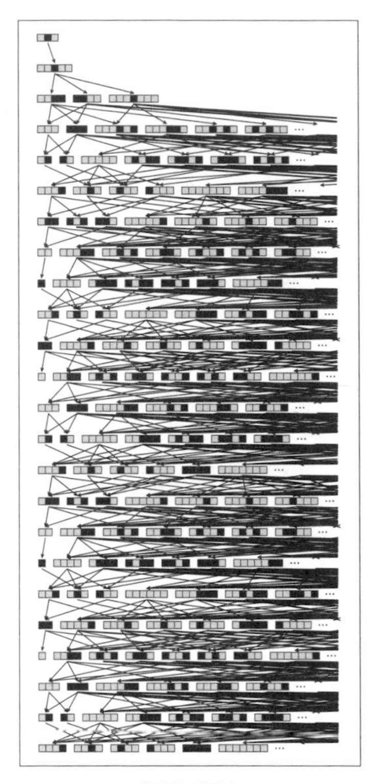

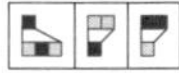

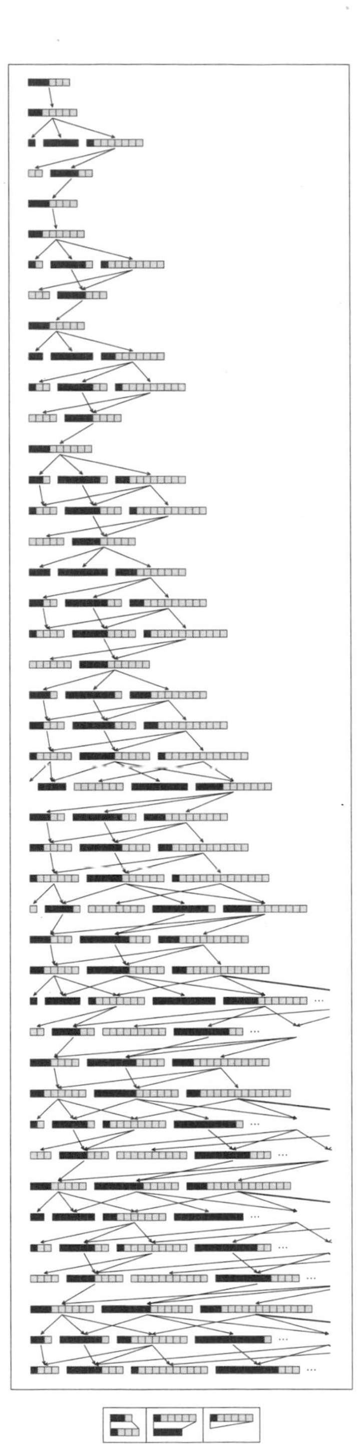

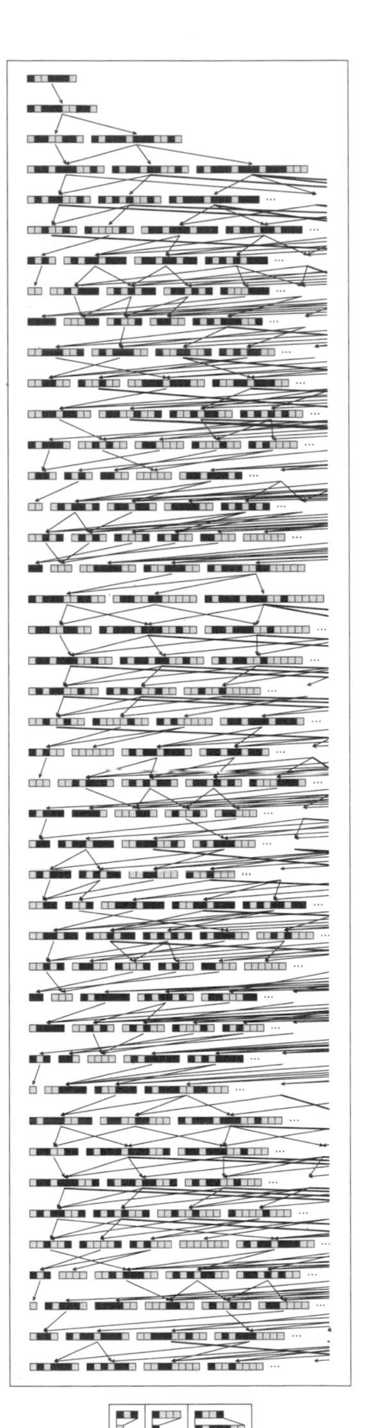

Three examples of multiway systems that show the analog of long proofs. In each case a string consisting of a single white element is eventually generated—but this takes respectively 12, 28 and 34 steps to happen. The first multiway system actually generates all strings in the end (not least since it yields the lemmas ■ → ■■ and ■ → □)—and in fact strings of length $n > 2$ appear after at most $2n + 7$ steps. The second multiway system generates only the $n + 1$ strings where black comes before white—and all of these strings appear after at most $7n$ steps. The third multiway system generates a complicated collection of strings; the numbers of lengths up to 8 are 1, 2, 4, 8, 14, 22, 34, 45. All the strings generated have an even number of black elements.

to state will also be easy to prove. But experience suggests that this is far from correct. And indeed there are all sorts of well-known examples—such as Fermat's Last Theorem and the Four-Color Theorem—in which a theorem that is easy to state seems to require a proof that is immensely long.

So is there an analog of this in multiway systems? It turns out that often there is, and it is that even though a string may be short it may nevertheless take a great many steps to reach.

If the rules for a multiway system always increase string length then it is inevitable that any given string that is ever going to be generated must appear after only a limited number of steps. But if the rules can both increase and decrease string length the story is quite different, as the picture on the facing page illustrates. And often one finds that even a short string can take a rather large number of steps to produce.

But are all these steps really necessary? Or is it just that the rule one has used is somehow inefficient, and there are other rules that generate the short strings much more quickly?

Certainly one can take the rules for any multiway system and add transformations that immediately generate particular short strings. But the crucial point is that like so many other systems I have discussed in this book there are many multiway systems that I suspect are computationally irreducible—so that there is no way to shortcut their evolution, and no general way to generate their short strings quickly.

And what I believe is that essentially the same phenomenon operates in almost every area of mathematics. Just like in multiway systems, one can always add axioms to make it easier to prove particular theorems. But I suspect that ultimately there is almost always computational irreducibility, and this makes it essentially inevitable that there will be short theorems that only allow long proofs.

In the previous section we saw that computational irreducibility tends to make infinite questions undecidable. So for example the question of whether a particular string will ever be generated in the evolution of a multiway system—regardless of how long one waits—is in general undecidable. And similarly it can be undecidable whether any proof—regardless of length—exists for a specific result in a mathematical system with particular axioms.

So what are the implications of this?

Probably the most striking arise when one tries to apply traditional ideas of logic—and particularly notions of true and false.

The way I have set things up, one can find all the statements that can be proved true in a particular axiom system just by starting with an expression that represents "true" and then using the rules of the axiom system, as in the picture on the facing page.

In a multiway system, one can imagine identifying "true" with a string consisting of a single black element. And this would mean that every string in networks like the ones below should correspond to a statement that can be proved true in the axiom system used.

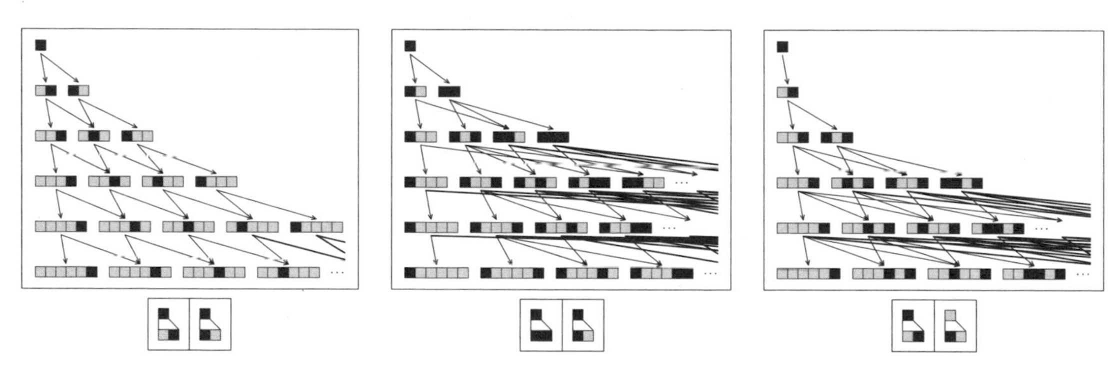

Multiway systems starting from a single black element that represents TRUE. All strings that appear can be thought of as statements that are true according to the axioms represented by the multiway system rules. One can take negation to be the operation that interchanges black and white. This then means that the first multiway system represents an inconsistent axiom system, since on step 2, both and its negation appear. The other two multiway systems are consistent, so that they never generate both a string and its negation. The third one, however, is incomplete, since for example it never generates either or its negation . The second one, however, is both complete and consistent: it generates all strings that begin with , but none that begin with .

But is this really reasonable? In traditional logic there is always an operation of negation which takes any true statement, and makes it into a false one, and vice versa. And in a multiway system, one possible way negation might work is just to reverse the colors of the elements in a string. But this then leads to a problem in the first picture above.

For the picture implies that both and its negation can be proved to be true statements. But this cannot be correct. And so what

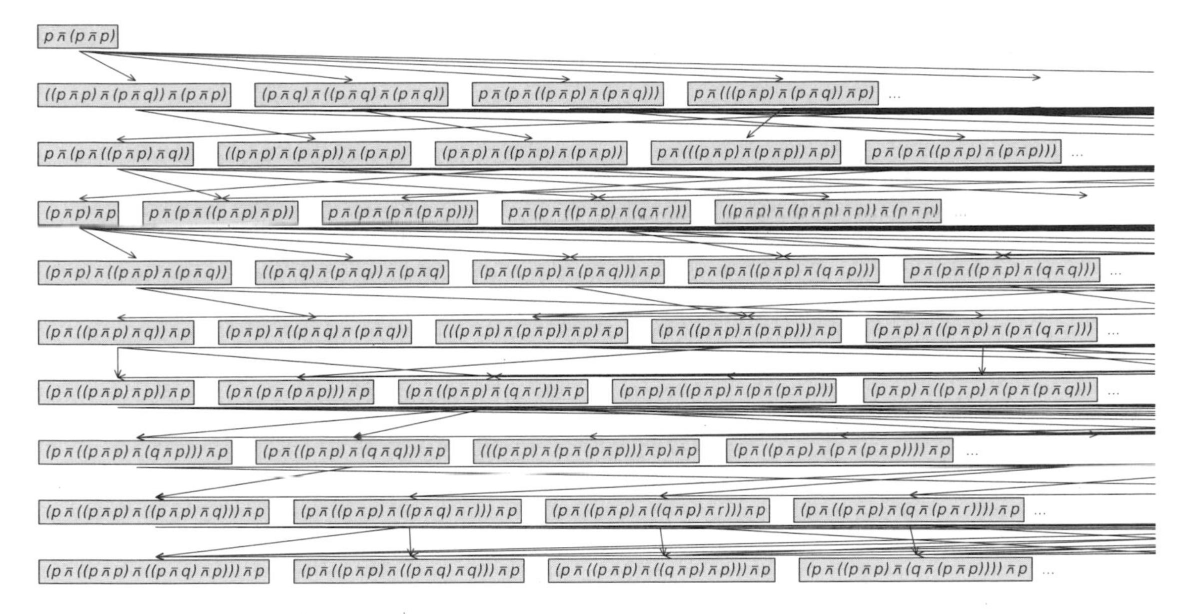

The network of statements that can be proved true using the axiom system for logic from page 775. *p ⊼ (p ⊼ p)* is the simplest representation for TRUE when logic is set up using the NAND operator ⊼. Each arrow indicates an equivalence established by applying a single axiom. On each row only statements that have not appeared before are given. The statements are sorted so that the simplest are first. Note that some fairly simple statements do not show up for at least several rows. The total number of statements on successive rows grows faster than exponentially; for the first few it is 1, 6, 91, 2180, 76138. If continued forever the network would eventually include all possible true statements (tautologies) of logic (see also page 818). Other simple axiom systems for logic like those on page 808 yield networks similar to the one shown.

this means is that with the setup used the underlying axiom system is inconsistent. So what about the other multiway systems on the facing page? At least with the strings one can see in the pictures there are no inconsistencies. But what about with longer strings? For the particular rules shown it is fairly easy to demonstrate that there are never inconsistencies. But in general it is not possible to do this, for after some given string has appeared, it can for example be undecidable whether the negation of that particular string ever appears.

So what about the axiom systems normally used in actual mathematics? None of those on pages 773 and 774 appear to be inconsistent. And what this means is that the set of statements that can be proved true will never overlap with the set that can be proved false.

But can every possible statement that one might expect to be true or false actually in the end be proved either true or false?

In the early 1900s it was widely believed that this would effectively be the case in all reasonable mathematical axiom systems. For at the time there seemed to be no limit to the power of mathematics, and no end to the theorems that could be proved.

But this all changed in 1931 when Gödel's Theorem showed that at least in any finitely-specified axiom system containing standard arithmetic there must inevitably be statements that cannot be proved either true or false using the rules of the axiom system.

This was a great shock to existing thinking about the foundations of mathematics. And indeed to this day Gödel's Theorem has continued to be widely regarded as a surprising and rather mysterious result.

But the discoveries in this book finally begin to make it seem inevitable and actually almost obvious. For it turns out that at some level it can be viewed as just yet another consequence of the very general Principle of Computational Equivalence.

So what is the analog of Gödel's Theorem for multiway systems? Given the setup on page 780 one can ask whether a particular multiway system is complete in the sense that for every possible string the system eventually generates either that string or its negation.

And one can see that in fact the third multiway system is incomplete, since by following its rules one can never for example generate either □□ or its negation ■■. But what if one extends the rules by adding more transformations, corresponding to more axioms? Can one always in the end make the system complete?

If one is not quite careful, one will generate too many strings, and inevitably get inconsistencies where both a string and its negation appear, as in the second picture on the facing page. But at least if one only has to worry about a limited number of steps, it is always possible to set things up so as to get a system that is both complete and consistent, as in the third picture on the facing page.

And in fact in the particular case shown on the facing page it is fairly straightforward to find rules that make the system always complete and consistent. But knowing how to do this requires having behavior that is in a sense simple enough that one can foresee every aspect of it.

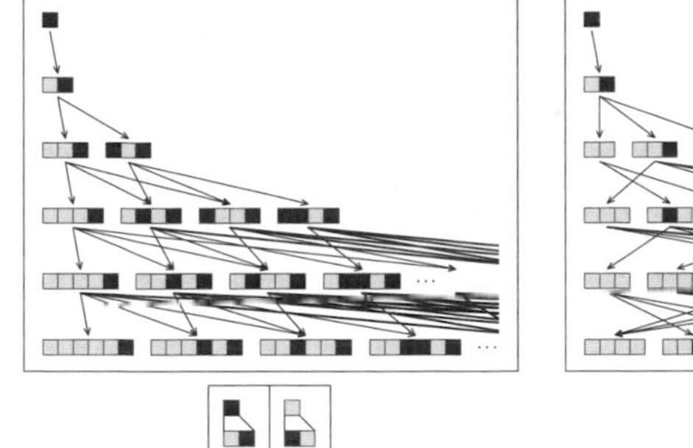

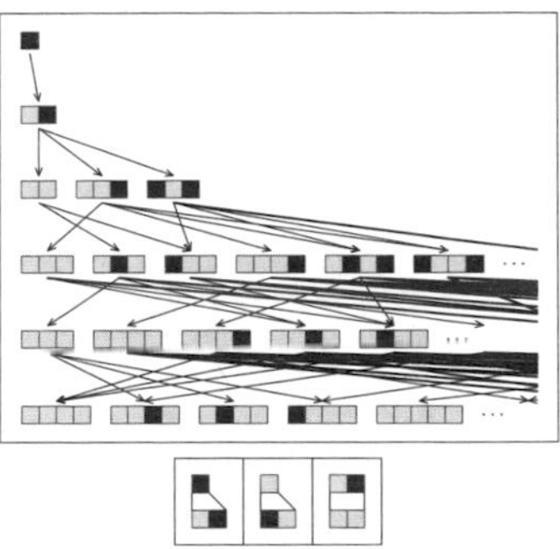

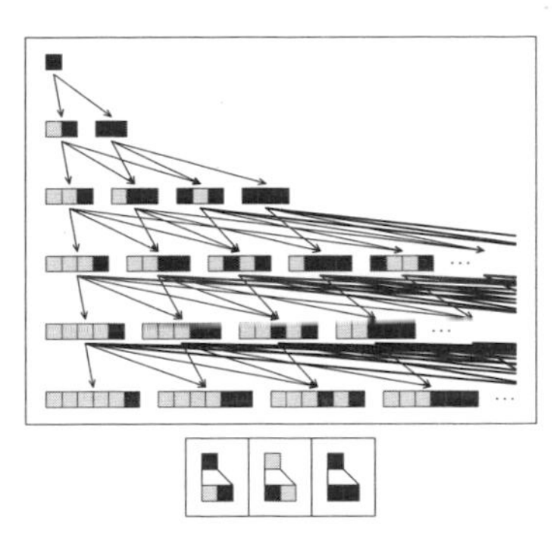

The effect of adding transformations to the rules for a multiway system. The first multiway system is incomplete, in the sense that for some strings, it generates neither the string nor its negation. The second multiway system yields more strings—but introduces inconsistency, since it can generate both and its negation . The third multiway system is however both complete and consistent: for every string it eventually generates either that string or its negation.

Yet if a system is computationally irreducible this will inevitably not be possible. For at any point the system will always in effect be able to do more things that one did not expect. And this means that in general one will not be able to construct a finite set of axioms that can be guaranteed to lead to ultimate completeness and consistency.

And in fact it turns out that as soon as the question of whether a particular string can ever be reached is undecidable it immediately follows that there must be either incompleteness or inconsistency. For to say that such a question is undecidable is to say that it cannot in general be answered by any procedure that is guaranteed to finish.

But if one had a system that was complete and consistent then it is easy to come up with such a procedure: one just runs the system until either one reaches the string one is looking for or one reaches its negation. For the completeness of the system guarantees that one must always reach one or the other, while its consistency implies that reaching one allows one to conclude that one will never reach the other.

So the result of this is that if the evolution of a multiway system is computationally irreducible—so that questions about its ultimate behavior are undecidable—the system cannot be both complete and consistent. And if one assumes consistency then it follows that there must be strings where neither the string nor its negation can be reached—corresponding to the fact that statements must exist that cannot be proved either true or false from a given set of axioms.

But what does it take to establish that such incompleteness will actually occur in a specific system?

The basic way to do it is to show that the system is universal.

But what exactly does universality mean for something like an axiom system? In effect what it means is that any question about the behavior of any other universal system can be encoded as a statement in the axiom system—and if the answer to the question can be established by watching the evolution of the other universal system for any finite number of steps then it must also be able to be established by giving a proof of finite length in the axiom system.

So what axiom systems in mathematics are then universal?

Basic logic is not, since at least in principle one can always determine the truth of any statement in this system by the finite—if perhaps exponentially long—procedure of trying all possible combinations of truth values for the variables that appear in it.

And essentially the same turns out to be the case for pure predicate logic, in which one just formally adds "for all" and "there exists" constructs. But as soon as one also puts in an abstract function or relation with more than one argument, one gets universality.

And indeed the basis for Gödel's Theorem is the result that the standard axioms for basic integer arithmetic support universality.

Set theory and several other standard axiom systems can readily be made to reproduce arithmetic, and are therefore also universal. And the same is true of group theory and other algebraic systems like ring theory.

If one puts enough constraints on the axioms one uses, one can eventually prevent universality—and in fact this happens for commutative group theory, and for the simplified versions of both real algebra and geometry on pages 773 and 774.

But of the axiom systems actually used in current mathematics research every single one is now known to be universal.

From page 773 we can see that many of these axiom systems can be stated in quite simple ways. And in the past it might have seemed hard to believe that systems this simple could ever be universal, and thus in a sense be able to emulate essentially any system.

But from the discoveries in this book this now seems almost inevitable. And indeed the Principle of Computational Equivalence implies that beyond some low threshold almost any axiom system should be expected to be universal.

So how does universality actually work in the case of arithmetic?

One approach is illustrated in the picture on the next page. The idea is to set up an arithmetic statement that can be proved true if the evolution of a cellular automaton from a given initial condition makes a given cell be a given color at a given step, and can be proved false if it does not.

By changing numbers in this arithmetic statement one can then in effect sample different aspects of the cellular automaton evolution. And with the cellular automaton being a universal one such as rule 110 this implies that the axioms of arithmetic can support universality.

Such universality then implies Gödel's Theorem and shows that there must exist statements about arithmetic that cannot ever be proved true or false from its normal axioms.

So what are some examples of such statements?

The original proof of Gödel's Theorem was based on considering the particular self-referential statement "this statement is unprovable".

At first it does not seem obvious that such a statement could ever be set up as a statement in arithmetic. But if it could then one can see that it would immediately follow that—as the statement says—it cannot be proved, since otherwise there would be an inconsistency.

And in fact the main technical difficulty in the original proof of Gödel's Theorem had to do with showing—by doing what amounted to establishing the universality of arithmetic—that the statement could indeed meaningfully be encoded as a statement purely in arithmetic.

But at least with the original encoding used, the statement would be astronomically long if written out in the notation of page 773. And from this result, one might imagine that unprovability would never be relevant in any practical situation in mathematics.

But does one really need to have such a complicated statement in order for it to be unprovable from the axioms of arithmetic?

$$(-3x_6+x_7+x_8)^2+(2^{1+x_3(1+x_1+2x_3)}x_2-2x_4-x_{10}+x_{11})^2+(-2x_8-x_9+x_{10}+x_{11})^2+(1-2^{(1+x_3)(x_1+2x_3)}+x_4+x_{12})^2+$$
$$(1-2^{x_1}+x_2+x_{13})^2+(1-2^{x_1}+x_5+x_{14})^2+(-x_4+2^{x_3}x_5+2^{x_1+2x_3}x_6+2^{x_1+x_3}x_{15}+x_{16})^2+(1-2^{x_3}+x_{15}+x_{17})^2+$$
$$(1-2^{x_3}+x_{16}+x_{18})^2+(-x_6-2x_7+x_9+x_{19})^2+(-(2+2^{x_6})^{x_6}+(1+2^{x_6})^{x_7}(1+2x_{20}+(1+2^{x_6})x_{21})+x_{22})^2+(1-(1+2^{x_6})^{x_7}+x_{22}+x_{23})^2+$$
$$(1-2^{x_6}+2x_{20}+x_{24})^2+(-(2+4^{x_6})^{2x_6}+(1+4^{x_6})^{x_7}(1+2x_{25}+(1+4^{x_6})x_{26})+x_{27})^2+(1-(1+4^{x_6})^{x_7}+x_{27}+x_{28})^2+$$
$$(1-4^{x_6}+2x_{25}+x_{29})^2+(-(2+2^{x_8})^{x_8}+(1+2^{x_8})^{x_6}(1+2x_{30}+(1+2^{x_8})x_{31})+x_{32})^2+(1-(1+2^{x_8})^{x_6}+x_{32}+x_{33})^2+$$
$$(1-2^{x_8}+2x_{30}+x_{34})^2+(-(2+2^{x_8})^{x_8}+(1+2^{x_8})^{2x_6}(1+2x_{35}+(1+2^{x_8})x_{36})+x_{37})^2+(1-(1+2^{x_8})^{2x_6}+x_{37}+x_{38})^2+$$
$$(1-2^{x_8}+2x_{35}+x_{39})^2+(-(2+2^{x_6})^{x_6}+(1+2^{x_6})^{x_9}(1+2x_{40}+(1+2^{x_6})x_{41})+x_{42})^2+(1-(1+2^{x_6})^{x_9}+x_{42}+x_{43})^2+$$
$$(1-2^{x_6}+2x_{40}+x_{44})^2+(-(2+4^{x_7})^{2x_7}+(1+4^{x_7})^{x_9}(1+2x_{45}+(1+4^{x_7})x_{46})+x_{47})^2+(1-(1+4^{x_7})^{x_9}+x_{47}+x_{48})^2+$$
$$(1-4^{x_7}+2x_{45}+x_{49})^2+(-(2+2^{x_{19}})^{x_{19}}+(1+2^{x_{19}})^{x_6}(1+2x_{50}+(1+2^{x_{19}})x_{51})+x_{52})^2+(1-(1+2^{x_{19}})^{x_6}+x_{52}+x_{53})^2+$$
$$(1-2^{x_{19}}+2x_{50}+x_{54})^2+(-(2+2^{x_{19}})^{x_{19}}+(1+2^{x_{19}})^{2x_7}(1+2x_{55}+(1+2^{x_{19}})x_{56})+x_{57})^2+(1-(1+2^{x_{19}})^{2x_7}+x_{57}+x_{58})^2+$$
$$(1-2^{x_{19}}+2x_{55}+x_{59})^2+(-(2+2^{x_9})^{x_9}+(1+2^{x_9})^{x_{10}}(1+2x_{60}+(1+2^{x_9})x_{61})+x_{62})^2+(1-(1+2^{x_9})^{x_{10}}+x_{62}+x_{63})^2+(1-2^{x_9}+2x_{60}+x_{64})^2+$$
$$(-(2+4^{x_8})^{2x_8}+(1+4^{x_8})^{x_{10}}(1+2x_{65}+(1+4^{x_8})x_{66})+x_{67})^2+(1-(1+4^{x_8})^{x_{10}}+x_{67}+x_{68})^2+(1-4^{x_8}+2x_{65}+x_{69})^2+$$
$$(-(2+2^{x_{11}})^{x_{11}}+(1+2^{x_{11}})^{x_9}(1+2x_{70}+(1+2^{x_{11}})x_{71})+x_{72})^2+(1-(1+2^{x_{11}})^{x_9}+x_{72}+x_{73})^2+(1-2^{x_{11}}+2x_{70}+x_{74})^2+$$
$$(-(2+2^{x_{11}})^{x_{11}}+(1+2^{x_{11}})^{2x_8}(1+2x_{75}+(1+2^{x_{11}})x_{76})+x_{77})^2+(1-(1+2^{x_{11}})^{2x_8}+x_{77}+x_{78})^2+(1-2^{x_{11}}+2x_{75}+x_{79})^2=0$$

x_1 (initial width)	1
x_2 (initial state)	1
x_3 (steps)	1
x_4 (evolution)	22
x_5	1
x_6	2
x_7	0
x_8	6
x_9	0
x_{10}	0
x_{11}	12
x_{12}	41
x_{13}	0
x_{14}	0
x_{15}	1
⋮	

x_1 (initial width)	1
x_2 (initial state)	1
x_3 (steps)	2
x_4 (evolution)	4508
x_5	1
x_6	140
x_7	8
x_8	412
x_9	0
x_{10}	0
x_{11}	824
x_{12}	28259
x_{13}	0
x_{14}	0
x_{15}	3
⋮	

x_1 (initial width)	1
x_2 (initial state)	1
x_3 (steps)	3
x_4 (evolution)	17177704
x_5	1
x_6	134200
x_7	2096
x_8	400504
x_9	32
x_{10}	32
x_{11}	801008
x_{12}	251257751
x_{13}	0
x_{14}	0
x_{15}	6
⋮	

x_1 (initial width)	1
x_2 (initial state)	1
x_3 (steps)	4
x_4 (evolution)	1105983545840
x_5	1
x_6	2160124112
x_7	8437888
x_8	6471934448
x_9	32768
x_{10}	32768
x_{11}	12943868896
x_{12}	34078388542991
x_{13}	0
x_{14}	0
x_{15}	15
⋮	

x_1 (initial width)	3
x_2 (initial state)	5
x_3 (steps)	4
x_4 (evolution)	1409438147512048
x_5	7
x_6	688202220464
x_7	940049184
x_8	2063666612208
x_9	805306880
x_{10}	805306880
x_{11}	4127333224416
x_{12}	34619358871451919
x_{13}	2
x_{14}	0
x_{15}	13
⋮	

$x_4 = 22 =$

0 1 0
1 1 0

$x_4 = 4508 =$

0 0 1 0 0
0 1 1 0 0
1 1 1 0 0

$x_4 = 17177704 =$

0 0 0 1 0 0 0
0 0 1 1 0 0 0
0 1 1 1 0 0 0
1 1 0 1 0 0 0

$x_4 = 1105983545840 =$

0 0 0 0 1 0 0 0 0
0 0 0 1 1 0 0 0 0
0 0 1 1 1 0 0 0 0
0 1 1 0 1 0 0 0 0
1 1 1 1 1 0 0 0 0

$x_4 = 1409438147512048 =$

0 0 0 0 1 0 1 0 0 0 0
0 0 0 1 1 1 1 0 0 0 0
0 0 1 1 0 0 1 0 0 0 0
0 1 1 1 0 1 1 0 0 0 0
1 1 0 1 1 1 1 0 0 0 0

Universality in arithmetic, illustrated by an integer equation whose solutions in effect emulate the rule 110 universal cellular automaton from Chapter 11. The equation has many solutions, but all of them satisfy the constraint that the variables x_1 through x_4 must encode possible initial conditions and evolution histories for rule 110. If one fills in fixed values for x_1, x_2 and x_3, then only one value for x_4 is ever possible—corresponding to the evolution history of rule 110 for x_3 steps starting from a width x_1 initial condition given by the digit sequence of x_2. In general any statement about the possible behavior of rule 110 can be encoded as a statement in arithmetic about solutions to the equation. So for example if one fills in values for x_1, x_2 and x_4, but not x_3, then the statement that the equation has no solution for any x_3 corresponds to a statement that rule 110 can never exhibit certain behavior, even after any number of steps. But the universality of rule 110 implies that such statements must in general be undecidable. So from this it follows that in at least some instances the axioms of arithmetic can never be used to give a finite proof of whether or not the statement is true. The construction shown here can be viewed as providing a simple proof of Gödel's Theorem on the existence of unprovable statements in arithmetic. Note that the equation shown is a so-called exponential Diophantine one, in which some variables appear in exponents. At the cost of considerably more complication—and using for example 2154 variables—it is possible to avoid this. The equation above can however already be viewed as capturing the essence of what is needed to demonstrate the general unsolvability of Diophantine equations and Hilbert's Tenth Problem.

Over the past seventy years a few simpler examples have been constructed—mostly with no obviously self-referential character.

But usually these examples have involved rather sophisticated and obscure mathematical constructs—most often functions that are somehow set up to grow extremely rapidly. Yet at least in principle there should be examples that can be constructed based just on statements that no solutions exist to particular integer equations.

If an integer equation such as $x^2 = y^3 + 12$ has a definite solution such as $x = 47$, $y = 13$ in terms of particular finite integers then this fact can certainly be proved using the axioms of arithmetic. For it takes only a finite calculation to check the solution, and this very calculation can always in effect be thought of as a proof.

But what if the equation has no solutions? To test this explicitly one would have to look at an infinite number of possible integers. But the point is that even so, there can still potentially be a finite mathematical proof that none of these integers will work.

And sometimes the proof may be straightforward—say being based on showing that one side of the equation is always odd while the other is always even. In other cases the proof may be more difficult—say being based on establishing some large maximum size for a solution, then checking all integers up to that size.

And the point is that in general there may in fact be absolutely no proof that can be given in terms of the normal axioms of arithmetic.

So how can one see this?

The picture on the facing page shows that one can construct an integer equation whose solutions represent the behavior of a system like a cellular automaton. And the way this works is that for example one variable in the equation gives the number of steps of evolution, while another gives the outcome after that number of steps.

So with this setup, one can specify the number of steps, then solve for the outcome after that number of steps. But what if for example one instead specifies an outcome, then tries to find a solution for the number of steps at which this outcome occurs?

If in general one was able to tell whether such a solution exists then it would mean that one could always answer the question of whether, say, a particular pattern would ever die out in the evolution of a given cellular automaton. But from the discussion of the previous section we know that this in general is undecidable.

So it follows that it must be undecidable whether a given integer equation of some particular general form has a solution. And from the arguments above this in turn implies that there must be specific integer equations that have no solutions but where this fact cannot be proved from the normal axioms of arithmetic.

So how ultimately can this happen?

At some level it is a consequence of the involvement of infinity. For at least in a universal system like arithmetic any question that is entirely finite can in the end always be answered by a finite procedure.

But what about questions that somehow ask, say, about infinite numbers of possible integers? To have a finite way to address questions like these is often in the end the main justification for setting up typical mathematical axiom systems in the first place.

For the point is that instead of handling objects like integers directly, axiom systems can just give abstract rules for manipulating statements about them. And within such statements one can refer, say, to infinite sets of integers just by a symbol like s.

And particularly over the past century there have been many successes in mathematics that can be attributed to this basic kind of approach. But the remarkable fact that follows from Gödel's Theorem is that whatever one does there will always be cases where the approach must ultimately fail. And it turns out that the reason for this is essentially the phenomenon of computational irreducibility.

For while simple infinite quantities like $1/0$ or the total number of integers can readily be summarized in finite ways—often just by using symbols like ∞ and $\aleph_0$—the same is not in general true of all infinite processes. And in particular if an infinite process is computationally irreducible then there cannot in general be any useful finite summary of what it does—since the existence of such a summary would imply computational reducibility.

So among other things this means that there will inevitably be questions that finite proofs based on axioms that operate within ordinary computational systems will never in general be able to answer.

And indeed with integer equations, as soon as one has a general equation that is universal, it typically follows that there will be specific instances in which the absence of solutions—or at least of solutions of some particular kind—can never be proved on the basis of the normal axioms of arithmetic.

For several decades it has been known that universal integer equations exist. But the examples that have actually been constructed are quite complicated—like the one on page 786—with the simplest involving 9 variables and an immense number of terms.

Yet from the discoveries in this book I am quite certain that there are vastly simpler examples that exist—so that in fact there are in the end rather simple integer equations for which the absence of solutions can never be proved from the normal axioms of arithmetic.

If one just starts looking at sequences of integer equations—as on the next page—then in the very simplest cases it is usually fairly easy to tell whether a particular equation will have any solutions. But this rapidly becomes very much more difficult. For there is often no obvious pattern to which equations ultimately have solutions and which do not. And even when equations do have solutions, the integers involved can be quite large. So, for example, the smallest solution to $x^2 = 61 y^2 + 1$ is $x = 1766319049$, $y = 226153980$, while the smallest solution to $x^3 + y^3 = z^3 + 2$ is $x = 1214928$, $y = 3480205$, $z = 3528875$.

Integer equations such as $a x + b y + c z = d$ that have only linear dependence on any variable were largely understood even in antiquity. Quadratic equations in two variables such as $x^2 = a y^2 + b$ were understood by the 1800s. But even equations such as $x^2 = a y^3 + b$ were not properly understood until the 1980s. And with equations that have higher powers or more variables questions of whether solutions exist quickly end up being unsolved problems of number theory.

It has certainly been known for centuries that there are questions about integer equations and other aspects of number theory that are easy to state, yet seem very hard to answer. But in practice it has almost

Equation	Solution	
$2x+3y=1$	□	
$2x+3y=2$	□	
$2x+3y=3$	□	
$2x+3y=4$	□	
$2x+3y=5$	$x=1$	$y=1$
$2x+3y=6$	□	
$2x+3y=7$	$x=2$	$y=1$
$2x+3y=8$	$x=1$	$y=2$
$2x+3y=9$	$x=3$	$y=1$
$2x+3y=10$	$x=2$	$y=2$
$2x+3y=11$	$x=1$	$y=3$
$2x+3y=12$	$x=3$	$y=2$
$2x+3y=13$	$x=2$	$y=3$
$2x+3y=14$	$x=1$	$y=4$
$2x+3y=15$	$x=3$	$y=3$

Equation	Solution	
$x^2=y^2+1$	□	
$x^2=y^2+2$	□	
$x^2=y^2+3$	$x=2$	$y=1$
$x^2=y^2+4$	□	
$x^2=y^2+5$	$x=3$	$y=2$
$x^2=y^2+6$	□	
$x^2=y^2+7$	$x=4$	$y=3$
$x^2=y^2+8$	$x=3$	$y=1$
$x^2=y^2+9$	$x=5$	$y=4$
x^2-y^2+10	□	
$x^2=y^2+11$	$x=6$	$y=5$
$x^2=y^2+12$	$x=4$	$y=2$
$x^2=y^2+13$	$x=7$	$y=6$
$x^2=y^2+14$	□	
$x^2=y^2+15$	$x=4$	$y=1$
$x^2=y^2+16$	$x=5$	$y=3$

Equation	Solution	
$x^2=y^2+1$	□	
$x^2=2y^2+1$	$x=3$	$y=2$
$x^2=3y^2+1$	$x=2$	$y=1$
$x^2=4y^2+1$	□	
$x^2=5y^2+1$	$x=9$	$y=4$
$x^2=6y^2+1$	$x=5$	$y=2$
$x^2=7y^2+1$	$x=8$	$y=3$
$x^2=8y^2+1$	$x=3$	$y=1$
$x^2=9y^2+1$	□	
$x^2=10y^2+1$	$x=19$	$y=6$
$x^2=11y^2+1$	$x=10$	$y=3$
$x^2=12y^2+1$	$x=7$	$y=2$
$x^2=13y^2+1$	$x=649$	$y=180$
$x^2=14y^2+1$	$x=15$	$y=4$
$x^2=15y^2+1$	$x=4$	$y=1$
$x^2=16y^2+1$	□	
$x^2=17y^2+1$	$x=33$	$y=8$
$x^2=18y^2+1$	$x=17$	$y=4$
$x^2=19y^2+1$	$x=170$	$y=39$
$x^2=20y^2+1$	$x=9$	$y=2$

Equation	Solution	
$x^2=y^3-20$	$x=14$	$y=6$
$x^2=y^3-19$	$x=18$	$y=7$
$x^2=y^3-18$	$x=3$	$y=3$
$x^2=y^3-17$	□	
$x^2=y^3-16$	□	
$x^2=y^3-15$	$x=7$	$y=4$
$x^2=y^3-14$	□	
$x^2=y^3-13$	$x=70$	$y=17$
$x^2=y^3-12$	□	
$x^2=y^3-11$	$x=4$	$y=3$
$x^2=y^3-10$	□	
$x^2=y^3-9$	□	
$x^2=y^3-8$	□	
$x^2=y^3-7$	$x=1$	$y=2$
$x^2=y^3-6$	□	
$x^2=y^3-5$	□	
$x^2=y^3-4$	$x=2$	$y=2$
$x^2=y^3-3$	□	
$x^2=y^3-2$	$x=5$	$y=3$
$x^2=y^3-1$	□	
$x^2=y^3$	$x=1$	$y=1$
$x^2=y^3+1$	$x=3$	$y=2$
$x^2=y^3+2$	□	
$x^2=y^3+3$	$x=2$	$y=1$
$x^2=y^3+4$	□	
$x^2=y^3+5$	□	
$x^2=y^3+6$	□	
$x^2=y^3+7$	□	
$x^2=y^3+8$	$x=3$	$y=1$
$x^2=y^3+9$	$x=6$	$y=3$
$x^2=y^3+10$	□	
$x^2=y^3+11$	□	
$x^2=y^3+12$	$x=47$	$y=13$
$x^2=y^3+13$	□	
$x^2=y^3+14$	□	
$x^2=y^3+15$	$x=4$	$y=1$
$x^2=y^3+16$	□	
$x^2=y^3+17$	$x=5$	$y=2$
$x^2=y^3+18$	$x=19$	$y=7$
$x^2=y^3+19$	$x=12$	$y=5$
$x^2=y^3+20$	□	

Equation	Solution	
$x^2=y^3+1$	$x=3$	$y=2$
$x^2=2y^3+1$	□	
$x^2=3y^3+1$	$x=2$	$y=1$
$x^2=4y^3+1$	□	
$x^2=5y^3+1$	□	
$x^2=6y^3+1$	$x=7$	$y=2$
$x^2=7y^3+1$	□	
$x^2=8y^3+1$	$x=3$	$y=1$
$x^2=9y^3+1$	□	
$x^2=10y^3+1$	$x=9$	$y=2$

Equation	Solution	
$x^3=y^4-20xy-1$	$x=10$	$y=7$
$x^3=y^4-19xy-1$	$x=3$	$y=4$
$x^3=y^4-18xy-1$	$x=75$	$y=26$
$x^3=y^4-17xy-1$		
$x^3=y^4-16xy-1$		
$x^3=y^4-15xy-1$	$x=624$	$y=125$
$x^3=y^4-14xy-1$		
$x^3=y^4-13xy-1$		
$x^3=y^4-12xy-1$	$x=3$	$y=2$
$x^3=y^4-11xy-1$		
$x^3=y^4-10xy-1$		
$x^3=y^4-9xy-1$	$x=80$	$y=27$
$x^3=y^4-8xy-1$	$x=12$	$y=7$
$x^3=y^4-7xy-1$	$x=1$	$y=2$
$x^3=y^4-6xy-1$	$x=15$	$y=8$
$x^3=y^4-5xy-1$		
$x^3=y^4-4xy-1$	$x=30$	$y=13$
$x^3=y^4-3xy-1$		
$x^3=y^4-2xy-1$		
$x^3=y^4-xy-1$		
$x^3=y^4-1$	□	
$x^3=y^4+xy-1$	$x=1$	$y=1$
$x^3=y^4+2xy-1$	$x=3$	$y=2$
$x^3=y^4+3xy-1$	$x=5$	$y=3$
$x^3=y^4+4xy-1$	$x=2$	$y=1$
$x^3=y^4+5xy-1$		
$x^3=y^4+6xy-1$		
$x^3=y^4+7xy-1$		
$x^3=y^4+8xy-1$	$x=20$	$y=9$
$x^3=y^4+9xy-1$	$x=3$	$y=1$
$x^3=y^4+10xy-1$		
$x^3=y^4+11xy-1$	$x=5$	$y=2$
$x^3=y^4+12xy-1$		
$x^3=y^4+13xy-1$		
$x^3=y^4+14xy-1$		
$x^3=y^4+15xy-1$		
$x^3=y^4+16xy-1$	$x=4$	$y=1$
$x^3=y^4+17xy-1$		
$x^3=y^4+18xy-1$	$x=8$	$y=3$
$x^3=y^4+19xy-1$		
$x^3=y^4+20xy-1$		

Equation	Solution	
$x^2=y^5+3$	$x=2$	$y=1$
$x^2=y^5+y+3$	$x=2537$	$y=23$
$x^2=y^5+2y+3$		
$x^2=y^5+3y+3$		
$x^2=y^5+4y+3$		
$x^2=y^5+5y+3$	$x=3$	$y=1$
$x^2=y^5+6y+3$		
$x^2=y^5+7y+3$	$x=7$	$y=2$
$x^2=y^5+8y+3$		
$x^2=y^5+9y+3$		

Equation	Solution		
$x^3+y^3=z^2+1$	$x=1$	$y=1$	$z=1$
$x^3+y^3=z^2+2$	$x=107$	$y=232$	$z=3703$
$x^3+y^3=z^2+3$	$x=1$	$y=3$	$z=5$
$x^3+y^3=z^2+4$	$x=5$	$y=12$	$z=43$
$x^3+y^3=z^2+5$	$x=1$	$y=2$	$z=2$
$x^3+y^3=z^2+6$	$x=7$	$y=24$	$z=119$
$x^3+y^3=z^2+7$	$x=2$	$y=2$	$z=3$
$x^3+y^3=z^2+8$	$x=1$	$y=2$	$z=1$
$x^3+y^3=z^2+9$	$x=3$	$y=7$	$z=19$
$x^3+y^3=z^2+10$	$x=2$	$y=3$	$z=5$

Equation	Solution		
$x^3+y^3=z^3-20$	$x=107$	$y=137$	$z=156$
$x^3+y^3=z^3-19$	$x=14$	$y=16$	$z=19$
$x^3+y^3=z^3-18$	$x=1$	$y=2$	$z=3$
$x^3+y^3=z^3-17$	$x=103$	$y=111$	$z=135$
$x^3+y^3=z^3-16$	$x=10$	$y=12$	$z=14$
$x^3+y^3=z^3-15$	$x=262$	$y=265$	$z=332$
$x^3+y^3=z^3-14$	□		
$x^3+y^3=z^3-13$	□		
$x^3+y^3=z^3-12$	$x=5725013$	$y=9019406$	$z=9730705$
$x^3+y^3=z^3-11$	$x=2$	$y=2$	$z=3$
$x^3+y^3=z^3-10$	$x=3$	$y=3$	$z=4$
$x^3+y^3=z^3-9$	$x=52$	$y=216$	$z=217$
$x^3+y^3=z^3-8$	$x=16$	$y=12$	$z=18$
$x^3+y^3=z^3-7$	$x=605809$	$y=680316$	$z=812918$
$x^3+y^3=z^3-6$	$x=1$	$y=1$	$z=2$
$x^3+y^3=z^3-5$	□		
$x^3+y^3=z^3-4$	□		
$x^3+y^3=z^3-3$			
$x^3+y^3=z^3-2$	$x=5$	$y=6$	$z=7$
$x^3+y^3=z^3-1$	$x=6$	$y=8$	$z=9$
$x^3+y^3=z^3$	□		
$x^3+y^3=z^3+1$	$x=1$	$y=2$	$z=2$
$x^3+y^3=z^3+2$	$x=1214928$	$y=3480205$	$z=3528875$
$x^3+y^3=z^3+3$	$x=4$	$y=4$	$z=5$
$x^3+y^3=z^3+4$	□		
$x^3+y^3=z^3+5$	□		
$x^3+y^3=z^3+6$	$x=10529$	$y=60248$	$z=60355$
$x^3+y^3=z^3+7$	$x=32$	$y=104$	$z=105$
$x^3+y^3=z^3+8$	$x=1$	$y=2$	$z=1$
$x^3+y^3=z^3+9$	$x=2097$	$y=11305$	$z=11329$
$x^3+y^3=z^3+10$	$x=130$	$y=141$	$z=171$
$x^3+y^3=z^3+11$	$x=297$	$y=619$	$z=641$
$x^3+y^3=z^3+12$	$x=7$	$y=10$	$z=11$
$x^3+y^3=z^3+13$	□		
$x^3+y^3=z^3+14$	□		
$x^3+y^3=z^3+15$	$x=2$	$y=2$	$z=1$
$x^3+y^3=z^3+16$	$x=2429856$	$y=6960410$	$z=7057750$
$x^3+y^3=z^3+17$	$x=25$	$y=50$	$z=52$
$x^3+y^3=z^3+18$	$x=94$	$y=101$	$z=123$
$x^3+y^3=z^3+19$	$x=26$	$y=76$	$z=77$
$x^3+y^3=z^3+20$	$x=1$	$y=3$	$z=2$

universally been assumed that with the continued development of mathematics any of these questions could in the end be answered.

However, what Gödel's Theorem shows is that there must always exist some questions that cannot ever be answered using the normal axioms of arithmetic. Yet the fact that the few known explicit examples have been extremely complicated has made this seem somehow fundamentally irrelevant for the actual practice of mathematics.

But from the discoveries in this book it now seems quite certain that vastly simpler examples also exist. And it is my strong suspicion that in fact of all the current unsolved problems seriously studied in number theory a fair fraction will in the end turn out to be questions that cannot ever be answered using the normal axioms of arithmetic.

If one looks at recent work in number theory, most of it tends to be based on rather sophisticated methods that do not obviously depend only on the normal axioms of arithmetic. And for example the elaborate proof of Fermat's Last Theorem that has been developed may make at least some use of axioms that come from fields like set theory and go beyond the normal ones for arithmetic.

But so long as one stays within, say, the standard axiom systems of mathematics on pages 773 and 774, and does not in effect just end up implicitly adding as an axiom whatever result one is trying to prove, my strong suspicion is that one will ultimately never be able to go much further than one can purely with the normal axioms of arithmetic.

And indeed from the Principle of Computational Equivalence I strongly believe that in general undecidability and unprovability will start to occur in practically any area of mathematics almost as soon as one goes beyond the level of questions that are always easy to answer.

But if this is so, why then has mathematics managed to get as far as it has? Certainly there are problems in mathematics that have remained unsolved for long periods of time. And I suspect that many of these will in fact in the end turn out to involve undecidability and

◀ Smallest solutions for various sequences of integer (or so-called Diophantine) equations. □ indicates that it can be proved that no solution exists. A blank indicates that I know only that no solution exists below a billion. Methods for resolving some of the equations in the first column were known in antiquity; all had been resolved by the 1800s. Practical methods for resolving the so-called elliptic curve equations in the second column were developed only in the 1980s. No general methods are yet known for most of the other equations given—and some classes of them may in fact show undecidability.

unprovability. But the issue remains why such phenomena have not been much more obvious in everyday work in mathematics.

At some level I suspect the reason is quite straightforward: it is that like most other fields of human inquiry mathematics has tended to define itself to be concerned with just those questions that its methods can successfully address. And since the main methods traditionally used in mathematics have revolved around doing proofs, questions that involve undecidability and unprovability have inevitably been avoided.

But can this really be right? For at least in the past century mathematics has consistently given the impression that it is concerned with questions that are somehow as arbitrary and general as possible.

But one of the important conclusions from what I have done in this book is that this is far from correct. And indeed for example traditional mathematics has for the most part never even considered most of the kinds of systems that I discuss in this book—even though they are based on some of the very simplest rules possible.

So how has this happened? The main point, I believe, is that in both the systems it studies and the questions it asks mathematics is much more a product of its history than is usually realized.

And in fact particularly compared to what I do in this book the vast majority of mathematics practiced today still seems to follow remarkably closely the traditions of arithmetic and geometry that already existed even in Babylonian times.

It is a fairly recent notion that mathematics should even try to address arbitrary or general systems. For until not much more than a century ago mathematics viewed itself essentially just as providing a precise formulation of certain aspects of everyday experience—mainly those related to number and space.

But in the 1800s, with developments such as non-Euclidean geometry, quaternions, group theory and transfinite numbers it began to be assumed that the discipline of mathematics could successfully be applied to any abstract system, however arbitrary or general.

Yet if one looks at the types of systems that are actually studied in mathematics they continue even to this day to be far from as general as possible. Indeed at some level most of them can be viewed as having been arrived at by the single rather specific approach of starting from some known set of theorems, then trying to find systems that are progressively more general, yet still manage to satisfy these theorems.

And given this approach, it tends to be the case that the questions that are considered interesting are ones that revolve around whatever theorems a system was set up to satisfy—making it rather likely that these questions can themselves be addressed by similar theorems, without any confrontation with undecidability or unprovability.

But what if one looks at other kinds of systems?

One of the main things I have done in this book is in a sense to introduce a new approach to generalization in which one considers systems that have simple but completely arbitrary rules—and that are not set up with any constraint about what theorems they should satisfy.

But if one has such a system, how does one decide what questions are interesting to ask about it? Without the guidance of known theorems, the obvious thing to do is just to look explicitly at how the system behaves—perhaps by making some kind of picture.

And if one does this, then what I have found is that one is usually immediately led to ask questions that run into phenomena like undecidability. Indeed, from my experiments it seems that almost as soon as one leaves behind the constraints of mathematical tradition undecidability and unprovability become rather common.

As the picture on the next page indicates, it is quite straightforward to set up an axiom system that deals with logical statements about a system like a cellular automaton. And within such an axiom system one can ask questions such as whether the cellular automaton will ever behave in a particular way after any number of steps.

But as we saw in the previous section, such questions are in general undecidable. And what this means is that there will inevitably be cases of them for which no proof of a particular answer can ever be given within whatever axiom system one is using.

So from this one might conclude that as soon as one looks at cellular automata or other kinds of systems beyond those normally studied in mathematics it must immediately become effectively impossible to make progress using traditional mathematical methods.

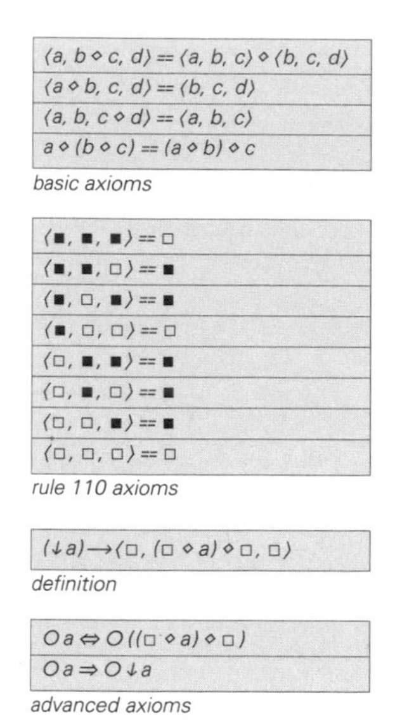

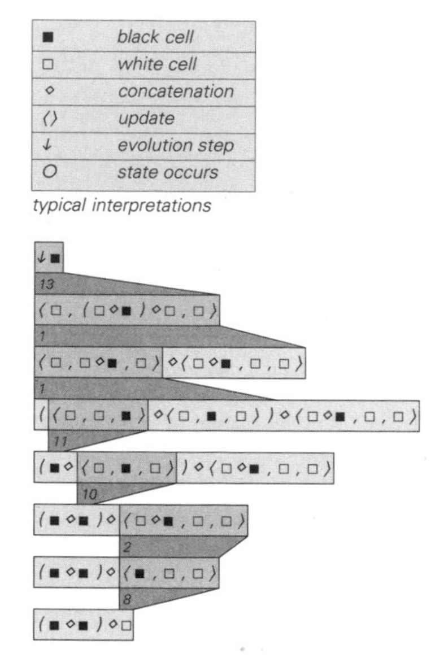

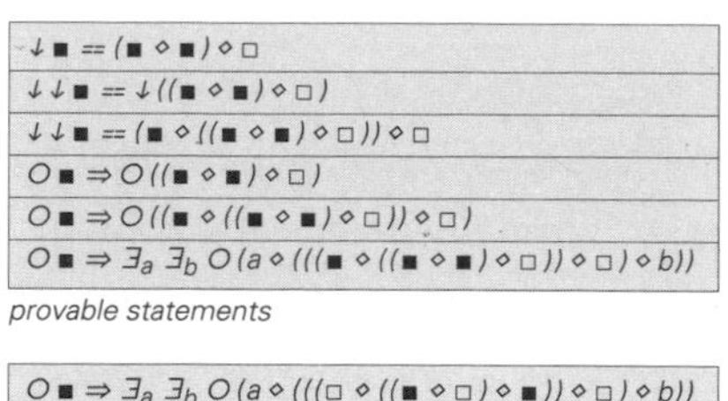

An axiom system for statements about the rule 110 cellular automaton. The top statement above makes the assertion that the outcome after one step of evolution from a single black cell has a particular form. A proof of this statement is shown to the left. All the statements in the top block above can be proved true from the axiom system. The statement at the bottom, however, cannot be proved either true or false. The axioms given are set up using predicate logic.

But in fact, in the fifteen years or so since I first emphasized the importance of cellular automata all sorts of traditional mathematical work has actually been done on them. So how has this been possible?

The basic point is that the work has tended to concentrate on particular aspects of cellular automata that are simple enough to avoid undecidability and unprovability. And typically it has achieved this in one of two ways: either by considering only very specific cases that have been observed or constructed to be simple, or by looking at things in so much generality that only rather simple properties ever survive.

So for example when presented with the 256 elementary cellular automaton patterns shown on page 55 mathematicians in my experience have two common responses: either to single out specific patterns that have a simple repetitive or perhaps nested form, or to generalize and look not at individual patterns, but rather at aggregate properties obtained say by evolving from all possible initial conditions.

And about questions that concern, for example, the structure of a pattern that looks to us complex, the almost universal reaction is that such questions can somehow not be of any real mathematical interest.

Needless to say, in the framework of the new kind of science in this book, such questions are now of great interest. And my results suggest that if one is ever going to study many important phenomena that occur in nature one will also inevitably run into them. But to traditional mathematics they seem uninteresting and quite alien.

As I said above, it is at some level not surprising that questions will be considered interesting in a particular field only if the methods of that field can say something useful about them. But this I believe is ultimately why there have historically been so few signs of undecidability or unprovability in mathematics. For any kinds of questions in which such phenomena appear are usually not amenable to standard methods of mathematics based on proof, and as a result such questions have inevitably been viewed as being outside what should be considered interesting for mathematics.

So how then can one set up a reasonable idealization for mathematics as it is actually practiced? The first step—much as I discussed earlier in this section—is to think not so much about systems that might be described by mathematics as about the internal processes associated with proof that go on inside mathematics.

A proof must ultimately be based on an axiom system, and one might have imagined that over the course of time mathematics would have sampled a wide range of possible axiom systems. But in fact in its historical development mathematics has normally stuck to only rather few such systems—each one corresponding essentially to some identifiable field of mathematics, and most given on pages 773 and 774.

So what then happens if one looks at all possible simple axiom systems—much as we looked, say, at all possible simple cellular automata earlier in this book? To what extent does what one sees capture the features of mathematics? With axiom systems idealized as multiway systems the pictures on the next page show some results.

In some cases the total number of theorems that can ever be proved is limited. But often the number of theorems increases rapidly with the length of proof—and in most cases an infinite number of theorems can eventually be proved. And given experience with mathematics an obvious question to ask in such cases is to what extent the system is consistent, or complete, or both.

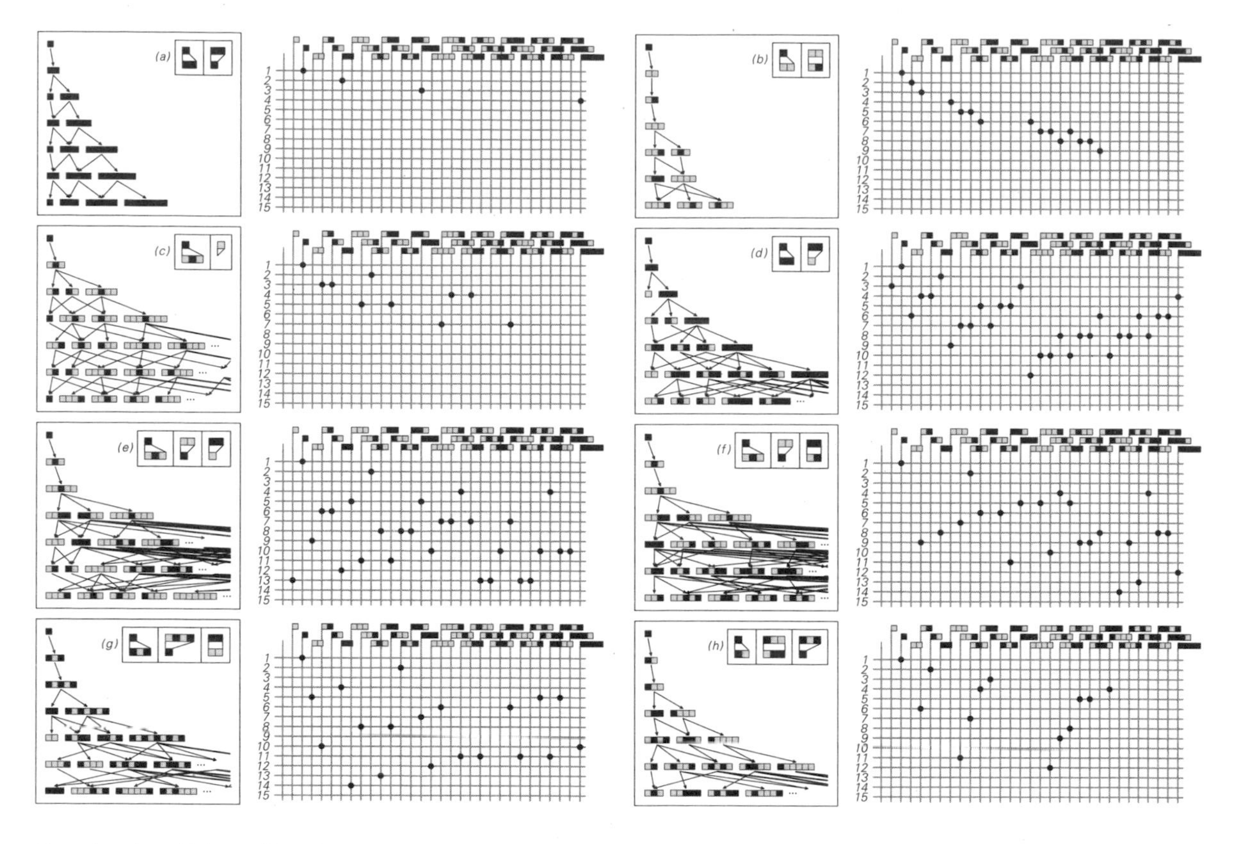

Plots showing which possible strings get generated in the first 15 steps of evolution in various multiway systems. Each string that is generated can be thought of as a theorem derived from the set of axioms represented by the rules of the multiway system. A dot shows at which step a given string first appears—and indicates the shortest proof of the theorem that string represents. In most cases, many strings are never produced—so that there are many possible statements that simply do not follow from the axioms given. Thus for example in first case shown only strings containing nothing but black elements are ever produced.

But to formulate such a question in a meaningful way one needs a notion of negation. In general, negation is just some operation that takes a string and yields another, giving back the original if it is applied a second time. Earlier in this section we discussed cases in which negation simply reverses the color of each element in a string. And as a generalization of this one can consider cases in which negation can be any operation that preserves lengths of strings.

And in this case it turns out that the criterion for whether a system is complete and consistent is simply that exactly half the possible strings of a given length are eventually generated if one starts from the string representing "true".

For if more than half the strings are generated, then somewhere both a string and its negation would have to appear, implying that the system must be inconsistent. And similarly, if less than half the strings are generated, there must be some string for which neither that string nor its negation ever appear, implying that the system is incomplete.

The pictures on the next page show the fractions of strings of given lengths that are generated on successive steps in various multiway systems. In general one might have to wait an arbitrarily large number of steps to find out whether a given string will ever be generated. But in practice after just a few steps one already seems to get a reasonable indication of the overall fraction of strings that will ever be generated.

And what one sees is that there is a broad distribution: from cases in which very few strings can be generated—corresponding to a very incomplete axiom system—to cases in which all or almost all strings can be generated—corresponding to a very inconsistent axiom system.

So where in this distribution do the typical axiom systems of ordinary mathematics lie? Presumably none are inconsistent. And a few—like basic logic and real algebra—are both complete and consistent, so that in effect they lie right in the middle of the distribution. But most are known to be incomplete. And as we discussed above, this is inevitable as soon as universality is present.

But just how incomplete are they? The answer, it seems, is typically not very. For if one looks at axiom systems that are widely used in mathematics they almost all tend to be complete enough to prove at least a fair fraction of statements either true or false.

So why should this be? I suspect that it has to do with the fact that in mathematics one usually wants axiom systems that one can think of as somehow describing definite kinds of objects—about which one then expects to be able to establish all sorts of definite statements.

And certainly if one looks at the history of mathematics most basic axiom systems have been arrived at by starting with objects—such as finite integers or finite sets—then trying to find collections of axioms that somehow capture the relevant properties of these objects.

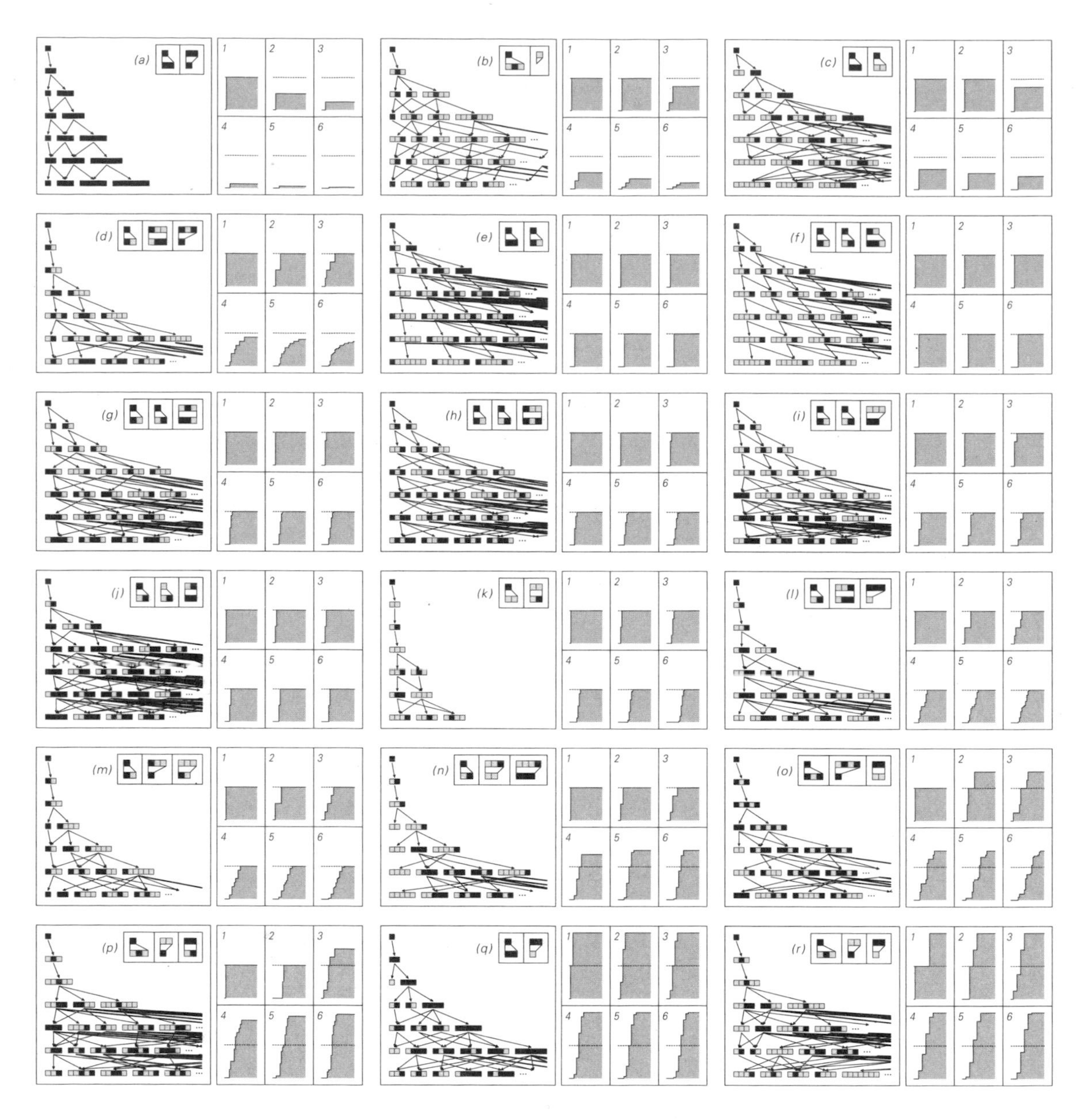

Examples of multiway systems that generate different fractions of possible strings, and in effect range from being highly incomplete to highly inconsistent. The plots show what fraction of strings of a given length have been produced by each of the first 25 steps in the evolution of each multiway system. If less than half the strings of a given length are ever produced, this means that there must be some strings where neither the string nor its negation can be proved, indicating incompleteness. But if more than half the strings are produced, there must be cases where both a string and its negation can be proved, indicating inconsistency. Rules (f) through (i), however, produce exactly half the strings of any given length, and can be considered complete and consistent.

But one feature is that normally the resulting axiom system is in a sense more general than the objects one started from. And this is why for example one can often use the axiom system to extrapolate to infinite situations. But it also means that it is not clear whether the axiom system actually describes only the objects one wants—or whether for example it also describes all sorts of other quite different objects.

One can think of an axiom system—say one of those listed on pages 773 and 774—as giving a set of constraints that any object it describes must satisfy. But as we saw in Chapter 5, it is often possible to satisfy a single set of constraints in several quite different ways.

And when this happens in an axiom system it typically indicates incompleteness. For as soon as there are just two objects that both satisfy the constraints but for which there is some statement that is true about one but false about the other it immediately follows that at least this statement cannot consistently be proved true or false, and that therefore the axiom system must be incomplete.

One might imagine that if one were to add more axioms to an axiom system one could always in the end force there to be only one kind of object that would satisfy the constraints of the system. But as we saw earlier, as soon as there is universality it is normally impossible to avoid incompleteness. And if an axiom system is incomplete there must inevitably be different kinds of objects that satisfy its constraints. For given any statement that cannot be proved from the axioms there must be distinct objects for which it is true, and for which it is false.

If an axiom system is far from complete—so that a large fraction of statements cannot be proved true or false—then there will typically be many different kinds of objects that are easy to specify and all satisfy the constraints of the system but for which there are fairly obvious properties that differ. But if an axiom system is close to complete—so that the vast majority of statements can be proved true or false—then it is almost inevitable that the different kinds of objects that satisfy its constraints must differ only in obscure ways.

And this is presumably the case in the standard axiom system for arithmetic from page 773. Originally this axiom system was intended to describe just ordinary integers. But Gödel's Theorem showed that it is incomplete, so that there must be more than one kind of object that can satisfy its constraints. Yet it is rather close to being complete—since as we saw earlier one has to go through at least millions of statements before finding ones that it cannot prove true or false.

And this means that even though there are objects other than the ordinary integers that satisfy the standard axioms of arithmetic, they are quite obscure—in fact, so much so that none have ever yet actually been constructed with any real degree of explicitness. And this is why it has been reasonable to think of the standard axiom system of arithmetic as being basically just about ordinary integers.

But if instead of this standard axiom system one uses the reduced axiom system from page 773—in which the usual axiom for induction has been weakened—then the story is quite different. There is again incompleteness, but now there is much more of it, for even statements as simple as $x + y = y + x$ and $x + 0 = x$ cannot be proved true or false from the axioms. And while ordinary integers still satisfy all the constraints, the system is sufficiently incomplete that all sorts of other objects with quite different properties also do. So this means that the system is in a sense no longer about any very definite kind of mathematical object—and presumably that is why it is not used in practice in mathematics.

At this juncture it should perhaps be mentioned that in their raw form quite a few well-known axiom systems from mathematics are actually also far from complete. An example of this is the axiom system for group theory given on page 773. But the point is that this axiom system represents in a sense just the beginning of group theory. For it yields only those theorems that hold abstractly for any group.

Yet in doing group theory in practice one normally adds axioms that in effect constrain one to be dealing say with a specific group rather than with all possible groups. And the result of this is that once again one typically has an axiom system that is at least close to complete.

In basic arithmetic and also usually in fields like group theory the underlying objects that one imagines describing can at some level be manipulated—and understood—in fairly concrete ways. But in a field like set theory this is less true. Yet even in this case an attempt has historically been made to get an axiom system that somehow describes definite kinds of objects. But now the main way this has been done is by progressively adding axioms so as to get closer to having a system that is complete—with only a rather vague notion of just what underlying objects one is really expecting to describe.

In studying basic processes of proof multiway systems seem to do well as minimal idealizations. But if one wants to study axiom systems that potentially describe definite objects it seems to be somewhat more convenient to use what I call operator systems. And indeed the version of logic used on page 775—as well as many of the axiom systems on pages 773 and 774—are already set up essentially as operator systems.

The basic idea of an operator system is to work with expressions such as $(p \circ q) \circ ((q \circ r) \circ p)$ built up using some operator $\circ$, and then to consider for example what equivalences may exist between such expressions. If one has an operator whose values are given by some finite table then it is always straightforward to determine whether expressions are equivalent. For all one need do, as in the pictures at the top of the next page, is to evaluate the expressions for all possible values of each variable, and then to see whether the patterns of results one gets are the same.

And in this way one can readily tell, for example, that the first operator shown is idempotent, so that $p \circ p = p$, while both the first two operators are associative, so that $(p \circ q) \circ r = p \circ (q \circ r)$, and all but the third operator are commutative, so that $p \circ q = q \circ p$. And in principle one can use this method to establish any equivalence that exists between any expressions with an operator of any specific form.

But the crucial idea that underlies the traditional approach to mathematical proof is that one should also be able to deduce such results just by manipulating cxpressions in purely symbolic form, using the rules of an axiom system, without ever having to do anything like filling in explicit values of variables.

And one advantage of this approach is that at least in principle it allows one to handle operators—like those found in many areas of mathematics—that are not based on finite tables. But even for operators given by finite tables it is often difficult to find axiom systems that can successfully reproduce all the results for a particular operator.

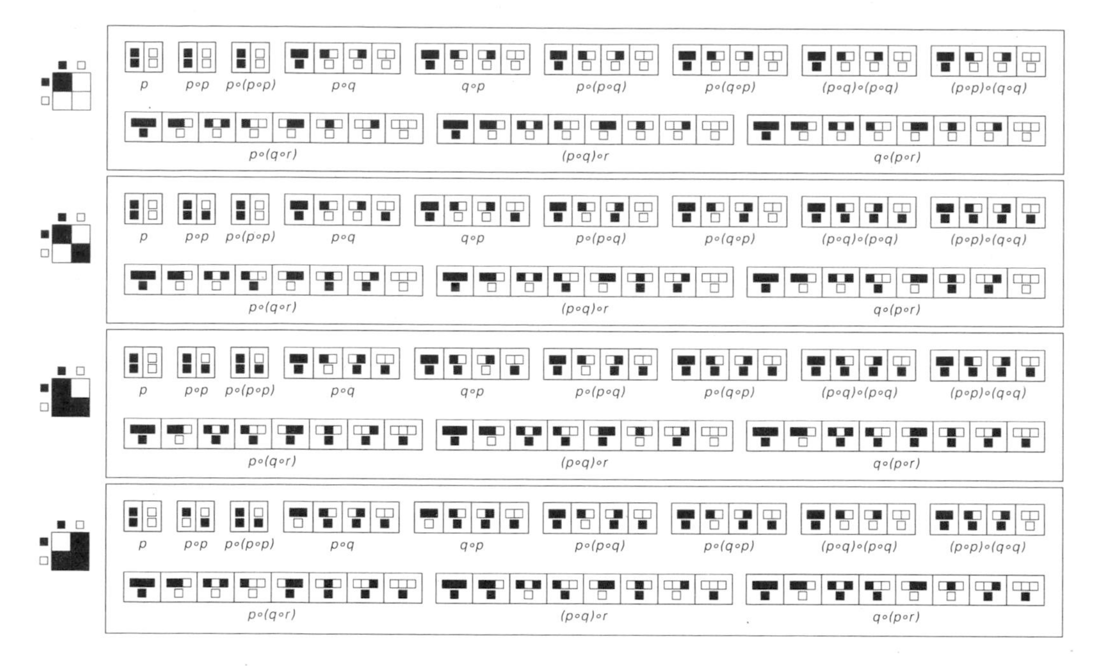

Values of expressions obtained by using operators of various forms. For each expression the sequence of values for every possible combination of values of variables is shown. Two expressions are equivalent when this sequence of values is the same. With black and white interpreted as TRUE and FALSE, the forms of operators shown here correspond respectively to AND, EQUAL, IMPLIES and NAND. (The first argument to each operator is shown on the left; the second on top.) The arrays of values generated can be thought of as being like truth tables.

With the way I have set things up, any axiom system is itself just a collection of equivalence results. So the question is then which equivalence results need to be included in the axiom system in order that all other equivalence results can be deduced just from these.

In general this can be undecidable—for there is no limit on how long even a single proof might need to be. But in some cases it turns out to be possible to establish that a particular set of axioms can successfully generate all equivalence results for a given operator—and indeed the picture at the top of the facing page shows examples of this for each of the four operators in the picture above.

So if two expressions are equivalent then by applying the rules of the appropriate axiom system it must be possible to get from one to the other—and in fact the picture on page 775 shows an example of how

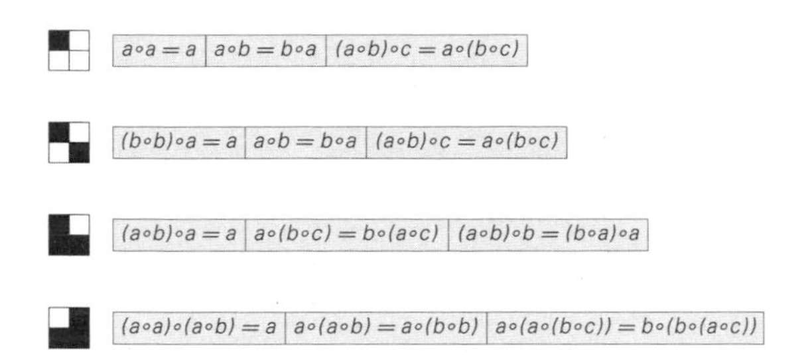

Axiom systems that can be used to derive all the equivalences between expressions that involve operators with the forms shown. Each axiom can be applied in either direction—as in the picture on page 775, with each variable standing for any expression, as in a *Mathematica* pattern. The operators shown are AND, EQUAL, IMPLIES and NAND. They yield respectively junctional, equivalential, implicational and full propositional or sentential calculus (ordinary logic).

this can be done for the fourth axiom system above. But if one removes just a single axiom from any of the axiom systems above then it turns out that they no longer work, and for example they cannot establish the equivalence result stated by whichever axiom one has removed.

In general one can think of axioms for an operator system as giving constraints on the form of the operator. And if one is going to reproduce all the equivalences that hold for a particular form then these constraints must in effect be such as to force that form to occur.

So what happens in general for arbitrary axiom systems? Do they typically force the operator to have a particular form, or not?

The pictures on the next two pages show which forms of operators are allowed by various different axiom systems. The successive blocks of results in each case give the forms allowed with progressively more possible values for each variable.

Indicated by stars near the bottom of the picture are the four axiom systems from the top of this page. And for each of these only a limited number of forms are allowed—all of which ultimately turn out to be equivalent to just the single forms shown on the facing page.

But what about other axiom systems? Every axiom system must allow an operator of at least some form. But what the pictures on the next two pages show is that the vast majority of axiom systems actually allow operators with all sorts of different forms.

And what this means is that these axiom systems are in a sense not really about operators of any particular form. And so in effect they are also far from complete—for they can prove only equivalence results that hold for every single one of the various operators they allow.

a = a
(19683)
(4294967296)
b = a
(0)
a∘a = a
(729)
(16777216)
a∘b = a
b∘a = a
b∘b = a
a∘b = b∘a
(1048576)
a∘(a∘a) = a
(3375)
(157351936)
(a∘a)∘a = a
a∘(a∘b) = a
(27)
(10000)
a∘(b∘a) = a
(136)
(46121)
a∘(b∘b) = a
(298)
(1147649)
b∘(a∘a) = a
(10)
b∘(a∘b) = a
(18)
b∘(b∘a) = a
(64)
(a∘a)∘b = a
(a∘b)∘a = a
(a∘b)∘b = a
(b∘a)∘a = a
(b∘a)∘b = a
(b∘b)∘a = a
b∘(b∘b) = a
(b∘b)∘b = a
(a∘a)∘(a∘a) = a
(2916)
(167772160)
a∘((a∘a)∘b) = a
(146)
(168780)
(a∘a)∘(a∘b) = a
(216)
(335008)
(a∘b)∘(b∘c) = a
(100)
1
(a∘b)∘c = a∘(b∘c)
(113)
(3492)
((b∘b)∘a)∘(a∘b) = a
(21)
(1022)
((b∘(a∘a))∘a)∘b = a
(16)
2
b∘(c∘(a∘(b∘c))) = a
(a∘b)∘(a∘(b∘c)) = a
(262)

Forms of a binary operator satisfying the constraints of a series of different axiom systems. The successive blocks of results in each case show forms of the operator allowed with 2, 3 and 4 possible elements. Note that with 3 and 4 elements, only forms inequivalent under interchange of element labels are shown. Representations of notable systems in mathematics are: (1) semigroup theory, (2) commutative group theory, (3) basic logic, (4) commutative semigroup theory, (5) squag theory, (6) group theory, (7) junctional calculus, (8) equivalential calculus and (9) implicational calculus. In each case the operator forms shown correspond to possible semigroups, commutative groups, systems of logic (Boolean algebras), etc. with 2, 3 and 4 possible elements. The operator forms shown can be thought of as giving multiplication tables. In model theory, these forms are usually called the models of an axiom system.

So if one makes a list of all possible axiom systems—say starting with the simplest—where in such a list should one expect to see axiom systems that correspond to traditional areas of mathematics?

Most axiom systems as they are given in typical textbooks are sufficiently complicated that they will not show up at all early. And in fact the only immediate exception is the axiom system $\{(a \circ b) \circ c = a \circ (b \circ c)\}$ for what are known as semigroups—which ironically are usually viewed as rather advanced mathematical objects.

But just how complicated do the axiom systems for traditional areas of mathematics really need to be? Often it seems that they can be vastly simpler than their textbook forms. And so, for example, as page 773 indicates, interpreting the $\circ$ operator as division, $\{a \circ (b \circ (c \circ (a \circ b))) = c\}$ is known to be an axiom system for commutative group theory, and $\{a \circ ((((a \circ a) \circ b) \circ c) \circ (((a \circ a) \circ a) \circ c)) = b\}$ for general group theory.

So what about basic logic? How complicated an axiom system does one need for this? Textbook discussions of logic mostly use axiom systems at least as complicated as the first one on page 773. And such axiom systems not only involve several axioms—they also normally involve three separate operators: AND ($\wedge$), OR ($\vee$) and NOT ($\neg$).

But is this in fact the only way to formulate logic?

As the picture below shows, there are 16 different possible operators that take two arguments and allow two values, say true and false. And of these AND, OR and NOT are certainly the most commonly used in both everyday language and most of mathematics.

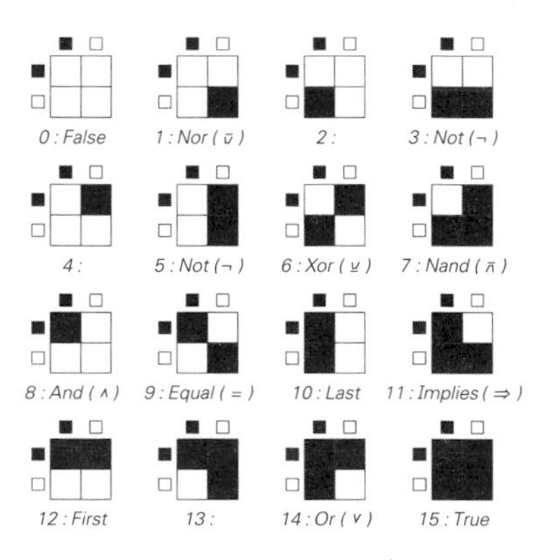

Logical functions of two arguments and their common names. Black stands for TRUE; white for FALSE. AND, OR, NOT, and IMPLIES are widely used in traditional logic. EQUAL (if and only if) is common in more mathematical settings, while XOR is widespread in discrete mathematics. NAND and NOR are mostly used only in circuit design and in a few foundational studies of logic. The first argument for each function appears on the left in the picture; the second argument on top. The functions are numbered like 2-neighbor analogs of the cellular automaton rules of page 53.

But at least at a formal level, logic can be viewed simply as a theory of functions that take on two possible values given variables with two possible values. And as we discussed on page 616, any such function can be represented as a combination of AND, OR and NOT.

But the table below demonstrates that as soon as one goes beyond the familiar traditions of language and mathematics there are other operators that can also just as well be used as primitives. And indeed it has been known since before 1900 that both NAND and NOR on their own work—a fact I already used on pages 617 and 775.

0	¬ a ∧ a	1	¬ (a ∨ b)	2	¬ a ∧ b	3	¬ a
4	¬ b ∧ a	5	¬ b	6	¬ (a ∧ b) ∧ (a ∨ b)	7	¬ (a ∧ b)
8	a ∧ b	9	a ∧ b ∨ ¬ (a ∨ b)	10	b	11	¬ a ∨ b
12	a	13	¬ b ∨ a	14	a ∨ b	15	¬ a ∨ a

And (∧) Or (∨) Not (¬)

0	¬ a ∧ a	1	¬ a ∧ ¬ b	2	¬ a ∧ b	3	¬ a
4	¬ b ∧ a	5	¬ b	6	¬ (¬ a ∧ ¬ b) ∧ ¬ (a ∧ b)	7	¬ (a ∧ b)
8	a ∧ b	9	¬ (¬ a ∧ b) ∧ ¬ (¬ b ∧ a)	10	b	11	¬ (¬ b ∧ a)
12	a	13	¬ (¬ a ∧ b)	14	¬ (¬ a ∧ ¬ b)	15	¬ (¬ a ∧ a)

And (∧) Not (¬)

0	¬ (¬ a ∨ a)	1	¬ (a ∨ b)	2	¬ (¬ b ∨ a)	3	¬ a
4	¬ (¬ a ∨ b)	5	¬ b	6	¬ (¬ a ∨ b) ∨ ¬ (¬ b ∨ a)	7	¬ a ∨ ¬ b
8	¬ (¬ a ∨ ¬ b)	9	¬ (¬ a ∨ ¬ b) ∨ ¬ (a ∨ b)	10	b	11	¬ a ∨ b
12	a	13	¬ b ∨ a	14	a ∨ b	15	¬ a ∨ a

Or (∨) Not (¬)

0	¬ (a ⇒ a)	1	¬ (¬ a ⇒ b)	2	¬ (b ⇒ a)	3	¬ a
4	¬ (a ⇒ b)	5	¬ b	6	(a ⇒ b) ⇒ ¬ (b ⇒ a)	7	a ⇒ ¬ b
8	¬ (a ⇒ ¬ b)	9	¬ ((a ⇒ b) ⇒ ¬ (b ⇒ a))	10	b	11	a ⇒ b
12	a	13	b ⇒ a	14	¬ a ⇒ b	15	a ⇒ a

Implies (⇒) Not (¬)

0	a ⊻ a	1	(a ⇒ b) ⊻ b	2	((a ⇒ b) ⇒ b) ⊻ a	3	(a ⇒ a) ⊻ a
4	((a ⇒ b) ⇒ b) ⊻ b	5	(a ⇒ a) ⊻ b	6	a ⊻ b	7	(a ⇒ b) ⊻ a
8	(((a ⇒ b) ⇒ b) ⊻ a) ⊻ b	9	((a ⇒ a) ⊻ a) ⊻ b	10	b	11	a ⇒ b
12	a	13	b ⇒ a	14	(a ⇒ b) ⇒ b	15	a ⇒ a

Xor (⊻) Implies (⇒)

0	a∘a	1	a∘(a ∘̄ b)	2	a∘b	3	a∘(a ∘̄ a)
4	b∘a	5	b∘(a ∘̄ a)	6	a∘b ∘̄ (b ∘̄ a)	7	a∘b ∘̄ b
8	(a∘b)∘b	9	(a∘b)∘(b ∘̄ a)	10	b	11	b ∘̄ a
12	a	13	a ∘̄ b	14	a ∘̄ (a ∘̄ b)	15	a ∘̄ a

2 (∘) 13 (∘̄)

0	((a ⊼ a) ⊼ a) ⊼ ((a ⊼ a) ⊼ a)	1	((a ⊼ a) ⊼ (b ⊼ b)) ⊼ ((a ⊼ a) ⊼ a)	2	((a ⊼ a) ⊼ a) ⊼ ((a ⊼ a) ⊼ b)	3	a ⊼ a
4	((a ⊼ a) ⊼ a) ⊼ ((a ⊼ b) ⊼ a)	5	b ⊼ b	6	((a ⊼ a) ⊼ b) ⊼ ((a ⊼ b) ⊼ a)	7	a ⊼ b
8	(a ⊼ b) ⊼ (a ⊼ b)	9	((a ⊼ a) ⊼ (b ⊼ b)) ⊼ (a ⊼ b)	10	b	11	(a ⊼ b) ⊼ a
12	a	13	(a ⊼ a) ⊼ b	14	(a ⊼ a) ⊼ (b ⊼ b)	15	(a ⊼ a) ⊼ a

Nand (⊼)

0	(a ⊽ a) ⊽ a	1	a ⊽ b	2	(a ⊽ b) ⊽ a	3	a ⊽ a
4	(a ⊽ a) ⊽ b	5	b ⊽ b	6	((a ⊽ a) ⊽ (b ⊽ b)) ⊽ (a ⊽ b)	7	((a ⊽ a) ⊽ (b ⊽ b)) ⊽ ((a ⊽ a) ⊽ a)
8	(a ⊽ a) ⊽ (b ⊽ b)	9	((a ⊽ a) ⊽ b) ⊽ ((a ⊽ b) ⊽ a)	10	b	11	((a ⊽ a) ⊽ a) ⊽ ((a ⊽ a) ⊽ b)
12	a	13	((a ⊽ a) ⊽ a) ⊽ ((a ⊽ b) ⊽ a)	14	(a ⊽ b) ⊽ (a ⊽ b)	15	((a ⊽ a) ⊽ a) ⊽ ((a ⊽ a) ⊽ a)

Nor (⊽)

Functions that can be used to formulate logic. In each case the minimal combinations of primitive functions necessary to reproduce each of the 16 logical functions of two arguments is given. From these any possible logical function with any number of arguments can be obtained. Most textbook treatments of logic use AND, OR, and NOT as primitive functions. NAND and NOR are the only primitive functions that work on their own.

So this means that logic can be set up using just a single operator. But how complicated an axiom system does it then need? The first box in the picture below shows that the direct translation of the standard textbook AND, OR, NOT axiom system from page 773 is very complicated.

(a) $(a\circ b)\circ(a\circ b) = (b\circ a)\circ(b\circ a)$ | $(a\circ a)\circ(b\circ b) = (b\circ b)\circ(a\circ a)$ | $(a\circ((b\circ b)\circ((b\circ b)\circ(b\circ b))))\circ(a\circ((b\circ b)\circ((b\circ b)\circ(b\circ b)))) = a$ | $(a\circ a)\circ(((b\circ(b\circ b))\circ(b\circ(b\circ b)))\circ((b\circ(b\circ b))\circ(b\circ(b\circ b)))) = a$ | $a\circ b = ((a\circ b)\circ(a\circ b))\circ((a\circ b)\circ(a\circ b))$ | $(a\circ((b\circ b)\circ(c\circ c)))\circ(a\circ((b\circ b)\circ(c\circ c))) = (((a\circ b)\circ(a\circ b))\circ((a\circ b)\circ(a\circ b)))\circ(((a\circ c)\circ(a\circ c))\circ((a\circ c)\circ(a\circ c)))$ | $(a\circ a)\circ(((b\circ c)\circ(b\circ c))\circ((b\circ c)\circ(b\circ c))) = (((a\circ a)\circ(b\circ b))\circ((a\circ a)\circ(c\circ c)))\circ(((a\circ a)\circ(b\circ b))\circ((a\circ a)\circ(c\circ c)))$

(b) $(a\circ a)\circ(a\circ a) = a$ | $a\circ b = b\circ a$ | $a\circ((b\circ c)\circ(b\circ c)) = b\circ((a\circ c)\circ(a\circ c))$ | $(a\circ b)\circ(a\circ(b\circ b)) = a$

(c) $(a\circ a)\circ(a\circ a) = a$ | $a\circ(b\circ(b\circ b)) = a\circ a$ | $(a\circ(b\circ c))\circ(a\circ(b\circ c)) = ((b\circ b)\circ a)\circ((c\circ c)\circ a)$

(d) $(a\circ a)\circ(a\circ b) = a$ | $a\circ(a\circ b) = a\circ(b\circ b)$ | $a\circ(a\circ(b\circ c)) = b\circ(b\circ(a\circ c))$

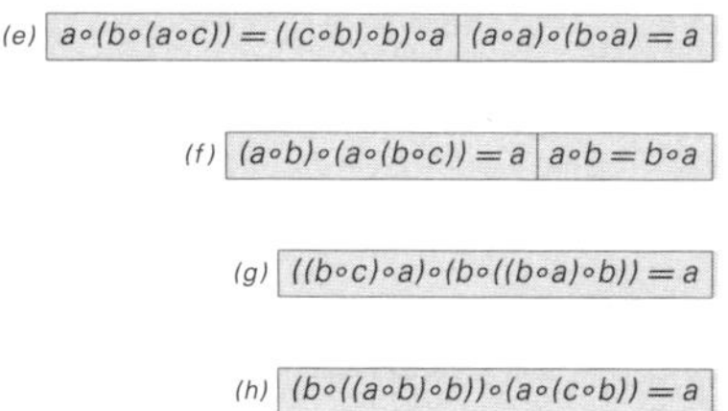

Axiom systems for basic logic (propositional calculus) formulated in terms of NAND ($\bar{\wedge}$). The number of operators that occur in these axiom systems is respectively 94, 17, 17, 13, 9, 6, 6, 6. System (a) is a translation of the standard textbook one given on page 773 in terms of AND, OR and NOT. (b) is based on the Robbins axioms from page 773. (c) is the Sheffer axiom system. (e) is the Meredith axiom system. The other axiom systems were found for this book. (d) was used on page 775. (g) and (h) are as short as is possible. Each axiom system given applies equally well to NOR as well as NAND.

But boxes (b) and (c) show that known alternative axiom systems for logic reduce the size of the axiom system by about a factor of ten. And some further reduction is achieved by manipulating the resulting axioms—leading to the axiom system used above and given in box (d).

But can one go still further? And what happens for example if one just tries to search simple axiom systems for ones that work?

One can potentially test axiom systems by seeing what operators satisfy their constraints, as on page 805. The first non-trivial axiom system that even allows the NAND operator is $\{(a\circ a)\circ(a\circ a) = a\}$. And the first axiom system for which NAND and NOR are the only operators allowed that involve 2 possible values is $\{((b\circ b)\circ a)\circ(a\circ b) = a\}$.

But if one now looks at operators involving 3 possible values then it turns out that this axiom system allows ones not equivalent to NAND

and Nor. And this means that it cannot successfully reproduce all the results of logic. Yet if any axiom system with just a single axiom is going to be able to do this, the axiom must be of the form $\{\ldots = a\}$.

With up to 6 Nands and 2 variables none of the 16,896 possible axiom systems of this kind work even up to 3-value operators. But with 6 Nands and 3 variables, 296 of the 288,684 possible axiom systems work up to 3-value operators, and 100 work up to 4-value operators.

And of the 25 of these that are not trivially equivalent, it then turns out that the two given as (g) and (h) on the facing page can actually be proved as on the next two pages to be axiom systems for logic—thus showing that in the end quite remarkable simplification can be achieved relative to ordinary textbook axiom systems.

If one looks at axiom systems of the form $\{\ldots = a,\ a\circ b = b\circ a\}$ the first one that one finds that allows only Nand and Nor with 2-value operators is $\{(a\circ a)\circ(a\circ a) = a,\ a\circ b = b\circ a\}$. But as soon as one uses a total of just 6 Nands, one suddenly finds that out of the 3402 possibilities with 3 variables 32 axiom systems equivalent to case (f) above all end up working all the way up to at least 4-value operators. And in fact it then turns out that (f) indeed works as an axiom system for logic.

So what this means is that if one were just to go through a list of the simplest few thousand axiom systems one would already be quite likely to find one that represents logic.

In human intellectual history logic has had great significance. But if one looks just at axiom systems is there anything obviously special about the ones for logic? My guess is that unless one asks about very specific details there is really not—and that standard logic is in a sense distinguished in the end only by its historical context.

One feature of logic is that its axioms effectively describe a single specific operator. But it turns out that there are all sorts of other axioms that also do this. I gave three examples on page 803, and in the picture on the right I give two more very simple examples. Indeed, given many forms of operator there are always axiom systems that can be found to describe it.

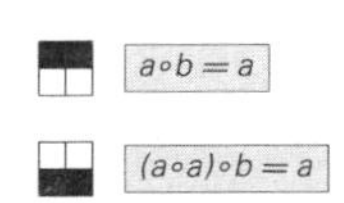

Axiom systems that reproduce equivalence results for the forms of operators shown.

L1 (a ((a a) a)) (a ((a a) a))
= A (((((a a) a) (a ((a a) a))) (a ((a a) a))) (((a a) a) ((((a a) a) (a ((a a) a))) ((a a) a)))) (a ((a a) a))
= A (((((a a) a) (a ((a a) a))) (a ((a a) a))) (((a a) a) (a ((a a) a)))) (a ((a a) a))
= A (((((a a) a) (a ((a a) a))) (a ((a a) a))) a) (a ((a a) a))
= A ((a (a ((a a) a))) a) (a ((a a) a))
= A a

L2 (a a) ((a ((a a) a)) a)
= A (a a) ((a ((a a) a)) (((a ((a a) a)) a) (a ((a a) a))))
= L1 (a a) ((a ((a a) a)) (((a ((a a) a)) ((a ((a a) a)) (a ((a a) a)))) (a ((a a) a))))
= L1 (((a ((a a) a)) (a ((a a) a))) ((a ((a a) a)) (a ((a a) a)))) ((a ((a a) a)) (((a ((a a) a)) ((a ((a a) a)) (a ((a a) a)))) (a ((a a) a))))
= A (a ((a a) a)) (a ((a a) a))
= L1 a

L3 (a b) ((a a) (((a a) b) (a a)))
= L2 (((a a) ((a ((a a) a)) a)) b) ((a a) (((a a) b) (a a)))
= A b

L4 ((a ((a b) a)) d) (b ((b d) b))
= A ((a ((a b) a)) d) (b ((b d) (((a c) b) (a ((a b) a)))))
= A ((a ((a b) a)) d) (b (((((a c) b) (a ((a b) a))) d) (((a c) b) (a ((a b) a)))))
= A ((a ((a b) a)) d) ((((a c) b) (a ((a b) a))) (((((a c) b) (a ((a b) a))) d) (((a c) b) (a ((a b) a)))))
= L3 (((((a c) b) (a ((a b) a))) ((((a c) b) ((a c) b)) (((((a c) b) ((a c) b)) (a ((a b) a))) (((a c) b) ((a c) b))))) d) ((((a c) b) (a ((a b) a))) (((((a c) b) (a ((a b) a))) d) (((a c) b) (a ((a b) a)))))
= A d

L5 (a ((a a) a)) ((a ((a a) a)) a)
= A (a ((a a) a)) ((a ((a a) a)) (((a ((a a) a)) a) (a ((a a) a))))
= A (((((a a) a) (((a a) a) (a ((a a) a))) ((a a) a))) (a ((a a) a))) (((a a) a) ((((a a) a) (a ((a a) a))) ((a a) a)))) ((a ((a a) a)) (((a ((a a) a)) a) (a ((a a) a))))
= A (((((a a) a) (((a a) a) (a ((a a) a))) ((a a) a))) (a ((a a) a))) (((a a) a) (a ((a a) a)))) ((a ((a a) a)) (((a ((a a) a)) a) (a ((a a) a))))
= A (((((a a) a) (((a a) a) (a ((a a) a))) ((a a) a))) (a ((a a) a))) a) ((a ((a a) a)) (((a ((a a) a)) a) (a ((a a) a))))
= A (((((a a) a) (((a a) a) (a ((a a) a))) ((a a) a))) (a ((a a) a))) (((((a a) a) (((a a) a) (a ((a a) a))) ((a a) a))) (a ((a a) a))) (((a a) a) ((((a a) a) (a ((a a) a))) ((a a) a)))) a) ((a ((a a) a)) (((a ((a a) a)) a) (a ((a a) a))))
= L4 a

L6 (a ((a a) a)) (a (a ((a a) a)))
= L1 (a ((a a) a)) (((a ((a a) a)) (a ((a a) a))) (a ((a a) a)))
= A ((((a ((a a) a)) ((a ((a a) a)) a)) (a ((a a) a))) ((a ((a a) a)) (((a ((a a) a)) (a ((a a) a))) (a ((a a) a))))) ((((a ((a a) a)) ((a ((a a) a)) a)) ((((a ((a a) a)) ((a ((a a) a)) a)) ((a ((a a) a)) (((a ((a a) a)) (a ((a a) a))) (a ((a a) a))))) ((a ((a a) a)) ((a ((a a) a)) a))))
= A (a ((a a) a)) (((a ((a a) a)) ((a ((a a) a)) a)) ((((a ((a a) a)) ((a ((a a) a)) a)) ((a ((a a) a)) (((a ((a a) a)) (a ((a a) a))) (a ((a a) a))))) ((a ((a a) a)) ((a ((a a) a)) a))))
= L5 (a ((a a) a)) (a ((((a ((a a) a)) ((a ((a a) a)) a)) ((a ((a a) a)) (((a ((a a) a)) (a ((a a) a))) (a ((a a) a))))) ((a ((a a) a)) ((a ((a a) a)) a))))
= L6 (a ((a a) a)) (a ((a ((a a) a)) (((a ((a a) a)) (a ((a a) a))) (a ((a a) a))))) ((a ((a a) a)) ((a ((a a) a)) a))))
= L1 (a ((a a) a)) (a ((a ((a ((a a) a)) (a (a ((a a) a))))) ((a ((a a) a)) ((a ((a a) a)) a))))
= L1 (a ((a a) a)) (a ((a ((a ((a a) a)) (((a ((a a) a)) (a ((a a) a))) (a ((a a) a))))) ((a ((a a) a)) ((a ((a a) a)) a))))
= L4 (a ((a a) a)) (a (((((a ((a a) a)) a) (a ((a a) a))) ((a ((a a) a)) (((a ((a a) a)) (a ((a a) a))) (a ((a a) a))))) ((a ((a a) a)) ((a ((a a) a)) a))))
= A (a ((a a) a)) (a ((a ((a a) a)) ((a ((a a) a)) ((a ((a a) a)) a))))

L7 (a ((a a) a)) (a (a ((a a) a)))
= L6 (a ((a a) a)) (a ((a ((a a) a)) ((a ((a a) a)) ((a ((a a) a)) a))))
= L5 (a ((a a) a)) (a ((a ((a a) a)) a))
= A (a ((a a) a)) (a (((((a a) a) (a ((a a) a))) ((a a) (((a a) (a ((a a) a))) (a a)))) a))
= A (a ((a a) a)) (a ((a ((a a) (((a a) (a ((a a) a))) (a a)))) a))
= L3 ((a (a ((a a) a))) ((a a) (((a a) (a ((a a) a))) (a a)))) (a ((a ((a a) (((a a) (a ((a a) a))) (a a)))) a))
= A (a a) (((a a) (a ((a a) a))) (a a))

L8 ((a a) (((a a) (a ((a a) a))) (a a))) ((a a) (((a a) (a ((a a) a))) (a a)))
= L7 ((a ((a a) a)) (a (a ((a a) a)))) ((a ((a a) a)) (a (a ((a a) a))))
= L1 ((a ((a a) a)) (((a ((a a) a)) (a ((a a) a))) (a ((a a) a)))) ((a ((a a) a)) (((a ((a a) a)) (a ((a a) a))) (a ((a a) a))))
= L1 a ((a a) a)

L9 (a ((a a) a)) (a (a ((a a) a)))
= L1 (a ((a a) a)) (((a ((a a) a)) (a ((a a) a))) (a ((a a) a)))
= L8 (((a ((a a) a)) (a ((a a) a))) ((((a ((a a) a)) (a ((a a) a))) ((a ((a a) a)) (((a ((a a) a)) (a ((a a) a))) (a ((a a) a))))) ((a ((a a) a)) (a ((a a) a))))) (((a ((a a) a)) (a ((a a) a))) ((((a ((a a) a)) (a ((a a) a))) ((a ((a a) a)) (((a ((a a) a)) (a ((a a) a))) (a ((a a) a))))) ((a ((a a) a)) (a ((a a) a)))))
= L1 (a ((((a ((a a) a)) (a ((a a) a))) ((a ((a a) a)) (((a ((a a) a)) (a ((a a) a))) (a ((a a) a))))) ((a ((a a) a)) (a ((a a) a))))) (a ((((a ((a a) a)) (a ((a a) a))) ((a ((a a) a)) (((a ((a a) a)) (a ((a a) a))) (a ((a a) a))))) ((a ((a a) a)) (a ((a a) a)))))

L10 (a a) (((a a) (a ((a a) a))) (a a))
= L7 (a ((a a) a)) (a (a ((a a) a)))
= L9 (a ((((a ((a a) a)) (a ((a a) a))) ((a ((a a) a)) (((a ((a a) a)) (a ((a a) a))) (a ((a a) a))))) ((a ((a a) a)) (a ((a a) a))))) (a ((((a ((a a) a)) (a ((a a) a))) ((a ((a a) a)) (((a ((a a) a)) (a ((a a) a))) (a ((a a) a))))) ((a ((a a) a)) (a ((a a) a)))))
= L1 (a ((a ((a ((a a) a)) (((a ((a a) a)) (a ((a a) a))) (a ((a a) a))))) ((a ((a a) a)) (a ((a a) a))))) (a ((a ((a ((a a) a)) (((a ((a a) a)) (a ((a a) a))) (a ((a a) a))))) ((a ((a a) a)) (a ((a a) a)))))

L11 (a a) (((a a) (a ((a a) a))) (a a))
= L10 (a ((a ((a ((a a) a)) (((a ((a a) a)) (a ((a a) a))) (a ((a a) a))))) ((a ((a a) a)) (a ((a a) a))))) (a ((a ((a ((a a) a)) (((a ((a a) a)) (a ((a a) a))) (a ((a a) a))))) ((a ((a a) a)) (a ((a a) a)))))
= L1 (a ((a ((a ((a a) a)) (a (a ((a a) a))))) ((a ((a a) a)) (a ((a a) a))))) (a ((a ((a ((a a) a)) (a (a ((a a) a))))) ((a ((a a) a)) (a ((a a) a)))))
= L7 (a ((a ((a a) (((a a) (a ((a a) a))) (a a)))) ((a ((a a) a)) (a ((a a) a))))) (a ((a ((a a) (((a a) (a ((a a) a))) (a a)))) ((a ((a a) a)) (a ((a a) a)))))

L12 (a a) (((a a) (a ((a a) a))) (a a))
= L11 (a ((a ((a a) (((a a) (a ((a a) a))) (a a)))) ((a ((a a) a)) (a ((a a) a))))) (a ((a ((a a) (((a a) (a ((a a) a))) (a a)))) ((a ((a a) a)) (a ((a a) a)))))
= A (a ((((a a) a) (a ((a a) a))) ((a a) (((a a) (a ((a a) a))) (a a)))) ((a ((a a) a)) (a ((a a) a))))) (a ((((a a) a) (a ((a a) a))) ((a a) (((a a) (a ((a a) a))) (a a)))) ((a ((a a) a)) (a ((a a) a)))))
= A (a ((a ((a a) a)) ((a ((a a) a)) (a ((a a) a))))) (a ((a ((a a) a)) ((a ((a a) a)) (a ((a a) a)))))
= L1 (a ((a ((a a) a)) a)) (a ((a ((a a) a)) a))

L13 (a ((a a) a)) (a (a ((a a) a)))
= L7 (a a) (((a a) (a ((a a) a))) (a a))
= L12 (a ((a ((a a) a)) a)) (a ((a ((a a) a)) a))

L14 (a ((a ((a a) a)) a)) (a ((a ((a a) a)) a))
= L13 (a ((a a) a)) (a (a ((a a) a)))
= L1 (a ((a a) a)) (((a ((a a) a)) (a ((a a) a))) (a ((a a) a)))
= A (((((a a) a) (((a a) a) (a ((a a) a))) ((a a) a))) (a ((a a) a))) (a ((a a) a))) ((a ((a ((a a) a)) (a ((a a) a)))) (a ((a a) a))) (((((a a) a) (((a a) a) (a ((a a) a))) ((a a) a))) (a ((a a) a))) ((a ((a a) a)) (((a a) a) (((a a) a) (a ((a a) a))) ((a a) a))) (a ((a a) a))) (((a a) a) (a ((a a) a))) ((a a) a))) (a ((a a) a))))
= A (((((a a) a) (((a a) a) (a ((a a) a))) ((a a) a))) ((a a) a))) (a ((a a) a))) (((a a) a) (((a a) a) (a ((a a) a))) ((a a) a))) (((a a) a) (a ((a a) a))) ((a a) a))) (a ((a a) a))) (((a a) a) (a ((a a) a))) ((a a) a))) (((a a) a) (a ((a a) a))) (((a a) a) (a ((a a) a))) ((a a) a))) (a ((a a) a))) ((((a ((a a) a)) (((a a) a) (a ((a a) a))) ((a a) a))) (a ((a a) a))))) (((a a) a) (((a a) a) (a ((a a) a))) ((a a) a))) (a ((a a) a)))
= L4 (a ((a a) a)) (((((a a) a) (((a a) a) (a ((a a) a))) ((a a) a))) (a ((a a) a))) (((((a a) a) (((a a) a) (a ((a a) a))) ((a a) a))) (a ((a a) a))) (((a a) a) (((a a) a) (a ((a a) a))) ((a a) a))) ((a a) a))) ((a a) a))) (a ((a a) a))))
= L4 (a ((a a) a)) (((((a a) a) (((a a) a) (a ((a a) a))) ((a a) a))) ((a a) a))) (((a a) a) (a ((a a) a))) ((a a) a))) (a ((a a) a)))
= A (a ((a a) a)) ((((a a) a) (a ((a a) a))) ((a a) a))) (((a a) a) ((((a a) a) (a ((a a) a))) ((a a) a))) (a ((a a) a)))
= A (a ((a a) a)) ((a ((a a) a)) ((a ((a a) a)) ((((a a) a) (a ((a a) a))) (a ((a a) a)))) (a ((a a) a)))
= A (a ((a a) a)) ((a ((a a) a)) ((a ((a a) a)) (((a a) a) (a ((a a) a))) (a ((a a) a)))))

L15 (a ((a ((a a) a)) a)) (a ((a ((a a) a)) a))
= L14 (a ((a a) a)) (a ((a ((a a) a)) ((a ((a a) a)) ((((a a) a) (a ((a a) a))) (a ((a a) a))))
= A (a ((a a) a)) ((a ((a ((a a) a)) ((a ((a a) a)) (a ((a a) a)))))
= L13 (a ((a a) a)) ((a ((a ((a a) a)) a)) (a ((a ((a a) a)) a)))

L16 (a ((a ((a a) a)) a)) (a ((a ((a a) a)) a))
= L15 (a ((a a) a)) ((a ((a ((a a) a)) a)) ((a ((a ((a a) a)) a)) (a ((a ((a a) a)) a))))
= L12 (a ((a a) a)) ((a ((a ((a a) a)) a)) ((a a) (((a a) (a ((a a) a))) (a a))))
= L3 (a ((a a) a)) (a ((a a) a))

L17 (a ((a a) a)) (a (a ((a a) a)))
= L13 (a ((a ((a a) a)) a)) (a ((a ((a a) a)) a))
= L16 (a ((a a) a)) (a ((a a) a))
= L1 a

L18 a ((a a) a)
= L1 ((a ((a a) a)) (((a ((a a) a)) (a ((a a) a))) (a ((a a) a)))) ((a ((a a) a)) (((a ((a a) a)) (a ((a a) a))) (a ((a a) a))))
= L16 ((a ((a a) a)) (((a ((a a) a)) (a ((a a) a))) (a ((a a) a)))) ((a ((a a) a)) (((a ((a a) a)) (a ((a a) a))) (a ((a a) a))))
= L1 ((a ((a a) a)) (((a ((a a) a)) (a ((a a) a))) (a ((a a) a)))) ((a ((a a) a)) (((a ((a a) a)) (a ((a a) a))) (a ((a a) a))))
= L17 ((a ((a a) a)) (a (a ((a a) a)))) ((a ((a a) a)) (a (a ((a a) a))))
= L17 a a

L19 (a a) (a (a a))
= L18 (a a) (a (a ((a a) a)))
= L18 (a ((a a) a)) (a (a ((a a) a)))
= L17 a

L20 a (a ((a ((a a) a)) a))
= L18 a (a ((a ((a a) a)) a))
= L2 ((a a) ((a ((a a) a)) a)) (a ((a ((a a) a)) a))
= A (a ((a a) a)) a

L21 a (a a)
= L18 a (a ((a a) a))
= L18 a (a ((a ((a a) a)) a))
= L20 (a ((a a) a)) a

L22 (a a) (a a)
= L18 (a ((a a) a)) (a ((a a) a))
= L1 a

T1 (a a) (a a)
= L22 a

L23 (a a) a
= L18 (a ((a a) a)) a
= L21 a (a a)

L24 ((a b) a) (a a)
= L18 ((a b) a) (a ((a a) a))
= A a

L25 a ((a ((a b) (a b))) a)
= A (((a b) ((a b) (a b))) (a ((a ((a b) (a b))) a))) ((a b) (((a b) (a ((a (a b)) (a b))) a))) (a b)))
= A ((a b) (a b)) ((a b) (((a b) (a ((a (a b)) a))) (a b)))
= L19 ((a b) (a b)) ((a b) (((((a b) (a b)) ((a b) ((a b) (a b)))) (a ((a ((a b) (a b))) a))) (a b)))
= A ((a b) (a b)) ((a b) (((((a b) (a b)) (((a b) (a b)) (a ((a (a b)) a)))) ((a b) (a b)))) (a ((a ((a b) (a b))) a))) (a b)))
= L24 ((a b) (a b)) ((a b) ((((((a b) (a b)) (((a b) (a b)) (a ((a (a b)) a)))) ((a b) (a b))))) ((((a b) (a b)) ((((a b) (a b)) (a ((a (a b)) a))) ((a b) (a b))))) ((((a b) (a b)) (((a b) (a b)) (a ((a (a b)) a)))) ((a b) (a b))))) (a ((a ((a b) (a b))) a))) (a b)))
= A ((a b) (a b)) ((a b) ((((a (a ((a (a b)) a)) ((((a b) (a b)) ((((a b) (a b)) (a ((a (a b)) a))) ((a b) (a b))))) (a ((a ((a b) (a b))) a))) (a b)))
= A ((a b) (a b)) ((a b) ((((a ((a (a b)) a)) ((((a b) (a b)) ((((a b) (a b)) (a ((a (a b)) a))) ((a b) (a b)))))) ((a b) (a b))) (a ((a ((a b) (a b))) a))) (a b)))
= L19 ((a b) (a b)) ((a b) ((((a ((a (a b)) a)) ((a b) (a b))) (a ((a ((a b) (a b))) a))) (a b)))
= A ((a b) (a b)) ((a b) (((a b) (a b)) (a b)))

L26 a ((a ((a b) (a b))) a)
= L25 ((a b) (a b)) ((a b) (((a b) (a b)) (a b)))
= A ((a b) (a b)) (((((a b) (a b)) (a b)) ((a b) (((a b) (a b)) (a b)))) (((a b) (a b)) ((((a b) (a b)) ((a b) (((a b) (a b)) (a b)))) ((a b) (a b)))))
= A ((a b) (a b)) ((a b) (((a b) (a b)) ((((a b) (a b)) ((a b) (((a b) (a b)) (a b)))) ((a b) (a b)))))
= L12 ((a b) (a b)) ((a b) (((a b) (((a b) (((a b) (a b)) (a b))) (a b))) ((a b) (((a b) (((a b) (a b)) (a b))) (a b)))))
= L16 ((a b) (a b)) ((a b) (((a b) (((a b) (a b)) (a b))) ((a b) (((a b) (a b)) (a b)))))
= L1 ((a b) (a b)) ((a b) (a b))

L27 a ((a ((a b) (a b))) a)
= L26 ((a b) (a b)) ((a b) (a b))
= L22 a b

L28 a
= L24 ((a b) a) (a a)
= L27 ((a b) a) ((((a b) a) ((((a b) a) (a a)) (((a b) a) (a a)))) ((a b) a))
= L24 ((a b) a) ((((a b) a) (a a)) ((a b) a))

L29 ((a b) a) (a ((a b) a))
= L24 ((a b) a) ((((a b) a) (a a)) ((a b) a))
= L28 a

L30 (a b) (a a)
= L27 (a ((a ((a b) (a b))) a)) (a a)
= L29 (a ((a ((a b) (a b))) a)) ((((a ((a b) (a b))) a) (a ((a ((a b) (a b))) a))) (((a ((a b) (a b))) a) (a ((a ((a b) (a b))) a))))
= L29 (a ((a ((a b) (a b))) a)) ((((a ((a b) (a b))) a) ((((a ((a b) (a b))) a) (a ((a ((a b) (a b))) a))) ((a ((a b) (a b))) a))) (((a ((a b) (a b))) a) ((((a ((a b) (a b))) a) (a ((a ((a b) (a b))) a))) ((a ((a b) (a b))) a))))
= A (((((a ((a b) (a b))) a) ((((a ((a b) (a b))) a) (a ((a ((a b) (a b))) a))) ((a ((a b) (a b))) a))) (a ((a ((a b) (a b))) a))) (((a ((a b) (a b))) a) ((((a ((a b) (a b))) a) (a ((a ((a b) (a b))) a))) ((a ((a b) (a b))) a)))) ((((a ((a b) (a b))) a) ((((a ((a b) (a b))) a) (a ((a ((a b) (a b))) a))) ((a ((a b) (a b))) a))) (((a ((a b) (a b))) a) ((((a ((a b) (a b))) a) (a ((a ((a b) (a b))) a))) ((a ((a b) (a b))) a))))
= L24 ((a ((a b) (a b))) a) ((((a ((a b) (a b))) a) (a ((a ((a b) (a b))) a))) ((a ((a b) (a b))) a))
= L29 ((a ((a b) (a b))) a) (a ((a ((a b) (a b))) a))
= L29 a

L31 b ((a b) (a b))
= L3 ((a b) ((a a) (((a a) b) (a a)))) ((a b) (a b))
= L30 a b

L32 a ((a b) (a b))
= L30 ((a b) (a a)) ((a b) (a b))
= L30 a b

L33 a ((a b) a)
= L32 a ((a ((a b) (a b))) a)
= L27 a b

L34 ((a b) c) (b c)
= L33 ((a b) c) (b ((b c) b))
= L33 ((a ((a b) a)) c) (b ((b c) b))
= L4 c

L35 b a
= L31 a ((b a) (b a))
= L33 a ((a ((b a) (b a))) a)
= L31 a ((b a) a)

L36 b (b (a b))
= L33 b (b (a ((a b) a)))
= L33 b (b ((b (a ((a b) a))) b))
= A (((a ((a b) a)) b) (a ((a b) a))) (b ((b (a ((a b) a))) b))
= A (((a ((a b) a)) (((a ((a b) a)) b) (a ((a b) a)))) (a ((a b) a))) (b ((b (a ((a b) a))) b))
= L4 a ((a b) a)
= L33 a b

L37 (a b) a
= L36 a (a ((a b) a))
= L33 a (a b)

L38 (b a) a
= L36 a (a ((b a) a))
= L35 a (b a)

L39 (b b) (b (a b))
= L38 (b b) ((a b) b)
= L33 (b b) ((a b) (((a b) b) (a b)))
= L3 (b b) ((a b) (((a b) ((a b) ((a a) (((a a) b) (a a))))) (a b)))
= L3 (((a b) ((a a) (((a a) b) (a a)))) ((a b) ((a a) (((a a) b) (a a))))) ((a b) (((a b) ((a b) ((a a) (((a a) b) (a a))))) (a b)))
= A (a b) ((a a) (((a a) b) (a a)))
= L3 b

L40 a
= L39 (a a) (a ((b a) a))
= L38 (a a) (a (a (b a)))
= L36 (a a) (b a)

L41 a
= L39 (a a) (a ((a b) a))
= L33 (a a) (a b)

L42 b a
= L41 b ((a a) (a b))

= [L33] b((aa)(a((ab)a)))
= [L33] b((aa)(((aa)(a((ab)a)))(aa)))
= [A] (((aa)b)(a((ab)a)))((aa)(((aa)(a((ab)a)))(aa)))
= [A] a((ab)a)
= [L33] ab

[L43] (ac)((ab)c)
= [L42] ((ab)c)(ac)
= [L33] ((ab)c)(a((ac)a))
= [A] c

[L44] (bc)((ab)c)
= [L42] ((ab)c)(bc)
= [L34] c

[L45] (ba)((ac)b)
= [L42] (ab)((ac)b)
= [L43] b

[L46] (ba)(a(bc))
= [L42] (ba)((bc)a)
= [L43] a

[L47] bc
= [L46] ((ab)(bc))((bc)((ab)c))
= [L44] ((ab)(bc))c
= [L42] c((ab)(bc))

[L48] ab
= [L45] ((ab)(ca))(((ca)b)(ab))
= [L42] ((ab)(ca))((ab)((ca)b))
= [L44] ((ab)(ca))b

[L49] b((ab)(ca))
= [L42] ((ab)(ca))b
= [L48] ab

[L50] (ab)c
= [L43] (a((ab)c))((ac)((ab)c))
= [L43] (a((ab)c))c
= [L42] c(a((ab)c))

[L51] a(b(ab))
= [L42] a((ab)b)
= [L42] ((ab)b)a
= [L50] a((ab)(((ab)b)a))
= [L42] a((ba)(((ab)b)a))
= [L44] aa

[L52] (ba)(ab)
= [L50] (ab)(b((ba)(ab)))
= [L47] (ab)(ab)

[L53] (aa)((ba)(ba))
= [L44] (aa)((ba)(((aa)(ba))((((aa)(ba))(aa))(ba))))
= [L50] (aa)((((aa)(ba))(aa))(ba))
= [L42] (aa)((ba)(((aa)(ba))(aa)))
= [L42] (aa)((ba)((aa)((aa)(ba))))
= [L40] (aa)((ba)((aa)a))
= [L42] (aa)((ba)(a(aa)))
= [L47] a(aa)

[L54] ((ab)(ab))((ab)(ab))
= [L52] ((ba)(ab))((ba)(ab))
= [L52] ((ab)(ba))((ba)(ab))
= [L52] ((ab)(ba))((ab)(ab))

[L55] ab
= [L22] ((ab)(ab))((ab)(ab))
= [L54] ((ab)(ba))((ab)(ab))
= [L52] ((ba)(ba))((ab)(ab))

[L56] a(b(bb))
= [L53] a((bb)((ab)(ab)))
= [L42] a(((ab)(ab))(bb))
= [L40] a(((ab)(ab))(((bb)(ab))((bb)(ab))))
= [L53] a((ab)((ab)(ab)))
= [L42] a(((ab)(ab))(ab))
= [L32] a(((ab)(ab))(a((ab)(ab))))
= [L51] aa

[T2] a(b(bb))
= [L56] aa

[L57] ((aa)(((ab)(ab))c))((aa)(((ab)(ab))c))
= [L56] ((aa)(((ab)(d(dd)))c))((aa)(((ab)(d(dd)))c))
= [L56] ((a(d(dd)))(((ab)(d(dd)))c))((a(d(dd)))(((ab)(d(dd)))c))
= [L56] ((a(d(dd)))(((ab)(d(dd)))c))(d(dd))
= [L42] (d(dd))((a(d(dd)))(((ab)(d(dd)))c))
= [L42] (d(dd))((((ab)(d(dd)))c)(a(d(dd))))
= [L42] ((((ab)(d(dd)))c)(a(d(dd))))(d(dd))
= [L46] ((((ab)(d(dd)))c)(a(d(dd))))(((ab)(d(dd)))((d(dd))((ab)(d(dd)))))
= [L33] ((((ab)(d(dd)))c)(a((a(d(dd)))a)))(((ab)(d(dd)))((d(dd))((ab)(d(dd)))))
= [A] ((((ab)(d(dd)))c)(a((a(d(dd)))a)))(((ab)(d(dd)))((((ab)(d(dd)))(a((a(d(dd)))a)))((ab)(d(dd)))))
= [A] a((a(d(dd)))a)
= [L33] a(d(dd))

[L58] (bb)(((bc)(bc))d)
= [L22] (((bb)(((bc)(bc))d))((bb)(((bc)(bc))d)))(((bb)(((bc)(bc))d))((bb)(((bc)(bc))d)))
= [L57] (b(a(aa)))(b(a(aa)))
= [L56] (bb)(bb)

[L59] (aa)(((ab)(ab))c)
= [L58] (aa)(aa)
= [L22] a

[L60] a
= [L58] (aa)(((a((ba)(cb)))(a((ba)(cb))))((d((a((ba)(cb)))(a((ba)(cb)))))(ed)))
= [L49] (aa)(d((a((ba)(cb)))(a((ba)(cb)))))
= [L49] (aa)(d((ba)(ba)))

[L61] c((ac)(((ab)c)((ab)c)))
= [L42] c((((ab)c)((ab)c))(ac))
= [L42] ((((ab)c)((ab)c))(ac))c
= [L46] ((((ab)c)((ab)c))(ac))(((ab)c)(c((ab)c)))
= [L33] ((((ab)c)((ab)c))(a((ac)a)))(((ab)c)(c((ab)c)))
= [A] ((((ab)c)((ab)c))(a((ac)a)))(((ab)c)((((ab)c)(a((ac)a)))((ab)c)))
= [A] a((ac)a)
= [L33] ac

[L62] (aa)b
= [L61] b(((aa)b)((((aa)(((ac)(ac))d))b)(((aa)(((ac)(ac))d))b)))
= [L59] b(((aa)b)((ab)(ab)))
= [L50] b(((ab)(ab))((aa)(((aa)b)((ab)(ab)))))
= [L45] b(((ab)(ab))((((ab)((bc)a))((ab)((bc)a)))(((aa)b)((ab)(ab)))))
= [L59] b(ab)

[L63] a(ab)
= [L42] a(ba)
= [L62] (bb)a

[L64] a(bc)
= [L45] ((a(bc))c)((ca)(a(bc)))
= [L42] ((a(bc))c)((ca)((bc)a))
= [L44] ((a(bc))c)a

[L65] a(bc)
= [L64] ((a(bc))c)a
= [L42] a((a(bc))c)
= [L42] a(c(a(bc)))

[L66] ac
= [L59] ((ac)(ac))((((ac)(ca))((ac)(ca)))b)
= [L52] ((ac)(ac))((((ca)(ca))((ca)(ca)))b)
= [L22] ((ac)(ac))((ca)b)

[L67] (ab)(ab)
= [L59] (((ab)(ab))((ab)(ab)))((((ab)(ab))((ba)(ba)))(((ab)(ab))((ba)(ba))))c)
= [L55] (((ab)(ab))((ab)(ab)))(((ba)(ba))c)
= [L22] (ab)(((ba)(ba))c)

[L68] a(((bc)(ba))((bc)(ba)))
= [L42] (((bc)(ba))((bc)(ba)))a
= [L63] a(a((bc)(ba)))
= [L33] a(a((bc)(b((ba)b))))
= [L33] a(a((bc)(((bc)(b((ba)b)))(bc))))
= [A] a((((bc)a)(b((ba)b)))((bc)(((bc)(b((ba)b)))(bc))))
= [A] a(b((ba)b))
= [L33] a(ba)
= [L62] (bb)a

[L69] (bc)a
= [L22] (((bc)(bc))((bc)(bc)))a
= [L68] a(((((bc)(bc))((cb)(cb)))(((bc)(bc))a))((((bc)(bc))((cb)(cb)))(((bc)(bc))a)))
= [L55] a(((cb)(((bc)(bc))a))((cb)(((bc)(bc))a)))

[L70] (bc)a
= [L69] a(((cb)(((bc)(bc))a))((cb)(((bc)(bc))a)))
= [L67] a(((cb)(cb))((cb)(cb)))
= [L22] a(cb)

[L71] ((bc)(bc))a
= [L68] a((((bc)(cb))((bc)a))(((bc)(cb))((bc)a)))
= [L52] a((((cb)(cb))((bc)a))(((cb)(cb))((bc)a)))
= [L66] a((cb)(cb))

[L72] (ba)(((bc)a)((bc)a))
= [L3] (a((ba)(((bc)a)((bc)a))))((aa)(((aa)((ba)(((bc)a)((bc)a))))(aa)))
= [L33] (a((ba)(((bc)a)((bc)a))))((aa)((ba)(((bc)a)((bc)a))))
= [L61] (ba)((aa)((ba)(((bc)a)((bc)a))))

[L73] (ba)(((bc)a)((bc)a))
= [L72] (ba)((aa)((ba)(((bc)a)((bc)a))))
= [L60] (ba)a
= [L70] a(ab)
= [L63] (bb)a

[L74] (aa)c
= [L73] (ac)(((ab)c)((ab)c))
= [L50] (((ab)c)((ab)c))(a((ac)(((ab)c)((ab)c))))
= [L73] (((ab)c)((ab)c))(a((aa)c))
= [L22] (((ab)c)((ab)c))(((aa)(aa))((aa)c))
= [L41] (((ab)c)((ab)c))(aa)

[L75] (aa)((c(ab))(c(ab)))
= [L71] (((ab)c)((ab)c))(aa)
= [L74] (aa)c

[L76] (b(ac))(aa)
= [L22] (((b(ac))(b(ac)))((b(ac))(b(ac))))(aa)
= [L63] (aa)((aa)((b(ac))(b(ac))))
= [L75] (aa)((aa)b)

[L77] ((ab)(ab))(ca)
= [L75] ((ab)(ab))(((ca)((ab)c))((ca)((ab)c)))
= [L45] ((ab)(ab))(cc)

[L78] ((bc)(bc))a
= [L45] ((bc)(bc))((ab)((bc)a))
= [L70] (((bc)a)(ab))((bc)(bc))
= [L22] (((((bc)a)(ab))(((bc)a)(ab)))((((bc)a)(ab))(((bc)a)(ab))))((bc)(bc))
= [L63] ((bc)(bc))(((bc)(bc))((((bc)a)(ab))(((bc)a)(ab))))
= [L76] (((((bc)a)(ab))(((bc)a)(ab)))((bc)(ab)))((bc)(bc))
= [L70] ((bc)(bc))(((bc)(ab))((((bc)a)(ab))(((bc)a)(ab))))
= [L73] ((bc)(bc))(((bc)(bc))(ab))
= [L63] ((ab)(ab))((bc)(bc))
= [L47] ((ab)(ab))((c((ab)(bc)))(c((ab)(bc))))
= [L75] ((ab)(ab))c
= [L71] c((ba)(ba))

[L79] a(c((ab)(ab)))
= [L42] a(c((ba)(ba)))
= [L78] a(((bc)(bc))a)
= [L62] (((bc)(bc))((bc)(bc)))a
= [L22] (bc)a

[L80] a((ba)c)
= [L70] (c(ba))a
= [L79] a((ba)((ac)(ac)))
= [L42] a(((ac)(ac))(ba))
= [L77] a(((ac)(ac))(bb))
= [L78] a(c((a(bb))(a(bb))))
= [L79] ((bb)c)a

[L81] ((ca)(ab))((ca)(ab))
= [L40] ((((ca)(ab))((ca)(ab)))(((ca)(ab))((ca)(ab))))((aa)(((ca)(ab))((ca)(ab))))
= [L75] ((((ca)(ab))((ca)(ab)))(((ca)(ab))((ca)(ab))))((aa)(ca))
= [L22] ((ca)(ab))((aa)(ca))
= [L70] ((aa)(ca))((ab)(ca))
= [L40] a((ab)(ca))
= [L70] ((ca)(ab))a
= [L79] a((ab)((a(ca))(a(ca))))
= [L70] a(((a(ca))(a(ca)))(ba))
= [L77] a(((a(ca))(a(ca)))(bb))
= [L78] a((ca)((a(bb))(a(bb))))
= [L79] ((bb)(ca))a

[T3] ((bb)a)((cc)a)
= [L42] ((bb)a)(a(cc))
= [L42] (a(cc))((bb)a)
= [L22] (((aa)(aa))(cc))((bb)a)
= [L80] ((bb)a)(((aa)((bb)a))(cc))
= [L70] ((cc)((aa)((bb)a)))((bb)a)
= [L81] (((aa)((bb)a))(((bb)a)c))(((aa)((bb)a))(((bb)a)c))
= [L40] (a(((bb)a)c))(a(((bb)a)c))
= [L80] ((((bb)(bb))c)a)((((bb)(bb))c)a)
= [L42] ((c((bb)(bb)))a)((c((bb)(bb)))a)
= [L40] ((((c((bb)(bb)))(c((bb)(bb))))((bb)(c((bb)(bb)))))a)((((c((bb)(bb)))(c((bb)(bb))))((bb)(c((bb)(bb)))))a)
= [L60] ((((c((bb)(bb)))(c((bb)(bb))))b)a)((((c((bb)(bb)))(c((bb)(bb))))b)a)
= [L42] ((b((c((bb)(bb)))(c((bb)(bb)))))a)((b((c((bb)(bb)))(c((bb)(bb)))))a)
= [L78] ((((cb)(cb))((bb)(bb)))a)((((cb)(cb))((bb)(bb)))a)
= [L40] ((((cb)(cb))((b((bb)(cb)))(b((bb)(cb)))))a)((((cb)(cb))((b((bb)(cb)))(b((bb)(cb)))))a)
= [L65] ((((cb)(cb))((b((cb)(b((bb)(cb)))))(b((cb)(b((bb)(cb)))))))a)((((cb)(cb))((b((cb)(b((bb)(cb)))))(b((cb)(b((bb)(cb)))))))a)
= [L75] ((((cb)(cb))b)a)((((cb)(cb))b)a)
= [L78] ((b((cb)(cb)))a)((b((cb)(cb)))a)
= [L31] ((cb)a)((cb)a)
= [L70] (a(bc))(a(bc))

A proof that the axiom system $\{((b\circ c)\circ a)\circ(b\circ((b\circ a)\circ b)) = a\}$ given as example (g) on page 808 can reproduce the Sheffer axiom system (c), and is thus a complete axiom system for logic. The proof involves taking the original axiom [A] and using it to establish a sequence of lemmas [Ln], from which it is eventually possible to prove the three Sheffer axioms [Tn]. In each part of the proof each line can be obtained from the previous one just as on page 775 by applying the axiom or lemma indicated. Explicit ⊼ operators have been omitted to allow expressions to be printed more compactly. The proof shown takes a total of 343 steps, and involves intermediate expressions with as many as 128 NANDS. It is quite possible that the proof could be considerably shortened. Note that any proof can always be recast without lemmas, but will usually then be much longer.

So what about patterns of theorems? Does logic somehow stand out when one looks at these? The picture below shows which possible simple equivalence theorems hold in systems from page 805.

And comparing with page 805 one sees that typically the more forms of operator are allowed by the constraints of an axiom system, the fewer equivalence results hold in that axiom system.

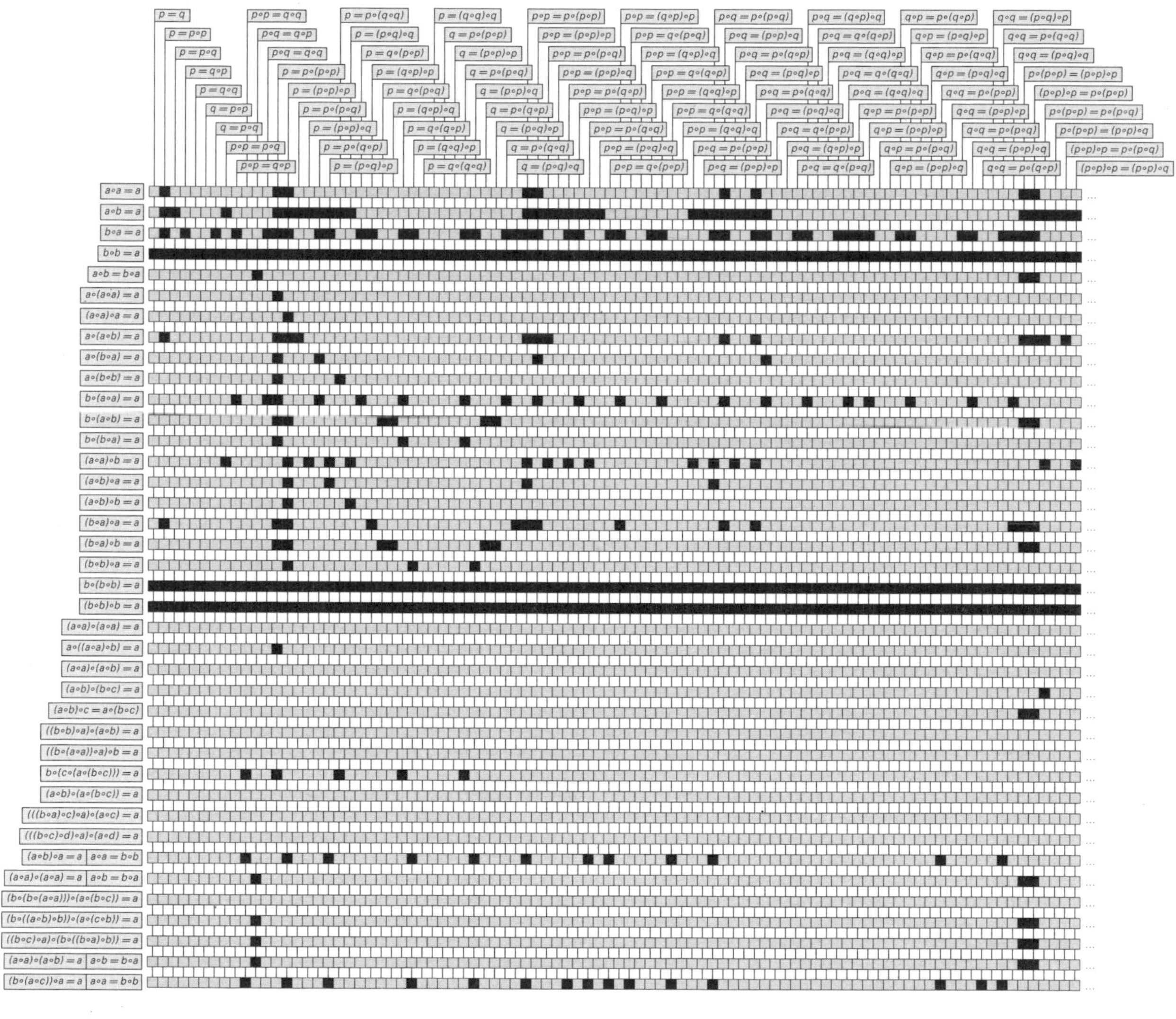

Theorems that can be proved on the basis of simple axiom systems from page 805. A black square indicates that a particular theorem holds in a particular axiom system. In general the question of whether a given theorem holds is undecidable, but the particular theorems given here happen to be simple enough that results for them can with some effort be established with certainty.

So what happens if essentially just a single form of operator is allowed? The pictures below show results for the 16 forms from page 806, and among these one sees that logic yields the fewest theorems.

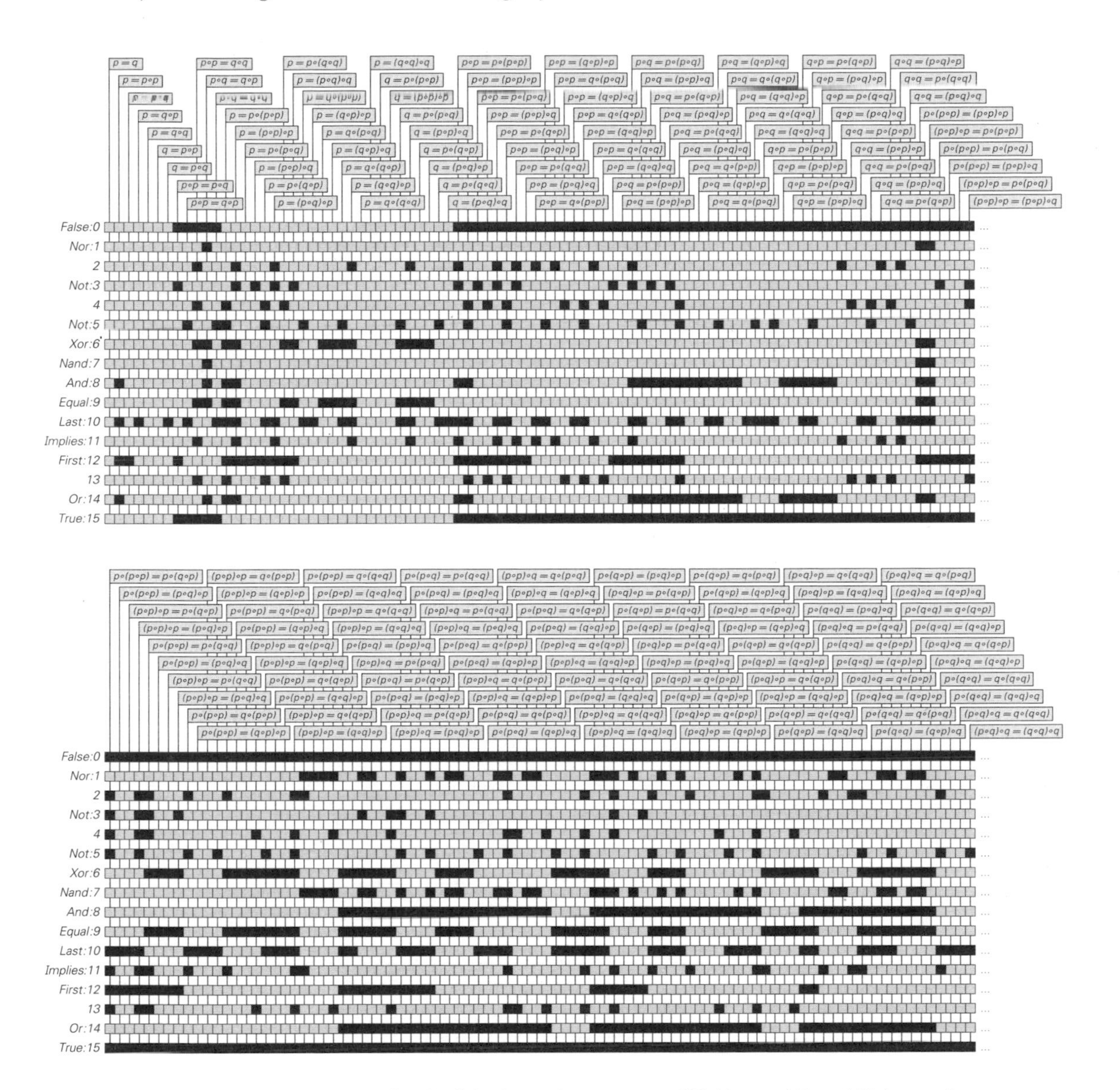

Theorems that hold with operators of each of the forms shown on page 806. NAND and NOR yield the smallest number of theorems.

But if one considers for example analogs of logic for variables with more than two possible values, the picture below shows that one immediately gets systems with still fewer theorems.

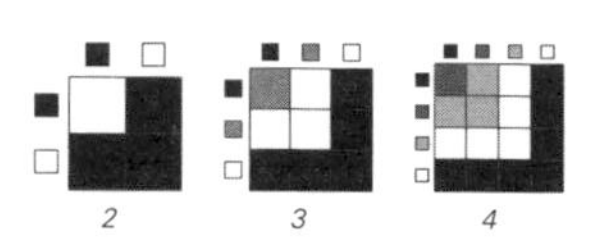

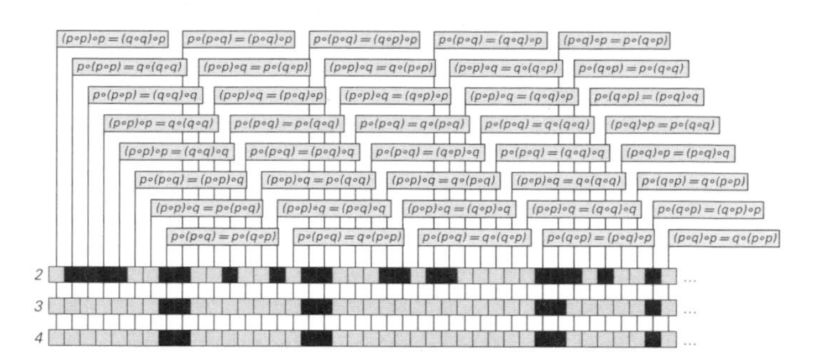

Theorems in analogs of logic that allow different numbers of truth values. Statements like $p = \neg\neg p$ do not hold in general with more than 2 truth values.

So what about proofs? Is there something about these that is somehow special in the case of ordinary logic?

In the axiom systems on page 803 the typical lengths of proofs seem to increase from one system to the next, so that they end up being longest for the last axiom system, which corresponds to logic.

But if one picks a different axiom system for logic—say one of the others on page 808—then the length of a particular proof will usually change. But since one can always just start by proving the new axioms, the change can only be by a fixed amount. And as it turns out, even the simplest axiom system (f) given on page 808 seems to allow fairly short proofs of at least most short theorems.

But as one tries to prove progressively longer theorems it appears that whatever axiom system one uses for logic the lengths of proofs can increase as fast as exponentially. A crucial point, however, is that for theorems of a given length there is always a definite upper limit on the length of proof needed. Yet once again this is not something unique to logic. Indeed, it turns out that this must always be the case for any axiom system—like those on page 803—that ends up allowing essentially only operators of a single form.

So what about other axiom systems?

The very simplest ones on pages 805 and 812 seem to yield proofs that are always comparatively short. But when one looks at axiom systems that are even slightly more complicated the proofs of anything but the simplest results can get much longer—making it in practice often difficult to tell whether a given result can actually even be proved at all.

And this is in a sense just another example of the same basic phenomenon that we already saw early in this section in multiway systems, and that often seems to occur in real mathematics: that even if a theorem is short to state, its proof can be arbitrarily long.

And this I believe is ultimately a reflection of the Principle of Computational Equivalence. For the principle suggests that most axiom systems whose consequences are not obviously simple will tend to be universal. And this means that they will exhibit computational irreducibility and undecidability—and will allow no general upper limit to be placed on how long a proof could be needed for any given result.

As I discussed earlier, most of the common axiom systems in traditional mathematics are known to be universal—basic logic being one of the few exceptions. But one might have assumed that to achieve their universality these axiom systems would have to be specially set up with all sorts of specific sophisticated features.

Yet from the results of this book—as embodied in the Principle of Computational Equivalence—we now know that this is not the case, and that in fact universality should already be rather common even among very simple axiom systems, like those on page 805.

And indeed, while operator systems and multiway systems have many superficial differences, I suspect that when it comes to universality they work very much the same. So in either idealization, one should not have to go far to get axiom systems that exhibit universality—just like most of the ones in traditional mathematics.

But once one has reached an axiom system that is universal, why should one in a sense ever have to go further? After all, what it means for an axiom system to be universal is that by setting up a suitable encoding it must in principle be possible to make that axiom system reproduce any other possible axiom system.

But the point is that the kinds of encodings that are normally used in mathematics are in practice rather limited. For while it is common, say, to take a problem in geometry and reformulate it as a problem in algebra, this is almost always done just by setting up a direct translation between the objects one is describing—usually in effect just by renaming the operators used to manipulate them.

Yet to take full advantage of universality one must consider not only translations between objects but also translations between complete proofs. And if one does this it is indeed perfectly possible, say, to program arithmetic to reproduce any proof in set theory. In fact, all one need do is to encode the axioms of set theory in something like the arithmetic equation system of page 786.

But with the notable exception of Gödel's Theorem these kinds of encodings are not normally used in mathematics. So this means that even when universality is present realistic idealizations of mathematics must still distinguish different axiom systems.

So in the end what is it that determines which axiom systems are actually used in mathematics? In the course of this section I have discussed a few criteria. But in the end history seems to be the only real determining factor. For given almost any general property that one can pick out in axiom systems like those on pages 773 and 774 there typically seem to be all sorts of operator and multiway systems—often including some rather simple ones—that share the exact same property.

So this leads to the conclusion that there is in a sense nothing fundamentally special about the particular axiom systems that have traditionally been used in mathematics—and that in fact there are all sorts of other axiom systems that could perfectly well be used as foundations for what are in effect new fields of mathematics—just as rich as the traditional ones, but without the historical connections.

So what about existing fields of mathematics? As I mentioned earlier in this section, I strongly believe that even within these there are fundamental limitations that have implicitly been imposed on what has actually been studied. And most often what has happened is that there are only certain kinds of questions or statements that have been considered of real mathematical interest.

The picture on the facing page shows a rather straightforward version of this. It lists in order a large number of theorems from basic logic, highlighting just those few that are considered interesting enough by typical textbooks of logic to be given explicit names.

1 $a = a \land a$	2 $a = a \lor a$	$a \land a = a \lor a$	3 $a \land b = b \land a$	4 $a \lor b = b \lor a$	5 $a = \neg\neg a$
$a \land a = \neg\neg a$	$a \lor a = \neg\neg a$	$\neg a = \neg(a \land a)$	$\neg a = \neg(a \lor a)$	$a = (a \land a) \land a$	$a = (a \lor a) \land a$
$a = a \land (a \land a)$	$a = a \land (a \lor a)$	$a = a \land a \lor a$	$a = (a \lor a) \lor a$	$a = a \lor a \land a$	$a = a \lor (a \lor a)$
6 $a = a \land (a \lor b)$	7 $a = a \lor a \land b$	$a = (a \lor b) \land a$	$a = a \land (b \lor a)$	$a = a \land b \lor a$	$a = a \lor b \land a$
$a = (b \lor a) \land a$	$a = b \land a \lor a$	$\neg a = \neg a \land \neg a$	$\neg a = \neg a \lor \neg a$	$\neg a \land a = a \land \neg a$	$\neg a \lor a = a \lor \neg a$
$\neg(a \land a) = \neg(a \lor a)$	$\neg\neg a = (a \land a) \land a$	$\neg\neg a = (a \lor a) \land a$	$\neg\neg a = a \land (a \land a)$	$\neg\neg a = a \land (a \lor a)$	$\neg\neg a = a \land a \lor a$
$\neg\neg a = (a \lor a) \lor a$	$\neg\neg a = a \lor a \land a$	$\neg\neg a = a \lor (a \lor a)$	$\neg\neg a = a \land (a \lor b)$	$\neg\neg a = a \lor a \land b$	$\neg\neg a = (a \lor b) \land a$
$\neg\neg a = a \land (b \lor a)$	$\neg\neg a = a \land b \lor a$	$\neg\neg a = a \lor b \land a$	8 $\neg a \land a = \neg b \land b$	$a \land \neg a = \neg b \land b$	$\neg a \land a = b \land \neg b$
$a \land \neg a = b \land \neg b$	9 $\neg a \lor a = \neg b \lor b$	$a \lor \neg a = \neg b \lor b$	$\neg a \lor a = b \lor \neg b$	$a \lor \neg a = b \lor \neg b$	$\neg\neg a = (b \lor a) \land a$
$\neg\neg a = b \land a \lor a$	$a \land \neg b = \neg b \land a$	$\neg a \land b = b \land \neg a$	$a \lor \neg b = \neg b \lor a$	$\neg a \lor b = b \lor \neg a$	$\neg(a \land b) = \neg(b \land a)$
$\neg(a \lor b) = \neg(b \lor a)$	$a \land a = (a \land a) \land a$	$a \lor a = (a \land a) \land a$	$a \land a = (a \lor a) \land a$	$a \lor a = (a \lor a) \land a$	$a \land a = a \land (a \land a)$
$a \lor a = a \land (a \land a)$	$a \land a = a \land (a \lor a)$	$a \lor a = a \land (a \lor a)$	$a \land a = a \land a \lor a$	$a \lor a = a \land a \lor a$	$a \land a = (a \lor a) \lor a$
$a \lor a = (a \lor a) \lor a$	$a \land a = a \lor a \land a$	$a \lor a = a \lor a \land a$	$a \land a = a \lor (a \lor a)$	$a \lor a = a \lor (a \lor a)$	$a \land a = a \land (a \lor b)$
$a \lor a = a \land (a \lor b)$	$a \land a = a \lor a \land b$	$a \lor a = a \lor a \land b$	$a \land a = (a \lor b) \land a$	$a \lor a = (a \lor b) \land a$	$a \land a = a \land (b \lor a)$
$a \lor a = a \land (b \lor a)$	$a \land a = a \land b \lor a$	$a \lor a = a \land b \lor a$	$a \land a = a \lor b \land a$	$a \lor a = a \lor b \land a$	$a \land a = (b \lor a) \land a$
$a \lor a = (b \lor a) \land a$	$a \land a = b \land a \lor a$	$a \lor a = b \land a \lor a$	$a \land b = (a \land a) \land b$	$a \land b = (a \lor a) \land b$	$a \land b = a \land (a \land b)$
$a \lor b = a \land a \lor b$	$a \lor b = (a \lor a) \lor b$	$a \lor b = a \lor (a \lor b)$	$a \land b = (a \land b) \land a$	$a \land b = a \land (b \land a)$	$a \lor b = (a \lor b) \lor a$
$a \lor b = a \lor (b \lor a)$	$a \land b = (a \land b) \land b$	$a \land b = a \land (b \land b)$	$a \land b = a \land (b \lor b)$	$a \lor b = (a \lor b) \lor b$	$a \lor b = a \lor b \land b$
$a \lor b = a \lor (b \lor b)$	$a \land b = (b \land a) \land a$	$a \land b = b \land (a \land a)$	$a \land b = b \land (a \lor a)$	$a \lor b = (b \lor a) \lor a$	$a \lor b = b \lor a \land a$
$a \lor b = b \lor (a \lor a)$	$a \land b = (b \land a) \land b$	$a \land b = b \land (a \land b)$	$a \lor b = (b \lor a) \lor b$	$a \lor b = b \lor (a \lor b)$	$a \land b = (b \land b) \land a$
$a \land b = (b \lor b) \land a$	$a \land b = b \land (b \land a)$	$a \lor b = b \land b \lor a$	$a \lor b = (b \lor b) \lor a$	$a \lor b = b \lor (b \lor a)$	$\neg(a \land a) = \neg a \land \neg a$
$\neg(a \lor a) = \neg a \land \neg a$	$\neg(a \land a) = \neg a \lor \neg a$	$\neg(a \lor a) = \neg a \lor \neg a$	10 $\neg(a \lor b) = \neg a \land \neg b$	$\neg(a \land b) = \neg a \lor \neg b$	$\neg(a \lor b) = \neg b \land \neg a$
$\neg(a \land b) = \neg b \lor \neg a$	$\neg a \land \neg a = \neg a \lor \neg a$	$\neg a \land \neg b = \neg b \land \neg a$	$\neg a \lor \neg b = \neg b \lor \neg a$	$(a \land a) \land a = (a \lor a) \land a$	$(a \land a) \land a = a \land (a \land a)$
$(a \lor a) \land a = a \land (a \land a)$	$(a \land a) \land a = a \land (a \lor a)$	$(a \lor a) \land a = a \land (a \lor a)$	$a \land (a \land a) = a \land (a \lor a)$	$(a \land a) \land a = a \land a \lor a$	$(a \lor a) \land a = a \land a \lor a$
$a \land (a \land a) = a \land a \lor a$	$a \land (a \lor a) = a \land a \lor a$	$(a \land a) \land a = (a \lor a) \lor a$	$(a \lor a) \land a = (a \lor a) \lor a$	$a \land (a \land a) = (a \lor a) \lor a$	$a \land (a \lor a) = (a \lor a) \lor a$
$a \land a \lor a = (a \lor a) \lor a$	$(a \land a) \land a = a \lor a \land a$	$(a \lor a) \land a = a \lor a \land a$	$a \land (a \land a) = a \lor a \land a$	$a \land (a \lor a) = a \lor a \land a$	$a \land a \lor a = a \lor a \land a$
$(a \lor a) \lor a = a \lor a \land a$	$(a \land a) \land a = a \lor (a \lor a)$	$(a \lor a) \land a = a \lor (a \lor a)$	$a \land (a \land a) = a \lor (a \lor a)$	$a \land (a \lor a) = a \lor (a \lor a)$	$a \land a \lor a = a \lor (a \lor a)$
$(a \lor a) \lor a = a \lor (a \lor a)$	$a \lor a \land a = a \lor (a \lor a)$	$(a \land a) \land a = a \land (a \lor b)$	$(a \lor a) \land a = a \land (a \lor b)$	$a \land (a \land a) = a \land (a \lor b)$	$a \land (a \lor a) = a \land (a \lor b)$
$a \land a \lor a = a \land (a \lor b)$	$(a \lor a) \lor a = a \land (a \lor b)$	$a \lor a \land a = a \land (a \lor b)$	$a \lor (a \lor a) = a \land (a \lor b)$	$(a \land a) \land a = a \lor a \land b$	$(a \lor a) \land a = a \lor a \land b$
$a \land (a \land a) = a \lor a \land b$	$a \land (a \lor a) = a \lor a \land b$	$a \land a \lor a = a \lor a \land b$	$(a \lor a) \lor a = a \lor a \land b$	$a \lor a \land a = a \lor a \land b$	$a \lor (a \lor a) = a \lor a \land b$

⋮ *50 lines*

$a \land b \lor b = b \land (b \lor a)$	$(a \lor b) \lor b = b \land b \lor a$	$a \lor b \land b = b \land b \lor a$	$a \lor (b \lor b) = b \land b \lor a$	$(a \lor b) \lor b = (b \lor b) \lor a$	$a \lor b \land b = (b \lor b) \lor a$
$a \lor (b \lor b) = (b \lor b) \lor a$	$(a \lor b) \land b = b \lor b \land a$	$a \land b \lor b = b \lor b \land a$	$(a \lor b) \lor b = b \lor (b \lor a)$	$a \lor b \land b = b \lor (b \lor a)$	$a \lor (b \lor b) = b \lor (b \lor a)$
$(a \lor b) \land b = (b \land b) \land b$	$a \land b \lor b = (b \land b) \land b$	$(a \lor b) \land b = (b \lor b) \land b$	$a \land b \lor b = (b \lor b) \land b$	$(a \lor b) \land b = b \land (b \land b)$	$a \land b \lor b = b \land (b \land b)$
$(a \lor b) \land b = b \land (b \lor b)$	$a \land b \lor b = b \land (b \lor b)$	$(a \lor b) \land b = b \land b \lor b$	$a \land b \lor b = b \land b \lor b$	$(a \lor b) \land b = (b \lor b) \lor b$	$a \land b \lor b = (b \lor b) \lor b$
$(a \lor b) \land b = b \lor b \land b$	$a \land b \lor b = b \lor b \land b$	$(a \lor b) \land b = b \lor (b \lor b)$	$a \land b \lor b = b \lor (b \lor b)$	$(a \lor b) \land b = b \land (b \lor c)$	$a \land b \lor b = b \land (b \lor c)$
$(a \lor b) \land b = b \lor b \land c$	$a \land b \lor b = b \lor b \land c$	$(a \lor b) \land b = (b \lor c) \land b$	$a \land b \lor b = (b \lor c) \land b$	$(a \lor b) \land b = b \land (c \lor b)$	$a \land b \lor b = b \land (c \lor b)$
$(a \lor b) \land b = b \land c \lor b$	$a \land b \lor b = b \land c \lor b$	$(a \lor b) \land b = b \lor c \land b$	$a \land b \lor b = b \lor c \land b$	$(a \lor b) \land b = (c \lor b) \land b$	$a \land b \lor b = (c \lor b) \land b$
$(a \lor b) \land b = c \land b \lor b$	$a \land b \lor b = c \land b \lor b$	11 $(a \land b) \land c = a \land (b \land c)$	12 $(a \lor b) \lor c = a \lor (b \lor c)$	$(a \land b) \land c = (a \land c) \land b$	$a \land (b \land c) = (a \land c) \land b$
$(a \land b) \land c = a \land (c \land b)$	$a \land (b \land c) = a \land (c \land b)$	$a \land (b \lor c) = a \land (c \lor b)$	$(a \lor b) \lor c = (a \lor c) \lor b$	$a \lor (b \lor c) = (a \lor c) \lor b$	$a \lor b \land c = a \lor c \land b$
$(a \lor b) \lor c = a \lor (c \lor b)$	$a \lor (b \lor c) = a \lor (c \lor b)$	$(a \land b) \land c = (b \land a) \land c$	$a \land (b \land c) = (b \land a) \land c$	$(a \lor b) \land c = (b \lor a) \land c$	$(a \land b) \land c = b \land (a \land c)$

⋮ *392 lines*

$a \land (b \lor c) = (a \land a) \land (c \lor b)$	$a \land (b \lor c) = (a \lor a) \land (c \lor b)$	$a \lor b \land c = a \land a \lor c \land b$	$a \lor b \land c = (a \lor a) \lor c \land b$	$(a \lor b) \lor c = a \land a \lor (c \lor b)$
$(a \lor b) \lor c = (a \lor a) \lor (c \lor b)$	$a \lor (b \lor c) = a \land a \lor (c \lor b)$	$a \lor (b \lor c) = (a \lor a) \lor (c \lor b)$	$(a \land b) \land c = (a \land b) \land (a \land c)$	$a \land (b \land c) = (a \land b) \land (a \land c)$
13 $a \land (b \lor c) = a \land b \lor a \land c$	14 $a \lor b \land c = (a \lor b) \land (a \lor c)$	$(a \lor b) \lor c = (a \lor b) \lor (a \lor c)$	$a \lor (b \lor c) = (a \lor b) \lor (a \lor c)$	$(a \land b) \land c = (a \land b) \land (b \land c)$
$a \land (b \land c) = (a \land b) \land (b \land c)$	$(a \lor b) \lor c = (a \lor b) \lor (b \lor c)$	$a \lor (b \lor c) = (a \lor b) \lor (b \lor c)$	$(a \land b) \land c = (a \land b) \land (c \land a)$	$a \land (b \land c) = (a \land b) \land (c \land a)$

⋮

The theorems of basic logic written out in order of increasing complexity. Those considered interesting enough to name in typical textbooks are highlighted. The theorems are respectively: (1), (2) idempotence (laws of tautology) of AND and OR, (3), (4) commutativity of AND and OR, (5) law of double negation, (6), (7) absorption (redundancy) laws, (8) law of noncontradiction (definition of FALSE), (9) law of excluded middle (definition of TRUE), (10) de Morgan's law, (11), (12) associativity of AND and OR, (13), (14) distributive laws. With the exception of the second distributive law, it turns out that the highlighted theorems are exactly the ones that cannot be derived from preceding theorems in the list. The distributive laws appear at positions 2813 and 2814 in the list; it takes a long proof to obtain the second one from preceding theorems.

But what determines which theorems these will be? One might have thought that it would be purely a matter of history. But actually looking at the list of theorems it always seems that the interesting ones are in a sense those that show the least unnecessary complication.

And indeed if one starts from the beginning of the list one finds that most of the theorems can readily be derived from simpler ones earlier in the list. But there are a few that cannot—and that therefore provide in a sense the simplest statements of genuinely new information. And remarkably enough what I have found is that these theorems are almost exactly the ones highlighted on the previous page that have traditionally been identified as interesting.

So what happens if one applies the same criterion in other settings? The picture below shows as an example theorems from the formulation of logic discussed above based on NAND.

$a \barwedge b = b \barwedge a$	**$a = (a \barwedge a) \barwedge (a \barwedge a)$**	**$a = (a \barwedge a) \barwedge (a \barwedge b)$**	$a = (a \barwedge a) \barwedge (b \barwedge a)$
$a = (a \barwedge b) \barwedge (a \barwedge a)$	$a = (b \barwedge a) \barwedge (a \barwedge a)$	$(a \barwedge a) \barwedge a = a \barwedge (a \barwedge a)$	**$(a \barwedge a) \barwedge a = (b \barwedge b) \barwedge b$**
$a \barwedge (a \barwedge a) = (b \barwedge b) \barwedge b$	$(a \barwedge a) \barwedge a = b \barwedge (b \barwedge b)$	$a \barwedge (a \barwedge a) = b \barwedge (b \barwedge b)$	$a \barwedge (a \barwedge b) = (a \barwedge b) \barwedge a$
$a \barwedge (a \barwedge b) = a \barwedge (b \barwedge a)$	**$(a \barwedge a) \barwedge b = (a \barwedge b) \barwedge b$**	$a \barwedge (a \barwedge b) = a \barwedge (b \barwedge b)$	$a \barwedge (a \barwedge b) = (b \barwedge a) \barwedge a$
$(a \barwedge a) \barwedge b = b \barwedge (a \barwedge a)$	$(a \barwedge a) \barwedge b = (b \barwedge a) \barwedge b$	$(a \barwedge a) \barwedge b = b \barwedge (a \barwedge b)$	$a \barwedge (a \barwedge b) = (b \barwedge b) \barwedge a$
$(a \barwedge a) \barwedge b = b \barwedge (b \barwedge a)$	$(a \barwedge b) \barwedge a = a \barwedge (b \barwedge a)$	$(a \barwedge b) \barwedge a = a \barwedge (b \barwedge b)$	$a \barwedge (b \barwedge a) = a \barwedge (b \barwedge b)$
$(a \barwedge b) \barwedge a = (b \barwedge a) \barwedge a$	$a \barwedge (b \barwedge a) = (b \barwedge a) \barwedge a$	$(a \barwedge b) \barwedge a = (b \barwedge b) \barwedge a$	$a \barwedge (b \barwedge a) = (b \barwedge b) \barwedge a$
$a \barwedge (b \barwedge b) = (b \barwedge a) \barwedge a$	$(a \barwedge b) \barwedge b = b \barwedge (a \barwedge a)$	$(a \barwedge b) \barwedge b = (b \barwedge a) \barwedge b$	$(a \barwedge b) \barwedge b = b \barwedge (a \barwedge b)$
$a \barwedge (b \barwedge b) = (b \barwedge b) \barwedge a$	$(a \barwedge b) \barwedge b = b \barwedge (b \barwedge a)$	$a \barwedge (b \barwedge c) = a \barwedge (c \barwedge b)$	$(a \barwedge b) \barwedge c = (b \barwedge a) \barwedge c$
$a \barwedge (b \barwedge c) = (b \barwedge c) \barwedge a$	$(a \barwedge b) \barwedge c = c \barwedge (a \barwedge b)$	$a \barwedge (b \barwedge c) = (c \barwedge b) \barwedge a$	$(a \barwedge b) \barwedge c = c \barwedge (b \barwedge a)$
$(a \barwedge a) \barwedge (a \barwedge a) = (a \barwedge a) \barwedge (a \barwedge b)$	$(a \barwedge a) \barwedge (a \barwedge a) = (a \barwedge a) \barwedge (b \barwedge a)$	$(a \barwedge a) \barwedge (a \barwedge a) = (a \barwedge b) \barwedge (a \barwedge a)$	$(a \barwedge a) \barwedge (a \barwedge a) = (b \barwedge a) \barwedge (a \barwedge a)$
$(a \barwedge a) \barwedge (a \barwedge b) = (a \barwedge a) \barwedge (a \barwedge c)$	$(a \barwedge a) \barwedge (a \barwedge b) = (a \barwedge a) \barwedge (b \barwedge a)$	$(a \barwedge a) \barwedge (a \barwedge b) = (a \barwedge a) \barwedge (c \barwedge a)$	$(a \barwedge a) \barwedge (a \barwedge b) = (a \barwedge b) \barwedge (a \barwedge a)$

⋮ *118 lines*

$a \barwedge ((a \barwedge b) \barwedge b) = (c \barwedge (a \barwedge a)) \barwedge a$	$a \barwedge ((a \barwedge b) \barwedge b) = ((c \barwedge a) \barwedge c) \barwedge a$	$a \barwedge ((a \barwedge b) \barwedge b) = (c \barwedge (a \barwedge c)) \barwedge a$	$(a \barwedge (a \barwedge b)) \barwedge b = (c \barwedge (b \barwedge b)) \barwedge b$
$(a \barwedge (a \barwedge b)) \barwedge b = ((c \barwedge b) \barwedge c) \barwedge b$	$(a \barwedge (a \barwedge b)) \barwedge b = (c \barwedge (b \barwedge c)) \barwedge b$	$a \barwedge ((a \barwedge b) \barwedge b) = (c \barwedge (c \barwedge a)) \barwedge a$	$(a \barwedge (a \barwedge b)) \barwedge b = (c \barwedge (c \barwedge b)) \barwedge b$
$a \barwedge ((a \barwedge b) \barwedge b) = ((c \barwedge c) \barwedge c) \barwedge a$	$a \barwedge ((a \barwedge b) \barwedge b) = (c \barwedge (c \barwedge c)) \barwedge a$	$(a \barwedge (a \barwedge b)) \barwedge b = ((c \barwedge c) \barwedge c) \barwedge b$	$(a \barwedge (a \barwedge b)) \barwedge b = (c \barwedge (c \barwedge c)) \barwedge b$
$a \barwedge (a \barwedge (b \barwedge c)) = a \barwedge (a \barwedge (c \barwedge b))$	$(a \barwedge (a \barwedge b)) \barwedge c = ((a \barwedge b) \barwedge a) \barwedge c$	$(a \barwedge (a \barwedge b)) \barwedge c = (a \barwedge (b \barwedge a)) \barwedge c$	$a \barwedge ((a \barwedge b) \barwedge c) = a \barwedge ((b \barwedge a) \barwedge c)$
$((a \barwedge a) \barwedge b) \barwedge c = ((a \barwedge b) \barwedge b) \barwedge c$	$(a \barwedge (a \barwedge b)) \barwedge c = (a \barwedge (b \barwedge b)) \barwedge c$	**$a \barwedge ((a \barwedge b) \barwedge c) = a \barwedge ((b \barwedge b) \barwedge c)$**	$a \barwedge ((a \barwedge b) \barwedge c) = ((a \barwedge b) \barwedge c) \barwedge a$
$a \barwedge (a \barwedge (b \barwedge c)) = (a \barwedge (b \barwedge c)) \barwedge a$	$a \barwedge (a \barwedge (b \barwedge c)) = a \barwedge ((b \barwedge c) \barwedge a)$	$a \barwedge (a \barwedge (b \barwedge c)) = ((a \barwedge b) \barwedge c) \barwedge c$	$(a \barwedge (a \barwedge b)) \barwedge c = (a \barwedge (b \barwedge c)) \barwedge c$
$a \barwedge ((a \barwedge b) \barwedge c) = a \barwedge ((b \barwedge c) \barwedge c)$	$a \barwedge ((a \barwedge b) \barwedge c) = a \barwedge (c \barwedge (a \barwedge b))$	$a \barwedge (a \barwedge (b \barwedge c)) = (a \barwedge (c \barwedge b)) \barwedge a$	$a \barwedge (a \barwedge (b \barwedge c)) = a \barwedge ((c \barwedge b) \barwedge a)$
$a \barwedge ((a \barwedge b) \barwedge c) = a \barwedge (c \barwedge (b \barwedge a))$	$a \barwedge (a \barwedge (b \barwedge c)) = ((a \barwedge c) \barwedge b) \barwedge b$	$a \barwedge ((a \barwedge b) \barwedge c) = a \barwedge (c \barwedge (b \barwedge b))$	$((a \barwedge a) \barwedge b) \barwedge c = ((a \barwedge c) \barwedge b) \barwedge c$
$(a \barwedge (a \barwedge b)) \barwedge c = (a \barwedge (c \barwedge b)) \barwedge c$	$a \barwedge ((a \barwedge b) \barwedge c) = a \barwedge ((c \barwedge b) \barwedge c)$	$a \barwedge ((a \barwedge b) \barwedge c) = a \barwedge (c \barwedge (b \barwedge c))$	$a \barwedge ((a \barwedge b) \barwedge c) = a \barwedge (c \barwedge (c \barwedge b))$

⋮

The theorems of logic formulated in terms of NAND. Theorems which cannot be derived from ones earlier in the list are highlighted. The last highlighted theorem is 539th in the list. No later theorems would be highlighted since the ones shown form a complete axiom system from which any theorem of logic can be derived. The last highlighted theorem is however an example of one that follows from the axioms, but is hard to prove.

Now there is no particular historical tradition to rely on. But the criterion nevertheless still seems to agree rather well with judgements a human might make. And much as in the picture on page 817, what one sees is that right at the beginning of the list there are several theorems that are identified as interesting. But after these one has to go a long way before one finds other ones.

So if one were to go still further, would one eventually find yet more? It turns out that with the criterion we have used one would not. And the reason is that just the six theorems highlighted already happen to form an axiom system from which any possible theorem about NANDs can ultimately be derived.

And indeed, whenever one is dealing with theorems that can be derived from a finite axiom system the criterion implies that only a finite number of theorems should ever be considered interesting—ending as soon as one has in a sense got enough theorems to be able to reproduce some formulation of the axiom system.

But this is essentially like saying that once one knows the rules for a system nothing else about it should ever be considered interesting. Yet most of this book is concerned precisely with all the interesting behavior that can emerge even if one knows the rules for a system.

And the point is that if computational irreducibility is present, then there is in a sense all sorts of information about the behavior of a system that can only be found from its rules by doing an irreducibly large amount of computational work. And the analog of this in an axiom system is that there are theorems that can be reached only by proofs that are somehow irreducibly long.

So what this suggests is that a theorem might be considered interesting not only if it cannot be derived at all from simpler theorems but also if it cannot be derived from them except by some long proof. And indeed in basic logic the last theorem identified as interesting on page 817—the distributivity of OR—is an example of one that can in principle be derived from earlier theorems, but only by a proof that seems to be much longer than other theorems of comparable size.

In logic, however, all proofs are in effect ultimately of limited length. But in any axiom system where there is universality—and thus undecidability—this is no longer the case, and as I discussed above I suspect that it will actually be quite common for there to be all sorts of short theorems that have only extremely long proofs.

No doubt many such theorems are much too difficult ever to prove in practice. But even if they could be proved, would they be considered interesting? Certainly they would provide what is in essence new information, but my strong suspicion is that in mathematics as it is currently practiced they would only rarely be considered interesting.

And most often the stated reason for this would be that they do not seem to fit into any general framework of mathematical results, but instead just seem like isolated random mathematical facts.

In doing mathematics, it is common to use terms like difficult, powerful, surprising and deep to describe theorems. But what do these really mean? As I mentioned above, any field of mathematics can at some level be viewed as a giant network of statements in which the connections correspond to theorems. And my suspicion is that our intuitive characterizations of theorems are in effect just reflections of our perception of various features of the structure of this network.

And indeed I suspect that by looking at issues such as how easy a given theorem makes it to get from one part of a network to another it will be possible to formalize many intuitive notions about the practice of mathematics—much as earlier in this book we were able to formalize notions of everyday experience such as complexity and randomness.

Different fields of mathematics may well have networks with characteristically different features. And so, for example, what are usually viewed as more successful areas of pure mathematics may have more compact networks, while areas that seem to involve all sorts of isolated facts—like elementary number theory or theory of specific cellular automata—may have sparser networks with more tendrils.

And such differences will be reflected in proofs that can be given. For example, in a sparser network the proof of a particular theorem may not contain many pieces that can be used in proving other theorems. But in a more compact network there may be intermediate definitions and concepts that can be used in a whole range of different theorems.

Indeed, in an extreme case it might even be possible to do the analog of what has been done, say, in the computation of symbolic integrals, and to set up some kind of uniform procedure for finding a proof of essentially any short theorem.

And in general whenever there are enough repeated elements within a single proof or between different proofs this indicates the presence of computational reducibility. Yet while this means that there is in effect less new information in each theorem that is proved, it turns out that in most areas of mathematics these theorems are usually the ones that are considered interesting.

The presence of universality implies that there must at some level be computational irreducibility—and thus that there must be theorems that cannot be reached by any short procedure. But the point is that mathematics has tended to ignore these, and instead to concentrate just on what are in effect limited patches of computational reducibility in the network of all possible theorems.

Yet in a sense this is no different from what has happened, say, in physics, where the phenomena that have traditionally been studied are mostly just those ones that show enough computational reducibility to allow analysis by traditional methods of theoretical physics.

But whereas in physics one has only to look at the natural world to see that other more complex phenomena exist, the usual approaches to mathematics provide almost no hint of anything analogous.

Yet with the new approach based on explicit experimentation used in this book it now becomes quite clear that phenomena such as computational irreducibility occur in abstract mathematical systems.

And indeed the Principle of Computational Equivalence implies that such phenomena should be close at hand in almost every direction: it is merely that—despite its reputation for generality—mathematics has in the past implicitly tended to define itself to avoid them.

So what this means is that in the future, when the ideas and methods of this book have successfully been absorbed, the field of mathematics as it exists today will come to be seen as a small and surprisingly uncharacteristic sample of what is actually possible.

Implications for Mathematics and Its Foundations

■ **History.** Babylonian and Egyptian mathematics emphasized arithmetic and the idea of explicit calculation. But Greek mathematics tended to focus on geometry, and increasingly relied on getting results by formal deduction. For being unable to draw geometrical figures with infinite accuracy this seemed the only way to establish anything with certainty. And when Euclid around 330 BC did his work on geometry he started from 10 axioms (5 "common notions" and 5 "postulates") and derived 465 theorems. Euclid's work was widely studied for more than two millennia and viewed as a quintessential example of deductive thinking. But in arithmetic and algebra—which in effect dealt mostly with discrete entities—a largely calculational approach was still used. In the 1600s and 1700s, however, the development of calculus and notions of continuous functions made use of more deductive methods. Often the basic concepts were somewhat vague, and by the mid-1800s, as mathematics became more elaborate and abstract, it became clear that to get systematically correct results a more rigid formal structure would be needed.

The introduction of non-Euclidean geometry in the 1820s, followed by various forms of abstract algebra in the mid-1800s, and transfinite numbers in the 1880s, indicated that mathematics could be done with abstract structures that had no obvious connection to everyday intuition. Set theory and predicate logic were proposed as ultimate foundations for all of mathematics (see note below). But at the very end of the 1800s paradoxes were discovered in these approaches. And there followed an increasing effort—notably by David Hilbert—to show that everything in mathematics could consistently be derived just by starting from axioms and then using formal processes of proof.

Gödel's Theorem showed in 1931 that at some level this approach was flawed. But by the 1930s pure mathematics had already firmly defined itself to be based on the notion of doing proofs—and indeed for the most part continues to do so even today (see page 859). In recent years, however, the increasing use of explicit computation has made proof less important, at least in most applications of mathematics.

■ **Models of mathematics.** Gottfried Leibniz's notion in the late 1600s of a "universal language" in which arguments in mathematics and elsewhere could be checked with logic can be viewed as an early idealization of mathematics. Starting in 1879 with his "formula language" (*Begriffsschrift*) Gottlob Frege followed a somewhat similar direction, suggesting that arithmetic and from there all of mathematics could be built up from predicate logic, and later an analog of set theory. In the 1890s Giuseppe Peano in his *Formulario* project organized a large body of mathematics into an axiomatic framework involving logic and set theory. Then starting in 1910 Alfred Whitehead and Bertrand Russell in their *Principia Mathematica* attempted to derive many areas of mathematics from foundations of logic and set theory. And although its methods were flawed and its notation obscure this work did much to establish the idea that mathematics could be built up in a uniform way.

Starting in the late 1800s, particularly with the work of Gottlob Frege and David Hilbert, there was increasing interest in so-called metamathematics, and in trying to treat mathematical proofs like other objects in mathematics. This led in the 1920s and 1930s to the introduction of various idealizations for mathematics—notably recursive functions, combinators, lambda calculus, string rewriting systems and Turing machines. All of these were ultimately shown to be universal (see page 784) and thus in a sense capable of reproducing any mathematical system. String rewriting systems—as studied particularly by Emil Post—are close to the multiway systems that I use in this section (see page 938).

Largely independent of mathematical logic the success of abstract algebra led by the end of the 1800s to the notion that any mathematical system could be represented in algebraic terms—much as in the operator systems of this section. Alfred Whitehead to some extent captured this in his 1898 *Universal Algebra*, but it was not until the 1930s that the theory of structures emphasized commonality in the axioms for different fields of mathematics—an idea taken further in the 1940s by category theory (and later by topos theory). And following the work of the Bourbaki group beginning at the end of the 1930s it has become almost universally accepted that structures together with set theory are the appropriate framework for all of pure mathematics.

But in fact the *Mathematica* language released in 1988 is now finally a serious alternative. For while it emphasizes calculation rather than proof its symbolic expressions and transformation rules provide an extremely general way to represent mathematical objects and operations—as for example the notes to this book illustrate.

(See also page 1176.)

■ **Page 773 · Axiom systems.** In the main text I argue that there are many consequences of axiom systems that are quite independent of their details. But in giving the specific axiom systems that have been used in traditional mathematics one needs to take account of all sorts of fairly complicated details.

As indicated by the tabs in the picture, there is a hierarchy to axiom systems in traditional mathematics, with those for basic and predicate logic, for example, being included in all others. (Contrary to usual belief my results strongly suggest however that the presence of logic is not in fact essential to many overall properties of axiom systems.)

As discussed in the main text (see also page 1155) one can think of axioms as giving rules for transforming symbolic expressions—much like rules in *Mathematica*. And at a fundamental level all that matters for such transformations is the structure of expressions. So notation like $a + b$ and $a \times b$, while convenient for interpretation, could equally well be replaced by more generic forms such as $f[a, b]$ or $g[a, b]$ without affecting any of the actual operation of the axioms.

My presentation of axiom systems generally follows the conventions of standard mathematical literature. But by making various details explicit I have been able to put all axiom systems in forms that can be used almost directly in *Mathematica*. Several steps are still necessary though to get the actual rules corresponding to each axiom system. First, the definitions at the top of page 774 must be used to expand out various pieces of notation. In basic logic I use the notation $u = v$ to stand for the pair of rules $u \to v$ and $v \to u$. (Note that $=$ has the precedence of $\to$ not $==$.) In predicate logic the tab at the top specifies how to construct rules (which in this case are often called rules of inference, as discussed on page 1155). $x\_ \land y\_ \to x\_$ is the *modus ponens* or detachment rule (see page 1155). $x\_ \to \forall_{y\_}\, x\_$ is the generalization rule. $x\_ \to x\_ \land \# \,\&$ is applied to the axioms given to get a list of rules. Note that while $=$ in basic logic is used in the underlying construction of rules, $==$ in predicate logic is just an abstract operator with properties defined by the last two axioms given.

As is typical in mathematical logic, there are some subtleties associated with variables. In the axioms of basic logic literal variables like a must be replaced with patterns like $a\_$ that can stand for any expression. A rule like $a\_ \land (b\_ \lor \neg b\_) \to a$ can then immediately be applied to part of an expression using *Replace*. But to apply a rule like $a\_ \to a \land (b \lor \neg b)$ requires in effect choosing some new expression for b (see page 1155). And one way to represent this process is just to have the pattern $a\_ \to a\_ \land (b\_ \lor \neg b\_)$ and then to say that any actual rule that can be used must match this pattern. The rules given in the tab for predicate logic work the same way. Note, however, that in predicate logic the expressions that appear on each side of any rule are required to be so-called well-formed formulas (WFFs) consisting of variables (such as a) and constants (such as 0 or $\emptyset$) inside any number of layers of functions (such as $+$, $\cdot$, or Δ) inside a layer of predicates (such as $==$ or $\in$) inside any number of layers of logical connectives (such as $\land$ or $\Rightarrow$) or quantifiers (such as $\forall$ or $\exists$). (This setup is reflected in the grammar of the *Mathematica* language, where the operator precedences for functions are higher than for predicates, which are in turn higher than for quantifiers and logical connectives—thus yielding for example few parentheses in the presentation of axiom systems here.)

In basic logic any rule can be applied to any part of any expression. But in predicate logic rules can be applied only to whole expressions, always in effect using *Replace[expr, rules]*. The axioms below (devised by Matthew Szudzik as part of the development of this book) set up basic logic in this way.

$(a \vee b) \vee c = (b \vee a) \vee c$	$a \vee ((b \wedge c) \wedge d) = a \vee ((c \wedge b) \wedge d)$
$((a \vee b) \vee c) \vee d = (a \vee (b \vee c)) \vee d$	$a \vee (((b \wedge c) \wedge d) \wedge e) = a \vee ((b \wedge (c \wedge d)) \wedge e)$
$a \vee (b \wedge (c \wedge \neg c)) = a$	$a \vee (b \wedge (c \vee \neg c)) = a \vee b$
$a \vee (b \vee (c \wedge d)) = a \vee ((b \vee c) \wedge (b \vee d))$	$a \vee (b \wedge (c \vee d)) = a \vee ((b \wedge c) \vee (b \wedge d))$
$a \vee (b \wedge \neg (c \vee d)) = a \vee (b \wedge (\neg c \wedge \neg d))$	$a \vee (b \wedge \neg (c \wedge d)) = a \vee (b \wedge (\neg c \vee \neg d))$
$a \vee (b \wedge c) = a \vee (b \wedge \neg \neg c)$	

▪ **Basic logic.** The formal study of logic began in antiquity (see page 1099), with verbal descriptions of many templates for valid arguments—corresponding to theorems of logic—being widely known by medieval times. Following ideas of abstract algebra from the early 1800s, the work of George Boole around 1847 introduced the notion of representing logic in a purely symbolic and algebraic way. (Related notions had been considered by Gottfried Leibniz in the 1680s.) Boole identified *1* with *True* and *0* with *False*, then noted that theorems in logic could be stated as equations in which *Or* is roughly *Plus* and *And* is *Times*—and that such equations can be manipulated by algebraic means. Boole's work was progressively clarified and simplified, notably by Ernst Schröder, and by around 1900, explicit axiom systems for Boolean algebra were being given. Often they included most of the 14 highlighted theorems of page 817, but slight simplifications led for example to the "standard version" of page 773. (Note that the duality between *And* and *Or* is no longer explicit here.) The "Huntington version" of page 773 was given by Edward Huntington in 1933, along with

$$\{(\neg \neg a) = a,\ (a \vee \neg (b \vee \neg b)) = a,\ (\neg (\neg (a \vee \neg b) \vee \neg (a \vee \neg c))) = (a \vee \neg (b \vee c))\}$$

The "Robbins version" was suggested by Herbert Robbins shortly thereafter, but only finally proved correct in 1996 by William McCune using automated theorem proving (see page 1157). The "Sheffer version" based on *Nand* (see page 1173) was given by Henry Sheffer in 1913. The shorter version was devised by David Hillman as part of the development of this book. The shortest version is discussed on page 808. (See also page 1175.)

In the main text each axiom defines an equivalence between expressions. The tradition in philosophy and mathematical logic has more been to take axioms to be true statements from which others can be deduced by the *modus ponens* inference rule $\{x, x \Rightarrow y\} \rightarrow y$ (see page 1155). In 1879 Gottlob Frege used his diagrammatic notation to set up a symbolic representation for logic on the basis of the axioms

$$\{a \Rightarrow (b \Rightarrow a),\ (a \Rightarrow (b \Rightarrow c)) \Rightarrow ((a \Rightarrow b) \Rightarrow (a \Rightarrow c)),\ (a \Rightarrow (b \Rightarrow c)) \Rightarrow (b \Rightarrow (a \Rightarrow c)),\ (a \Rightarrow b) \Rightarrow ((\neg b) \Rightarrow (\neg a)),\ (\neg \neg a) \Rightarrow a,\ a \Rightarrow (\neg \neg a)\}$$

Charles Peirce did something similar at almost the same time, and by 1900 this approach to so-called propositional or sentential calculus was well established. (Alfred Whitehead and Bertrand Russell used an axiom system based on *Or* and *Not* in their original 1910 edition of *Principia Mathematica*.) In 1948 Jan Łukasiewicz found the single axiom version

$$\{((a \Rightarrow (b \Rightarrow a)) \Rightarrow ((((\neg c) \Rightarrow (d \Rightarrow (\neg e))) \Rightarrow ((c \Rightarrow (d \Rightarrow f)) \Rightarrow ((e \Rightarrow d) \Rightarrow (e \Rightarrow f)))) \Rightarrow g)) \Rightarrow (h \Rightarrow g)\}$$

equivalent for example to

$$\{((\neg a) \Rightarrow (b \Rightarrow (\neg c))) \Rightarrow ((a \Rightarrow (b \Rightarrow d)) \Rightarrow ((c \Rightarrow b) \Rightarrow (c \Rightarrow d))),\ a \Rightarrow (b \Rightarrow a)\}$$

It turns out to be possible to convert any axiom system that works with *modus ponens* (and supports the properties of $\Rightarrow$) into a so-called equational one that works with equivalences between expressions by using

```
Module[{a}, Join[Thread[axioms == a ⇒ a],
    {((a ⇒ a) ⇒ b) == b, ((a ⇒ b) ⇒ b) == (b ⇒ a) ⇒ a}]]
```

An analog of *modus ponens* for *Nand* is $\{x, x \bar{\wedge} (y \bar{\wedge} z)\} \rightarrow z$, and with this Jean Nicod found in 1917 the single axiom

$$\{(a \bar{\wedge} (b \bar{\wedge} c)) \bar{\wedge} ((e \bar{\wedge} (e \bar{\wedge} e)) \bar{\wedge} ((d \bar{\wedge} b) \bar{\wedge} ((a \bar{\wedge} d) \bar{\wedge} (a \bar{\wedge} d))))\}$$

which was highlighted in the 1925 edition of *Principia Mathematica*. In 1931 Mordechaj Wajsberg found the slightly simpler

$$\{(a \bar{\wedge} (b \bar{\wedge} c)) \bar{\wedge} (((d \bar{\wedge} c) \bar{\wedge} ((a \bar{\wedge} d) \bar{\wedge} (a \bar{\wedge} d))) \bar{\wedge} (a \bar{\wedge} (a \bar{\wedge} b)))\}$$

Such an axiom system can be converted to an equational one using

```
Module[{a}, With[{t = a ⊼ (a ⊼ a), i = #1 ⊼ (#2 ⊼ #2) &},
    Join[Thread[axioms == t], {i[t ⊼ (b ⊼ c), c] == t,
        i[t, b] == b, i[i[a, b], b] == i[i[b, a], a]}]]]
```

but then involves 4 axioms.

The question of whether any particular statement in basic logic is true or false is always formally decidable, although in general it is NP-complete (see page 768).

▪ **Predicate logic.** Basic logic in effect concerns itself with whole statements (or "propositions") that are each either *True* or *False*. Predicate logic on the other hand takes into account how such statements are built up from other constructs—like those in mathematics. A simple statement in predicate logic is $\forall_x (\forall_y x == y) \vee \forall_x (\exists_y (\neg x == y))$, where $\forall$ is "for all" and $\exists$ is "there exists" (defined in terms of $\forall$ on page 774)—and this particular statement can be proved *True* from the axioms. In general statements in predicate logic can contain arbitrary so-called predicates, say $p[x]$ or $r[x, y]$, that are each either *True* or *False* for given x and y. When predicate logic is used as part of other axiom systems, there are typically axioms which define properties of the predicates. (In real algebra, for example, the predicate $>$ satisfies $a > b \Rightarrow a \neq b$.) But in pure predicate logic the predicates are not assumed to have any particular properties.

Notions of quantifiers like $\forall$ and $\exists$ were already discussed in antiquity, particularly in the context of syllogisms. The first explicit formulation of predicate logic was given by Gottlob Frege in 1879, and by the 1920s predicate logic had become widely accepted as a basis for mathematical axiom systems. (Predicate logic has sometimes also been used as a model for general reasoning—and particularly in the 1980s was the basis for several initiatives in artificial intelligence. But for the most part it has turned out to be too rigid to capture directly typical everyday reasoning processes.)

Monadic pure predicate logic—in which predicates always take only a single argument—reduces in effect to basic logic and is not universal. But as soon as there is even one arbitrary predicate with two arguments the system becomes universal (see page 784). And indeed this is the case even if one considers only statements with quantifiers $\forall \exists \forall$. (The system is also universal with one two-argument function or two one-argument functions.)

In basic logic any statement that is true for all possible assignments of truth values to variables can always be proved from the axioms of basic logic. In 1930 Kurt Gödel showed a similar result for pure predicate logic: that any statement that is true for all possible explicit values of variables and all possible forms of predicates can always be proved from the axioms of predicate logic. (This is often called Gödel's Completeness Theorem, but is not related to completeness of the kind I discuss on page 782 and elsewhere in this section.)

In discussions of predicate logic there is often much said about scoping of variables. A typical issue is that in, say, $\forall_x (\exists_y (\neg x == y))$, x and y are dummy variables whose specific names are not supposed to be significant; yet the names become significant if, say, x is replaced by y. In *Mathematica* most such issues are handled automatically. The axioms for predicate logic given here follow the work of Alfred Tarski in 1962 and use properties of $==$ to minimize issues of variable scoping.

(See also higher-order logics on page 1167.)

▪ **Arithmetic.** Most of the Peano axioms are straightforward statements of elementary facts about arithmetic. The last axiom is a schema (see page 1156) that states the principle of mathematical induction: that if a statement is valid for $a = 0$, and its validity for $a = b$ implies its validity for $a = b + 1$, then it follows that the statement must be valid for all a. Induction was to some extent already used in antiquity—for example in Euclid's proof that there are always larger primes. It began to be used in more generality in the 1600s. In effect it expresses the idea that the integers form a single ordered sequence, and it provides a basis for the notion of recursion.

In the early history of mathematics arithmetic with integers did not seem to need formal axioms, for facts like $x + y == y + x$ appeared to be self-evident. But in 1861 Hermann Grassmann showed that such facts could be deduced from more basic ones about successors and induction. And in 1891 Giuseppe Peano gave essentially the Peano axioms listed here (they were also given slightly less formally by Richard Dedekind in 1888)—which have been used unchanged ever since. (Note that in second-order logic—and effectively set theory— $+$ and $\times$ can be defined just in terms of Δ; see page 1160. In addition, as noted by Julia Robinson in 1948 it is possible to remove explicit mention of $+$ even in the ordinary Peano axioms, using the fact that if $c == a + b$ then $(\Delta a \times c) \times (\Delta b \times c) == \Delta(c \times c) \times (\Delta a \times b)$. Axioms 3, 4 and 6 can then be replaced by $a \times b == b \times a$, $a \times (b \times c) == (a \times b) \times c$ and $(\Delta a) \times (\Delta a \times b) == \Delta a \times (\Delta b \times (\Delta a))$. See also page 1163.)

The proof of Gödel's Theorem in 1931 (see page 1158) demonstrated the universality of the Peano axioms. It was shown by Raphael Robinson in 1950 that universality is also achieved by the Robinson axioms for reduced arithmetic (usually called Q) in which induction—which cannot be reduced to a finite set of ordinary axioms (see page 1156)—is replaced by a single weaker axiom. Statements like $x + y == y + x$ can no longer be proved in the resulting system (see pages 800 and 1169).

If any single one of the axioms given for reduced arithmetic is removed, universality is lost. It is not clear however exactly what minimal set of axioms is needed, for example, for the existence of solutions to integer equations to be undecidable (see page 787). (It is known, however, that essentially nothing is lost even from full Peano arithmetic if for example one drops axioms of logic such as $\neg \neg a = a$.)

A form of arithmetic in which one allows induction but removes multiplication was considered by Mojzesz Presburger in 1929. It is not universal, although it makes statements of size n potentially take as many as about 2^{2^n} steps to prove (though see page 1143).

The Peano axioms for arithmetic seem sufficient to support most of the whole field of number theory. But if as I believe there are fairly simple results that are unprovable from these axioms it may in fact be necessary to extend the Peano axioms to make certain kinds of progress even in practical number theory. (See also page 1166.)

▪ **Algebraic axioms.** Axioms like $a \circ (b \circ c) = (a \circ b) \circ c$ can be used in at least three ways. First, as equations which can be manipulated—like the axioms of basic logic—to establish whether expressions are equal. Second, as on page 773, as statements to be added to the axioms of predicate logic to yield results that hold for every possible system described by the axioms (say every possible semigroup). And third, as definitions of sets whose properties can be studied—and compared—using set theory. High-school algebra typically treats axioms as equations. More advanced algebra often uses predicate logic, but implicitly uses set theory whenever it addresses for example mappings between objects. Note that as discussed on page 1159 how one uses algebraic axioms can affect issues of universality and undecidability. (See also page 1169.)

▪ **Groups.** Groups have been used implicitly in the context of geometrical symmetries since antiquity. In the late 1700s specific groups began to be studied explicitly, mainly in the context of permutations of roots of polynomials, and notably by Evariste Galois in 1831. General groups were defined by Arthur Cayley around 1850 and their standard axioms became established by the end of the 1800s. The alternate axioms given in the main text are the shortest known. The first for ordinary groups was found by Graham Higman and Bernhard Neumann in 1952; the second by William McCune (using automated theorem proving) in 1992. For commutative (Abelian) groups the first alternate axioms were found by Alfred Tarski in 1938; the second by William McCune (using automated theorem proving) in 1992. In this case it is known that no shorter axioms are possible. (See page 806.) Note that in terms of the $\bar{\circ}$ operator $1 == a \bar{\circ} a$, $\overline{b} == (a \bar{\circ} a) \bar{\circ} b$, and $a \cdot b == a \bar{\circ} ((a \bar{\circ} a) \bar{\circ} b)$. Ordinary group theory is universal; commutative group theory is not (see page 1159).

▪ **Semigroups.** Despite their simpler definition, semigroups have been much less studied than groups, and there have for example been about 7 times fewer mathematical publications about them (and another 7 times fewer about monoids). Semigroups were defined by Jean-Armand de Séguier in 1904, and beginning in the late 1920s a variety of algebraic results about them were found. Since the 1940s they have showed up sporadically in various areas of mathematics—notably in connection with evolution processes, finite automata and category theory.

▪ **Fields.** With $\oplus$ being $+$ and $\otimes$ being $\times$ rational, real and complex numbers are all examples of fields. Ordinary integers lack inverses under $\times$, but reduction modulo a prime p gives a finite field. Since the 1700s many examples of fields have arisen, particularly in algebra and number theory. The general axioms for fields as given here emerged around the end of the 1800s. Shorter versions can undoubtedly be found. (See page 1168.)

▪ **Rings.** The axioms given are for commutative rings. With $\oplus$ being $+$ and $\otimes$ being $\times$ the integers are an example. Several examples of rings arose in the 1800s in number theory and algebraic geometry. The study of rings as general algebraic structures became popular in the 1920s. (Note that from the axioms of ring theory one can only expect to prove results that hold for any ring; to get most results in number theory, for example, one needs to use the axioms of arithmetic, which are intended to be specific to ordinary integers.) For non-commutative rings the last axiom given is replaced by $(a \oplus b) \otimes c == a \otimes c \oplus b \otimes c$. Non-commutative rings already studied in the 1800s include quaternions and square matrices.

▪ **Other algebraic systems.** Of algebraic systems studied in traditional mathematics the vast majority are special cases of either groups, rings or fields. Probably the most common other examples are those based on lattice theory. Standard axioms for lattice theory are ($\wedge$ is usually called meet, and $\vee$ join)

$$\{a \wedge b == b \wedge a,\ a \vee b == b \vee a,\ (a \wedge b) \wedge c == a \wedge (b \wedge c),\\ (a \vee b) \vee c == a \vee (b \vee c),\ a \wedge (a \vee b) == a,\ a \vee a \wedge b == a\}$$

Boolean algebra (basic logic) is a special case of lattice theory, as is the theory of partially ordered sets (of which the causal networks in Chapter 9 are an example). The shortest single axiom currently known for lattice theory has *LeafCount* 79 and involves 7 variables. But I suspect that in fact a *LeafCount* less than about 20 is enough.

(See also page 1171.)

▪ **Real algebra.** A notion of real numbers as measures of space or quantity has existed since antiquity. The development of basic algebra gave a formal way to represent operations on such numbers. In the late 1800s there were efforts—notably by Richard Dedekind and Georg Cantor—to set up a general theory of real numbers relying only on basic concepts about integers—and these efforts led to set theory. For purely algebraic questions of the kind that might arise in high-school algebra, however, one can use just the axioms given here. These add to field theory several axioms for ordering, as well as the axiom at the bottom expressing a basic form of continuity (specifically that any polynomial which changes sign must have a zero). With these axioms one can prove results about real polynomials, but not about arbitrary mathematical functions, or integers. The axioms were shown to be complete by Alfred Tarski in the 1930s. The proof was based on setting up a procedure that could in principle resolve any set of real polynomial equations or inequalities. This is now in practice done by *Simplify* and other functions in *Mathematica* using methods of cylindrical algebraic decomposition invented in the 1970s—which work roughly by finding a succession of points of change using *Resultant*. (Note that with n variables the number of steps needed can increase like 2^{2^n}.) (See the note about real analysis below.)

■ **Geometry.** Euclid gave axioms for basic geometry around 300 BC which were used with fairly little modification for more than 2000 years. In the 1830s, however, it was realized that the system would remain consistent even if the so-called parallel postulate was modified to allow space to be curved. Noting the vagueness of Euclid's original axioms there was then increasing interest in setting up more formal axiom systems for geometry. The best-known system was given by David Hilbert in 1899—and by describing geometrical figures using algebraic equations he showed that it was as consistent as the underlying axioms for numbers.

The axioms given here are illustrated below. They were developed by Alfred Tarski and others in the 1940s and 1950s. (Unlike Hilbert's axioms they require only first-order predicate logic.) The first six give basic properties of betweenness of points and congruence of line segments. The second- and third-to-last axioms specify that space has two dimensions; they can be modified for other dimensions. The last axiom is a schema that asserts the continuity of space. (The system is not finitely axiomatizable.)

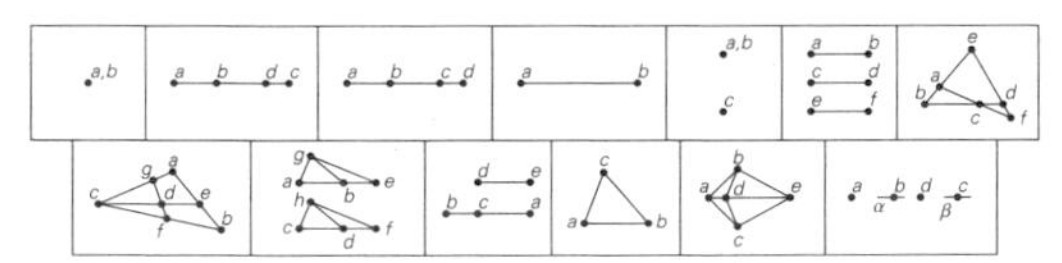

The axioms given can prove most of the results in an elementary geometry textbook—indeed all results that are about geometrical figures such as triangles and circles specified by a fixed finite number of points, but which do not involve concepts like area. The axioms are complete and consistent—and thus not universal. They can however be made universal if axioms from set theory are added.

■ **Category theory.** Developed in the 1940s as a way to organize constructs in algebraic topology, category theory works at the level of whole mathematical objects rather than their elements. In the basic axioms given here the variables represent morphisms that correspond to mappings between objects. (Often morphisms are shown as arrows in diagrams, and objects as nodes.) The axioms specify that when morphisms are composed their domains and codomains must have appropriately matching types. Some of the methodology of category theory has become widely used in mathematics, but until recently the basic theory itself was not extensively studied—and its axiomatic status remains unclear. Category theory can be viewed as a formalization of operations on abstract data types in computer languages—though unlike in *Mathematica* it normally requires that functions take a single bundle of data as an argument.

■ **Set theory.** Basic notions of finite set theory have been used since antiquity—though became widespread only after their introduction into elementary mathematics education in the 1960s. Detailed ideas about infinite sets emerged in the 1880s through the work of Georg Cantor, who found it useful in studying trigonometric series to define sets of transfinite numbers of points. Several paradoxes associated with infinite sets were quickly noted—a 1901 example due to Bertrand Russell being to ask whether a set containing all sets that do not contain themselves in fact contains itself. To avoid such paradoxes Ernst Zermelo in 1908 suggested formalizing set theory using the first seven axioms given in the main text. (The axiom of infinity, for example, was included to establish that an infinite set such as the integers exists.) In 1922 Abraham Fraenkel noted that Zermelo's axioms did not support certain operations that seemed appropriate in a theory of sets, leading to the addition of Thoralf Skolem's axiom of replacement, and to what is usually called Zermelo-Fraenkel set theory (ZF). (The replacement axiom formally makes the subset axiom redundant.) The axiom of choice was first explicitly formulated by Zermelo in 1904 to capture the idea that in a set all elements can be ordered, so that the process of transfinite induction is possible (see page 1160). The non-constructive character of the axiom of choice has made it always remain somewhat controversial. It has arisen in many different guises and been useful in proving theorems in many areas of mathematics, but it has seemingly peculiar consequences such as the Banach-Tarski result that a solid sphere can be divided into six pieces (each a non-measurable set) that can be reassembled into a solid sphere twice the size. (The nine axioms with the axiom of choice are usually known as ZFC.) The axiom of regularity (or axiom of foundation) formulated by John von Neumann in 1929 explicitly forbids sets which for example can be elements of themselves. But while this axiom is convenient in simplifying work in set theory it has not been found generally useful in mathematics, and is normally considered optional at best.

A few additional axioms have also arisen as potentially useful. Most notable is the Continuum Hypothesis discussed on page 1127, which was proved independent of ZFC by Paul Cohen in 1963. (See also page 1166.)

Note that by using more complicated axioms the only construct beyond predicate logic needed to formulate set theory is ∈ . As discussed on page 1176, however, one cannot avoid axiom schemas in the formulation of set theory given here. (The von Neumann-Bernays-Gödel formulation does avoid these, but at the cost of introducing additional objects more general than sets.)

(See also page 1160.)

▪ **General topology.** The axioms given define properties of open sets of points in spaces—and in effect allow issues like connectivity and continuity to be discussed in terms of set theory without introducing any explicit distance function.

▪ **Real analysis.** The axiom given is Dedekind's axiom of continuity, which expresses the connectedness of the set of real numbers. Together with set theory it allows standard results about calculus to be derived. But as well as ordinary real numbers, these axioms allow non-standard analysis with constructs such as explicit infinitesimals (see page 1172).

▪ **Axiom systems for programs.** (See pages 794 and 1168.)

▪ **Page 775 · Implementation.** Given the axioms in the form

```
s[1] = (a_ ⊼ a_) ⊼ (a_ ⊼ b_) → a;
s[2, x_] := b_ → (b ⊼ b) ⊼ (b ⊼ x); s[3] =
  a_ ⊼ (a_ ⊼ b_) → a ⊼ (b ⊼ b); s[4] = a_ ⊼ (b_ ⊼ b_) → a ⊼ (a ⊼ b);
s[5] = a_ ⊼ (a_ ⊼ (b_ ⊼ c_)) → b ⊼ (b ⊼ (a ⊼ c));
```

the proof shown here can be represented by

```
{{s[2, b], {2}}, {s[4], {}}, {s[2, (b ⊼ b) ⊼ ((a ⊼ a) ⊼ (b ⊼ b))],
    {2, 2}}, {s[1], {2, 2, 1}}, {s[2, b ⊼ b], {2, 2, 2, 2, 2, 2}},
  {s[5], {2, 2, 2}}, {s[2, b ⊼ b], {2, 2, 2, 2, 2, 1}},
  {s[1], {2, 2, 2, 2, 2}}, {s[3], {2, 2, 2}},
  {s[1], {2, 2, 2, 2}}, {s[4], {2, 2, 2}}, {s[5], {}},
  {s[2, a], {2, 2, 1}}, {s[1], {2, 2}}, {s[3], {}}, {s[1], {2}}}
```

and applied using

```
FoldList[Function[{u, v},
    MapAt[Replace[#, v[[1]]] &, u, {v[[2]]}]], a ⊼ b, proof]
```

▪ **Page 776 · Proof structures.** The proof shown is in a sense based on very low-level steps, each consisting of applying a single axiom from the original axiom system. But in practical mathematics it is usual for proofs to be built up in a more hierarchical fashion using intermediate results or lemmas. In the way I set things up lemmas can in effect be introduced as new axioms which can be applied repeatedly during a proof. And in the case shown here if one first proves the lemma

```
(a ⊼ (a ⊼ (b ⊼ ((a ⊼ a) ⊼ c)))) = (b ⊼ a)
```

and treats it as rule 6, then the main proof can be shortened:

When one just applies axioms from the original axiom system one is in effect following a single line of steps. But when one proves a lemma one is in effect on a separate branch, which only merges with the main proof when one uses the lemma. And if one has nested lemmas one can end up with a proof that is in effect like a tree. (Repeated use of a single lemma can also lead to cycles.) Allowing lemmas can in extreme cases probably make proofs as much as exponentially shorter. (Note that lemmas can also be used in multiway systems.)

In the way I have set things up one always gets from one step in a proof to the next by taking an expression and applying some transformation rule to it. But while this is familiar from algebraic mathematics and from the operation of *Mathematica* it is not the model of proofs that has traditionally been used in mainstream mathematical logic. For there one tends to think not so much about transforming expressions as about taking collections of true statements (such as equations $u == v$), and using so-called rules of inference to deduce other ones. Most often there are two basic rules of inference: *modus ponens* or detachment which uses the logic result $(x \wedge x \Rightarrow y) \Rightarrow y$ to deduce the statement y from statements x and $x \Rightarrow y$, and substitution, which takes statements x and y and deduces $x /. p \to y$, where p is a logical variable in x (see page 1151). And with this approach axioms enter merely as initial true statements, leaving rules of inference to generate successive steps in proofs. And instead of being mainly linear sequences of results, proofs instead become networks in which pairs of results are always combined when *modus ponens* is used. But it is still always in principle possible to convert any proof to a purely sequential one—though perhaps at the cost of having exponentially many more steps.

■ **Substitution strategies.** With the setup I am using each step in a proof involves transforming an expression like $u = v$ using an expression like $s = t$. And for this to happen s or t must match some part w of u or v. The simplest way this can be achieved is for s or t to reproduce w when its variables are replaced by appropriate expressions. But in general one can make replacements not only for variables in s and t, but also for ones in w. And in practice this often makes many more matches possible. Thus for example the axiom $a \circ a = a$ cannot be applied directly to $(p \circ q) \circ (p \circ r) = q \circ r$. But after the replacement $r \to q$, $a \circ a$ matches $(p \circ q) \circ (p \circ r)$ with $a \to p \circ q$, yielding the new theorem $p \circ q = q \circ q$. These kinds of substitutions are used in the proof on page 810. One approach to finding them is so-called paramodulation, which was introduced around 1970 in the context of automated theorem-proving systems, and has been used in many such systems (see page 1157). (Such substitutions are not directly relevant to *Mathematica*, since it transforms expressions rather than theorems or equations. But when I built SMP in 1981, its semantic pattern matching mechanism did use essentially such substitutions.)

■ **One-way transformations.** As formulated in the main text, axioms define two-way transformations. One can also set up axiom systems based on one-way transformations (as in multiway systems). For basic logic, examples of this were studied in the mid-1900s, and with the transformations thought of as rules of inference they were sometimes known as "axiomless formulations".

■ **Axiom schemas.** An axiom like $a + 0 == a$ is a single well-formed formula in the sense of page 1150. But sometimes one needs infinite collections of such individual axioms, and in the main text these are represented by axiom schemas given as *Mathematica* patterns involving objects like $x\_$. Such schemas are taken to stand for all individual axioms that match the patterns and are well-formed formulas. The induction axiom in arithmetic is an example of a schema. (See the note on finite axiomatizability on page 1176.) Note that as mentioned on page 1150 all the axioms given for basic logic should really be thought of as schemas.

■ **Reducing axiom details.** Traditional axiom systems have many details not seen in the basic structure of multiway systems. But in most cases these details can be avoided—and in the end the universality of multiway systems implies that they can always be made to emulate any axiom system.

Traditional axiom systems tend to be based on operator systems (see page 801) involving general expressions, not just strings. But any expression can always be written as a string using something like *Mathematica FullForm*. (See also page 1169.) Traditional axiom systems also involve symbolic variables, not just literal string elements. But by using methods like those for combinators on page 1121 explicit mention of variables can always be eliminated.

■ **Proofs in practice.** At some level the purpose of a proof is to establish that something is true. But in the practice of modern mathematics proofs have taken on a broader role; indeed they have become the primary framework for the vast majority of mathematical thinking and discourse. And from this perspective the kinds of proofs given on pages 810 and 811—or typically generated by automated theorem proving—are quite unsatisfactory. For while they make it easy at a formal level to check that certain statements are true, they do little at a more conceptual level to illuminate why this might be so. And indeed the kinds of proofs normally considered most mathematically valuable are ones that get built up in terms of concepts and constructs that are somehow expected to be as generally applicable as possible. But such proofs are inevitably difficult to study in a uniform and systematic way (though see page 1176). And as I argue in the main text, it is in fact only for the rather limited kinds of mathematics that have historically been pursued that such proofs can be expected to be sufficient. For in general proofs can be arbitrarily long, and can be quite devoid of what might be considered meaningful structure.

Among practical proofs that show signs of this (and whose mathematical value is thus often considered controversial) most have been done with aid of computers. Examples include the Four-Color Theorem (coloring of maps), the optimality of the Kepler packing (see page 986), the completeness of the Robbins axiom system (see page 1151) and the universality of rule 110 (see page 678).

In the past it was sometimes claimed that using computers is somehow fundamentally incompatible with developing mathematical understanding. But particularly as the use of *Mathematica* has become more widespread there has been increasing recognition that computers can provide crucial raw material for mathematical intuition—a point made rather forcefully by the discoveries in this book. Less well recognized is the fact that formulating mathematical ideas in a *Mathematica* program is at least as effective a way to produce clarity of thinking and understanding as formulating a traditional proof.

■ **Page 778 · Properties.** The second rule shown has the property that black elements always appear before white, so that strings can be specified just by the number of elements of each color that they contain—making the rule one of the sorted type discussed on page 937, based on the difference vector *{{2, -1}, {-1, 3}, {-4, -1}}*. The question of whether a given string can be generated is then analogous to finding whether there is a solution with certain positivity properties to a set of linear Diophantine equations.

■ **Page 781 · NAND tautologies.** At each step every possible transformation rule in the axioms is applied wherever it can. New expressions are also created by replacing each possible variable with $x \bar{\wedge} y$, where x and y are new variables, and by setting every possible pair of variables equal in turn. The longest tautology at step t is

Nest[(# ⊼ #) ⊼ (# ⊼ p_t) &, p ⊼ (p ⊼ p), t - 1]

whose *LeafCount* grows like 3^t. The distribution of sizes of statements generated at each step is shown below.

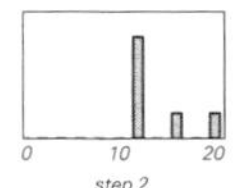

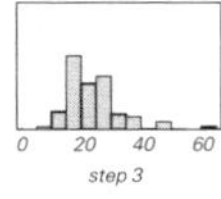

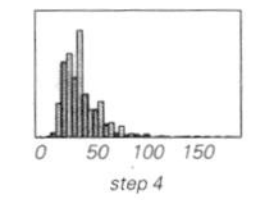

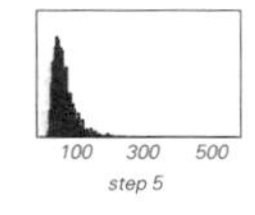

Even with the same underlying axioms the tautologies are generated in a somewhat different order if one uses a different strategy—say one based on paramodulation (see page 1156). Pages 818 and 1175 discuss the sequence of all NAND theorems listed in order of increasing complexity.

■ **Proof searching.** To find a proof of some statement $p = q$ in a multiway system one can always in principle just start from p, evolve the system until it first generates q, then pick out the sequence of strings on the path from p to q. But doing this will usually involve building up a vast network of strings. And although at some level computational irreducibility and NP completeness (see page 766) imply that in general only a limited amount of this computational work can be saved, there are in practice quite often important optimizations that can be made. For finding a proof of $p = q$ is like searching for a path satisfying the constraint of going from p to q. And just like in the systems based on constraints in Chapter 5 one can usually do at least somewhat better than just to look at every possible path in turn.

For a start, in generating the network of paths one only ever need keep a single path that leads to any particular string; just like in many of my pictures of multiway systems one can in effect always drop any duplicate strings that occur. One might at first imagine that if p and q are both short strings then one could also drop any very long strings that are produced. But as we have seen, it is perfectly possible for long intermediate strings to be needed to get from p to q. Still, it is often reasonable to weight things so that at least at first one looks at paths that involve only shorter strings.

In the most direct approach, one takes a string and at each step just applies the underlying rules or axioms of the multiway system. But as soon as one knows that there is a path from a string u to a string v, one can also imagine applying the rule $u \to v$ to any string—in effect like a lemma. And one can choose which lemmas to try first by looking for example at which involve the shortest or commonest strings.

It is often important to minimize the number of lemmas one has to keep. Sometimes one can do this by reducing every lemma—and possibly every string—to some at least partially canonical form. One can also use the fact that in a multiway system if $u \to v$ and $r \to s$ then $u <> r \to v <> s$.

If one wants to get from p to q the most efficient thing is to use properties of q to avoid taking wrong turns. But except in systems with rather simple structure this is usually difficult to achieve. Nevertheless, one can for example always in effect work forwards from p, and backwards from q, seeing whether there is any overlap in the sets of strings one gets.

■ **Automated theorem proving.** Since the 1950s a fair amount of work has been done on trying to set up computer systems that can prove theorems automatically. But unlike systems such as *Mathematica* that emphasize explicit computation none of these efforts have ever achieved widespread success in mathematics. And indeed given my ideas in this section this now seems not particularly surprising.

The first attempt at a general system for automated theorem proving was the 1956 Logic Theory Machine of Allen Newell and Herbert Simon—a program which tried to find proofs in basic logic by applying chains of possible axioms. But while the system was successful with a few simple theorems the searches it had to do rapidly became far too slow. And as the field of artificial intelligence developed over the next few years it became widely believed that what would be needed was a general system for imitating heuristics used in human thinking. Some work was nevertheless still done on applying results in mathematical logic to speed up the search process. And in 1963 Alan Robinson suggested the idea of resolution theorem proving, in which one constructs *¬ theorem ∨ axioms*, then typically writes this in conjunctive normal form and repeatedly applies rules like $(\neg p \vee q) \wedge (p \vee q) \to q$ to try to reduce it to *False*, thereby proving given *axioms* that *theorem* is *True*. But after early enthusiasm it became clear that this approach could not be expected to make theorem proving easy—a point emphasized by the discovery of NP completeness in the early 1970s. Nevertheless, the approach was used with some success, particularly in proving that various mechanical and other engineering systems would behave as intended—although by the mid-1980s such verification was more often done by systematic Boolean function methods (see page 1097). In the 1970s simple versions of the resolution method were incorporated into logic programming languages such as Prolog, but little in the way of mathematical theorem proving was done with them. A notable system under development since the 1970s is the Boyer-Moore theorem prover Nqthm, which uses resolution together with methods related to induction to try to find proofs of statements in a version of LISP. Another family of systems under development at Argonne National Laboratory since the 1960s are intended to find proofs in pure operator (equational) systems (predicate logic with equations). Typical of this effort was the Otter system started in the mid-1980s, which uses the resolution method, together with a variety of ad hoc strategies that are mostly versions of the general ones for multiway systems in the previous note. The development of so-called unfailing completion algorithms (see page 1037) in the late 1980s made possible much more systematic automated theorem provers for pure operator systems—with a notable example being the Waldmeister system developed around 1996 by Arnim Buch and Thomas Hillenbrand.

Ever since the 1970s I at various times investigated using automated theorem-proving systems. But it always seemed that extensive human input—typically from the creators of the system—was needed to make such systems actually find non-trivial proofs. In the late 1990s, however, I decided to try the latest systems and was surprised to find that some of them could routinely produce proofs hundreds of steps long with little or no guidance. Almost any proof that was easy to do by hand almost always seemed to come out automatically in just a few steps. And the overall ability to do proofs—at least in pure operator systems—seemed vastly to exceed that of any human. But as page 810 illustrates, long proofs produced in this way tend to be difficult to read—in large part because they lack the higher-level constructs that are typical in proofs created by humans. As I discuss on page 821, such lack of structure is in some respects inevitable. But at least for specific kinds of theorems in specific areas of mathematics it seems likely that more accessible proofs can be created if each step is allowed to involve sophisticated computations, say as done by *Mathematica*.

▪ Proofs in *Mathematica*. Most of the individual built-in functions of *Mathematica* I designed to be as predictable as possible—applying transformations in definite ways and using algorithms that are never of fundamentally unknown difficulty. But as their names suggest *Simplify* and *FullSimplify* were intended to be less predictable—and just to do what they can and then return a result. And in many cases these functions end up trying to prove theorems; so for example *FullSimplify[(a + b)/2 ≥ Sqrt[a b], a > 0 && b > 0]* must in effect prove a theorem to get the result *True*.

▪ Page 781 · Truth and falsity. The notion that statements can always be classified as either true or false has been a common idealization in logic since antiquity. But in everyday language, computer languages and mathematics there are many ways in which this idealization can fail. An example is $x + y == z$, which cannot reasonably be considered either true or false unless one knows what x, y and z are. Predicate logic avoids this particular kind of case by implicitly assuming that what is meant is a general statement about all values of any variable—and avoids cases like the expression $x + y$ by requiring all statements to be well-formed formulas (see page 1150). In *Mathematica* functions like *TrueQ* and *IntegerQ* are set up always to yield *True* or *False*—but just by looking at the explicit structure of a symbolic expression.

Note that although the notion of negation seems fairly straightforward in everyday language it can be difficult to implement in computational or mathematical settings. And thus for example even though it may be possible to establish by a finite computation that a particular system halts, it will often be impossible to do the same for the negation of this statement. The same basic issue arises in the intuitionistic approach to mathematics, in which one assumes that any object one handles must be found by a finite construction. And in such cases one can set up an analog of logic in which one no longer takes $\neg \neg a = a$.

It is also possible to assume a specific number $k > 2$ of truth values, as on page 1175, or to use so-called modal logics.

(See also page 1167.)

▪ Page 782 · Gödel's Theorem. What is normally known as "Gödel's Theorem" (or "Gödel's First Incompleteness Theorem") is the centerpiece of the paper "On Undecidable Propositions of *Principia Mathematica* and Related Systems" published by Kurt Gödel in 1931. What the theorem shows is that there are statements that can be formulated within the standard axiom system for arithmetic but which cannot be proved true or false within that system. Gödel's paper does this first for the statement "this statement is unprovable", and much of the paper is concerned with showing how such a statement can be encoded within arithmetic. Gödel in effect does this by first converting the statement to one about recursive functions and then—by using tricks of number theory such as the beta function of page 1120—to one purely about arithmetic. (Gödel's main achievement is sometimes characterized as the "arithmetization of metamathematics": the discovery that concepts such as provability related to the processes of mathematics can be represented purely as statements in arithmetic.) (See page 784.)

Gödel originally based his theorem on Peano arithmetic (as discussed in the context of *Principia Mathematica*), but expected that it would in fact apply to any reasonable formal system for mathematics—and in later years considered that this had been established by thinking about Turing machines. He suggested that his results could be avoided if some form of transfinite hierarchy of formalisms could be used, and appears to have thought that at some level humans and mathematics do this (compare page 1167).

Gödel's 1931 paper came as a great surprise, although the issues it addressed were already widely discussed in the field of mathematical logic. And while the paper is at a technical level rather clear, it has never been easy for typical mathematicians to read. Beginning in the late 1950s its results began to be widely known outside of mathematics, and by the late 1970s Gödel's Theorem and various misstatements of it were often assigned an almost mystical significance. Self-reference was commonly seen as its central feature, and connections with universality and computation were usually missed. And with the belief that humans must somehow have intrinsic access to all truths in mathematics, Gödel's Theorem has been used to argue for example that computers can fundamentally never emulate human thinking.

The picture on page 786 can be viewed as a modern proof of Gödel's Theorem based on Diophantine equations.

In addition to what is usually called Gödel's Theorem, Kurt Gödel established a second incompleteness theorem: that the statement that the axioms of arithmetic are consistent cannot be proved by using those axioms (see page 1168). He also established what is often called the Completeness Theorem for predicate logic (see page 1152)—though here "completeness" is used in a different sense.

■ **Page 783 · Properties.** The first multiway system here generates all strings that end in □■; the third all strings that end in ■. The second system generates all strings where the second-to-last element is white, or the string ends with a run of black elements delimited by white ones.

■ **Page 783 · Essential incompleteness.** If a consistent axiom system is complete this means that any statement in the system can be proved true or false using its axioms, and the question of whether a statement is true can always be decided by a finite procedure. If an axiom system is incomplete then this means that there are statements that cannot be proved true or false using its axioms—and which must therefore be considered independent of those axioms. But even given this it is still possible that a finite procedure can exist which decides whether a given statement is true, and indeed this happens in the theory of commutative groups (see note below). But often an axiom system will not only be incomplete, but will also be what is called essentially incomplete. And what this means is that there is no finite set of axioms that can consistently be added to make the system complete. A consequence of this is that there can be no finite procedure that always decides whether a given statement is true—making the system what is known as essentially undecidable. (When I use the term "undecidable" I normally mean "essentially undecidable". Early work on mathematical logic sometimes referred to statements that are independent as being undecidable.)

One might think that adding rules to a system could never reduce its computational sophistication. And this is correct if with suitable input one can always avoid the new rules. But often these rules will allow transformations that in effect short-circuit any sophisticated computation. And in the context of axiom systems, adding axioms can be thought of as putting more constraints on a system—thus potentially in effect forcing it to be simpler. The result of all this is that an axiom system that is universal can stop being universal when more axioms are added to it. And indeed this happens when one goes from ordinary group theory to commutative group theory, and from general field theory to real algebra.

■ **Page 784 · Predicate logic.** The universality of predicate logic with a single two-argument function follows immediately from the result on page 1156 that it can be used to emulate any two-way multiway system.

■ **Page 784 · Algebraic axioms.** How universality works with algebraic axioms depends on how those axioms are being used (compare page 1153). What is said in the main text here assumes that they are being used as on page 773—with each variable in effect standing for any object (compare page 1169), and with the axioms being added to predicate logic. The first of these points means that one is concerned with so-called pure group theory—and with finding results valid for all possible groups. The second means that the statements one considers need not just be of the form $\ldots == \ldots$, but can explicitly involve logic; an example is Cayley's theorem

$$a \cdot x == a \cdot y \Rightarrow (x == y \wedge \exists_z\, a \cdot z == x) \wedge$$
$$a \cdot x == b \cdot x \Rightarrow a == b \wedge (a \cdot b) \cdot x == a \cdot (b \cdot x)$$

With this setup, Alfred Tarski showed in 1946 that any statement in Peano arithmetic can be encoded as a statement in group theory—thus demonstrating that group theory is universal, and that questions about it can be undecidable. This then also immediately follows for semigroup theory and monoid theory. It was shown for ring theory and field theory by Julia Robinson in 1949. But for commutative group theory it is not the case, as shown by Wanda Szmielew around 1950. And indeed there is a procedure based on quantifier elimination for determining in a finite number of steps whether any statement in commutative group theory can be proved. (Commutative group theory is thus a decidable theory. But as mentioned in the note above, it is not complete—since for example it cannot establish the theorem $a == b$ which states that a group has just one element. It is nevertheless not essentially incomplete—and for example adding the axiom $a == b$ makes it complete.) Real algebra is also not universal (see page 1153), and the same is for example true for finite fields—but not for arbitrary fields.

As discussed on page 1141, word problems for systems such as groups are undecidable. But to set up a word problem in general formally requires going beyond predicate logic, and including axioms from set theory. For a word problem relates not, say, to groups in general, but to a particular group, specified by relations between generators. Within predicate logic one can give the relations as statements, but in effect one cannot specify that no other relations hold. It turns out, however, that undecidability for word problems occurs in essentially the same places as universality for axioms with predicate logic. Thus, for example, the word problem is undecidable for groups and semigroups, but is decidable for commutative groups.

One can also consider using algebraic axioms without predicate logic—as in basic logic or in the operator systems of page 801. And one can now ask whether there is then universality. In the case of semigroup theory there is not. But certainly systems of this type can be universal—since for example they can be set up to emulate any multiway system. And it seems likely that the axioms of ordinary group theory are sufficient to achieve universality.

▪ Page 784 · Set theory. Any integer n can be encoded as a set using for example $Nest[Union[\#, \{\#\}] \&, \{\}, n]$. And from this a statement s in Peano arithmetic (with each variable explicitly quantified) can be translated to a statement in set theory by using

$$Replace[s, \{\forall_{a\_}\, b\_ \to \forall_a (a \in N \Rightarrow b), \exists_{a\_}\, b\_ \to \exists_a (a \in N \wedge b)\}, \{0, \infty\}]$$

and then adding the statements below to provide definitions (N is the set of non-negative integers, $\langle x, y, z\rangle$ is an ordered triple, and $\updownarrow a$ determines whether each triple in a set a is of the form $\langle x, y, f[x, y]\rangle$ specifying a single-valued function).

$$a = N \Leftrightarrow \forall_b ((\emptyset \in b \wedge \forall_c (c \in b \Rightarrow \cup\{c, \{c\}\} \in b)) \Rightarrow a \subseteq b)$$

$$a = \Delta b \Leftrightarrow a = \cup\{b, \{b\}\}$$

$$a = \langle b, c, d\rangle \Leftrightarrow a = \{\{\{\{b, c\}, \{c\}\}, d\}, \{d\}\}$$

$$\updownarrow a \Leftrightarrow (\forall_b \forall_c \forall_d (\langle b, c, d\rangle \in a \Rightarrow \forall_e (\langle b, c, e\rangle \in a \Rightarrow d = e)) \wedge \forall_b \forall_c ((b \in N \wedge c \in N) \Rightarrow \exists_d (d \in N \wedge \langle b, c, d\rangle \in a)))$$

$$a = b + c \Leftrightarrow \forall_d ((\updownarrow d \wedge \forall_e \forall_f \forall_g (\langle \Delta e, f, g\rangle \in d \Rightarrow \langle e, f, \Delta g\rangle \in d) \wedge \forall_f \forall_g (\langle \emptyset, f, g\rangle \in d \Rightarrow g = f)) \Rightarrow \langle b, c, a\rangle \in d)$$

$$a = b \times c \Leftrightarrow \forall_d ((\updownarrow d \wedge \forall_e \forall_f \forall_g (\langle \Delta e, f, g\rangle \in d \Rightarrow \langle e, f, f + g\rangle \in d) \wedge \forall_f \forall_g (\langle \emptyset, f, g\rangle \in d \Rightarrow g = \emptyset)) \Rightarrow \langle b, c, a\rangle \in d)$$

This means that set theory can be used to prove any statement that can be proved in Peano arithmetic. But it can also prove other statements—such as Goodstein's result (see note below), and the consistency of arithmetic (see page 1168). An important reason for this is that set theory allows not just ordinary induction over sequences of integers but also transfinite induction over arbitrary ordered sets (see below).

▪ Page 786 · Universal Diophantine equation. The equation is built up from ones whose solutions are set up to be integers that satisfy particular relations. So for example the equation $a^2 + b^2 == 0$ has solutions that are exactly those integers that satisfy the relation $a == 0 \wedge b == 0$. Similarly, assuming as in the rest of this note that all variables are non-negative, $b == a + c + 1$ has solutions that are exactly those integers that satisfy $a < b$, with c having some allowed value. From various number-theoretical results many relations can readily be encoded as integer equations:

$$(a == 0 \vee b == 0) \longleftrightarrow a\, b == 0$$

$$(a == 0 \wedge b == 0) \longleftrightarrow a + b == 0$$

$$a < b \longleftrightarrow b == a + c + 1$$

$$a == Mod[b, c] \longleftrightarrow (b == a + c\, d \wedge a < c)$$

$$a == Quotient[b, c] \longleftrightarrow (b == a\, c + d \wedge d < c)$$

$$a == Binomial[b, c] \longleftrightarrow With[\{n = 2^b + 1\}, (n + 1)^b == n^c\, (a + d\, n) + e \wedge e < n^c \wedge a < n]$$

$$a == b! \longleftrightarrow a == Quotient[c^b, Binomial[c, b]]$$

$$a == GCD[b, c] \longleftrightarrow (b\, c > 0 \wedge a\, d == b \wedge a\, e == c \wedge a + c\, f == b\, g)$$

$$a == Floor[b/c] \longleftrightarrow (a\, c + d == b \wedge d < c)$$

$$PrimeQ[a] \longleftrightarrow (GCD[(a - 1)!, a] == 1 \wedge a > 1)$$

$$a == BitAnd[c, d] \wedge b == BitOr[c, d] \longleftrightarrow (\sigma[c, a] \wedge \sigma[d, a] \wedge \sigma[b, c] \wedge \sigma[b, d] \wedge a + b == c + d) /. \sigma[x\_, y\_] \to Mod[Binomial[x, y], 2] == 1$$

where the last encoding uses the result on page 608. (Note that any variable a can be forced to be non-negative by including an equation $a == w^2 + x^2 + y^2 + z^2$, as on page 910.)

Given an integer a for which $IntegerDigits[a, 2]$ gives the cell values for a cellular automaton, a single step of evolution according say to rule 30 is given by

$$BitXor[a, 2\, BitOr[a, 2\, a]]$$

where (see page 871)

$$BitXor[x, y] == BitOr[x, y] - BitAnd[x, y]$$

and a is assumed to be padded with 0's at each end. The corresponding form for rule 110 is

$$BitXor[BitAnd[a, 2a, 4a], BitOr[2a, 4a]]$$

The final equation is then obtained from

$$\begin{aligned}\{&1 + x_4 + x_{12} == 2^{(1+x_3)(x_1+2x_3)},\ x_2 + x_{13} == 2^{x_1},\\ &1 + x_5 + x_{14} == 2^{x_1},\ 2^{x_3} x_5 + 2^{x_1+2x_3} x_6 + 2^{x_1+x_3} x_{15} + x_{16} == x_4,\\ &1 + x_{15} + x_{17} == 2^{x_3},\ 1 + x_{16} + x_{18} == 2^{x_3},\\ &2^{1+x_3(1+x_1+2x_3)}(-1 + x_2) - x_{10} + x_{11} == 2x_4,\\ &x_7 == BitAnd[x_6, 2x_6] \wedge x_8 == BitOr[x_6, 2x_6],\\ &x_9 == BitAnd[x_6, 2x_7] \wedge x_{19} == BitOr[x_6, 2x_7],\\ &x_{10} == BitAnd[x_9, 2x_8] \wedge x_{11} == BitOr[x_9, 2x_8]\}\end{aligned}$$

where x_1 through x_4 have the meanings indicated in the main text, and satisfy $x_i \geq 0$. Non-overlapping subsidiary variables are introduced for *BitOr* and *BitAnd*, yielding a total of 79 variables.

Note that it is potentially somewhat easier to construct Diophantine equations to emulate register machines—or arithmetic systems from page 673—than to emulate cellular automata, but exactly the same basic methods can be used.

In the universal equation in the main text variables appear in exponents. One can reduce such an exponential equation to a pure polynomial equation by encoding powers using integer equations. The simplest known way of doing this (see note below) involves a degree 8 equation with 60 variables:

$$\begin{aligned}&a == b^c \longleftrightarrow \alpha[d, 4 + b e, 1 + z] \wedge \alpha[f, e, 1 + z] \wedge\\ &\quad a == Quotient[d, f] \wedge \alpha[g, 4 + b, 1 + z] \wedge e == 16 g (1 + z)\end{aligned}$$

$$\begin{aligned}&\lambda[a\_, b\_, c\_] := Module[\{x\},\\ &\qquad 2a + x_1 == c \wedge (Mod[b - a, c] == 0 \vee Mod[b + a, c] == 0)]\end{aligned}$$

$$\begin{aligned}&\alpha[a\_, b\_, c\_] := Module[\{x\}, x_1^2 - b x_1 x_2 + x_2^2 == 1 \wedge\\ &\quad x_3^2 - b x_3 x_4 + x_4^2 == 1 \wedge 1 + x_4 + x_5 == x_3 \wedge Mod[x_3, x_1^2] ==\\ &\quad 0 \wedge 2x_4 + x_7 == b x_3 \wedge Mod[-b + x_8, x_7] == 0 \wedge\\ &\quad Mod[-2 + x_8, x_1] == 0 \wedge x_8 - x_{11} == 3 \wedge x_{12}^2 - x_8 x_{12} x_{13} +\\ &\quad x_{13}^2 == 1 \wedge 1 + 2a + x_{14} == x_1 \wedge \lambda[a, x_{12}, x_7] \wedge \lambda[c, x_{12}, x_1]]\end{aligned}$$

(This roughly uses the idea that solutions to Pell equations grow exponentially, so that for example $x^2 == 2y^2 + 1$ has solutions $With[\{u = 3 + 2\sqrt{2}\}, (u^n + u^{-n})/2]$.) From this representation of *Power* the universal equation can be converted to a purely polynomial equation with 2154 variables—which when expanded has 1683150 terms, total degree 16 (average per term 6.8), maximum coefficient 17827424 and *LeafCount* 16540206.

Note that the existence of universal Diophantine equations implies that any problem of mathematics—even, say, the Riemann Hypothesis—can in principle be formulated as a question about the existence of solutions to a Diophantine equation. It also means that given any specific enumeration of polynomials, there must be some universal polynomial u which if fed the enumeration number of a polynomial p, together with an encoding of the values of its variables, will yield the corresponding value of p as a solution to $u == 0$.

▪ **Hilbert's Tenth Problem.** Beginning in antiquity various procedures were developed for solving particular kinds of Diophantine equations (see page 1164). In 1900, as one of his list of 23 important mathematical problems, David Hilbert posed the problem of finding a single finite procedure that could systematically determine whether a solution exists to any specified Diophantine equation. The original proof of Gödel's Theorem from 1931 in effect involves showing that certain logical and other operations can be represented by Diophantine equations—and in the end Gödel's Theorem can be viewed as saying that certain statements about Diophantine equations are unprovable. The notion that there might be universal Diophantine equations for which Hilbert's Tenth Problem would be fundamentally unsolvable emerged in work by Martin Davis in 1953. And by 1961 Davis, Hilary Putnam and Julia Robinson had established that there are exponential Diophantine equations that are universal. Extending this to show that Hilbert's original problem about ordinary polynomial Diophantine equations is unsolvable required proving that exponentiation can be represented by a Diophantine equation, and this was finally done by Yuri Matiyasevich in 1969 (see note above).

By the mid-1970s, Matiyasevich had given a construction for a universal Diophantine equation with 9 variables—though with a degree of about 10^{15}. It had been known since the 1930s that any Diophantine equation can be reduced to one with degree 4—and in 1980 James Jones showed that a universal Diophantine equation with degree 4 could be constructed with 58 variables. In 1979 Matiyasevich also showed that universality could be achieved with an exponential Diophantine equation with many terms, but with only 3 variables. As discussed in the main text I believe that vastly simpler Diophantine equations can also be universal. It is even conceivable that a Diophantine equation with 2 variables could be universal: with one variable essentially being used to represent the program and input, and the other the execution history of the program—with no finite solution existing if the program does not halt.

▪ **Polynomial value sets.** Closely related to issues of solving Diophantine equations is the question of what set of positive values a polynomial can achieve when fed all possible positive integer values for its variables. A polynomial with a single variable must always yield either be a finite set, or a simple polynomial progression of values. But already the sequence of values for $x^2 y - x y^3$ or even $x(y^2 + 1)$ seem quite complicated. And for example from the fact that $x^2 == y^2 + (x y \pm 1)$ has solutions $Fibonacci[n]$ it follows that the positive values of $(2-(x^2-y^2-xy)^2)x$ are just *Fibonacci[n]* (achieved when *{x, y}* is *Fibonacci[{n, n-1}]*). This is the simplest polynomial giving *Fibonacci[n]*, and there are for example no polynomials with 2 variables, up to 4 terms, total degree less than 4, and integer coefficients between -2 and +2, that give any of 2^n, 3^n or *Prime[n]*. Nevertheless, from the representation for *PrimeQ* in the note above it has been shown that the positive values of a particular polynomial with 26 variables, 891 terms and total degree 97 are exactly the primes. (Polynomials with 42 variables and degree 5, and 10 variables and degree 10^{45}, are also known to work, while it is known that one with 2 variables cannot.) And in general the existence of a universal Diophantine equation implies that any set obtained by any finite computation must correspond to the positive values of some polynomial. The analog of doing a long computation to find a result is having to go to large values of variables to find a positive polynomial value. Note that one can imagine, say, emulating the evolution of a cellular automaton by having the tth positive value of a polynomial represent the tth step of evolution. That universality can be achieved just in the positive values of a polynomial is already remarkable. But I suspect that in the end it will take only a surprisingly simple polynomial, perhaps with just three variables and fairly low degree.

(See also page 1165.)

■ **Statements in Peano arithmetic.** Examples include:

- $\sqrt{2}$ is irrational:
  $\neg \exists_a (\exists_b (b \neq 0 \wedge a \times a == (\Delta\Delta 0) \times (b \times b)))$
- There are infinitely many primes of the form n^2+1:
  $\neg \exists_n (\forall_c (\exists_a (\exists_b (n+c) \times (n+c) + \Delta 0 == (\Delta\Delta a) \times (\Delta\Delta b))))$
- Every even number (greater than 2) is the sum of two primes (Goldbach's Conjecture; see page 135):
  $\forall_a (\exists_b (\exists_c ((\Delta\Delta 0) \times (\Delta\Delta a) == b + c \wedge \forall_d (\forall_e (\forall_f ((f == (\Delta\Delta d) \times (\Delta\Delta e) \vee f == \Delta 0) \Rightarrow (f \neq b \wedge f \neq c)))))))$

The last two statements have never been proved true or false, and remain unsolved problems of number theory. The picture shows spacings between n for which n^2+1 is prime.

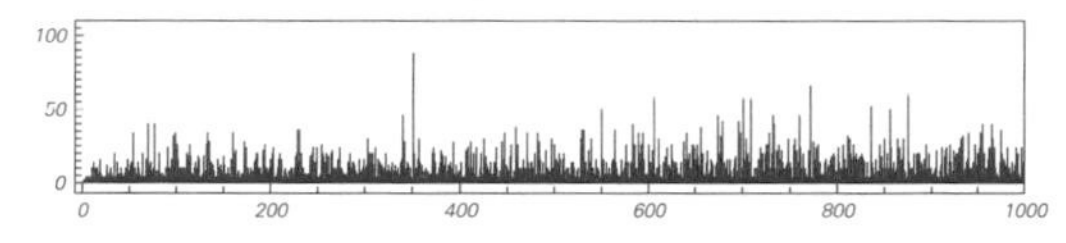

■ **Transfinite numbers.** For most mathematical purposes it is quite adequate just to have a single notion of infinity, usually denoted ∞. But as Georg Cantor began to emphasize in the 1870s, it is possible to distinguish different levels of infinity. Most of the details of this have not been widely used in typical mathematics, but they can be helpful in studying foundational issues. Cantor's theory of ordinal numbers is based on the idea that every integer must have a successor. The next integer after all of the ordinary ones—the first infinite integer—is given the name ω. In Cantor's theory $\omega+1$ is still larger (though $1+\omega$ is not), as are 2ω, ω^2 and ω^ω. Any arithmetic expression involving ω specifies an ordinal number—and can be thought of as corresponding to a set containing all integers up to that number. The ordinary axioms of arithmetic do not apply, but there are still fairly straightforward rules for manipulating such expressions. In general there are many different expressions that correspond to a given number, though there is always a unique Cantor normal form—essentially a finite sequence of digits giving coefficients of descending powers of ω. However, not all infinite integers can be represented in this way. The first one that cannot is ϵ_0, given by the limit $\omega^{\omega^{\omega^{\cdot}}}$, or effectively *Nest[ω^# &, ω, ω]*. ϵ_0 is the smallest solution to $\omega^\epsilon == \epsilon$. Subsequent solutions ($\epsilon_1$, ..., ϵ_ω, ..., ϵ_{ϵ_0}, ...) define larger ordinals, and one can go on until one reaches the limit $\epsilon_{\epsilon_{\epsilon}}$, which is the first solution to $\epsilon_\alpha == \alpha$. Giving this ordinal a name, one can then go on again, until eventually one reaches another limit. And it turns out that in general one in effect has to introduce an infinite sequence of names in order to be able to specify all transfinite integers. (Naming a single largest or "absolutely infinite" integer is never consistent, since one can always then talk about its successor.) As Cantor noted, however, even this only allows one to reach the lowest class of transfinite numbers—in effect those corresponding to sets whose size corresponds to the cardinal number $\aleph_0$. Yet as discussed on page 1127, one can also consider larger cardinal numbers, such as $\aleph_1$, considered in connection with the number of real numbers, and so on. And at least for a while the ordinary axioms of set theory can be used to study the sets that arise.

■ **Growth rates.** One can characterize most functions by their ultimate rates of growth. In basic mathematics these might be n, $2n$, $3n$, ... or n^2, n^3, ..., or 2^n, 3^n, ..., or 2^n, 2^{2^n}, $2^{2^{2^n}}$, ... To go further one begins by defining an analog to the Ackermann function of page 906:

```
f[1][n_] = 2 n; f[s_][n_] := Nest[f[s - 1], 1, n]
```

f[2][n] is then 2^n, *f[3]* is iterated power, and so on. Given this one can now form the "diagonal" function

```
f[ω][n_] := f[n][n]
```

and this has a higher growth rate than any of the *f[s][n]* with finite *s*. This higher growth rate is indicated by the transfinite index ω. And in direct analogy to the transfinite numbers discussed above one can then in principle form a hierarchy of functions using operations like

```
f[ω + s][n_] := Nest[f[ω + s - 1], 1, n]
```

together with diagonalization at limit ordinals. In practice, however, it gets more and more difficult to determine that the functions defined in this way actually in a sense halt and yield definite values—and indeed for $f[\epsilon_0]$ this can no longer be proved using the ordinary axioms of arithmetic (see below). Yet it is still possible to define functions with even more rapid rates of growth. An example is the so-called busy beaver function (see page 1144) that gives the maximum number of steps that it takes for any Turing machine of size *n* to halt when started from a blank tape. In general this function must grow faster than any computable function, and is not itself computable.

■ **Page 787 · Unprovable statements.** After the appearance of Gödel's Theorem a variety of statements more or less directly related to provability were shown to be unprovable in Peano arithmetic and certain other axiom systems. Starting in the 1960s the so-called method of forcing allowed certain kinds of statements in strong axiom systems—like the Continuum Hypothesis in set theory (see page 1155)—to be shown to be unprovable. Then in 1977 Jeffrey Paris and Leo Harrington showed that a variant of Ramsey's Theorem (see page 1068)—a statement that is much more directly mathematical—is also unprovable in Peano arithmetic. The approach they used was in essence based on thinking about growth rates—and since the 1970s almost all new examples of unprovability have been based on similar ideas. Probably the simplest is a statement shown to be unprovable in Peano arithmetic by Laurence Kirby and Jeff Paris in 1982: that certain sequences *g[n]* defined by Reuben Goodstein in 1944 are of limited length for all *n*, where

```
g[n_] := Map[First, NestWhileList[
    {f[#] - 1, Last[#] + 1} &, {n, 3}, First[#] > 0 &]]
f[{0, _}] = 0; f[{n_, k_}] := Apply[Plus, MapIndexed[#1
  k^f[{#2[[1]] - 1, k}] &, Reverse[IntegerDigits[n, k - 1]]]]
```

As in the pictures below, *g[1]* is *{1, 0}*, *g[2]* is *{2, 2, 1, 0}* and *g[3]* is *{3, 3, 3, 2, 1, 0}*. *g[4]* increases quadratically for a long time, with only element $3 \times 2^{402653211} - 2$ finally being 0. And the point is that in a sense *Length[g[n]]* grows too quickly for its finiteness to be provable in general in Peano arithmetic.

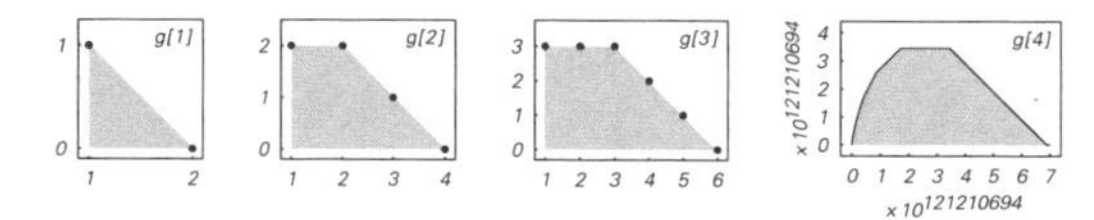

The argument for this as usually presented involves rather technical results from several fields. But the basic idea is roughly just to set up a correspondence between elements of *g[n]* and possible proofs in Peano arithmetic—then to use the fact that if one knew that *g[n]* always terminated this would establish the validity of all these proofs, which would in turn prove the consistency of arithmetic—a result which is known to be unprovable from within arithmetic.

Every possible proof in Peano arithmetic can in principle be encoded as an ordinary integer. But in the late 1930s Gerhard Gentzen showed that if proofs are instead encoded as ordinal numbers (see note above) then any proof can validly be reduced to a preceding one just by operations in logic. To cover all possible proofs, however, requires going up to the ordinal ϵ_0. And from the unprovability of consistency one can conclude that this must be impossible using the ordinary operation of induction in Peano arithmetic. (Set theory, however, allows transfinite induction—essentially induction on arbitrary sets—letting one reach such ordinals and thus prove the consistency of arithmetic.) In constructing *g[n]* the integer *n* is in effect treated like an ordinal number in Cantor normal form, and a sequence of numbers that should precede it are found. That this sequence terminates for all *n* is then provable in set theory, but not Peano arithmetic—and in effect *Length[g[n]]* must grow like $f[\epsilon_0][n]$.)

In general one can imagine characterizing the power of any axiom system by giving a transfinite number κ which specifies the first function $f[\kappa]$ (see note above) whose termination cannot be proved in that axiom system (or similarly how rapidly the first example of *y* must grow with *x* to prevent $\exists_y$ *p[x, y]* from being provable). But while it is known that in Peano arithmetic $\kappa = \epsilon_0$, quite how to describe the value of κ for, say, set theory remains unknown. And in general I suspect that there are a vast number of functions with simple definitions whose termination cannot be proved not just because they grow too quickly but instead for the more fundamental reason that their behavior is in a sense too complicated.

Whenever a general statement about a system like a Turing machine or a cellular automaton is undecidable, at least some instances of that statement encoded in an axiom system must be unprovable. But normally these tend to be complicated and not at all typical of what arise in ordinary mathematics. (See page 1167.)

■ **Encodings of arithmetic.** Statements in arithmetic are normally written in terms of *+*, *×* and *Δ* (and logical operations). But it turns out also to be possible to encode such statements in terms of other basic operations. This was for example done by Julia Robinson in 1949 with *Δ* (or *a + 1*) and *Mod[a, b] == 0*. And in the 1990s Ivan Korec and others showed that it could be done just with *Mod[Binomial[a + b, a], k]* with $k = 6$ or any product of primes—and that it could not be done with k a prime or prime power. These operations can be thought of as finding elements in nested Pascal's triangle patterns produced by k-color additive cellular automata. Korec showed that finding elements in the nested pattern produced by the $k = 3$ cellular automaton with rule *{{1, 1, 3}, {2, 2, 1}, {3, 3, 2}}[[#1, #2]] &* (compare page 886) was also enough.

■ **Page 788 · Infinity.** See page 1162.

■ **Page 789 · Diophantine equations.** If variables appear only linearly, then it is possible to use *ExtendedGCD* (see page 944) to find all solutions to any system of Diophantine equations—or to show that none exist. Particularly from the work of Carl Friedrich Gauss around 1800 there emerged a procedure to find solutions to any quadratic Diophantine equation in two variables—in effect by reduction to the Pell equation $x^2 == a y^2 + 1$ (see page 944), and then computing *ContinuedFraction*[$\sqrt{a}$]. The minimal solutions can be large; the largest ones for successive coefficient sizes are given below. (With size s coefficients it is for example known that the solutions must always be less than $(14 s)^{5 s}$.).

1	$1 + x + x^2 + y - x y == 0$	$\{x == 2, y == 7\}$
2	$1 + 2 x + 2 x^2 + 2 y + x y - 2 y^2 == 0$	$\{x == 687, y == 881\}$
3	$2 + 2 x + 3 x^2 + 3 y + x y - y^2 == 0$	$\{x == 545759, y == 1256763\}$
4	$-4 - x + 4 x^2 - y - 3 x y - 4 y^2 == 0$	$\{x == 251996202018, y == 174633485974\}$

There is a fairly complete theory of homogeneous quadratic Diophantine equations with three variables, and on the basis of results from the early and mid-1900s a finite procedure should in principle be able to handle quadratic Diophantine equations with any number of variables. (The same is not true of simultaneous quadratic Diophantine equations, and indeed with a vector x of just a few variables, a system $m . x^2 == a$ of such equations could quite possibly show undecidability.)

Ever since antiquity there have been an increasing number of scattered results about Diophantine equations involving higher powers. In 1909 Axel Thue showed that any equation of the form $p[x, y] == a$, where $p[x, y]$ is a homogeneous irreducible polynomial of degree at least 3 (such as $x^3 + x y^2 + y^3$) can have only a finite number of integer solutions. (He did this by formally factoring $p[x, y]$ into terms $x - \alpha_i y$, then looking at rational approximations to the algebraic numbers α_i.) In 1966 Alan Baker then proved an explicit upper bound on such solutions, thereby establishing that in principle they can be found by a finite search procedure. (The proof is based on having bounds for how close to zero *Sum*[α_i *Log*[α_j], i, j] can be for independent algebraic numbers α_k.) His bound was roughly *Exp*[$(c s)^{10^5}$]—but later work in essence reduced this, and by the 1990s practical algorithms were being developed. (Even with a bound of 10^{100}, rational approximations to real number results can quickly give the candidates that need to be tested.)

Starting in the late 1800s and continuing ever since a series of progressively more sophisticated geometric and algebraic views of Diophantine equations have developed. These have led for example to the 1993 proof of Fermat's Last Theorem and to the 1983 Faltings theorem (Mordell conjecture) that the topology of the algebraic surface formed by allowing variables to take on complex values determines whether a Diophantine equation has only a finite number of rational solutions—and shows for example that this is the case for any equation of the form $x^n == a y^n + 1$ with $n > 3$. Extensive work has been done since the early 1900s on so-called elliptic curve equations such as $x^2 == a y^3 + b$ whose corresponding algebraic surface has a single hole (genus 1). (A crucial feature is that given any two rational solutions to such equations, a third can always be found by a simple geometrical construction.) By the 1990s explicit algorithms for such equations were being developed—with bounds on solutions being found by Baker's method (see above). In the late 1990s similar methods were applied to superelliptic (e.g. $x^n == p[y]$) and hyperelliptic (e.g. $x^2 == p[y]$) equations involving higher powers, and it now at least definitely seems possible to handle any two-variable cubic Diophantine equation with a finite procedure. Knowing whether Baker's method can be made to work for any particular class of equations involves, however, seeing whether certain rather elaborate algebraic constructions can be done—and this may perhaps in general be undecidable. Most likely there are already equations of degree 4 where Baker's method cannot be used—perhaps ones like $x^3 == y^4 + x y + a$. But in recent years there have begun to be results by other methods about two-variable Diophantine equations, giving, for example, general upper bounds on the number of possible solutions. And although this has now led to the assumption that all two-variable Diophantine equations will eventually be resolved, based on the results of this book I would not be surprised if in fact undecidability and universality appeared in such equations—even perhaps at degree 4 with fairly small coefficients.

The vast majority of work on Diophantine equations has been for the case of two variables (or three for some homogeneous equations). No clear analog of Baker's method is known beyond two variables, and my suspicion is that with three variables undecidability and universality may already be present even in cubic equations.

As mentioned in the main text, proving that even simple specific Diophantine equations have no solutions can be very difficult. Obvious methods involve for example showing that no solutions exist for real variables, or for variables reduced modulo some n. (For quadratic equations Hasse's Principle implies that if no solutions exist for any n then there are no solutions for ordinary integers—but a cubic like $3x^3 + 4y^3 + 5z^3 == 0$ is a counterexample.) If one can find a bound on solutions—say by Baker's method—then one can also try to show that no values below this bound are actually solutions. Over the history of number theory the sophistication of equations for which proofs of no solutions can be given has gradually increased—though even now it is state of the art to show say that $x == y == 1$ is the only solution to $x^2 == 3y^4 - 2$.

Just as for all sorts of other systems with complex behavior, some idea of overall properties of Diophantine equations can be found on the basis of an approximation of perfect randomness. Writing equations in the form $p[x_1, x_2, \ldots, x_n] == 0$ the distribution of values of p will in general be complicated (see page 1161), but as a first approximation one can try taking it to be purely random. (Versions of this for large numbers of variables are validated by the so-called circle method from the early 1900s.) If p has total degree d then with $x_i < x$ the values of $Abs[p]$ will range up to about x^d. But with n variables the number of different cases sampled for $x_i < x$ will be x^n. The assumption of perfect randomness then suggests that for $d < n$, more and more cases with $p == 0$ will be seen as x increases, so that the equation will have an infinite number of solutions. For $d > n$, on the other hand, it suggests that there will often be no solutions, and that any solutions that exist will usually be small. In the boundary case $d == n$ it suggests that even for arbitrarily large x an average of about one solution should exist—suggesting that the smallest solution may be very large, and presumably explaining the presence of so many large solutions in the $n = d = 2$ and $n = d = 3$ examples in the main text. Note that even though large solutions may be rare when $d > n$ they must always exist in at least some cases whenever there is undecidability and universality in a class of equations. (See also page 1161.)

If one wants to enumerate all possible Diophantine equations there are many ways to do this, assigning different weights to numbers of variables, and sizes of coefficients and of exponents. But with several ways I have tried, it seems that of the first few million equations, the vast majority have no solutions—and this can in most cases be established by fairly elementary methods that are presumably within Peano arithmetic. When solutions do exist, most are fairly small. But as one continues the enumeration there are increasingly a few equations that seem more and more difficult to handle.

■ **Page 790 · Properties.** (All variables are assumed positive.)

- $2x + 3y == a$. There are $Ceiling[a/2] + Ceiling[2a/3] - (a + 1)$ solutions, the one with smallest x being $\{Mod[2a + 2, 3] + 1, 2 Floor[(2a + 2)/3] - (a + 2)\}$. Linear equations like this were already studied in antiquity. (Compare page 915.)
- $x^2 == y^2 + a$. Writing a in terms of distinct factors as $r s$, $\{r + s, r - s\}/2$ gives a solution if it yields integers—which happens when $Abs[a] > 4$ and $Mod[a, 4] \neq 2$.
- $x^2 == a y^2 + 1$ (Pell equation). As discussed on page 944, whenever a is not a perfect square, there are always an infinite number of solutions given in terms of $ContinuedFraction[\sqrt{a}]$. Note that even when the smallest solution is not very large, subsequent solutions can rapidly get large. Thus for example when $a = 13$, the second solution is already $\{842401, 233640\}$.
- $x^2 == y^3 + a$ (Mordell equation). First studied in the 1600s, a complete theory of this so-called elliptic curve equation was only developed in the late 1900s—using fairly sophisticated algebraic number theory. The picture below shows as a function of a the minimum x that solves the equation. For $a = 68$, the only solution is $x = 1874$; for $a = 1090$, it is $x = 149651610621$. The density of cases with solutions gradually thins out as a increases (for $0 < a \leq 10000$ there are 2468 such cases). There are always only a finite number of solutions (for $0 < a \leq 10000$ the maximum is 12, achieved for $a = 8900$).

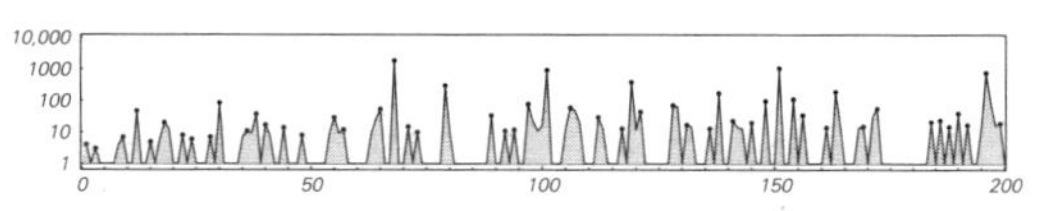

- $x^2 == a y^3 + 1$. Also an elliptic curve equation.
- $x^3 == y^4 + x y + a$. For most values of a (including specifically $a = 1$) the continuous version of this equation defines a surface of genus 3, so there are at most a finite number of integer solutions. (An equation of degree d generically defines a surface of genus $1/2 (d - 1)(d - 2)$.) Note that $x^3 == y^4 + a$ is equivalent to $x^3 == z^2 + a$ by a simple substitution.
- $x^2 == y^5 + a y + 3$. The second smallest solution to $x^2 == y^5 + 5y + 3$ is $\{45531, 73\}$. As for the equations above, there are always at most a finite number of integer solutions.
- $x^3 + y^3 == z^2 + a$. For the homogenous case $a = 0$ the complete solution was found by Leonhard Euler in 1756.

- $x^3 + y^3 == z^3 + a$. No solutions exist when $a = 9n \pm 4$; for $a = n^3$ or $2n^3$ infinite families of solutions are known. Particularly in its less strict form $x^3 + y^3 + z^3 == a$ with x, y, z positive or negative the equation was mentioned in the 1800s and again in the mid-1900s; computer searches for solutions were begun in the 1960s, and by the mid-1990s solutions such as {283059965, 2218888517, 2220422932} for the case $a = 30$ had been found. Any solution to the difficult case $x^3 + y^3 == z^3 - 3$ must have Mod[x, 9] == Mod[y, 9] == Mod[z, 9]. (Note that $x^2 + y^2 + z^2 == a$ always has solutions except when $a = 4^s(8n + 7)$, as mentioned on page 135.)

Large solutions. A few other 2-variable equations with fairly large smallest solutions are:

- $x^3 == 3y^3 - xy + 63$: {7149, 4957}
- $x^4 == y^3 + 2xy - 2y + 81$: {19674, 531117}
- $x^4 == 5y^3 + xy + y - 8x$: {69126, 1659072}

The equation $x^x y^y == z^z$ is known to have smallest non-trivial solution {2985984, 1679616, 4478976}.

Nearby powers. One can potentially find integer equations with large solutions but small coefficients by looking say for pairs of integer powers close in value. The pictures below show what happens if one computes x^m and y^n for many x and y, sorts these values, then plots successive differences. The differences are trivially zero when $x = s^n$, $y = s^m$. Often they are large, but surprisingly small ones can sometimes occur (despite various suggestions from the so-called ABC conjecture). Thus, for example, $5853886516781223^3 - 1641843$ is a perfect square, as found by Noam Elkies in 1998. (Another example is $55^5 - 22434^2 == 19$.)

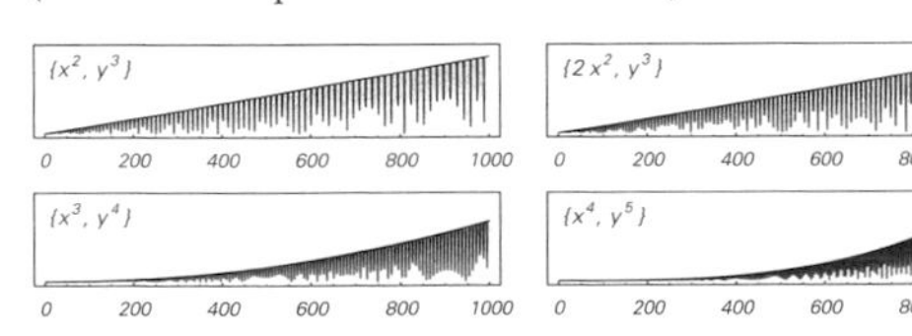

Page 791 · Unsolved problems. Problems in number theory that are simple to state (say in the notation of Peano arithmetic) but that so far remain unsolved include:

- Is there any odd number equal to the sum of its divisors? (Odd perfect number; 4th century BC) (See page 911.)
- Are there infinitely many primes that differ by 2? (Twin Prime Conjecture; 1700s?) (See page 909.)
- Is there a cuboid in which all edges and all diagonals are of integer length? (Perfect cuboid; 1719)
- Is there any even number which is not the sum of two primes? (Goldbach's Conjecture; 1742) (See page 135.)
- Are there infinitely many primes of the form $n^2 + 1$? (Quadratic primes; 1840s?) (See page 1162.)
- Are there infinitely many primes of the form $2^{2^n} + 1$? (Fermat primes; 1844)
- Are there no solutions to $x^m - y^n == 1$ other than $3^2 - 2^3 == 1$? (Catalan's Conjecture; 1844)
- Can every integer not of the form $9n \pm 4$ be written as $a^3 \pm b^3 \pm c^3$? (See note above.)
- How few n^{th} powers need be added to get any given integer? (Waring's Problem; 1770)

(See also Riemann Hypothesis on page 918.)

Page 791 · Fermat's Last Theorem. That $x^n + y^n == z^n$ has no integer solutions for $n > 2$ was suggested by Pierre Fermat around 1665. Fermat proved this for $n = 4$ around 1660; Leonhard Euler for $n = 3$ around 1750. It was proved for $n = 5$ and $n = 7$ in the early 1800s. Then in 1847 Ernst Kummer used ideas of factoring with algebraic integers to prove it for all $n < 37$. Extensions of this method gradually allowed more cases to be covered, and by the 1990s computers had effectively given proofs for all n up to several million. Meanwhile, many connections had been found between the general case and other areas of mathematics—notably the theory of elliptic curves. And finally around 1995, building on extensive work in number theory, Andrew Wiles managed to give a complete proof of the result. His proof is long and complicated, and relies on sophisticated ideas from many areas of mathematics. But while the statement of the proof makes extensive use of concepts from areas like set theory, it seems quite likely that in the end a version of it could be given purely in terms of Peano arithmetic. (By the 1970s it had for example been shown that many classic proofs with a similar character in analytic number theory could at least in principle be carried out purely in Peano arithmetic.)

Page 791 · More powerful axioms. If one looks for example at progressively more complicated Diophantine equations then one can expect that one will find examples where more and more powerful axiom systems are needed to prove statements about them. But my guess is that almost as soon as one reaches cases that cannot be handled by Peano arithmetic one will also reach cases that cannot be handled by set theory or even by still more powerful axiom systems.

Any statement that one can show is independent of the Peano axioms and at least not inconsistent with them one can potentially consider adding as a new axiom. Presumably it is best to add axioms that allow the widest range of new statements to be proved. But I strongly suspect that the set of statements that cannot be proved is somehow sufficiently fragmented that adding a few new axioms will actually make very little difference.

In set theory (see page 1155) a whole sequence of new axioms have historically been added to allow particular kinds of statements to be proved. And for several decades additional so-called large cardinal axioms have been discussed, that in effect state that sets exist larger than any that can be reached with the current axioms of set theory. (As discussed on page 816 any axiom system that is universal must in principle be able to prove any statement that can be proved in any axiom system—but not with the kinds of encodings normally considered in mathematical logic.)

It is notable, however, that if one looks at classic theorems in mathematics many can actually be derived from remarkably weak axioms. And indeed the minimal axioms needed to obtain most of mathematics as it is now practiced are probably much weaker than those on pages 773 and 774.

(If one considers for example theorems about computational issues such as whether Turing machines halt, then it becomes inevitable that to cover more Turing machines one needs more axioms—and to cover all possible machines one needs an infinite set of axioms, that cannot even be generated by any finite set of rules.)

▪ **Higher-order logics.** In ordinary predicate—or so called first-order—logic the objects x that $\forall_x$ and $\exists_x$ range over are variables of the kind used as arguments to functions (or predicates) such as `f[x]`. To set up second-order logic, however, one imagines also being able to use $\forall_f$ and $\exists_f$ where f is a function (say the head of `f[x]`). And then in third-order logic one imagines using $\forall_g$ and $\exists_g$ where g appears in `g[f][x]`.

Early formulations of axiom systems for mathematics made little distinction between first- and second-order logic. The theory of types used in *Principia Mathematica* introduced some distinction, and following the proof of Gödel's Completeness Theorem for first-order logic in 1930 (see page 1152) standard axiom systems for mathematics (as given on pages 773 and 774) began to be reformulated in first-order form, with set theory taking over many of the roles of second-order logic.

In current mathematics, second-order logic is sometimes used at the level of notation, but almost never in its full form beyond. And in fact with any standard computational system it can never be implemented in any explicit way. For even to enumerate theorems in second-order logic is in general impossible for a system like a Turing machine unless one assumes that an oracle can be added. (Note however that this is possible in Henkin versions of higher-order logic that allow only limited function domains.)

▪ **Truth and incompleteness.** In discussions of the foundations of mathematics in the early 1900s it was normally assumed that truth and provability were in a sense equivalent—so that all true statements could in principle be reached by formal processes of proof from fixed axioms (see page 782). Gödel's Theorem showed that there are statements that can never be proved from given axioms. Yet often it seemed inevitable just from the syntactic structure of statements (say as well-formed formulas) that each of them must at some level be either true or false. And this led to the widespread claim that Gödel's Theorem implies the existence of mathematical statements that are true but unprovable—with their negations being false but unprovable. Over the years this often came to be assigned a kind of mystical significance, mainly because it was implicitly assumed that somehow it must still ultimately be possible to know whether any given statement is true or false. But the Principle of Computational Equivalence implies that in fact there are all sorts of statements that simply cannot be decided by any computational process in our universe. So for example, it must in some sense be either true or false that a given Turing machine halts with given input—but according to the Principle of Computational Equivalence there is no finite procedure in our universe through which we can guarantee to know which of these alternatives is correct.

In some cases statements can in effect have default truth values—so that showing that they are unprovable immediately implies, say, that they must be true. An example in arithmetic is whether some integer equation has no solution. For if there were a solution, then given the solution it would be straightforward to give a proof that it is correct. So if it is unprovable that there is no solution, then it follows that there must in fact be no solution. And similarly, if it could be shown for example that Goldbach's Conjecture is unprovable then it would follow that it must be true, for if it were false then there would have to be a specific number which violates it, and this could be proved. Not all statements in mathematics have this kind of default truth value. And thus for example the Continuum Hypothesis in set theory is unprovable but could be either of true or false: it is just independent of the axioms of set theory. In computational systems, showing that it is unprovable that a given Turing machine halts with given input immediately implies that in fact it must not halt. But showing that it is unprovable whether a Turing machine halts with every input (a Π_2 statement in the notation of page 1139) does not immediately imply anything about whether this is in fact true or false.

▪ **Page 793 · Generalization in mathematics.** Systems that have evolved from the basic notion of numbers provide a characteristic example of the process of progressive generalization in mathematics. The main such systems and their dates of earliest known reasonably formalized use have been (see also page 901): positive integers (before 10,000 BC), rationals (3000 BC), square roots (2000 BC), other roots (1800 BC), all integers (600 AD, 1600s), decimals (950 AD), complex numbers (1500s, 1800s), polynomials (1591), infinitesimals (1635), algebraic numbers (1744), quaternions (1843), Grassmann algebra (1844), ideals (1844, 1871), octonions (~1845), Boolean algebra (1847), fields (1850s, 1871), matrices (1858), associative algebras (1870), axiomatic real numbers (1872), vectors (1881), transfinite ordinals (1883), transfinite cardinals (1883), operator calculus (1880s), Boolean algebras (1890), algebraic number fields (1893), rings (1897), p-adic numbers (1897), non-Archimedean fields (1899), q-numbers (1926), non-standard integers (1930s), non-standard reals (hyperreals) (1960), interval arithmetic (1968), fuzzy arithmetic (1970s), surreal numbers (1970s). New systems have usually been introduced in connection with extending the domains of particular existing operations. But in almost all cases the systems are set up so as to preserve as many theorems as possible—a notion that was for example made explicit in the Principle of Permanence discussed by George Peacock in 1830 and extended by Hermann Hankel in 1869.

▪ **Page 794 · Cellular automaton axioms.** The first 4 axioms are general to one-dimensional cellular automata. The next 8 are specific to rule 110. The final 3 work whenever patterns are embedded in a background of white cells. The universality of rule 110 presumably implies that the axiom system given is universal. (A complete proof would require handling various issues about boundary conditions.)

If the last 2 axioms are dropped any statement can readily be proved true or false essentially just by running rule 110 for a finite number of steps equal to the number of nested ↓ plus ⟨…⟩ in the statement. In practice, a large number of steps can however be required. As an example the statement

O■ ⇒ (∃*ₐ* (∃*_b* *O a* ◇ (■ ◇ (■ ◇ (■ ◇ (□ ◇
(■ ◇ (■ ◇ (■ ◇ (□ ◇ (■ ◇ (■ ◇ (■ ◇ (■ ◇ (■ ◇ (■ ◇ (■ ◇ (■ ◇ (□ ◇
(□ ◇ (□ ◇ (■ ◇ (□ ◇ (□ ◇ (■ ◇ (■ ◇ (□ ◇ (■ ◇ (■ ◇ (■ ◇ (■ ◇ (■ ◇
(□ ◇ (□ ◇ (□ ◇ (■ ◇ (□ ◇ *b*)))))))))))))))))))))))))))))))))))))

asserts that a particular localized structure occurs in the evolution of rule 110 from a single black cell. But page 38 shows that this happens for the first time after 2867 steps. (A proof of this without lemmas would probably have to be of length at least 32,910,300.)

The axioms as they are stated apply to any rule 110 evolution, regardless of initial conditions. One can establish that the statement at the bottom on the right cannot be proved either true or false from the axioms by showing that it is true for some initial conditions and false for others. Note from page 279 that the sequence □■□■□ cannot occur in rule 110 evolution except as an initial condition. So this means that the statement is false if the initial condition is ■ and true if the initial condition is □■□■□.

▪ **Practical programs.** Any equivalence between programs in a programming language can be thought of as a theorem. Simple examples in *Mathematica* include:

```
First[Prepend[p, q]] === q
Join[Join[p, q], r] === Join[p, Join[q, r]]
Partition[Union[p], 1] === Split[Union[p]]
```

One can set up axiom systems say by combining definitions of programming languages with predicate logic (as done by John McCarthy for LISP in 1963). And for programs whose structure is simple enough it has sometimes been possible to prove theorems useful for optimization or verification. But in the vast majority of cases this has been essentially impossible in practice. And I suspect that this is a reflection of widespread fundamental unprovability. In setting up programs with specific purposes there is inevitably some computational reducibility (see page 828). But I suspect that enough computational irreducibility usually remains to make unprovability common when one asks about all possible forms of behavior of the program.

▪ **Page 796 · Rules.** The examples shown here (roughly in order of increasing complexity) correspond respectively to cases (a), (k), (b), (q), (p), (r), (o), (d) on page 798.

▪ **Page 797 · Consistency.** Any axiom system that is universal can represent the statement that the system is consistent. But normally such a statement cannot be proved true or false within the system itself. And thus for example Kurt Gödel showed this in 1931 for Peano arithmetic (in his so-called second incompleteness theorem). In 1936, however, Gerhard Gentzen showed that the axioms of set theory imply the consistency of Peano arithmetic (see page 1160). In practical mathematics set theory is always taken to be consistent, but to set up a proof of this would require axioms beyond set theory.

▪ **Page 798 · Properties.** For most of the rules shown, there ultimately turn out to be quite easy characterizations of what strings can be produced.

- (a) At step t, the only new string produced is the one containing t black elements.

- (b) All strings of length n containing exactly one black cell are produced—after at most $2n-1$ steps.
- (c) All strings containing even-length runs of white cells are produced.
- (d) The set of strings produced is complicated. The last length 4 string produced is □□□■, after 16 steps; the last length 6 one is □□□□□■, after 26 steps.
- (e) All strings that begin with a black element are produced.
- (f) All strings that end with a white element but contain at least one black element, or consist of all white elements ending with black, are produced. Strings of length n take n steps to produce.
- (g) The same strings as in (f) are produced, but now a string of length n with m black elements takes $n+m-1$ steps.
- (h) All strings appear in which the first run of black elements is of length 1; a string of length n with m black elements appears after $n+m-1$ steps.
- (i) All strings containing an odd number of black elements are produced; a string of length n with m black cells occurs at step $n+m-1$.
- (j) All strings that end with a black element are produced.
- (k) Above length 1, the strings produced are exactly those starting with a white element. Those of length n appear after at most $3n-3$ steps.
- (l) The same strings as in (k) are produced, taking now at most $2n+1$ steps.
- (m) All strings beginning with a black element are produced, after at most $3n+1$ steps.
- (n) The set of strings produced is complicated, and seems to include many but not all that do not end with ■□.
- (o) All strings that do not end in ■□ are produced.
- (p) All strings are produced, except ones in which every element after the first is white. ■□□■ takes 14 steps.
- (q) All strings are produced, with a string of length n with m white elements taking $n+2m$ steps.
- (r) All strings are ultimately produced—which is inevitable after the lemmas ■ → ■■ and ■ → □ appear at steps 12 and 13. (See the first rule on page 778.)

■ **Page 800 · Non-standard arithmetic.** Goodstein's result from page 1163 is true for all ordinary integers. But since it is independent of the axioms of arithmetic there must be objects that still satisfy the axioms but for which it is false. It turns out however that any such objects must in effect be infinite. For any set of objects that satisfy the axioms of arithmetic must include all finite ordinary integers, since each of these can be reached just by using Δ repeatedly. And the axioms then turn out to imply that any additional objects must be larger than all these integers—and must therefore be infinite. But for any such truly infinite objects operations like $+$ and $\times$ cannot be computed by finite procedures, making it difficult to describe such objects in an explicit way. Ever since the work of Thoralf Skolem in 1933 non-standard models of arithmetic have been discussed, particularly in the context of ultrafilters and constructs like infinite trees. (See also page 1172.)

■ **Page 800 · Reduced arithmetic.** (See page 1152.) Statements that can be proved with induction but are not provable only with Robinson's axioms are: $x \neq \Delta x$; $x+y == y+x$; $x+(y+z) == (x+y)+z$; $0+x == x$; $\exists_x(\Delta x + y == z \Rightarrow y \neq z)$; $x \times y == y \times x$; $x \times (y \times z) == (x \times y) \times z$; $x \times (y+z) == x \times y + x \times z$.

■ **Page 800 · Generators and relations.** In the axiom systems of page 773, a single variable can stand for any element—much like a *Mathematica* pattern object such as x_. In studying specific instances of objects like groups one often represents elements as products of constants or generators, and then for example specifies the group by giving relations between these products. In traditional mathematical notation such relations normally look just like ordinary axioms, but in fact the variables that appear in them are now assumed to be literal objects—like x in *Mathematica* that are generically taken to be unequal. (Compare page 1159.)

■ **Page 801 · Comparison to multiway systems.** Operator systems are normally based on equations, while multiway systems are based on one-way transformations. But for multiway systems where each rule $p \to q$ is accompanied by its reverse $q \to p$, and such pairs are represented say by "AAB" ↔ "BBAA", an equivalent operator system can immediately be obtained either from

```
Apply[Equal,
  Map[Fold[#2[#1] &, x, Characters[#]] &, rules, {2}], {1}]
```

or from (compare page 1172)

```
Append[Apply[Equal,
    Map[(Fold[f, First[#], Rest[#]] &)[Characters[#]] &,
      rules, {2}], {1}], f[f[a, b], c] == f[a, f[b, c]]]
```

where now objects like "A" and "B" are treated as constants—essentially functions with zero arguments. With slightly more effort multiway systems with ordinary one-way rules can also be converted to operator systems. Converting from operator systems to multiway systems is more difficult, though ultimately always possible (see page 1156).

As discussed on page 898, one can set up operator evolution systems similar to symbolic systems (see page 103) that have essentially the same relationship to operator systems as sequential substitution systems do to multiway systems. (See also page 1172.)

■ **Page 802 · Operator systems.** One can represent the possible values of expressions like *f[f[p, q], p]* by rule numbers analogous to those used for cellular automata. Specifying an operator *f* (taken in general to have *n* arguments with *k* possible values) by giving the rule number *u* for *f[p, q, …]*, the rule number for an expression with variables *vars* can be obtained from

```
With[{m = Length[vars]}, FromDigits[
 Block[{f = Reverse[IntegerDigits[u, k, k^n]][[FromDigits[
    {##}, k] + 1]] &}, Apply[Function[Evaluate[vars], expr],
   Reverse[Array[IntegerDigits[# - 1, k, m] &, k^m]], {1}]], k]]
```

■ **Truth tables.** The method of finding results in logic by enumerating all possible combinations of truth values seems to have been rediscovered many times since antiquity. It began to appear regularly in the late 1800s, and became widely known after its use by Emil Post and Ludwig Wittgenstein in the early 1920s.

■ **Page 803 · Proofs of axiom systems.** One way to prove that an axiom system can reproduce all equivalences for a given operator is to show that its axioms can be used to transform any expression to and from a unique standard form. For then one can start with an expression, convert it to standard form, then convert back to any expression that is equivalent. We saw on page 616 that in ordinary logic there is a unique DNF representation in terms of *And*, *Or* and *Not* for any expression, and in 1921 Emil Post used essentially this to give the first proof that an axiom system like the first one on page 773 can completely reproduce all theorems of logic. A standard form in terms of *Nand* can be constructed essentially by direct translation of DNF; other methods can be used for the various other operators shown. (See also page 1175.)

Given a particular axiom system that one knows reproduces all equivalences for a given operator one can tell whether a new axiom system will also work by seeing whether it can be used to derive the axioms in the original system. But often the derivations needed may be very long—as on page 810. And in fact in 1948 Samuel Linial and Emil Post showed that in general the problem is undecidable. They did this in effect by arguing (much as on page 1169) that any multiway system can be emulated by an axiom system of the form on page 803, then knowing that in general it is undecidable whether a multiway system will ever reach some given result. (Note that if an axiom system does manage to reproduce logic in full then as indicated on page 814 its consequences can always be derived by proofs of limited length, if nothing else by using truth tables.)

Since before 1920 it has been known that one way to disprove the validity of a particular axiom system is to show that with $k > 2$ truth values it allows additional operators (see page 805). (Note that even if it works for all finite k this does not establish its validity.) Another way to do this is to look for invariants that should not be present—seeing if there are features that differ between equivalent expressions, yet are always left the same by transformations in the axiom system. (Examples for logic are axiom systems which never change the size of an expression, or which are of the form *{expr == a}* where *Flatten[expr]* begins or ends with *a*.)

■ **Junctional calculus.** Expressions are equivalent when *Union[Level[expr, {-1}]]* is the same, and this canonical form can be obtained from the axiom system of page 803 by flattening using $(a \circ b) \circ c == a \circ (b \circ c)$, sorting using $a \circ b == b \circ a$, and removing repeats using $a \circ a == a$. The operator can be either *And* or *Or* (8 or 14). With $k = 3$ there are 9 operators that yield the same results:

{13203, 15633, 15663, 16401,
 17139, 18063, 19539, 19569, 19599}

With $k = 4$ there are 3944 such operators (see below). No single axiom can reproduce all equivalences, since such an axiom must of the form *expr == a*, yet *expr* cannot contain variables other than *a*, and so cannot for example reproduce $a \circ b = b \circ a$.

■ **Equivalential calculus.** Expressions with variables *vars* are equivalent if they give the same results for

```
Mod[Map[Count[expr, #, {-1}] &, vars], 2]
```

With n variables, there are thus 2^n equivalence classes of expressions (compared to 2^{2^n} for ordinary logic). The operator can be either *Xor* or *Equal* (6 or 9). With $k = 3$ there are no operators that yield the same results; with $k = 4$ {458142180, 1310450865, 2984516430, 3836825115} work (see below). The shortest axiom system that works up to $k = 2$ is $\{(a \circ b) \circ a = b\}$. With *modus ponens* as the rule of inference, the shortest single-axiom system that works is known to be $\{(a \circ b) \circ ((c \circ b) \circ (a \circ c))\}$. Note that equivalential calculus describes the smallest non-trivial group, and can be viewed as an extremely minimal model of algebra.

■ **Implicational calculus.** With $k = 2$ the operator can be either 2 or 11 (*Implies*), with $k = 3$ {2694, 9337, 15980}, and with $k = 4$ any of 16 possibilities. (Operators exist for any k.) No single axiom, at least with up to 7 operators and 4 variables, reproduces all equivalences. With *modus ponens* as the rule of inference, the shortest single-axiom system that works is known to be $\{((a \circ b) \circ c) \circ ((c \circ a) \circ (d \circ a))\}$. Using the method of page 1151 this can be converted to the equational form

$$\{((a \circ b) \circ c) \circ ((c \circ a) \circ (d \circ a)) = d \circ d,$$
$$(a \circ a) \circ b = b,\ (a \circ b) \circ b = (b \circ a) \circ a\}$$

from which the validity of the axiom system in the main text can be established.

■ **Page 803 · Operators on sets.** There is always more than one operator that yields a given collection of equivalences. So for ordinary logic both *Nand* and *Nor* work. And with $k = 4$ any of the 12 operators

```
{1116699, 169585339, 290790239, 459258879,
 1090522958, 1309671358, 1430343258, 1515110058,
 2184380593, 2575151445, 2863760025, 2986292093}
```

also turn out to work. One can see why this happens by considering the analogy between operations in logic and operations on sets. As reflected in their traditional notations—and emphasized by Venn diagrams—*And* ($\wedge$), *Or* ($\vee$) and *Not* correspond directly to *Intersection* ($\cap$), *Union* ($\cup$) and *Complement*. If one starts from the single-element set *{1}* then applying *Union*, *Intersection* and *Complement* one always gets either *{}* or *{1}*. And applying *Complement[s, Intersection[a, b]]* to these two elements gives the same results and same equivalences as $a \bar{\wedge} b$ applied to *True* and *False*. But if one uses instead *s = {1, 2}* then starts with *{1}* and *{2}* one gets any of *{{}, {1}, {2}, {1, 2}}* and in general with *s = Range[n]* one gets any of the 2^n elements in the powerset

```
Distribute[Map[{{}, {#}} &, s], List, List, List, Join]
```

But applying *Complement[s, Intersection[a, b]]* to these elements still always produces the same equivalences as with $a \bar{\wedge} b$. Yet now $k = 2^n$. And so one therefore has a representation of Boolean algebra of size 2^n. For ordinary logic based on *Nand* it turns out that there are no other finite representations (though there are other infinite ones). But if one works, say, with *Implies* then there are for example representations of size 3 (see above). And the reason for this is that with *s = {1, 2}* the function *Union[Complement[s, a], b]* corresponding to $a \Rightarrow b$ only ever gets to the 3 elements *{{1}, {2}, {1, 2}}*. Indeed, in general with operators *Implies*, *And* and *Or* one gets to $2^n - 1$ elements, while with operators *Xor* and *Equal* one gets to *2^(2 Floor[n/2])* elements.

(One might think that one could force there only ever to be two elements by adding an axiom like $a == b \vee b == c \vee c == a$. But all this actually does is to force there to be only two objects analogous to *True* and *False*.)

■ **Page 805 · Implementation.** Given an axiom system in the form *{f[a, f[a, a]] = a, f[a, b] = f[b, a]}* one can find rule numbers for the operators *f[x, y]* with *k* values for each variable that are consistent with the axiom system by using

```
Module[{c, v}, c = Apply[Function,
  {v = Union[Level[axioms, {-1}]], Apply[And, axioms]}];
 Select[Range[0, k^(k^2) - 1], With[{u = IntegerDigits[#, k, k^2]},
   Block[{f}, f[x_, y_] := u[[-1 - k x - y]];
    Array[c, Table[k, {Length[v]}], 0, And]]] &]]
```

For $k = 4$ this involves checking nearly 16^4 or 4 billion cases, though many of these can often be avoided, for example by using analogs of the so-called Davis-Putnam rules. (In searching for an axiom system for a given operator it is in practice often convenient first to test whether each candidate axiom holds for the operator one wants.)

■ **Page 805 · Properties.** There are k^{k^2} possible forms for binary operators with k possible values for each argument. There is always at least some operator that satisfies the constraints of any given axiom system—though in a case like $a = b$ it has $k = 1$. Of the 274,499 axiom systems of the form *{... = a}* where ... involves $\circ$ up to 6 times, 32,004 allow only operators *{6, 9}*, while 964 allow only *{1, 7}*. The only cases of 2 or less operators that appear with $k = 2$ are *{{}, {10}, {12}, {1, 7}, {3, 12}, {5, 10}, {6, 9}, {10, 12}}*. (See page 1174.)

■ **Page 806 · Algebraic systems.** Operator systems can be viewed as algebraic systems of the kind studied in universal algebra (see page 1150). With a single two-argument operator (such as $\circ$) what one has is in general known as a groupoid (though this term means something different in topology and category theory); with two such operators a ringoid. Given a particular algebraic system, it is sometimes possible—as we saw on page 773—to reduce the number of operators it involves. But the number of systems that have traditionally been studied in mathematics and that are known to require only one 2-argument operator are fairly limited. In addition to basic logic, semigroups and groups, there are essentially only the rather obscure examples of semilattices, with axioms $\{a \circ (b \circ c) = (a \circ b) \circ c, a \circ b = b \circ a, a \circ a = a\}$, central groupoids, with axioms $\{(b \circ a) \circ (a \circ c) = b\}$, and squags (quasigroup representations of Steiner triple systems), with axioms $\{a \circ b = b \circ a, a \circ a = a, a \circ (a \circ b) = b\}$ or equivalently $\{a \circ ((b \circ (b \circ (((c \circ c) \circ d) \circ c))) \circ a) = d\}$. (Ordinary quasigroups are defined by $\{a \circ c = b, d \circ a = b\}$ with c, d unique for given a, b—so that their table is a Latin square; their axioms can be set up with 3 operators as $\{a \backslash a \circ b = b, a \circ b / b = a, a \circ (a \backslash b) = b, (a / b) \circ b = a\}$.)

Pages 773 and 774 indicate that most axiom systems in mathematics involve operators with at most 2 arguments (there are exceptions in geometry). (Constants such as *1* or $\emptyset$ can be viewed as 0-argument operators.) One can nevertheless generalize say to polyadic groups, with 3-argument composition and analogs of associativity such as

f[f[a, b, c], d, e] = f[a, f[b, c, d], e] = f[a, b, f[c, d, e]]

Another example is the cellular automaton axiom system of page 794; see also page 886. (A perhaps important generalization is to have expressions that are arbitrary networks rather than just trees.)

▪ **Symbolic systems.** By introducing constants (0-argument operators) and interpreting ∘ as function application one can turn any symbolic system such as *e[x][y] → x[x[y]]* from page 103 into an algebraic system such as $(e \circ a) \circ b = a \circ (a \circ b)$. Doing this for the combinator system from page 711 yields the so-called combinatory algebra $\{((s \circ a) \circ b) \circ c = (a \circ c) \circ (b \circ c),\ (k \circ a) \circ b = a\}$.

▪ **Page 806 · Groups and semigroups.** With k possible values for each variable, the forms of operators allowed by axiom systems for group theory and semigroup theory correspond to multiplication tables for groups and semigroups with k elements. Note that the first group that is not commutative (Abelian) is the group S3 with $k = 6$ elements. The total number of commutative groups with k elements is just

```
Apply[Times,
  Map[PartitionsP[Last[#]] &, FactorInteger[k]]]
```

(Relabelling of elements makes the number of possible operator forms up to $k!$ times larger.) (See also pages 945, 1153 and 1173.)

▪ **Forcing of operators.** Given a particular set of forms for operators one can ask whether an axiom system can be found that will allow these but no others. As discussed in the note on operators on sets on page 1171 some straightforwardly equivalent forms will always be allowed. And unless one limits the number of elements k it is in general undecidable whether a given axiom system will allow no more than a given set of forms. But even with fixed k it is also often not possible to force a particular set of forms. And as an example of this one can consider commutative group theory. The basic axioms for this allow forms of operators corresponding to multiplication tables for all possible commutative groups (see note above). So to force particular forms of operators would require setting up axioms satisfied only by specific commutative groups. But it turns out that given the basic axioms for commutative group theory any non-trivial set of additional axioms can always be reduced to a single axiom of the form $a^n == 1$ (where exponentiation is repeated application of ∘). Yet even given a particular number of elements k, there can be several distinct groups satisfying $a^n == 1$ for a given exponent n. (The groups can be written as products of cyclic ones whose orders correspond to the possible factors of n.) (Something similar is also known in principle to be true for general groups, though the hierarchy of axioms in this case is much more complicated.)

▪ **Model theory.** In model theory each form of operator that satisfies the constraints of a given axiom system is called a model of that axiom system. If there is only one inequivalent model the axiom system is said to be categorical—a notion discussed for example by Richard Dedekind in 1887. The Löwenheim-Skolem theorem from 1915 implies that any axiom system must always have a countable model. (For an operator system such a model can have elements which are simply equivalence classes of expressions equal according to the axioms.) So this means that even if one tries to set up an axiom system to describe an uncountable set—such as real numbers—there will inevitably always be extra countable models. Any axiom system that is incomplete must always allow more than one model. The model intended when the axiom system was originally set up is usually called the standard model; others are called non-standard. In arithmetic non-standard models are obscure, as discussed on page 1169. In analysis, however, Abraham Robinson pointed out in 1960 that one can potentially have useful non-standard models, in which there are explicit infinitesimals—much along the lines suggested in the original work on calculus in the late 1600s.

▪ **Pure equational logic.** Proofs in operator systems always rely on certain underlying rules about equality, such as the equivalence of $u == v$ and $v == u$, and of $u == v$ and $u == v$ /. $a \to b$. And as Garrett Birkhoff showed in 1935, any equivalence between expressions that holds for all possible forms of operator must have a finite proof using just these rules. (This is the analog of Gödel's Completeness Theorem from page 1152 for pure predicate logic.) But as soon as one introduces actual axioms that constrain the operators this is no longer true—and in general it can be undecidable whether or not a particular equivalence holds.

▪ **Multiway systems.** One can use ideas from operator systems to work out equivalences in multiway systems (compare page 1169). One can think of concatenation of strings as being an operator, in terms of which a string like "ABB" can be written *f[f[a, b], b]*. (The arguments to ∘ should strictly be distinct constants, but no equivalences are lost by allowing them to be general variables.) Assuming that the rules for a multiway system come in pairs $p \to q$, $q \to p$, like "AB" → "AAA", "AAA" → "AB", these can be written as statements about operators, like *f[a, b] == f[f[a, a], a]*. The basic properties of concatenation then also imply that *f[f[a, b], c] == f[a, f[b, c]]*. And this means that the possible forms for the operator ∘ correspond to possible semigroups. Given a particular such semigroup satisfying axioms derived from a multiway system, one can see whether the operator representations of particular strings are equal—and if they are not, then it follows that the strings can never be reached from each other through evolution of the multiway system. (Such operator representations are a rough analog for multiway systems of truth tables.) As an example, with the multiway system "AB" ↔ "BA" some possible forms of operators are shown below. (In this case these are the commutative semigroups. With $k = 2$, elements 6 out of the total of 8 possible semigroups appear; with $k = 3$, 63 out of 113, and with $k = 4$, 1140 out of 3492—all as shown on page 805.) (See also page 952.)

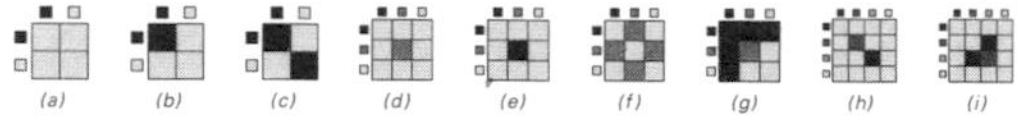

Taking ∘ to be each of these operators, one can work out a representation for any given string like "ABAA" by for example constructing the expression f[f[f[a, b], a], a] and finding its value for each of the k^2 possible pairs of values of a and b. Then for each successive operator, the sets of strings where the arrays of values are the same are as shown below.

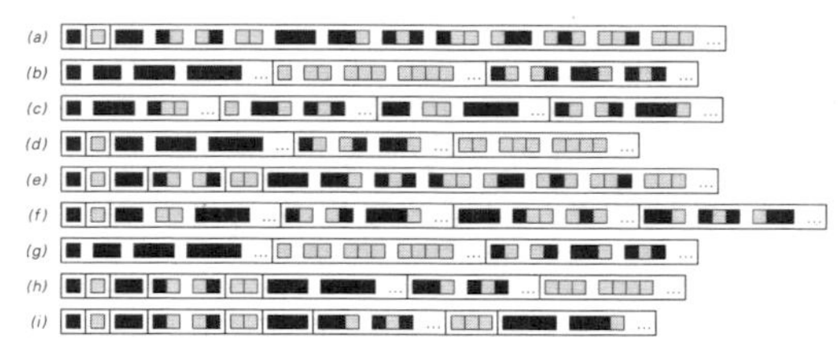

Ultimately the sets of strings equivalent under the multiway system are exactly those containing particular numbers of black and white elements. But as the pictures above suggest, only some of the distinctions between sets of strings are ever captured when any specific form for the operator is used.

Just as for operator systems, any bidirectional multiway system will allow a certain set of operators. (When there are multiple rules in the multiway system, tighter constraints are obtained by combining them with *And*.) And the pattern of results for simple multiway systems is roughly similar to those on page 805 for operator systems—although, for example, the associativity of concatenation makes it impossible for example to get the operators for *Nand* and basic logic.

▪ Page 806 · Logic in languages. Human languages always seem to have single words for AND, OR and NOT. A few have distinct words for OR and XOR: examples are Latin with *vel* and *aut* and Finnish with *vai* and *tai*. NOR is somewhat rare, though Dutch has *noch* and Old English *ne*. (Modern English has only the compound form *neither ... nor*.) But remarkably enough it appears that no ordinary language has a single word for NAND. The reason is not clear. Most people seem to find it difficult to think in terms of NAND (NAND is for example not associative, but then neither is NOR). And NAND on the face of it rarely seems useful in everyday situations. But perhaps these are just reflections of the historical fact that NAND has never been familiar from ordinary languages.

Essentially all computer languages support AND, OR and NOT as ways to combine logical statements; many support AND, OR and XOR as bitwise operations. Circuit design languages like Verilog and VHDL also support NAND, NOR and XNOR. (NAND is the operation easiest to implement with CMOS FETs—the transistors used in most current chips; it was also implemented by pentode vacuum tubes.) Circuit designers sometimes use the linguistic construct "*p* nand *q*".

The Laws of Form presented by George Spencer Brown in 1969 introduce a compact symbolic notation for NAND with any number of arguments and in effect try to develop a way of discussing NAND and reasoning directly in terms of it. (The axioms normally used are essentially the Sheffer ones from page 773.)

▪ Page 806 · Properties. Page 813 lists theorems satisfied by each function. {0, 1, 6, 7, 8, 9, 14, 15} are commutative (orderless) so that $a∘b = b∘a$, while {0, 6, 8, 9, 10, 12, 14, 15} are associative (flat), so that $a∘(b∘c) = (a∘b)∘c$. (Compare page 886.)

▪ Notations. Among those in current use are (highlighted ones are supported directly in *Mathematica*):

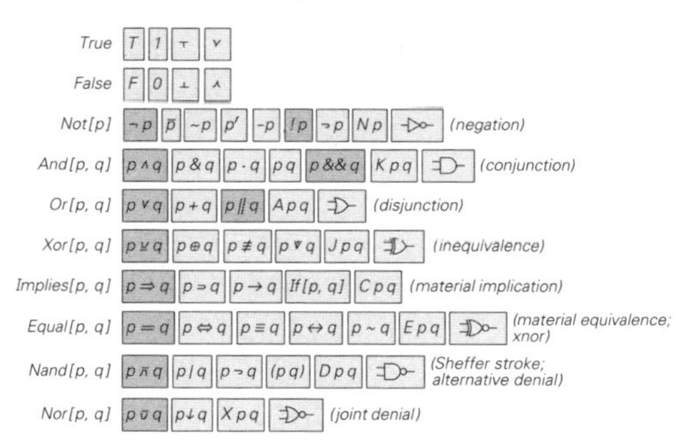

The grouping of terms is normally inferred from precedence of operators (typically ordered ==, ¬, ⊼, ∧, ⊻, ⊽, ∨, ⇒), or explicitly indicated by parentheses, function brackets, or sometimes nested underbars or dots. So-called Polish notation given second-to-last above avoids all explicit delimiters (see page 896).

▪ Page 807 · Universal logical functions. The fact that combinations of *Nand* or *Nor* are adequate to reproduce any logical function was noted by Charles Peirce around 1880, and became widely known after the work of Henry Sheffer in 1913. (See also page 1096.) *Nand* and *Nor* are the only 2-input functions universal in this sense. ({Equal} can for example reproduce only functions $\{9, 10, 12, 15\}$, $\{Implies\}$ only functions $\{10, 11, 12, 13, 14, 15\}$, and $\{Equal, Implies\}$ only functions $\{8, 9, 10, 11, 12, 13, 14, 15\}$.) For 3-input functions, corresponding to elementary cellular automaton rules, 56 of the 256 possibilities turn out to be universal. Of these, 6 are straightforward generalizations of *Nand* and *Nor*. Other universal functions include rules 1, 45 and 202 ($If[a == 1, b, c]$), but not 30, 60 or 110. For large n roughly $1/4$ of all n-input functions are universal. (See also page 1175.)

▪ **Page 808 · Searching for logic.** For axiom systems of the form $\{\ldots = a\}$ one finds:

	2 variables					3 variables				
number of $\circ$	2	3	4	5	6	2	3	4	5	6
total systems	4	16	80	448	2688	54	405	3402	30618	288684
allow $\bar{\wedge}$	0	5	44	168	1532	0	9	124	744	8764
allow only $\bar{\wedge}$ etc. for $k=2$	0	0	2	12	76	0	0	12	84	868
allow only $\bar{\wedge}$ etc. for $k\leq 3$	0	0	0	0	0	0	0	8	16	296
allow only $\bar{\wedge}$ etc. for $k\leq 4$	0	0	0	0	0	0	0	0	0	100

$\{((b\circ b)\circ a)\circ(a\circ b) = a\}$ allows the $k = 3$ operator 15552 for which the NAND theorem $(p\circ p)\circ q = (p\circ q)\circ q$ is not true. $\{(((b\circ a)\circ c)\circ a)\circ(a\circ c) = a\}$ allows the $k = 4$ operator 95356335 for which even $p\circ q = q\circ p$ is not true. Of the 100 cases that remain when $k = 4$, the 25 inequivalent under renaming of variables and reversing arguments of $\circ$ are

$\{(b\circ(b\circ(a\circ a)))\circ(a\circ(b\circ c))$,
$(b\circ(b\circ(a\circ a)))\circ(a\circ(c\circ b))$, $(b\circ(b\circ(a\circ b)))\circ(a\circ(b\circ c))$,
$(b\circ(b\circ(a\circ b)))\circ(a\circ(c\circ b))$, $(b\circ(b\circ(a\circ c)))\circ(a\circ(c\circ b))$,
$(b\circ(b\circ(b\circ a)))\circ(a\circ(b\circ c))$, $(b\circ(b\circ(b\circ a)))\circ(a\circ(c\circ b))$,
$(b\circ(b\circ(c\circ a)))\circ(a\circ(b\circ c))$, $(b\circ((a\circ b)\circ b))\circ(a\circ(b\circ c))$,
$(b\circ((a\circ b)\circ b))\circ(a\circ(c\circ b))$, $(b\circ((a\circ c)\circ b))\circ(a\circ(c\circ b))$,
$((b\circ c)\circ a)\circ(b\circ(b\circ(a\circ b)))$, $((b\circ c)\circ a)\circ(b\circ(b\circ(a\circ c)))$,
$((b\circ c)\circ a)\circ(b\circ((a\circ a)\circ b))$, $((b\circ c)\circ a)\circ(b\circ((a\circ b)\circ b))$,
$((b\circ c)\circ a)\circ(b\circ((a\circ c)\circ b))$, $((b\circ c)\circ a)\circ(b\circ((b\circ a)\circ b))$,
$((b\circ c)\circ a)\circ(b\circ((c\circ a)\circ b))$, $((b\circ c)\circ a)\circ(c\circ(c\circ(a\circ b)))$,
$((b\circ c)\circ a)\circ(c\circ(c\circ(a\circ c)))$, $((b\circ c)\circ a)\circ(c\circ((a\circ a)\circ c))$,
$((b\circ c)\circ a)\circ(c\circ((a\circ b)\circ c))$, $((b\circ c)\circ a)\circ(c\circ((a\circ c)\circ c))$,
$((b\circ c)\circ a)\circ(c\circ((b\circ a)\circ c))$, $((b\circ c)\circ a)\circ(c\circ((c\circ a)\circ c))\}$

Of these I was able in 2000—using automated theorem proving—to show that the ones given as (g) and (h) in the main text are indeed axiom systems for logic. (My proof essentially as found by Waldmeister is given on page 810.)

If one adds $a\circ b = b\circ a$ to any of the other 23 axioms above then in all cases the resulting axiom system can be shown to reproduce logic. But from any of the 23 axioms on their own I have never managed to derive $p\circ q = q\circ p$. Indeed, it seems difficult to derive much at all from them—though for example I have found a fairly short proof of $(p\circ p)\circ(p\circ q) = p$ from $\{(b\circ(b\circ(b\circ a)))\circ(a\circ(b\circ c)) = a\}$.

It turns out that the first of the 25 axioms allows the $k = 6$ operator 18857605376550238654534420036 and so cannot be logic. Axioms 3, 19 and 23 allow similar operators, leaving 19 systems as candidate axioms for logic.

It has been known since the 1940s that any axiom system for logic must have at least one axiom that involves more than 2 variables. The results above now show that 3 variables suffice. And adding more variables does not seem to help. The smallest axiom systems with more than 3 variables that work up to $k = 2$ are of the form $\{(((b\circ c)\circ d)\circ a)\circ(a\circ d) = a\}$. All turn out also to work at $k = 3$, but fail at $k = 4$. And with 6 NANDs (as in (g) and (h)) no system of the form $\{\ldots = a\}$ works even up to $k = 4$.

For axiom systems of the form $\{\ldots = a, a\circ b = b\circ a\}$:

	2 variables					3 variables				
number of $\circ$	4	5	6	7	8	4	5	6	7	8
total systems	4	16	80	448	2688	54	405	3402	30618	288684
allow $\bar{\wedge}$	0	5	44	168	1532	0	9	124	744	8764
allow only $\bar{\wedge}$ etc. for $k=2$	0	4	20	160	748	0	8	80	736	6248
allow only $\bar{\wedge}$ etc. for $k\leq 3$	0	0	0	64	16	0	0	32	416	2752
allow only $\bar{\wedge}$ etc. for $k\leq 4$	0	0	0	48	16	0	0	32	384	2368

With 2 variables the inequivalent cases that remain are

$\{(a\circ b)\circ(a\circ(b\circ(a\circ b)))$,
$(a\circ b)\circ(a\circ(b\circ(b\circ b)))$, $(a\circ(b\circ b))\circ(a\circ(b\circ(b\circ b)))\}$

but all of these allow the $k = 6$ operator

18857605376551254297384808884

and so cannot correspond to basic logic. With 3 variables, all 32 cases with 6 NANDs are equivalent to $(a\circ b)\circ(a\circ(b\circ c))$, which is axiom system (f) in the main text. With 7 NANDs there are 8 inequivalent cases:

$\{(a\circ a)\circ(b\circ(b\circ(a\circ c)))$, $(a\circ b)\circ(a\circ(b\circ(a\circ b)))$, $(a\circ b)\circ(a\circ(b\circ(a\circ c)))$,
$(a\circ b)\circ(a\circ(b\circ(b\circ b)))$, $(a\circ b)\circ(a\circ(b\circ(b\circ c)))$,
$(a\circ b)\circ(a\circ(b\circ(c\circ c)))$, $(a\circ b)\circ(a\circ(c\circ(a\circ c)))$, $(a\circ b)\circ(a\circ(c\circ(c\circ c)))\}$

and of these at least 5 and 6 can readily be proved to be axioms for logic.

Any axiom system must consist of equivalences valid for the operator it describes. But the fact that there are fairly few short such equivalences for *Nand* (see page 818) implies that there can be no axiom system for *Nand* with 6 or less NANDs except the ones discussed above.

▪ **Two-operator logic.** If one allows two operators then one can get standard logic if one of these operators is forced to be *Not* and the other is forced to be *And*, *Or* or *Implies*—or in fact any of operators 1, 2, 4, 7, 8, 11, 13, 14 from page 806.

A simple example that allows *Not* and either *And* or *Or* is the Robbins axiom system from page 773. Given the first two axioms (commutativity and associativity) it turns out that no shorter third axiom will work in this case (though ones such as $f[g[f[a, g[f[a, b]]]], g[g[b]]] = b$ of the same size do work).

Much as in the single-operator case, to reproduce logic two pairs of operators must be allowed for $k = 2$, none for $k = 3$, 12 for $k = 4$, and so on. Among single axioms, the shortest that works up to $k = 2$ is $(\neg(\neg(\neg b \vee a) \vee \neg(a \vee b))) = a$. The shortest that works up to $k = 3$ is $(\neg(\neg(a \vee b) \vee \neg b) \vee \neg(\neg a \vee a)) = b$. It is known, however, that at least 3 variables must appear in order to reproduce logic, and an example of a single axiom with 4 variables that has been found recently to work is $\{(\neg(\neg(c \vee b) \vee \neg a) \vee \neg(\neg(\neg d \vee d) \vee \neg a \vee c)) = a\}$.

▪ **Page 808 · History.** (See page 1151.) (c) was found by Henry Sheffer in 1913; (e) by Carew Meredith in 1967. Until this book, very little work appears to have been done on finding short axioms for logic in terms of *Nand*. Around 1949 Meredith found the axiom system

$$\{(a\circ(b\circ c))\circ(a\circ(b\circ c)) = ((c\circ a)\circ a)\circ((b\circ a)\circ a),\ (a\circ a)\circ(b\circ a) = a\}$$

In 1967 George Spencer Brown found (see page 1173)

$$\{(a\circ a)\circ((b\circ b)\circ b) = a,\ a\circ(b\circ c) = (((c\circ c)\circ a)\circ((b\circ b)\circ a))\circ(((c\circ c)\circ a)\circ((b\circ b)\circ a))\}$$

and in 1969 Meredith also gave the system

$$\{a\circ(b\circ(a\circ c)) = a\circ(b\circ(b\circ c)),\ (a\circ a)\circ(b\circ a) = a,\ a\circ b = b\circ a\}$$

▪ **Page 812 · Theorem distributions.** The picture below shows which of the possible theorems from page 812 hold for each of the numbered standard mathematical theories from page 805. The theorem close to the right-hand end valid in many cases is $(p\circ p)\circ p = p\circ(p\circ p)$. The lack of regularity in this picture can be viewed as a sign that it is difficult to tell which theorems hold, and thus in effect to do mathematics.

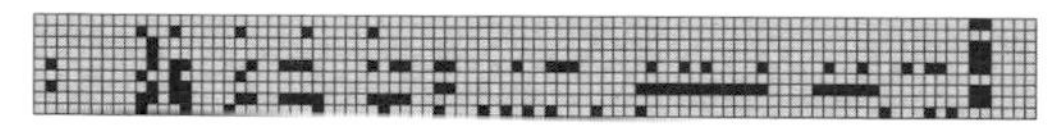

▪ **Page 814 · Multivalued logic.** As noted by Jan Łukasiewicz and Emil Post in the early 1920s, it is possible to generalize ordinary logic to allow k values *Range[0, 1, 1/(k - 1)]*, say with 0 being *False*, and 1 being *True*. Standard operations in logic can be generalized as *Not[a_] = 1 - a*, *And[a_, b_] = Min[a, b]*, *Or[a_, b_] = Max[a, b]*, *Xor[a_, b_] = Abs[a - b]*, *Equal[a_, b_] = 1 - Abs[a - b]*, *Implies[a_, b_] = 1 - UnitStep[a - b] (a - b)*. An alternative generalization for *Not* is *Not[a_] := Mod[(k - 1) a + 1, k]/(k - 1)*. The function *Nand[a_, b_] := Not[And[a, b]]* used in the main text turns out to be universal for any k. Axiom systems can be set up for multivalued logic, but they are presumably more complicated than for ordinary $k = 2$ logic. (Compare page 1171.)

The idea of intermediate truth values has been discussed intermittently ever since antiquity. Often—as in the work of George Boole in 1847—a continuum of values between 0 and 1 are taken to represent probabilities of events, and this is the basis for the field of fuzzy logic popular since the 1980s.

▪ **Page 814 · Proof lengths in logic.** As discussed on page 1170 equivalence between expressions can always be proved by transforming to and from canonical form. But with n variables a DNF-type canonical form can be of size 2^n—and can take up to at least 2^n proof steps to reach. And indeed if logic proofs could in general be done in a number of steps that increases only like a polynomial in n this would mean that the NP-complete problem of satisfiability could also be solved in this number of steps, which seems very unlikely (see page 768).

In practice it is usually extremely difficult to find the absolute shortest proof of a given logic theorem—and the exact length will depend on what axiom system is used, and what kinds of steps are allowed. In fact, as mentioned on page 1155, if one does not allow lemmas some proofs perhaps have to become exponentially longer. The picture below shows in each of the axiom systems from page 808 the lengths of the shortest proofs found by a version of Waldmeister (see page 1158) for all 582 equivalences (see page 818) that involve two variables and up to 3 NANDs on either side.

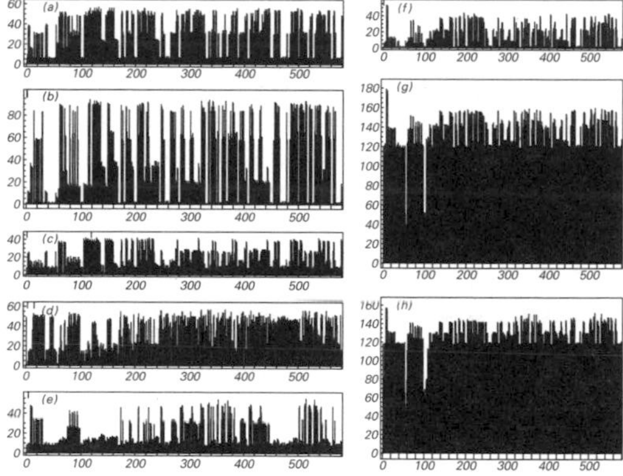

The longest of these are respectively $\{57, 94, 42, 57, 55, 53, 179, 157\}$ and occur for theorems

$$\{(((a \bar{\wedge} a) \bar{\wedge} b) \bar{\wedge} b) = (((a \bar{\wedge} b) \bar{\wedge} a) \bar{\wedge} a),\ (a \bar{\wedge} (a \bar{\wedge} (a \bar{\wedge} a))) = (a \bar{\wedge} ((a \bar{\wedge} b) \bar{\wedge} b)),\ (((a \bar{\wedge} a) \bar{\wedge} a) \bar{\wedge} a) = (((a \bar{\wedge} a) \bar{\wedge} b) \bar{\wedge} a),\ (((a \bar{\wedge} a) \bar{\wedge} b) \bar{\wedge} b) = (((a \bar{\wedge} b) \bar{\wedge} a) \bar{\wedge} a),\ (a \bar{\wedge} ((b \bar{\wedge} b) \bar{\wedge} a)) = (b \bar{\wedge} ((a \bar{\wedge} a) \bar{\wedge} b)),\ ((a \bar{\wedge} a) \bar{\wedge} a) = ((b \bar{\wedge} b) \bar{\wedge} b),\ ((a \bar{\wedge} a) \bar{\wedge} a) = ((b \bar{\wedge} b) \bar{\wedge} b),\ ((a \bar{\wedge} a) \bar{\wedge} a) = ((b \bar{\wedge} b) \bar{\wedge} b)\}$$

Note that for systems that do not already have it as an axiom, most theorems use the lemma $(a \bar{\wedge} b) = (b \bar{\wedge} a)$ which takes respectively $\{6, 1, 8, 49, 8, 1, 119, 118\}$ steps to prove.

▪ **Page 818 · NAND theorems.** The total number of expressions with n NANDs and s variables is: *Binomial[2 n, n]* $s^{n+1}/(n+1)$ (see page 897). With $s = 2$ and n from 0 to 7 the number of these *True* for all values of variables is $\{0, 0, 4, 0, 80, 108, 2592, 7296\}$, with the first few distinct ones being (see page 781)

$$\{(p \bar{\wedge} p) \bar{\wedge} p,\ (((p \bar{\wedge} p) \bar{\wedge} p) \bar{\wedge} p) \bar{\wedge} p,\ (((p \bar{\wedge} p) \bar{\wedge} p) \bar{\wedge} q) \bar{\wedge} q\}$$

The number of unequal expressions obtained is {2, 3, 3, 7, 10, 15, 12, 16} (compare page 1096), with the first few distinct ones being

{p, p ⊼ p, p ⊼ q, (p ⊼ p) ⊼ p, (p ⊼ q) ⊼ p, (p ⊼ p) ⊼ q}

Most of the axioms from page 808 are too long to appear early in the list of theorems. But those of system (d) appear at positions {3, 15, 568} and those of (e) at {855, 4}.

(See also page 1096.)

▪ **Page 819 · Finite axiomatizability.** It is known that the axiom systems (such as Peano arithmetic and set theory) given with axiom schemas on pages 773 and 774 can be set up only with an infinite number of individual axioms. But because such axioms can be described by schemas they must all have similar forms, so that even though the definition in the main text suggests that each corresponds to an interesting theorem these theorems are not in a sense independently interesting. (Note that for example the theory of specifically finite groups cannot be set up with a finite number even of schemas—or with any finite procedure for checking whether a given candidate axiom should be included.)

▪ **Page 820 · Empirical metamathematics.** One can imagine a network representing some field of mathematics, with nodes corresponding to theorems and connections corresponding to proofs, gradually getting filled in as the field develops. Typical rough characterizations of mathematical results—independent of their detailed history—might then end up being something like the following:

- lemma: short theorem appearing at an intermediate stage in a proof
- corollary: theorem connected by a short proof to an existing theorem
- easy theorem: theorem having a short proof
- difficult theorem: theorem having a long proof
- elegant theorem: theorem whose statement is short and somewhat unique
- interesting theorem (see page 817): theorem that cannot readily be deduced from earlier theorems, but is well connected
- boring theorem: theorem for which there are many others very much like it
- useful theorem: theorem that leads to many new ones
- powerful theorem: theorem that substantially reduces the lengths of proofs needed for many others
- surprising theorem: theorem that appears in an otherwise sparse part of the network of theorems
- deep theorem: theorem that connects components of the network of theorems that are otherwise far away
- important theorem: theorem that allows a broad new area of the network to be reached.

The picture below shows the network of theorems associated with Euclid's *Elements*. Each stated theorem is represented by a node connected to the theorems used in its stated proof. (Only the shortest connection from each theorem is shown explicitly.) The axioms (postulates and common notions) are given in the first column on the left, and successive columns then show theorems with progressively longer proofs. (Explicit annotations giving theorems used in proofs were apparently added to editions of Euclid only in the past few centuries; the picture below extends the usual annotations in a few cases.) The theorem with the longest proof is the one that states that there are only five Platonic solids.

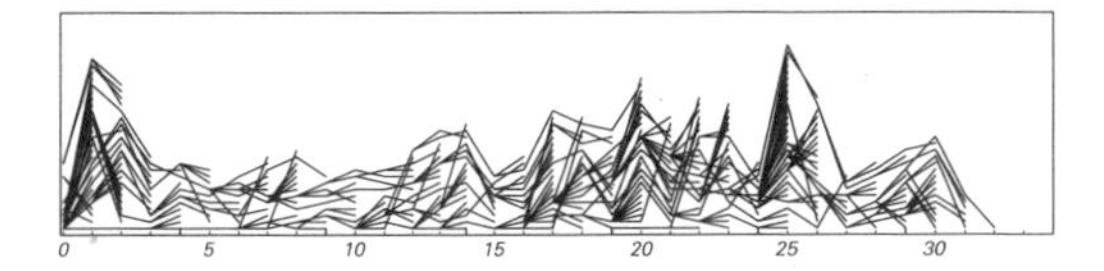

▪ **Speedups in other systems.** Multiway systems are almost unique in being able to be sped up just by adding "results" already derived in the multiway system. In other systems, there is no such direct way to insert such results into the rules for the system.

▪ **Character of mathematics.** Since at least the early 1900s several major schools of thought have existed:

- *Formalism* (e.g. David Hilbert): Mathematics studies formal rules that have no intrinsic meaning, but are relevant because of their applications or history.
- *Platonism* (e.g. Kurt Gödel): Mathematics involves trying to discover the properties of a world of ideal mathematical forms, of which we in effect perceive only shadows.
- *Logicism* (e.g. Gottlob Frege, Bertrand Russell): Mathematics is an elaborate application of logic, which is itself fundamental.
- *Intuitionism* (e.g. Luitzen Brouwer): Mathematics is a precise way of handling ideas that are intuitive to the human mind.

The results in this book establish a new point of view somewhere between all of these.

▪ **Invention versus discovery in mathematics.** One generally considers things invented if they are created in a somewhat arbitrary way, and discovered if they are identified by what seems like a more inexorable process. The results of this section thus strongly suggest that the basic directions taken by mathematics as currently practiced should mostly be considered invented, not discovered. The new kind of science that I describe in this book, however, tends to suggest forms of mathematics that involve discovery rather than invention.

▪ **Ordering of constructs.** One can deduce some kind of ordering among standard mathematical constructs by seeing how difficult they are to implement in various systems—such as cellular automata, Turing machines and Diophantine equations. My experience has usually been that addition is easiest, followed by multiplication, powers, Fibonacci numbers, perfect numbers and then primes. And perhaps this is similar to the order in which these constructs appeared in the early history of mathematics. (Compare page 640.)

▪ **Mathematics and the brain.** A possible reason for some constructs to be more common in mathematics is that they are somehow easier for human brains to manipulate. Typical human experience makes small positive integers and simple shapes familiar—so that all human brains are at least well adapted to such constructs. Yet of the limited set of people exposed to higher mathematics, different ones often seem to think in bizarrely different ways. Some think symbolically, presumably applying linguistic capabilities to algebraic or other representations. Some think more visually, using mechanical experience or visual memory. Others seem to think in terms of abstract patterns, perhaps sometimes with implicit analogies to musical harmony. And still others—including some of the purest mathematicians—seem to think directly in terms of constraints, perhaps using some kind of abstraction of everyday geometrical reasoning.

In the history of mathematics there are many concepts that seem to have taken surprisingly long to emerge. And sometimes these are ones that people still find hard to grasp. But they often later seem quite simple and obvious—as with many examples in this book.

It is sometimes thought that people understand concepts in mathematics most easily if they are presented in the order in which they arose historically. But for example the basic notion of programmability seems at some level quite easy even for young children to grasp—even though historically it appeared only recently.

In designing *Mathematica* one of my challenges was to use constructs that are at least ultimately easy for people to understand. Important criteria for this in my experience include specifying processes directly rather than through constraints, the explicitness in the representation of input and output, and the existence of small, memorable, examples. Typically it seems more difficult for people to understand processes in which larger numbers of different things happen in parallel. (Notably, FoldList normally seems more difficult to understand than NestList.) Tree structures such as *Mathematica* expressions are fairly easy to understand. But I have never found a way to make general networks similarly easy, and I am beginning to suspect that they may be fundamentally difficult for brains to handle.

▪ **Page 821 · Frameworks.** Symbolic integration was in the past done by a collection of ad hoc methods like substitution, partial fractions, integration by parts, and parametric differentiation. But in *Mathematica* Integrate is now almost completely systematic, being based on structure theorems for finding general forms of integrals, and on general representations in terms of MeijerG and other functions. (In recognizing, for example, whether an expression involving a parameter can have a pole undecidable questions can in principle come up, but they seem rare in practice.) Proofs are essentially always still done in an ad hoc way—with a few minor frameworks like enumeration of cases, induction, and proof by contradiction (reductio ad absurdum) occasionally being used. (More detailed frameworks are used in specific areas; an example are ϵ-δ arguments in calculus.) But although still almost unknown in mainstream mathematics, methods from automated theorem proving (see page 1157) are beginning to allow proofs of many statements that can be formulated in terms of operator systems to be found in a largely systematic way (e.g. page 810). (In the case of Euclidean geometry—which is a complete axiom system—algebraic methods have allowed complete systematization.) In general, the more systematic the proofs in a particular area become, the less relevant they will typically seem compared to the theorems that they establish as true.

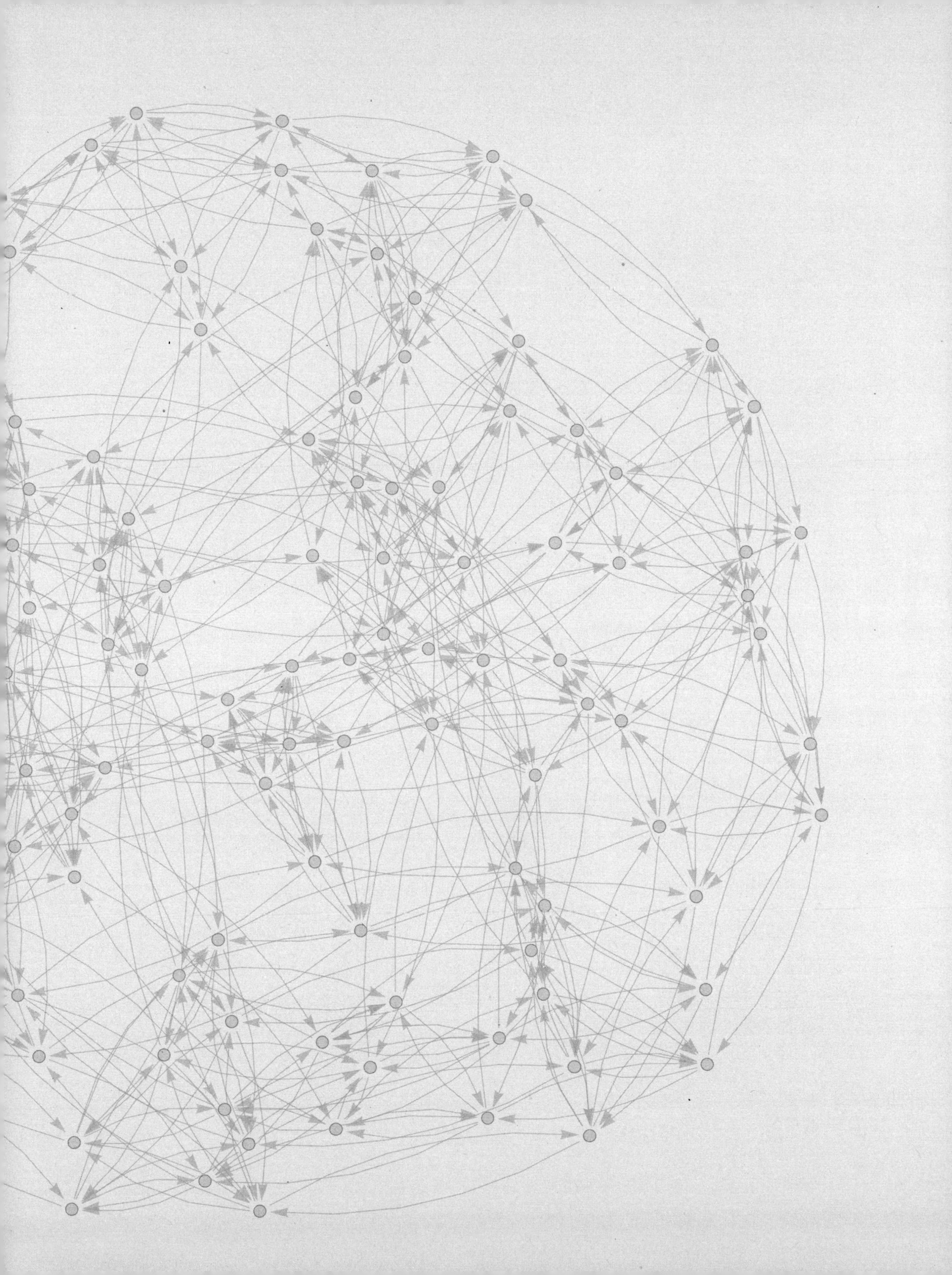

Index